Apomixis in Angiosperms

Apomixis is a difficult-to-analyse trait with a complex molecular basis and a substantive effect on the biology of a species. Thus, apomixis is an interesting characteristic for researchers and students working in different fields of plant science and agriculture and technological advances are enabling and making apomixis studies more common.

Apomixis in Angiosperms: Mechanisms, Occurrences, and Biotechnology provides a systematic introduction to the mechanisms and developmental types of apomixis along with an overview of alternative methodologies for identifying apomixis and a detailed reassessment of the occurrences of apomictic species among angiosperm families. Optional methods are illustrated with examples of all types of apomixis and biological levels of analysis, i.e. cells, ovules, seeds and offspring. Data on apomictic species are collected in tables along with information on ploidy, type of apomixis and references. Occurrences of apomixis are briefly discussed in phylogenetic and evolutionary contexts. An outline of the molecular basis of apomixis in plants is presented, together with prospects and challenges that remain for its biotechnological exploitation.

This book:

- Provides a systematic overview of the mechanisms of apomixis, developmental types and methodology for apomixis research.
- Reassessments of apomixis at the species level in angiosperm families.
- Contains tables summarizing relevant information on apomixis.
- Analyses occurrences of apomixis in phylogenetic and evolutionary contexts.
- Outlines the molecular basis and biotechnological perspective of apomixis breeding.

This book presents an accessible overview of apomixis research and a curated dataset of apomictic species. It serves as a reference book for students, researchers and citizen scientists interested in apomixis, as well as researchers, business innovators and entrepreneurs pursuing apomixis breeding. It can also be used as a textbook in graduate courses on plant reproduction.

Diego Hojsgaard Taxonomy and Evolutionary Biology, Leibniz Institute of Plant Genetics and Crop Plant Research, Gatersleben, Germany, and Faculty of Exact, Chemical and Natural Sciences, National University of Misiones, Posadas, Argentina.

Thammineni Pullaiah Department of Botany, Sri Krishnadevaraya University, Anantapur, India.

Apomixis in Angiosperms

Mechanisms, Occurrences, and Biotechnology

Diego Hojsgaard
Thammineni Pullaiah

CRC Press
Taylor & Francis Group
Boca Raton London New York

CRC Press is an imprint of the
Taylor & Francis Group, an **informa** business

Cover image: Mountain meadow at about 2100 m.a.s.l. with apomictic daisies and grasses near Santa Cruz (Tafí del Valle), northwestern Argentina in South America. Behind is the Cerro Pelao covered in clouds.

First edition published 2023
by CRC Press
6000 Broken Sound Parkway NW, Suite 300, Boca Raton, FL 33487-2742

and by CRC Press
4 Park Square, Milton Park, Abingdon, Oxon, OX14 4RN

CRC Press is an imprint of Taylor & Francis Group, LLC

ISBN: 978-0-367-54295-5 (hbk)
ISBN: 978-0-367-54296-2 (pbk)
ISBN: 978-1-003-08856-1 (ebk)

DOI: 10.1201/9781003088561

Typeset in Times
by SPi Technologies India Pvt Ltd (Straive)

Content

Preface

Advances in apomixis research over the past three decades have been remarkable, and the coming decades look auspicious. Increasingly easy access to molecular biology and flow cytometry methods, coupled with a plethora of open-source bioinformatics software for intuitive visualization, exploration and integration of biological data that require less and less programming knowledge, memory and processing power, is helping researchers characterize apomixis at many levels. Despite limited funding, apomixis scientists are currently exploring – with outstanding results – all major types of biological data for meta-analysis in molecular biology, ecology and evolutionary biology.

Outstanding in recent discoveries is the identification and functional characterization of genes that trigger components of apomixis or apomixis-like phenotypes. A noteworthy shift of paradigm is the growing recognition of apomixis as an evolutionary force that plays an active role in the maintenance and diversification of species. Remaining challenges concern the unresolved taxonomic consent on the criteria for delimiting species with apomixis, understanding the genetics underlying the ecological adaptation of apomicts and the dynamics of apomixis to sex reversals in nature, and unravelling the molecular basis for the occurrence of the trait.

Scientists investigating apomixis are now – more than ever – faced with the need to join forces and cooperate on different fronts to consolidate apomixis as a model in evolutionary studies and as a biotechnological tool in molecular breeding. Honest engagement, fair peer review and good data management and stewardship would be the keys to facilitating this path.

Sincere gratitude to those colleagues who helped in one way or another during the process of this book, whether by sharing invaluable ideas and discussions, assisting with some of the illustrations, embryological classifications or taxonomic identifications, or providing literature. We are indebted to Drs Silvia Arroyo-Leuenberger, Ana Caperta, John Carman, Julio Daviña, Timothy Dickinson, Angel Hojsgaard, Ana Honfi, Elvira Hörandl, Piyal Karunarathne, Eric Martínez, Patrik Mráz, Kristina Musiał, Camilo Quarin, Agostina Sassone, Jan Štěpánek and Peter van Dijk.

We also thank Jyotsna Jangra and Renu Apadhyay, from CRC Press, for their assistance in bringing this edition to print.

Diego Hojsgaard
Aschersleben, Germany

Thammineni Pullaiah
Anantapur, India

Authors

Diego Hojsgaard obtained his PhD (2010) in Natural Resources (Botany) from the National University of the Northeast, Argentina. He was a Post-Doctoral Fellow at Vienna University, Austria, during 2010–2011, before joining the University of Göttingen as a non-tenured Assistant Professor in 2012. Since 2016 he has acted as Visiting Professor at the National University of Misiones in Argentina. He has travelled extensively in Europe and South America, the United States, England and China and has visited universities and botanic gardens in about 16 countries. He is currently a Guest Scientist in the Taxonomy and Evolutionary Biology group at the Leibniz Institute of Plant Genetics and Crop Plant Research in Germany. Diego does research in plant evolutionary biology and genetics, focusing on understanding how species use different mechanisms to maximize the exploitation of genetic variability during adaptation and speciation. He is intrigued to understand the origin and evolution of apomixis in plants. Thus, his projects address topics in plant reproductive biology with an emphasis on apomixis, including ovule ontogeny and the origin of different types of female gametophytes, plant cytotype interactions (population dynamics) and natural distributions, extant genetic variability and the ecology of adaptation. He has published forty-three research articles, three book chapters and has edited one book on the *Molecular Basis of Apomixis in Plants*.

Dr T. Pullaiah is a former Professor at the Department of Botany at Sri Krishnadevaraya University in Andhra Pradesh, India, where he has taught for more than 35 years. He has held several positions at the university, including Dean, Faculty of Biosciences, Head of the Department of Botany, Head of the Department of Biotechnology and Member of Academic Senate. He was President of the Indian Botanical Society (2014), President of the Indian Association for Angiosperm Taxonomy (2013) and Fellow of Andhra Pradesh Akademi of Sciences. He was awarded the Panchanan Maheshwari Gold Medal, the Prof. P.C.Trivedi Medal, the Dr. G. Panigrahi Memorial Lecture Award of the Indian Botanical Society and Prof. Y.D. Tyagi Gold Medal of the Indian Association for Angiosperm Taxonomy, and a Best Teacher Award from the Government of Andhra Pradesh. Under his guidance, 54 students obtained their doctoral degrees. He has authored 52 books, edited 23 books and published over 330 research papers, including reviews and book chapters. His books include *Advances in Cell and Molecular Diagnostics*, *Camptothecin and Camptothecin Producing Plants*, *Ethnobotany of India* (5 volumes), *Global Biodiversity* (4 volumes), *Red Sanders: Silviculture and Conservation*, *Genetically Modified Crops* (2 volumes), *Monograph on Brachystelma and Ceropegia in India* and so on. He was also a member of Species Survival Commission of the International Union for Conservation of Nature (IUCN). Professor Pullaiah received his PhD from Andhra University, India, attended Moscow State University, Russia, and worked as Post-Doctoral Fellow during 1976–1978.

1 Introduction

Spermatophytes, or seed-bearing plants (from the Greek words *spérmatos*, meaning "seed", and *phytón*, meaning "plant"), comprise the gymnosperms and angiosperms, some of the most important and diverse organisms on Earth. A seed is a highly efficient dispersal unit, an embryonic plant enclosed in a protective outer covering that enhances survival and adaptation. Seeds can be produced sexually, by amphimixis, or asexually, by apomixis.

Amphimixis is derived from the Greek words *amphi* and *mixis*. Apomixis is derived from the Greek words *apo* and *mixis*. *Amphi* means "of both kinds", *apo* means "away from" and *mixis* is the "act of mixing or mingling". The term "amphimixis" was first introduced by Weismann (1891) to describe his observations in animals. The term "apomixis" was first coined by Haacke (1893) to describe a situation observed in a mammal (for details see Battaglia, 1991). For sexually differentiated plants, Winkler (1908, p. 300) defined amphimixis as follows: "Die Amphimixis ist die normale Art der geschlechtlichen Fortpflanzung, bei der also der Keim entsteht aus dem Verschmelzungsprodukt zweier Keimzellen, seien diese nun als Isogameten ausgebildet oder in Ei und Spermatozoon differenziert". (*Amphimixis is the normal mode of sexual reproduction in which the germ is formed from the fusion product of two germ cells, whether these are isogametes or differentiated into egg and spermatozoon*). Winkler (1908, pp. 302–303) also defined apomixis as follows: "Als Apomixis endlich bezeichnen wir den Ersatz der geschlechtlichen Fortpflanzung durch einen anderen, ungeschlechtlichen, nicht mit Kern- und Zellverschmelzung verbundenen Vermehrungsprozeß". (*Finally, we call apomixis the replacement of sexual reproduction by another, asexual, reproduction process not associated with nuclear and cell fusion*).

Winkler's definition of apomixis is broad and includes organisms in which asexual reproduction takes place through seeds (i.e. *agamospermy*, Täckholm, 1922), and organisms in which asexual reproduction is through vegetative structures such as bulbs, bulbils, tubers, runners and suckers. In seed plants, some authors consider vegetative propagation a form of clonal growth and not a mode of apomictic reproduction (e.g., Mogie, 1992), and most researchers use apomixis as a synonym for agamospermy.

For more historical aspects related mainly to the concepts and classifications of apomixis and the delimitations of developmental types, we recommend the detailed reviews published by Lakshmanan and Ambegaokar (1984), Nogler (1984), Naumova (1993) and Asker and Jerling (1992). For simplicity and following the terms most commonly used in the current literature, we will here use the classification and divisions of apomictic types suggested by Strasburger (1878), Webber (1940), Gustafsson (1946, 1947a,1947b) and Stebbins (1941, 1950) in relation to gametophytic and sporophytic apomixis (see Chapter 2). We believe that a seed-centred view that considers the clonal nature of the embryo originating directly from a single somatic cell (sporophytic apomixis) or through a gametophyte (gametophytic apomixis) simplifies the classification of the different types of apomixis and helps to resolve the issues raised by different authors (Battaglia, 1963, 1991; Crane, 1989; Solntzeva, 1989). A seed-centred view excludes cases of vegetative propagation (Figure 1.1), and therefore, apomixis can be used as a synonym for agamospermy, as it would refer to asexual reproduction through seeds. Unlike sexual reproduction, apomixis leads to the formation of clonal offspring that fully maintain the genetic constitution of the mother plant.

DOI: 10.1201/9781003088561-1

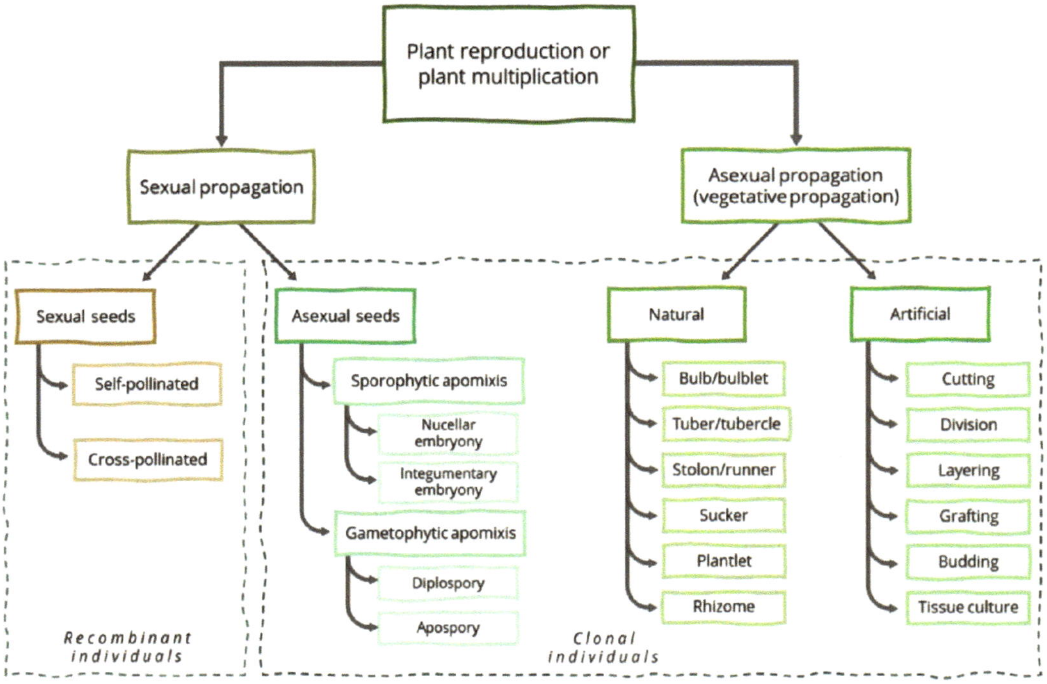

FIGURE 1.1 Alternative mechanisms of plant reproduction or multiplication.

In recent years, apomixis has gained interest in plant biology as an evolutionary force capable of promoting plant diversification and as a biotechnological challenge that is expected to revolutionize agriculture. Apomixis barely occurs in gymnosperms, where only one case has been reported recently (Pichot et al., 2000, 2001). In angiosperms, on the other hand, apomixis is much more frequent. However, as records of apomixis among angiospermous species date back to the nineteenth century and are scattered, it is not known exactly how many species exhibit this trait.

The present contribution aims to provide an outline of the main developmental pathways of asexual seeds, alternative methods to detect apomixis, a fairly comprehensive list of recorded species that have any of such pathways or recorded species that produce asexual seeds, an overview of its distribution among angiosperm families that will foster biogeographic and evolutionary studies. Last, a synopsis of the genetic knowledge on apomixis is provided, together with brief comments on emerging biotechnological applications that may allow the exploitation of apomixis as a plant molecular breeding tool in the medium term.

As apomixis involves cytological, embryological and genetic changes specific to the sexual reproduction, several methodologies can be used to effectively determine the formation of asexual seeds in any species (Chapter 3), without the need to know the type of apomixis. Following these premises about the undoubted record of apomictic cases, we reviewed all (or most) of the available literature on apomixis in grass species (Chapter 4), daisies (Chapter 5), roses (Chapter 6) and gathered all other families with reports of apomictic species (Chapter 7). A critical review of these studies allowed us to list all apomictic species and to include quarantine lists for doubtful cases, all accompanied by detailed observations. A first analysis of the occurrence of apomixis in angiosperms confirms a scattered distribution among angiosperms orders, with gametophytic apomixis being more frequent than sporophytic apomixis and associated to polyploidy, supporting the view that apomictic species play an active role in the evolution of genera and species complexes of different plant families (Chapter 8). One of the remaining challenges in understanding apomixis relates to the complex molecular basis of the trait, the elucidation of which will be necessary to advance its biotechnological exploitation in major crops (Chapter 9).

The reviewed and curated data present an up-to-date overview of the main aspects of apomixis research, as well as the latest advances towards apomixis breeding, which will help students and attract scientists from different fields, fostering broader evolutionary and biotechnological studies, necessary both to understand the origin and role of apomixis in nature and to harness its deployment for human benefit.

2 Sexual Development and Pathways to Apomixis

2.1 INTRODUCTION

Angiosperms are the most biodiverse group of plants. Flowering plants exhibit a remarkable variability in size and shape of floral traits, stamens, carpels and fruit developments (Simonini and Østergaard, 2019), all functional to the mechanism of reproduction, formation of diaspores (seeds) and dispersal.

From unisexual or bisexual flowers, with radial or bilateral shape, solitary or clustered in inflorescences, with bisporic or tetrasporic anthers, unilocular, bilocular or multilocular ovaries carrying several possible types of ovules (campylotropous, anatropous, amphitropous, orthotropous) and integuments (ategmic, unitegmic, bitegmic), in the nucellar tissue (tenuinucellate, crassinucellate), to numerous types of fruits (berries, drupes, pomes, capsules, nuts, caryopses), all are adapted to contribute to the reproduction and survival of the species. Such variability in reproductive organs synthesizes the evolutionary plasticity granted by the acquisition of a seed-syndrome as the innovation underlying angiosperms' reproductive success (Baroux and Grossniklaus, 2019).

Angiosperm flowers contain the cell lineage that undertakes the cycle of alternation of generations during the process of sexual reproduction to recreate new diaspores (new plant embryos contained in seeds). During this alternation of generations, specialized sporophytic cells in stamens and carpels, called mother cells, undergo meiosis and give rise to haploid spores, a process called sporogenesis. Meiosis is an extraordinarily complex and distinctive cell division involving one step of DNA synthesis and two nuclear divisions, reducing the genetic information to one half (Mirzaghaderi and Hörandl, 2016). Spore cells proliferate and differentiate into multicellular gametophytes containing the gametes (sperm and egg cells), a process called gametogenesis. The gametes then fuse to form the zygote and a new sporophyte in an embryo containing the body plant (Gifford and Foster, 1989; Drews and Koltunow, 2011).

Spore cells in angiosperms have different sizes (heterospory). The one produced in ovules (carpels) is the megaspore, and the one produced in anthers (stamens) is the microspore. The diploid megaspore and microspore mother cells give rise to haploid megaspores and microspores after megasporogenesis and microsporogenesis, respectively. Then, during megagametogenesis and microgametogenesis, megaspores and microspores develop into haploid female and male gametophytes, respectively. The female gametophyte is also called a megagametophyte or embryo sac. The male gametophyte is called a microgametophyte or pollen grain. The development of both gametophytes also shows variability among angiosperms (see below). Thus, flowers are highly specialized to ensure proper transfer of microgametophytes and delivery of male gametes for fertilization of female gametes and formation of seeds containing new sporophytic embryos. Seeds are specialized dispersal units consisting of an embryo, and endosperm and the seed coat.

The following is an overview of the developmental steps, plant structures, and organisms involved in the process of sexual reproduction and alternation of generations most frequently observed in angiosperms to understand reproductive alterations in apomicts. Additional details on specific developmental types or plant groups can be found elsewhere (e.g., Johri, 1984; Pandey and Chadha, 1997; Pullaiah and Febulaus, 2000; Batygina, 2002, 2005, 2009; Grossniklaus, 2019). Current technological advances in different areas of research have made it possible to identify genes and gene networks involved in each of the stages of reproduction. Seed formation is a tightly regulated process, with complex genetic and regulatory controls governing key steps (Gasser and Skinner,

DOI: 10.1201/9781003088561-2

2019; Hackenberg and Twell, 2019; Zhou and Dresselhaus, 2019). Here, attention will be given to patterns of sexual development important to understand apomictic developments.

2.2 THE SEXUAL PATHWAY OF SEED FORMATION

Female and male gametophytes of angiosperms are highly reduced and complex organisms, essential for plant reproduction (reviewed in Honys et al., 2006; Drews and Koltunow, 2011). They interact during the process of making a seed and coevolve to ensure reproductive success and species survival (Baroux and Grossniklaus, 2019).

2.2.1 The Female Gametophyte

So far, more than 15 distinct patterns of female gametophytes development have been described in angiosperms (Rutishauser, 1969; Battaglia, 1989; Friedman, 2006). The most frequently observed pattern is the Polygonum type, first described by Strasburger in 1879 in the dicot species *Polygonum divaricatum* L. (Strasburger, 1879; Maheshwari, 1950). This type is present in >70% of angiosperms and is exclusive in 53% of all plant families studied so far (Carman, 1997), including economically important plant families like Brassicaceae (*Arabidopsis* Heynh., *Capsella* Medik., *Brassica* L.), Fabaceae (beans, soybean), Malvaceae (cotton), Poaceae (maize, rice, wheat) and Solanaceae (pepper, tobacco, tomato, potato, *Petunia*), along with most apomictic species (Maheshwari, 1950; Willemse and van Went, 1984; Haig, 1990; Huang and Russell, 1992).

The development of the female gametophyte begins early in the development of the ovule with the differentiation of a sub-epidermal diploid (2*n*) archesporial cell that acquires a megaspore mother cell identity. This cell (also called female meiocyte or megasporocyte) is larger than other surrounding somatic cells, has a denser cytoplasm, has a larger nucleus and elongates before meiosis. Once the megaspore mother cell undergoes meiosis, it gives rise to four genetically recombinant haploid (*n*) megaspores. Subsequently, three megaspores degenerate by apoptosis and one, usually the closest to the chalaza, remains functional. This is called megaspore selection and the development of the female gametophyte from one megaspore is called monospory. Cases of bisporic or tetrasporic developments are known, in which the second or both meiotic divisions, respectively, are not followed by formation of a cell wall. In such cases, the development of the gametophyte starts from a megaspore with two or four haploid nuclei (Rutishauser, 1982).

During megasporogenesis and megaspore selection, megaspore cells accumulate callose in their cell wall, a β-1,3-glucan whose function is not yet well understood. Since callose seems to control molecular movement through plasmodesmata (Maule et al., 2012), it might regulate symplastic cell continuity and intercellular communication, and provide a physical barrier that isolates megaspore cells from the neighbouring tissue and promotes their demise. The functional megaspore is usually the one with less accumulated callose, due to its chalazal location and its late origin, and losses the callose as it begins its transition to gametogenesis (Drews and Koltunow, 2011).

During gametogenesis, the surviving megaspore in monosporic ovules enlarges oblongly and undergoes three consecutive mitosis, without cytokinesis, resulting in an eight-nucleate coenocyte with four identical haploid nuclei at each pole. In ovules exhibiting bisporic and tetrasporic developments, the gametogenesis is shortened to two or one mitotic divisions, respectively, and also results in an eight-nucleate gametophyte. During the last mitosis, a process of cellularization leads to the formation of phragmoplasts and cell plates between sister and non-sister nuclei, and cells develop cell walls (Webb and Gunning, 1994; Christensen et al., 1997). During this process, one nucleus from each pole in the developing gametophyte migrates toward the centre and the two can fuse. At this point, the female gametophyte is a multicellular structure consisting of three antipodal cells, a central cell with one homodiploid nuclei, two synergid cells and an egg cell.

During the processes of megaspore mother cell differentiation, megaspore selection and female gametophyte development, the cells and the coenocytic gametophyte become polarized at the cellular and subcellular levels (Willemse and van Went, 1984; Huang and Russell, 1992). The egg and synergid cells develop at the micropylar end of the gametophyte, whereas the antipodal cells develop at the chalazal end. The monosporic development of this female gametophyte with eight nuclei and seven cells is of the Polygonum type mentioned above (Figure 2.1). In certain species, such structural organization is quickly modified by cell death or growth events. In grasses in general, antipodal cells continue to grow mitotically and their number may vary (Huang and Russell, 1992; Drews et al., 1998). In *Arabidopsis*, on the other hand, antipodal cells undergo programmed cell death immediately after fertilization (Russell, 1993).

2.2.2 The Male Gametophyte

In bisexual flowers, the male gametophyte develops in anthers synchronously with the female one in ovules (McCormick, 2004). During microsporogenesis, diploid (2*n*) pollen mother cells differentiate, undergo meiosis and give rise to four haploid (*n*) recombinant microspores. Each microspore begins to increase in size and divides through an asymmetric mitosis that creates a large vegetative cell and a small generative cell (Borg et al., 2009). In many plants, the generative cell undergoes another mitosis before pollination and forms two sperm cells, but in most cases the generative cell completes its second mitosis during pollination and pollen tube growth (Friedman, 1999).

2.2.3 The Progamic Phase: From Pollination to Double Fertilization

The period between pollination and fertilization is called the progamic phase, a post-pollination period in which pollen grains interact with sporophyte-derived tissues and become sexually mature (Williams and Reese, 2019). Pollen landing on the stigma is followed by hydration and germination of a pollen tube that grows through the carpel's transmitting tissue guided by peptides secreted by the female gametophyte. The growing pollen tube first reaches the ovule, then the micropyle, and later penetrates the female gametophyte through the filiform apparatus of the synergid cells. The pollen tube burst releases the sperms, and the synergid cells degenerate (Dresselhaus and Coimbra, 2016). Then, following complex molecular signalling, one sperm cell moves and fuses with the egg cell to form a diploid (2*n*) zygote, and the second sperm cell fuses with the central cell to form a triploid (3*n*) primary endosperm nucleus (Sprunck, 2020). The time elapsed between pollination and the fertilization process is mediated by pollen performance traits, and in most angiosperm species, it takes about 3 hours or less (Mascarenhas, 1989).

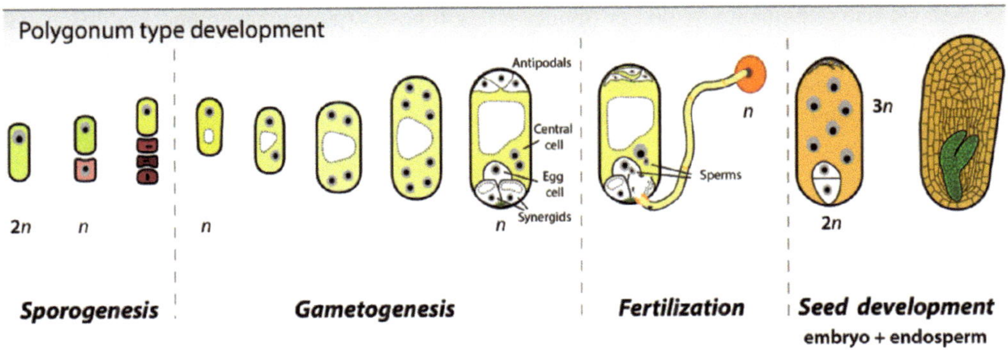

FIGURE 2.1 Polygonum type female gametophyte and seed development. During sporogenesis, the diploid (2*n*) megaspore mother cell produces four haploid (*n*) megaspores, and the chalazal one enters gametogenesis and gives rise to a reduced (*n*) female gametophyte carrying eight nuclei in seven cells. The egg and central cells are fertilized and develop into an embryo (green) and endosperm (brown), respectively, completing the formation of a viable seed (a nuclear endosperm from a dicot is here depicted).

2.2.4 The Making of a Seed: Embryo and Endosperm Formation

Post-fertilization, there is a maternal-to-zygotic transition where the zygotic genome is activated to regulate embryogenesis (Armenta-Medina and Gilmor, 2019). Yet, early embryogenesis displays maternal effects on the regulation (either sporophytic or gametophytic) of seed development. The zygote divides asymmetrically and forms a small apical cell and a large vacuolate basal cell. This basal–apical axis is central for cell fate determination and morphogenesis. While the apical cell lineage contributes most of the tissue types to the mature embryo, the basal cell lineage generates the suspensor and part of the root meristem (Kimata et al., 2016; Armenta-Medina and Gilmor, 2019). Further embryo growth can show species-specific patterns of cell divisions and hence display different embryogenesis stages (Natesh and Rau, 1984; Johri et al., 1992). In the dicot plant *Arabidopsis thaliana* (L.) Heynh., embryogenesis continues through a 2-4 cell stage, octant (8 cell), dermatogen (16 cell), globular (32 cell), heart (>1000 cell), and the torpedo stage in which epidermis, vascular, shoot and root meristems are defined (Armenta-Medina and Gilmor, 2019). Based on the patterns of cell divisions and contributions of basal and apical cell lineages, six patterns of embryogenesis are recognized in dicots: Crucifer/Onagrad type; Asterad type; Solanad type; Caryophyllad type, Chenopodiad type and the Piperad type (Johri et al., 2001). In monocots, the first embryo development stages up to the proembryo stage are similar to those in dicots (but see Zhao et al., 2017), and develop following a less predictable cell division pattern, lacking a typical suspensor, and showing a bent apical-basal axis. Later stages include the scutellar and coleoptilar stages which establish the main architecture of monocotyledonous embryos exhibiting a single cotyledon and an additional dorsoventral axis.

The primary endosperm nucleus divides several times mitotically, as well as its daughter nuclei, to constitute the endosperm. In most sexual angiosperms the endosperm is triploid and carries two maternal and one paternal genome ($2m{:}1p$). Even though violations of such maternal-to-paternal genome ratio lead to developmental defects (Scott et al., 1998; Erilova et al., 2009), species can tolerate changes in this ratio to some extent (e.g., Povilus et al., 2018; van Dijk et al. 2020), and in a few cases the endosperm might have a different ploidy like in *Butomopsis* Kunth (apparently holding a diploid endosperm; Johri, 1936) or in *Fritillaria* L. (pentaploid endosperm; Vijayaraghavan and Prabhakar, 1984). After the mitosis of the primary nucleus, the formation of a cell wall is used to classify the endosperm in cellular or nuclear types. An intermediary type exists and is called helobial. In the cellular type, the first and subsequent divisions are followed by cell wall formation. In the nuclear type, the nuclei undergo free-nuclear divisions and eventually become cellular. Micropylar and chalazal haustoria are observed in different plant groups. In the helobial type, the first mitosis produces two asymmetrical chambers, the micropylar one being usually the larger. Free-nuclear divisions occur in the micropylar chamber before cellularization, while the nucleus in the chalazal one remains undivided or divides a few times (Vijayaraghavan and Prabhakar, 1984). During late embryogenesis, the endosperm plays a crucial nourishing role that promotes embryo growth and germination. In mature seeds, the endosperm serves as a repository of polysaccharides, proteins, DNA and lipids. Finally, the endosperm has a morphogenetic role in the regulation of embryogenesis.

2.3 MAIN FEATURES OF SEXUAL REPRODUCTION

Overall, the main characteristics of sexual seed formation can be summarized as follows:

1 A megaspore mother cell differentiates, produces (three-)four haploid meiotic megaspores; one selected megaspore undergo mitosis and develops into a multicelled female gametophyte;
2 Fertilization of the haploid egg cell and central cell by two haploid sperm cells is required to reestablish the diploid sporophytic ploidy in the zygote and for proper embryo and endosperm developments;
3 Produced seeds and seedlings are genetically, physiologically and morphologically different.

2.4 APOMIXIS AND THE ALTERNATIVES TO SEXUAL SEEDS

Apomixis is the asexual formation of seeds (but see Chapter 1). Apomixis involves modifications in the normal development of a seed in which the embryo retains the same genetic information as the mother plant. Therefore, from a practical point of view, it basically implies a unique and singular feature: the formation of clonal seed embryos. However, from a biological standpoint, the making of a seed carrying a clonal embryo is more complicated. As stated above, the seed is a complex dispersal unit. Evolutionarily, the seed consists of not one but two embryos (Friedman, 2001), one of which evolved to have a transient nutritional function that ensures the survival of the pair and is called the endosperm. This surrogated embryo or endosperm is usually the only one fertilized in apomictic plants.

Clonal formation of the embryo in the seed can occur through two main developmental mechanisms. The first, the simplest and most frequent in nature (Hojsgaard et al., 2014a, but see Chapter 8), is through the acquisition of an embryonic fate by a somatic cell in the ovule and the development of an embryo carrying the maternal genetic information. This mechanism is called sporophytic apomixis or adventitious embryony and requires concurrent sexual development (endosperm) to produce a functional seed (see details below). The second mechanism parallels sexual development in many aspects, and involves the formation of unreduced female gametophytes; hence, it is called gametophytic apomixis. Under this mechanism, several alternative types of apomixis are observed in different plant species and families (Asker and Jerling, 1992; Carman, 1997; Hojsgaard et al., 2014a) and are subdivided into diplospory and apospory. In either case, the unreduced egg-cell in the gametophyte develops without fertilization into a clonal embryo and – as in sporophytic apomixis – in many cases requires the sexual development of the endosperm to produce a functional seed (see details below).

Even though the developmental pathways in sporophytic and gametophytic apomixis are evidently different, the fact that in both cases the sexual development (i.e., meiosis, gametogenesis, and the function of the female gametophyte) is partially altered or completely depleted, and that the asexual cell lineage arises from single cells that acquire a totipotent identity to form a clonal embryo (whether or not it undergoes gametogenesis), might suggest a possible connection between them. However, no molecular link between the sporophytic and gametophytic pathways has yet been verified. Perhaps it is now, in the era of "omics" technologies, that we will be able to answer this question by unveiling the molecular genetics of apomixis and whether or not there is a molecular connection – if any – between apomictic developmental alternatives. Still, both sporophytic and gametophytic apomixis probably share an evolutionary role, as these pathways possibly provide a selective survival advantage (see Chapter 8).

2.4.1 Sporophytic Apomixis

According to Strasburger (1879), any embryo arising from maternal sporophytic tissues (other than the female gametophyte) in the ovule is called an adventive embryo (Naumova, 1993). Within the ovule, only two sporophytic tissues are known to form adventive embryos: the nucellus and the integuments. Webber (1940) was the first to suggest classifying adventitious embryony into nucellar and integumentary embryony based on earlier observations. Thus, the formation of adventitious embryos in these maternal tissues occurs independently of the gametophytic development (either meiotic or apomeiotic) of the seed, reason for which is often associated with the presence of more than one embryo per seed (the normal sexual embryo plus one or more apomicts) and is called polyembryony. However, polyembryos can also arise from gametophytic cells other than the egg-cell, such as synergids or antipodal cells, or by cleavage of zygotic embryos (e.g., Secor and Russell, 1988; Aleza et al., 2010), or through the formation of two archesporial cells and simultaneous development of two meiotic female gametophytes (e.g., Duthie, 1915; Vinogradova, 2013). These cases in which the additional embryos are genetically different to the mother plant are not considered here. For a detailed analysis of the multiple origins of polyembryos see Batygina and Vinogradova (2007).

Polyembryony has been grouped into induced and autonomous (Lakshmanan and Ambegaokar, 1984). While in the first one adventitious embryo develop only after pollination (i.e., pseudogamously), in the second this happens independently of pollination (autonomously). However, according to Naumova (1993), initial cell differentiation and early embryo developments do not depend on pollination, fertilization or the presence of sexual embryos.

2.4.1.1 Initial Cell Growth and Embryo Formation

Initial cells of adventitious embryos are called primordial initial cells (Kobayashi et al., 1978), embryocytes (Naumova, 1993), adventitious embryo precursor cells (Bittencourt and Moraes, 2010) or simply initial cells. Such cells (like the aposporous initial cells in gametophytic apomicts) acquire a new, visually recognizable identity in the ovules during the flowering period – or shortly after – by having an enlarged (sometimes irregularly shaped) nucleus and nucleoli, and optically dense and/or vacuolized cytoplasm compared to surrounding nucellar or integumentary cells. Initial cells are usually located in the micropylar zone of the nucellus and in the inner integuments but can also appear in the lateral and chalazal zones of the nucellus (Wakana and Uemoto, 1987, 1988; Bittencourt and Moraes, 2010). During early mitotic growth, embryocytes undergo relevant structural changes such as vacuolization and loss of plasmodesmata (Naumova, 1993), likely aimed at shaping the embryogenesis programme without interference from surrounding signals.

Nucellar and integumentary embryos do not arise simultaneously, so they may present different stages of development (Figure 2.2). Initial growth is characterized by intense cell divisions and lack of polarity (i.e., in direction or sequence), creating multicellular structures with no defined shape. Since their growth is mostly irregular, a suspensor is frequently not observed (Lakshmanan and Ambegaokar, 1984; Teppner and Klein, 1985). With the gradual increase in cell number, a globular-like stage is reached in which the embryoderm is formed. Later development follows regular embryonic organogenesis, but the earliest stages show no analogy with any type of zygotic embryogenesis. According to Naumova (1993), this is due to an origin outside the female gametophyte and, therefore, free of the biochemical gradient of auxins, the close influence of the endosperm and hormones synthesized by the suspensor (Kawashima and Goldberg, 2010; Sundaresan and Alandete-Saez, 2010). This view is supported by cases where the initial cells grow by accessing the female gametophyte and develop suspensor-like structures and the pattern of zygotic embryogenesis can be determined (e.g., Bittencourt and Moraes, 2010), and could explain former observations (e.g., Rangaswamy, 1959). In unpollinated ovules without endosperm, the embryos are arrested at either globular or torpedo-like stages (Wakana and Uemoto, 1987; Naumova, 1993).

Embryos in mature seeds with this type of apomixis regularly exceed two and may reach up to 50 (Johri, 1984). In most cases, the number of polyembryonic seeds in species with sporophytic apomixis is higher than 50% (Kordyum, 1961; Salomão and Allem, 2001; Singh and Soule, 1963; Bittencourt and Moraes, 2010).

2.4.1.2 Sexual Development in Sporophytic Apomicts

Sexual development progresses normally in most species with adventitious embryony (Figure 2.2). The megaspore mother cell produces three-four megaspores after meiosis, one of which remains functional. Female gametogenesis in most species follows a monosporic Polygonum type development (Naumova, 1993), but cases with bisporic or tetrasporic female gametophytes development have been reported (Carano, 1926; Håkansson, 1953). Despite regular sexual development, deviations like arrest of meiosis during prophase, disruption of female gametophyte polarity, egg-cell degeneration or haploid parthenogenesis (the formation of embryos from unfertilized egg or synergid cells) have been observed in some species with adventitious embryony (e.g., Ganong, 1898a, 1898b, Hegelmaier, 1903; Naumova, 1993). Since similar deviations in female gametophyte development, fertilization and endospermogenesis have occasionally been observed in sexual species (Naumova, 1993), it is unclear whether there is a causal connection between such alterations of the normal sexual developmental programme and adventitious embryony.

Male meiosis and development of male gametophytes are usually normal, and plants display high fertility. However, in some cases plants with adventitious embryos showed drastic meiotic abnormalities and collapse of pollen grains leading to sterility (Naumova, 1993). Fertilization can progress normally, as can embryo and endosperm development. The first mitotic divisions in the zygote are asymmetric and formative, creating cells with distinct identity and crucial for the differentiation and development of sexual embryos (Kajala et al., 2014), unlike adventitious embryos. In a few species with adventitious embryony, the endosperm may develop autonomously without fertilization (Haberlandt, 1922; Edman, 1931; Gorham, 1953; Naumova, 1993). Autonomous endosperm development is known to occur in species exhibiting gametophytic apomixis (see Section 2.4.2), which might suggest that a similar stimulus or regulatory mechanism may underlie unfertilized endosperm formation.

2.4.1.3 Main Features of Sporophytic Apomixis

Overall, the main characteristics usually observed in species with adventitious embryony can be summarized as follows:

1. Somatic cells acquire an embryogenic identity, grow into an undifferentiated mass of cells first to later produce an embryo morphologically similar to zygotic embryos (with the exception of the usual absence of a suspensor) that can develop simultaneously in the ovule;
2. The development of adventitious embryos is irregular and can occur independently from fertilization, but final development and survival depend upon presence of endosperm in the seed (Orchidaceae, Podostemaceae and Trapaceae are exceptions where the function of the endosperm has been replaced by suspensor haustoria);
3. Sporophytic apomictic embryos are genetic copies of the mother plant and seedlings display morphological features alike zygotic seedlings.

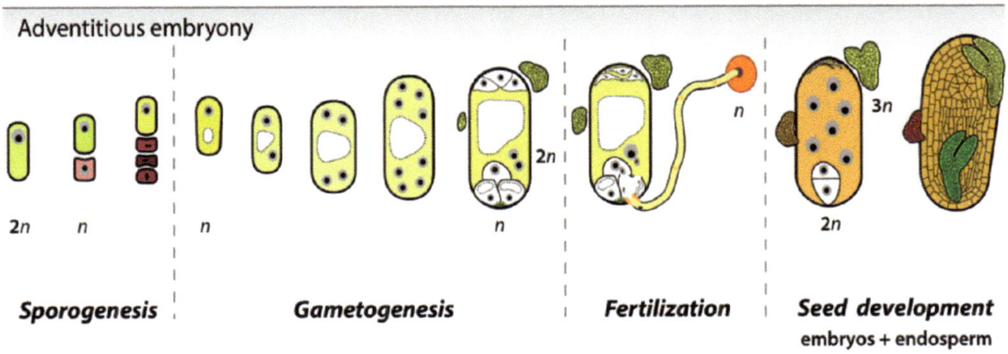

FIGURE 2.2 Female gametophyte and seed development in a plant exhibiting adventitious embryony. During maturation and fertilization of the female gametophyte (see references in Figure 2.1), initial cells in the nucellus or integuments acquire an embryogenic fate and grow to develop embryos carrying the maternal genetic configuration. The sexual development progress as in sexual species (here a monosporic Polygonum type development is represented) and fertilization of the reduced (*n*) female gametophyte is required to develop the embryo and endosperm and complete the formation of a viable seed (sexual embryo in green, viable adventitious embryo in yellow).

2.4.2 Gametophytic Apomixis

The term *gametophytic apomixis* was first introduced by Stebbins (1950) to represent those pathways of asexual seed formation developing through a gametophyte, and, hence, characterized by an alternation of generations. Although this point perfectly represents and serves to graphically separate the alternative developmental pathways among apomicts, we shall emphasize that from a genetic point of view there is no such alternance of generations *sensu stricto*, as new seed embryos are created from a single cell with a somatic genetic constitution.

The pathways of gametophytic apomixis are grouped into diplosporous and aposporous developments (although over time different authors have used different names; see e.g., Battaglia, 1991). In diplosporous species, modifications that occur during the meiotic cell division produce an unreduced megaspore that carries unrecombined chromosomes and develops into a female gametophyte. Thus, the female gametophyte arises from the meiotic germline. In aposporous species, the female gametophyte arises from a somatic cell in the nucellus of the ovule.

2.4.2.1 Diplosporous Developments

The development of an unreduced female gametophyte from the generative cell shows four different alternative developments: the bipolar Antennaria type, Taraxacum type and Ixeris type, and the monopolar Eragrostis type (Figure 2.3). These developments are named after the species on which they were first described.

2.4.2.1.1 The Antennaria Type

This type of diplospory was first described by Juel (1900) in *Antennaria alpina* (L.) Gaertn. In the development of the Antennaria type, meiosis in the megaspore mother cell is completely omitted (Figure 2.3). The megaspore mother cell, after a relatively long interphase, enlarges and undergoes a mitotic division. The two resulting nuclei move to opposite poles, separated by a vacuole, creating a bipolar female gametophyte. These two nuclei undergo two more mitotic divisions resulting in eight nuclei, four at each pole. Three of the micropylar end nuclei form the egg apparatus (constituted by an egg-cell and two synergid cells), and three chalazal end nuclei form antipodal cells. The other two nuclei, one from each pole of the gametophyte, function as polar nuclei in the central cell (Figure 2.3).

2.4.2.1.2 The Taraxacum Type

This type was first reported in *Taraxacum officinale* F.H. Wigg. by Juel (1904, 1905). In the Taraxacum type development, the megaspore mother cell enters the meiotic prophase, but due to a failure in the formation of the synaptonemal complex leading to asynapsis or desynapsis between homologous chromosomes, chiasma formation fails (asyndesis) along with chromosome pairing. The resulting univalents remain dispersed over the spindle and the first meiotic division becomes a restitution nucleus (Figure 2.3). The second meiotic division then continues with an unreduced number of chromosomes. Meiosis II is regular and produces a dyad of unreduced cells of which the chalazal dyad remains functional while the micropylar cell degenerates. The nucleus of the functional dyad cell undergoes three mitotic divisions which, after cellularization, give rise to a female gametophyte carrying eight nuclei and seven cells.

2.4.2.1.3 The Ixeris Type

The Ixeris type was first described by Okabe (1932) in *Ixeridium dentatum* (Thunb.) Tzvelev [≡*Ixeris dentata* (Thunb.) Nakai]. As in the Taraxacum type, here meiosis I gives rise to a restitution nucleus and meiosis II is regular (Figure 2.3). However, there is no wall formation after meiosis II and a single binucleate spore is formed. This spore undergoes two mitotic divisions and develops an unreduced female gametophyte with eight nuclei and seven cells.

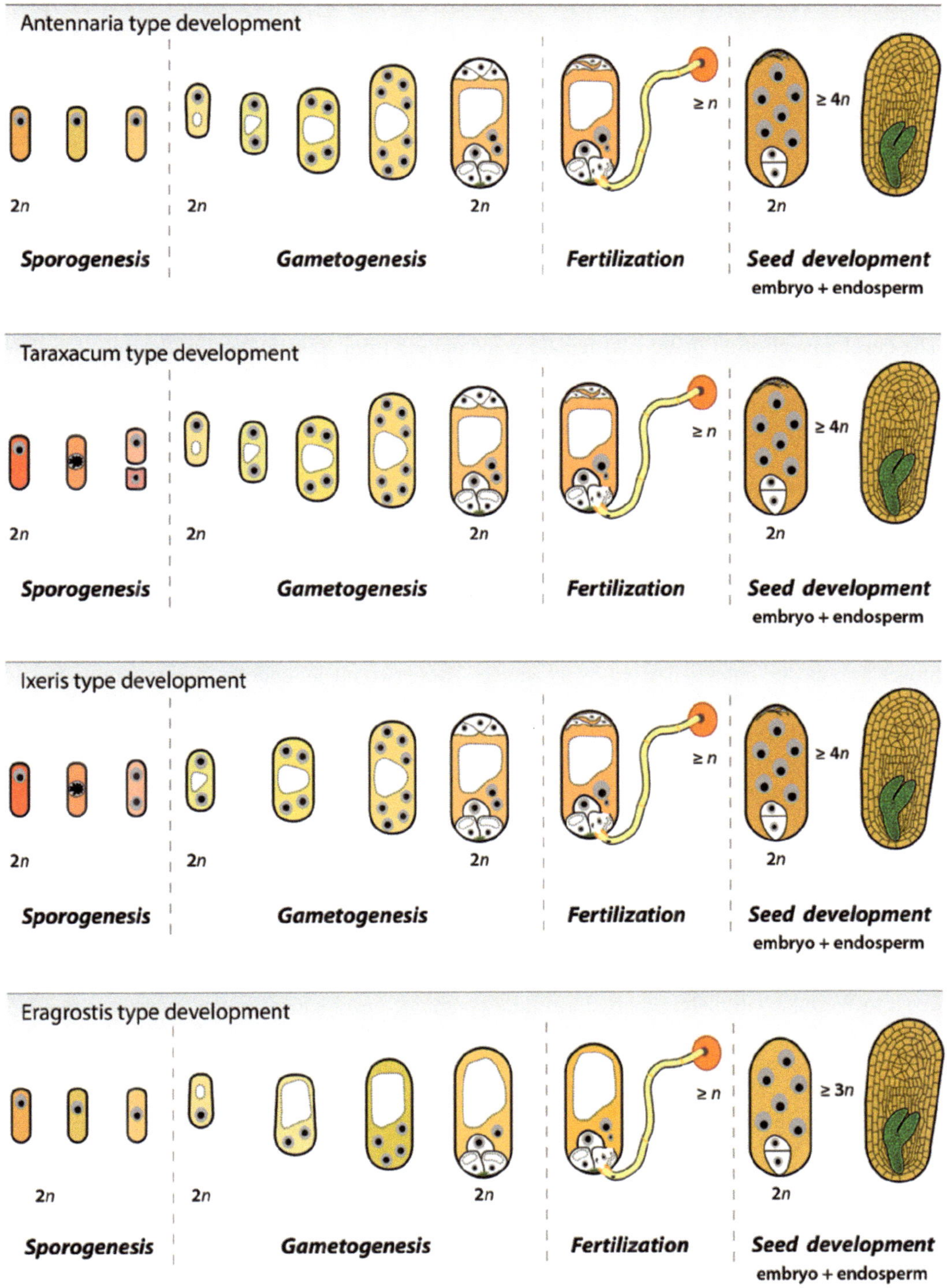

FIGURE 2.3 Different diplosporous developments are observed in angiosperms. In the Antennaria type the diploid (2*n*) megaspore mother cell does not undergo meiosis, enters gametogenesis and gives rise to an unreduced (2*n*) female gametophyte carrying eight nuclei in seven cells. In the Taraxacum type, the diploid (2*n*) megaspore mother cell enters meiosis, but chromosome pairing fails, and a restitution nucleus is formed during the first (reductional) division. The second (equational) division proceeds as usual and produces two diploid megaspores, the chalazal enters gametogenesis and gives rise to an unreduced (2*n*) female gametophyte carrying eight nuclei in seven cells. The Ixeris type differs from the Taraxacum type in the lack of cytokinesis after the meiotic restitution (see Chapter 5). In the Eragrostis type there is no polarization, and the enlarging megaspore produces a female gametophyte with four unreduced nuclei and four cells. In all cases, the egg cell develops an embryo by parthenogenesis and the central cell develops the endosperm after fertilization (pseudogamy) or autonomously.

2.4.2.1.4 The Eragrostis Type

The Eragrostis type was described by Liebenberg (1961, in Vorster and Liebenberg, 1984) and Streetman (1963, in Crane, 2001) for different *Eragrostis* species. This type of female gametophyte development starts as the Antennaria type, but there is no polarization and two mitotic divisions in the enlarging megaspore mother cell produce a monopolar gametophyte with four unreduced nuclei and four cells at the micropylar pole, an egg cell, two synergids and the central cell with one polar nucleus (2*n*).

2.4.2.2 Aposporous Developments

The development of an unreduced female gametophyte from somatic cells in the nucellus shows different alternative developments: the bipolar Hieracium type and the monopolar Panicum type, Setaria type and Paspalum type (Figure 2.4). Similarly, such developments are named after the species in which they were first described.

2.4.2.2.1 The Hieracium Type

This type of apospory was first discovered by Rosenberg (1906, 1907) in species of *Hieracium* subgenus *Pilosella*. Nucellar cells surrounding the megaspore(s) cell(s) enlarge and function as aposporous initials cells. These diploid initial cells acquire a new identity, their cytoplasm becomes optically dense and vacuolated, enlarge their nucleus and undergo three mitotic divisions. The spindle of the first mitosis is usually parallel to the nucellus and produces two nuclei, each of which moves to opposite poles separated by a vacuole. These two nuclei undergo two more mitotic divisions resulting in eight nuclei, four at each pole. After cellularization, the female gametophyte has eight nuclei and seven cells, organized into a three-celled egg apparatus towards the micropyle, three antipodal cells towards the chalaza and two diploid nuclei that usually fuse into a tetraploid secondary nucleus in the central cell.

2.4.2.2.2 The Panicum Type

Even though this type of aposporous development is called Panicum type, it was Narayan (1951) who reported it for the first time in *Cenchrus longisetus* M.C.Johnst., *C. orientalis* (Rich.) Morrone and *C. setaceus* (Forssk.) Morrone [sub *Pennisetum villosum* Fresen, *P. orientale* Rich. and *P. setaceum* (Forssk.) Chiov., respectively]. Since Narayan's work was in his PhD thesis, credit was given to Warmke, who later reported a similar apomictic development in *Panicum maximum* Jacq. (Warmke 1952, 1954). Battaglia (1991) pointed out the priority of Narayan's findings by providing a detailed analysis and suggesting this type should be renamed Pennisetum type. However, giving the wide acceptance of the name Panicum type and the current taxonomic transfer of *Pennisetum* species (including the taxa studied by Narayan) to *Cenchrus* (based on monophyly and name priority; Chemisquy et al., 2010), it seems appropriate – at least at the moment – to retain the name Panicum type.

In this type, like in the Hieracium type, one or more somatic cells acquire a new identity and function as initial of apospory cells. These cells enlarge, become vacuolated, and their nucleus is positioned at the micropylar end. After the first mitosis, the two nuclei remain in the same (micropylar) pole. These two nuclei undergo one more mitotic division, resulting in four nuclei, which then form a mature unreduced female gametophyte carrying a three-celled egg apparatus plus a diploid polar nucleus. Antipodal cells in this type are absent. In some species showing this type of development, the occurrence of female gametophytes with a two-celled egg apparatus (bearing one synergid cell) and two nuclei in the central cell has been observed (e.g., Ozias-Akins et al., 2003).

2.4.2.2.3 Variations to the Panicum Type: The Setaria Type and Paspalum Type

There are at least two types of aposporous development in which the female gametophyte lacks antipodal cells but maintains binucleate central cells. Even though these developments could be

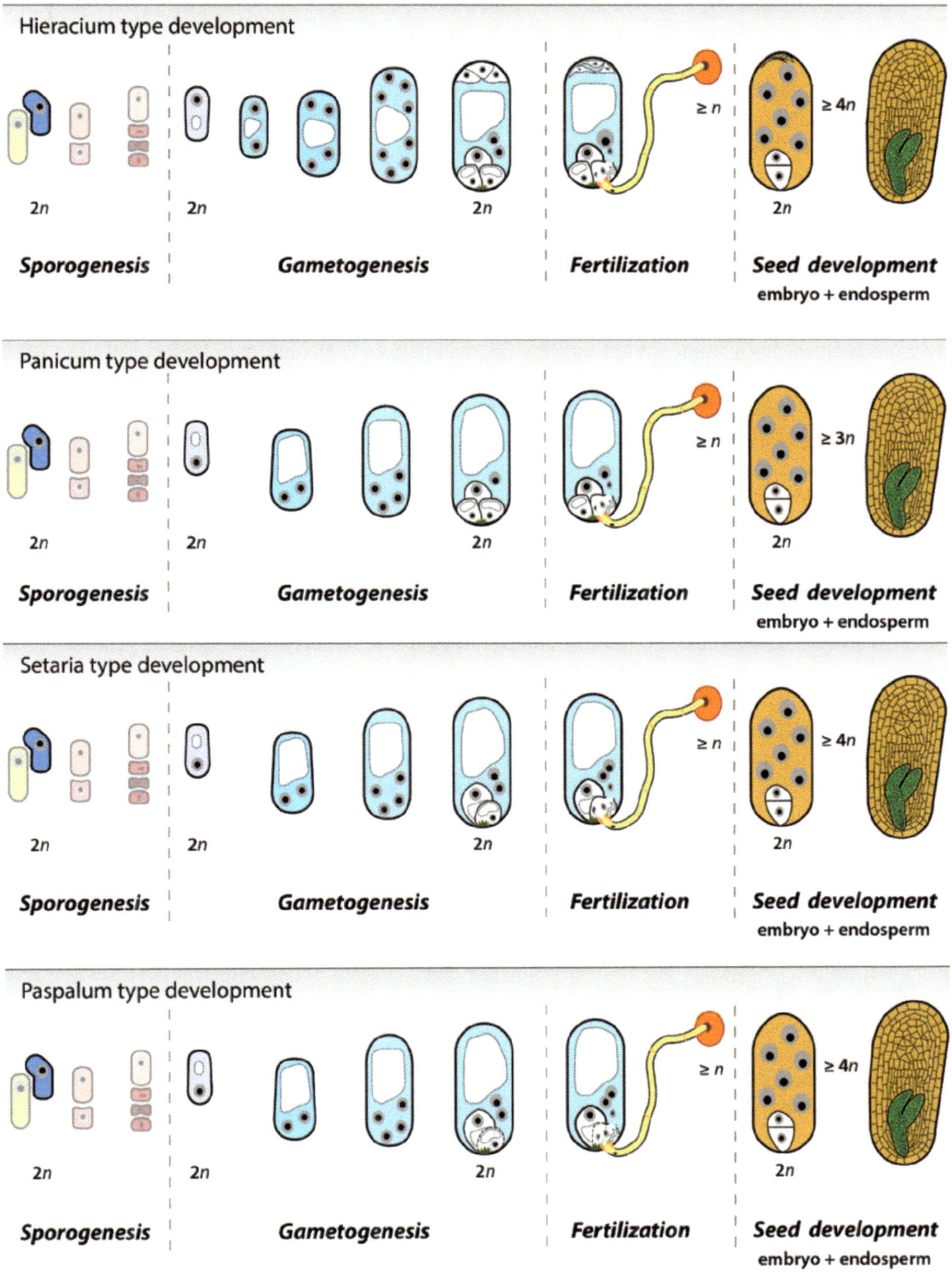

FIGURE 2.4 Different aposporous developments are observed in angiosperms. In the Hieracium type a nucellar (2*n*) cell acquire a gametogenetic fate and divides mitotically to create a coenocyte first and then an unreduced (2*n*) female gametophyte carrying eight nuclei in seven cells. The egg cell develops an embryo by parthenogenesis and the central cell develops the endosperm after fertilization. In the Panicum type, the nucellar (2*n*) cell enters gametogenesis but after the first mitosis nuclei remain in the same pole, and a second mitosis produces a coenocyte with four micropylar nuclei that became cellularized. The unreduced (2*n*) female gametophyte carries four nuclei in four cells. The egg cell develops a parthenogenetic embryo and the central cell with one unreduced nucleus develops the endosperm after fertilization. In the Setaria type and Paspalum type, gametogenesis in the somatic cell progress with variations from the Panicum type, both in number of mitosis and cellularization, and the central cell receives two unreduced nuclei.

considered as variations of the Panicum type, we refer to them separately here because (1) they occur consistently and are characteristic of species of certain genera of grasses, and (2) their anatomical differences allow the detection of apomixis by flow cytometry (see Chapter 3) and are therefore highly relevant from a practical point of view.

The first one was described in species of *Bothriochloa* Kuntze and *Dichanthium* Willemet (Celarier and Harlan, 1957) as a less frequent type of female gametophyte, secondary to the common Panicum type gametophyte characteristic of these species (see Chapter 4). Perhaps this is the reason why this type of female gametophyte development was originally considered a variant of the four-nucleate Panicum type gametophyte. This type of development involves a monopolar development similar to that followed by aposporous cells in the Panicum type, except that after cellularization the resulting gametophyte has three (instead of four) cells and two (instead of one) nuclei forming the central cell. This subtle difference has drastic consequences in embryological and cytological studies aimed at recognizing apomixis (see Chapter 3). In rare occasions, a fifth nucleus might be present by adding the second synergid cell to the egg apparatus and keeping the central cell binucleate. Yet, its origin is not clear. Almost simultaneously to Celarier and Harlan (1957), it was discovered that this type of female gametophyte development occurs exclusively in ovules of *Setaria* species (Emery, 1957), in which no other type of female gametophyte has been observed. The specific and exclusive occurrence of this type of development in certain species excludes the thesis that this type of development is simply a spontaneous variation of the Panicum type and point to a possible functional basis for the occurrence of such anatomical modification. Since all these species are pseudogamous, the relative contributions of the maternal and paternal genomes to the primary endosperm nucleus probably have a yet to be unveiled role for the progression and development of the endosperm and in the establishment or maintenance of apomixis in these species (but see below). Hence, it seems appropriate to name this development as Setaria type (Figure 2.4).

The second type of development in which the unreduced female gametophyte lacks antipodal cells and maintains a binucleate central cell is the one found in *Paspalum* L. In most aposporous species of this genus, the polarity of the growing unreduced female gametophyte is not strictly monopolar as in the Panicum type and Setaria type, and due to an unstable pattern of cell divisions and cell differentiation, a common situation in mature gametophytes is the presence of two cells (synergids are absent) and three nuclei, as well as gametophytes with three cells and four nuclei, or sometimes four cells and five nuclei (Quarin, pers. com.). This type of development is named Paspalum type (Chen et al., 2000; Ortiz et al., 2013). Since such female gametophyte anatomy contradicts the C^4 cellularization hypothesis that requires the presence of at least four nuclei in the micropylar pole (Battaglia, 1991), and because synergid cell death has been observed in other grasses with Panicum type development (e.g., Rathi et al., 2020), the case of *Paspalum* is interesting and might help to understand the role of polarization in the establishment of functional gradients and cell differentiation within the female gametophyte. A more detailed analysis of the ontogeny of this type of female gametophyte will provide clues to the nature of this aposporous development.

2.4.2.3 Sexual Development and Embryo and Endosperm Formations in Diplosporous and Aposporous Apomicts

In many apomicts, male sporogenesis and gametogenesis is normal and follows the same developmental pattern described above for sexuals. Abnormal male meiosis has been observed in several apomicts, especially among daisies and roses. Although a strict correlation between apomictic development and abnormalities in pollen mother cells is not generally observed (Stebbins, 1941; Asker and Jerling, 1992), in different species, gametophytic apomixis (mostly diplospory) is associated with the production of unreduced pollen (e.g., by synapsis failure and first division restitution in *Boechera* Á. Löve & D. Löve). The progamic phase and delivery of male gametes are apparently not altered. The developmental changes with the greatest consequences for seed formation in gametophytic apomicts occur in the ovule.

From a developmental point of view, in diplosporous species sexuality is impaired by apomictic development, as sporogenesis in the germline is modified to produce unreduced spores. In aposporous species this does not occur, as each pathway arises from different cell lineages. Hence, the germline may produce meiotic spores and form reduced female gametophytes, and simultaneously, initial of apospory cells could develop into unreduced gametophytes (Asker and Jerling, 1992). However, from a genetic point of view, gametophytic apomixis is associated with drastic genetic changes (see e.g. Sharbel et al., 2010; Pellino et al., 2013; Schmidt et al., 2014; Galla et al., 2019) that substantially alter normal development and deplete meiosis and sexual reproduction to a minimal level.

Such genetic and regulatory changes cause heterochronic developments to the normal sexual pathway, as has been shown by different studies comparing sexual and apomictic developments (either diplosporous or aposporous) (Asker and Jerling, 1992; Leblanc et al., 1995a, 1995b, 1995c; Koltunow et al., 1998; Grimanelli et al., 2003; Bradley et al., 2007; Galla et al., 2011; Sharma et al., 2014). While in some aposporous species the concurrence of sexual and apomictic developments within the ovule appear to accelerate meiosis (Carman et al. 2011), in others it was clearly delayed or arrested (Hojsgaard et al., 2013, 2014b, Soliman et al., 2019).

In many species, apomictic (unreduced) gametophytes display a modification from the normal double fertilization observed in sexual ovules to a single fertilization event. This process is called pseudogamy, wherein only the central cell is fertilized as a requirement for proper development of the endosperm, while the unreduced ($2n$) egg cell develops the seed embryo by parthenogenesis, i.e., without fertilization by one of the male gametes (Figures 2.3 and 2.4). Even though *centrogamy* might better refer to this event (see Solntzeva, 2003), here we will use the term *pseudogamy* owing to its wider acceptance. In some cases, the egg cell may initiate the embryo formation before the central cell is fertilized, but embryos are usually arrested at globular stage and seed completion occurs only after fertilization of the central cell. In a number of species, unreduced gametophytes develop into a functional seed autonomously, i.e., without fertilization of the egg cell or central cell. However, autonomous endosperm development may occasionally require pollination by compatible pollen for proper endosperm development (Bicknell et al., 2003).

In all species with gametophytic apomixis, development of the endosperm and the embryo has not been studied as intensively as all previous developmental stages (Asker and Jerling, 1992). A few studies have shown that they follow patterns similar to those observed in sexual ovules of the same (or closely related) species. However, there is a crucial difference in endosperm formation in these apomicts compared to their sexual relatives. The central cell in most apomictic species is generally insensitive to changes in maternal and paternal genome ratios, a critical barrier to hybridization and polyploidization in sexual species (see, e.g., Köhler et al., 2010; Lafon-Placette and Köhler, 2014). Thus, most species with gametophytic apomixis accept deviations in maternal:paternal genome ratio, from 4:0 in species with autonomous endosperm development to a range between 1:1 and 8:1 in pseudogamous species. Exceptions are species exhibiting the Panicum type development (e.g., *Panicum* spp., *Cenchrus/Pennisetum* spp.) in which the contribution of 2 maternal:1 paternal genomes to the primary endosperm nucleus is maintained as in sexual species.

Overall, even when meiosis and sexual development can proceed and produce sexual seeds in ovules of apomictic plants where diplosporous or aposporous apomixis development fails, in most flowers, seeds develop from unreduced female gametophytes (e.g., Hojsgaard et al., 2013; Hand et al., 2015). Gametophytic apomixis is intrinsically associated with high production of asexual seeds and low formation of sexual seeds. Consequently, most – if not all – species exhibiting gametophytic apomixis are facultative apomicts. Even though cases of obligate apomicts (e.g., in *Zeuxine* sp., Kant and Verma, 2012; *Paspalum* spp., Hojsgaard et al., 2013; *Potentilla* spp., Asker, 1971, Dobeš et al., 2015) have been recorded, they are based on temporary assessments (usually during one flowering season) of flowers/seeds from single or a few individuals. Therefore, a similar situation is unlikely to occur under natural conditions, where apomictic plants are subjected to variable environmental stressors. The observed influence of environmental factors on expression of reproductive pathways and the genotypic variability found in natural populations (see e.g. Klatt et al., 2016; Karunarathne et al., 2020a) support the view that obligate apomixis occurs only as a temporary reproductive expression.

2.4.2.4 Main Features of Gametophytic Apomixis

1. Meiosis is bypassed (apomeiosis) and a spore-like cell forms an unreduced female gametophyte;
2. Parthenogenesis in the unreduced egg cell simplifies fertilization to a single event (absent in species with autonomous development) required for endosperm and seed development;
3. Gametophytic apomictic embryos are genetic copies of the mother plant and seedlings have similar morphological features to zygotic seedlings.

2.5 COMMON FACTS AMONG APOMICTS

Despite the developmental variability observed among the different types of apomixis, two basic functional changes in the sexual development can be highlighted:

- The acquisition of a parthenogenetic or embryogenic identity (these terms may represent a similar – or the same – molecular mechanism occurring in cells with different ontogeny, either somatic or germline).
- The requirement of endospermogenesis to complete seed development.

2.6 OTHER DEVIATIONS FROM SEXUAL REPRODUCTION

In rare cases, variations other than those described above have been observed and have sometimes been included within apomixis by different authors (e.g., as non-recurrent types). These include cases of haploid parthenogenesis, apogamy (or apogamety), semigamy (or hemigamy), androgenesis, etc. These cases are relevant to point out the high flexibility of reproductive programmes and the possible functional variations that each reproductive module can incorporate. Comparative modular analyses among organisms exhibiting different transitions from sexuality to asexuality may provide useful information to unravel the molecular nature of apomixis and its evolutionary relevance (Hojsgaard and Schartl, 2021). However, since these types include the process of meiosis and the formation of reduced female gametophytes, some authors do not recognize them within apomixis (e.g., Heslop-Harrison, 1959). Here, the mentioned deviations from the standard steps of sexual reproduction will not be considered in detail. We may add that, due to different biological and evolutionary reasons, such cases (e.g., haploid parthenogenesis, apogamety) cannot be permanently fixed in a plant as regular methods of embryonic development unless highly specific conditions are met (see e.g. semigamy, Battaglia, 1981). Since these cases have been described elsewhere and their general relevance in nature – according to their frequency – is limited, they will not be presented here. For more specific descriptions and discussions we recommend the reader to consider works of Rosenberg (1930), Stebbins (1941), Gustafsson (1946, 1947a, 1947b), Nygren (1967), Johri (1984), Battaglia (1991), Asker and Jerling (1992), Naumova (1993), Pullaiah and Febulaus (2000), Johri et al. (2001), Talent (2009).

3 Identifying Apomixis in Nature

3.1 INTRODUCTION

Understanding apomixis require the use of specific techniques to identify and quantify cytological, genetic and physiological changes happening during the emergence of apomixis in plant ovules, and the formation of seeds and offspring. Apomixis can start at different times during development of the ovule, i.e., throughout megasporogenesis, megagametogenesis or embryogenesis and endospermogenesis. Even though the developmental type of apomixis within a species is constant, in some cases it may vary in the same species or in the same individual. In single individuals, apomixis is usually facultative and the frequency of apomixis versus sexual seeds is genotype-dependent and influenced by the environment (see Karunarathne et al., 2020a and references therein).

Lack of good screening techniques for the identification of apomixis and deviations in the sexual reproductive programme or reproductive modules was formerly recognized as a hurdle to apomixis research (e.g., Asker, 1980). Today, several methods have been developed or adapted for plants which allow the screening of apomixis throughout large samples at different developmental steps. Any phenotypic manifestation of apomixis (Chapter 2) can be used to suspect or infer its occurrence in plants. Thus, apomixis can be identified and quantified by several alternative procedures used to screen ovules, the formation of unreduced gametes, the constitution of seed tissues, or by the features of a plant's progeny or natural siblings. This means that apomixis can be evaluated at different levels during a plant's reproduction and life cycle, and in each case, the use of distinct techniques is required. In some cases, a simple "pollen exclusion test" could be enough to detect signs or the presence of apomixis. Yet, to confirm apomixis *per se* and its developmental type, in most cases the use of more than one technique is required (Hanna and Bashaw, 1987; Leblanc and Mazzucato, 2001; Sherwood, 2001). A combination of cytological and genetic tests (see below) is in general recommended to detect apomixis, to identify its type of development and expressivity, or to determine its penetrance in a sample of natural or experimental siblings. We present below the most used methods for detecting apomixis in plant species as well as residual levels of sexuality or cases of incomplete apomixis in otherwise known apomictic plant species.

3.2 DETECTING APOMIXIS IN FLOWERS AND OVULES

Apomixis *per se* occurs during the formation of gametes and their interaction during the formation of a zygote and the primary endosperm cell as starting point for the development of a seed. Thus, identification of the type of apomixis in a species can be achieved only by analysing these processes.

3.2.1 Embryology and the Identification of Apomixis Types

In any species, systematic examination of anatomical features throughout development of an ovule is needed to thoroughly understand the cytological basis of sexuality and its developmental changes during the emergence of apomixis. Thus, embryological evaluation of diagnostic stages is the only method to accurately assess the type of apomixis and reproductive development in a certain plant genotype.

In sporophytic apomixis (Chapter 2, Section 2.4.1), the presence of proembryos or embryo-like structures developing directly from somatic cells in the ovule indicates adventive embryony

DOI: 10.1201/9781003088561-3

(Naumova, 1993). Sporophytic apomixis can be detected in ovules with mature gametophytes, during or just before blooming. Apomictic female gametophytes are rarely formed in species with adventitious embryony, but the meiotic gametophytes must be formed and fertilized to develop the endosperm tissue of the seed.

In gametophytic apomixis (Chapter 2, Section 2.4.2), to classify the type of apomictic development, it is necessary to recognize the normal meiotic development during megasporogenesis and megagametogenesis and its alterations. Key developmental steps for proper assessments of apomixis and sexuality are the beginning and end of meiosis, and the end of gametogenesis. Among apomictic species, most sexual cytotypes (ca. 90%) exhibit a monosporic meiotic development of gametophytes of the Polygonum type (Carman, 1997). The rest of the genera and species have bisporic or tetrasporic types of gametophyte developments. The type of apomictic development is often stable in each species being only one of the alternative types described in Chapter 2. In few species, co-occurrence of different types of apomixis (apospory and diplospory) in single individuals or closely related individuals had been observed (e.g., Bonilla and Quarin, 1997; Carman et al., 2019). Despite the relevance of embryological evaluations, such analysis requires expensive instruments and certain experience to recognize cell types, developmental stages and deviations from the normal mechanism of sexual seed development in each targeted material. Typically, it is widely accepted to evaluate one or few selected genotypes and extrapolate such observations to the species. However, the occurrence of developmental variations and alternative types of gametophyte developments can be identified only through detailed embryological analyses of an appropriate number of ovules during key steps of development and numerous genotypes. Between five and ten ovules at each key reproductive step would be enough to have an idea of the overall pattern of development and to define the type of apomictic development. Depending upon the developmental complexity (i.e., apospory or diplospory, type of diplospory, type of female gametophyte, zygotic or parthenogenetic embryo and endosperm developments, type of development of the male gametophytes), the number of pistils and stamens examined to accurately determine or quantify the type of apomixis can vary.

In species with monopolar apospory like the Panicum type or Setaria type, meiotic and apomictic female gametophytes can be structurally differentiated and the analysis of as few as 15 pistils during pre-anthesis or anthesis would be enough to determine a relative proportion of sexual and apomictic gametophytes (see e.g. Karunarathne et al., 2020a). In species with bipolar apospory of Hieracium type where usually only one female gametophyte develops after meiosis and both meiotic and apomictic gametophytes are structurally similar, proportions for each reproductive type have to be estimated at an earlier developmental stage, in ovules around the end of meiosis and the beginning of gametogenesis. At such stages it is possible to observe and evaluate the viability of both the meiotic megaspore cell and initial of apospory cells in single ovules. However, while ovules of apomicts in some genera often show a single germline and one or few aposporous initial cells (e.g., *Ranunculus* spp., Hojsgaard et al., 2014b), in other cases the presence of multiple secondary archesporial cells besides initial of apospory cells complicates proper reproductive evaluation (e.g., *Crataegus* spp., Phipps and Muniyamma 1980; *Sorbus* spp., Liljefors, 1953).

In diplosporous species, proper understanding of the alteration of the events of meiosis is critical and must be examined as accurately as possible. Even though the lack of a linear (triad or) tetrad of spores resulting from meiosis is characteristic of diplospory (Gustafsson, 1946; Nogler, 1984), it is not always easy to visualize the meiotic events and identify all the products of meiosis. Detailed analysis of (apo)meiosis is particularly important in species with bipolar diplospory of Antennaria type, Taraxacum type and Ixeris type, in which mature apomictic female gametophytes are structurally similar to meiotic ones and apomixis can be quantified only through the analysis of the events during megasporogenesis. In species with monopolar diplospory of Eragrostis type where mature apomictic and meiotic female gametophytes are structurally distinct (e.g., Meier et al., 2011), analysis of a few pistils would be enough to determine relative proportions of each type of gametophytes.

3.2.1.1 Embryological Methods

The most used methods for embryological analyses are of two types: those involving the sectioning and staining of ovules, and those using whole-mount ovules and tissue-clearing methods. In either case, the sampling of pistils at the appropriate developmental stage can be standardized through the relative size of flower buds in dioecious or monoecious plants with pistillate flowers, or through the stage of development of the male gametophytes in monoecious plants with hermaphrodite flowers. In bisexual flowers, the development of the male gametophyte is synchronized or slightly faster than the female development.

Sectioning of paraffin or resin embedded pistils and staining with safranin-fast green or other dyes (Gurr, 1965) is recommended to evaluate cellular details of the reproductive development in a species (e.g., the presence of structural elements in cells, such as sugars or proteins, or processes such as the fusion of nuclei in the central cell or during fertilization). These methods commonly use bright field microscopy and are technically demanding and time-consuming (Johansen, 1940; Ruzin, 1999). Preparing paraffin or resin sections is laborious, and interpretations of multiple consecutive sections may be difficult.

Clearing methods are a faster and easier alternative to sectioning methods and therefore allow the characterization of larger samples, although the overall quality and amount of visual information is limited. Intact pistils can be cleared in different media (mostly based on modified lactophenol or aromatic esters) designed to match the refractive index of organelles or cell walls (Crane, 2001), and hence, ovules are observed in three dimensions. Structures with the same refractive index as the media become invisible, but changes in orientation and cellular components generate variations in the refractive index which turn certain aspects of the specimen visible (Herr, 1971; Young et al., 1979; Crane, 2001). The selection of a clearing medium will depend on the specimen's thickness, the refractive index of the cells/structures, and the preservation of the structural complexity of the cell (number of organelles and DNA content) which is affected by the type and duration of fixation. A detailed list of clearing media and their refractive indexes is given in Crane (2001), together with a list of advantages and disadvantages associated with each method, recipes and alternatives for fixation in organic solvents. Even when clearing methods usually use Nomarsky interference-contrast microscopy, less expensive alternatives are available (Ruzin, 1999; Crane, 2001).

3.2.1.2 Callose Deposition

In angiosperms, the differentiation of a megaspore (and microspore) mother cell(s) and the formation of a (triad or) tetrad of megaspores (or microspores) by meiosis is associated with the synthesis and deposition of callose (the β-1,3 glucan, Chapter 2) around meiocytes (Rodkiewicz, 1970), and to the altered expression of arabinogalactan proteins in the cellular membrane (Pennell and Roberts, 1990). So, callose likely has a role as a signalling barrier in megaspores destined to die, but callose is also degraded by β-1,3-glucanases (Chen and Kim, 2009) and the released monosaccharides may support the development of the female gametophyte after functional specification of the megaspore and during the initiation of gametogenesis. In addition, callose serves as a marker, distinguishing the germline from the somatic tissue and ovules showing mono- and bisporic patterns of megasporogenesis (Rodkiewicz, 1970; Russell, 1979; Tucker and Koltunow, 2014).

In diplosporous species, where the normal progression of meiosis is modified, callose deposition does not occur or is disturbed, and therefore, it has been used as a cytological marker in different studies. Lack of callose deposition has been documented in *Anthosachne rectiseta* (Nees) Barkworth & S.W.L. Jacobs [≡*Elymus rectisetus* (Nees) Á. Löve & Connor; Carman et al., 1991], in several *Tripsacum* species (Leblanc et al., 1995a), in *Eragrostis curvula* (Schrad.) Nees (Peel et al., 1997a, 1997b), in *Poa nemoralis* L. and *P. palustris* L. (Naumova et al., 1999), reported in meiotic mutants of alfalfa (*Medicago sativa* L.) that produce unreduced megagametophytes (Barcaccia et al., 1996) and in apomictic *Antennaria* hybrids (Carman, 2007). In other diplosporous species like *Taraxacum atricapillum* Sonck (Musiał et al., 2015) and possibly in *Chondrilla juncea* L. (a comparison to the meiotic pattern is missing; Musiał and Kościńska-Pająk, 2017), in *Paspalum minus* L. (Bonilla and Quarin, 1997) and even in *Eragrostis curvula* (Meier et al., 2011, using other genotypes than those evaluated by Peel et al., 1997a, 1997b), callose had been observed in elongated diplosporous megaspore mother cells and dyads showing thin walls and heterogeneous distribution, indicating a deficient or altered deposition of the polysaccharide.

Aposporous species do also show disturbances in callose deposition. Abnormal patterns have been observed in aposporous species like *Poa pratensis* L. (Naumova et al., 1993), in *Panicum maximum* Jacq. (Naumova and Willemse, 1995), in *Cenchrus squamulatus* (Fresen.) Morrone (≡*Pennisetum squamulatum* Fresen.) and *Cenchrus ciliaris* L. (Peel et al., 1997a, 1997b). A few studies in aposporic species from *Hieracium* L. have revealed functional aspects of the callose. As in species of the genera *Poa* L., *Cenchrus* L. (=*Pennisetum* Rich.) and *Brachiaria* (Trin.) Griseb. (Peel et al., 1997a, 1997b; Dusi and Willemse, 1999), Tucker et al. (2001) did not detect callose deposition on aposporous initial cells of *Hieracium*, or changes in callose deposition patterns in cells undergoing megasporogenesis, and an expected spatial-temporal expression of β-1,3-glucanase gene (*HpGluc*) encoding a putative callose-degrading enzyme was not observed. In a recent study using sequencing of laser captured cells and immunolocalization, Juranić et al. (2018) found an enrichment of genes involved in the synthesis of β-glucans in aposporous initial cells together with genes linked to pathways involved in functional megaspore specification, megaspore selection and megaspore death. They showed that callose appears in the aposporous initial cells during functional megaspore selection and remains in those cells which do not divide after megaspore degeneration or is removed in those expanding cells (either meiotic or apomeiotic) that enter the first mitosis of gametogenesis. Moreover, arabinogalactan proteins co-localize with callose and serve as markers of aposporous initial cell specification (Juranić et al., 2018). Since callose might control cell-to-cell communication, callose deposition in cell walls and cell contact junctions might regulate the activation and growth of just one aposporous initial into a female gametophyte.

Callose is detected in ovules by staining the tissues in Aniline Blue solution (Martin, 1959) and using epifluorescence microscopy at near UV excitation wavelength (365nm) (Ruzin, 1999). Depending upon the size of the ovule, the use of alternative sectioning or clearing procedures is recommended. See clearing options and staining procedures in Ruzin (1999) and Crane (2001).

3.2.1.3 The Auxin Test for Parthenogenesis

Auxin is a phytohormone with key endogenous signalling roles controlling initiation and development of flower organs (stamens and gynoecia) (Robert et al., 2015), and is essential for effective fertilization and seed development (Figueiredo and Köhler, 2018).

In a time where experimental techniques for analysis of apomixis were scarce, Matzk (1991) presented a method to screen parthenogenesis competence in plants using auxin. In this approach, different sexual and apomictic species, including sexual plants with parthenogenetic capacity, were treated with synthetic auxin hormones shortly before, during or immediately after anthesis. These hormones included the 3,6-dichloro-2-methoxybenzoic acid (DIC), the 2,4-dichlorophenoxyacetic acid (2,4-D), the 2,4,5-Trichlorophenoxy acetic acid (2,4,5-T), and the *p*-chlorophenoxy acetic acid (CPAA). The application of any of those hormones on flowers induced the development of grains carrying mature embryos and no endosperm only in parthenogenetic competent genotypes. In auxin-treated plants the fertility of the male gametophytes was not affected, but cellular changes initiated shortly after auxin application (before pollination) blocked the fertilization of the female gametophyte and induced sexual female sterility (Matzk, 1991). In natural conditions, the unreduced egg cell in apomictic ovules might develop parthenogenetically into a proembryo, but endosperm initiation (either through pseudogamy or autonomously) is needed to avoid proembryo's abortion and form a viable seed.

This approach was named "auxin test" or "auxin-induced parthenocarpy", whereby the application of auxins seems to replace the effect of the endosperm in plants presenting parthenogenesis-like genes, and to promote the development of the embryo in unfertilized ovules or in absence of endosperm. The auxin test has been successfully used to identify parthenogenesis competence in genera such as *Bromus* spp., *Festuca* spp., *Hordeum* spp., *Hypericum* spp., *Lolium* spp., *Poa* spp., and *Triticum* spp. (in the wheat-salmon system) (Matzk, 1991; Mazzucato et al., 1996; Barcaccia et al., 1998; Matzk, 2007; Niemann et al., 2012) and in *Dichanthium annulatum* (Forssk.) Stapf (Gupta et al., 1999).

Even though the method is useful to screen either apomictic species for sexual plants or sexual species for parthenogenetic individuals (e.g., Matzk, 1989), it has been replaced by other methodologies (e.g., by flow cytometry seed screen; see below) and in the last years only a few studies had used the auxin test for characterizing apomixis (e.g., Niemann et al., 2012). Moreover, in apomictic species where the development of a mature embryo depends on the nutritive support of the endosperm (e.g., in *Boechera* spp.) or in sexual relatives where auxin (2,4-D) can trigger central cell division and autonomous endosperm development in unfertilized ovules (Figueiredo et al., 2015), the auxin test might not be informative.

3.2.2 Testing the Autonomous Development of Seeds

Control of pollination and formation of seeds without the contribution of the pollen is an indication for the occurrence of apomixis. In fact, parthenogenetic development of seeds (apomixis) was first suspected (discovered) in a single female plant from the dioecious species *Alchornea aquifolia* (Js.Sm.) Domin (=*Coelebogyne ilicifolia* Js.Sm.) cultivated at Kew gardens which produced seeds in absence of male plants (Smith, 1841). Later, Strasburger (1878) confirmed that the seeds of *A. aquifolia* develop through sporophytic apomixis (see Chapter 7).

3.2.2.1 Pollen Exclusion Tests

Pollinator exclusion and hand pollination experiments had been extensively used to identify the pollination requirements of a plant (Pattemore, 2017). Bags and netting with different pore sizes are useful to exclude pollinators and evaluate its consequence on seed and fruit set.

In sexual plants, since male gametophytes are required to fertilize female gametes and for further development of the ovule into a seed, pollen exclusion experiments will lead to ovule abortion and complete lack of seeds. A similar case happens in apomictic plants with pseudogamous development. Yet, well-documented isolation and/or emasculation experiments (see below) assuring the exclusion of pollen can test for the development of seeds in absence of the male contribution to fertilization, and thus, it can provide good evidence of autonomous apomictic development of seeds (Richards, 1997).

In dioecious plants, female individuals can be easily isolated from male individuals and potential pollinators by isolating the plant or flower. In monoecious plants with separate flowers, female flowers bagged before anthesis can prove the formation of seeds without pollen. In hermaphrodite flowers, emasculation experiments using forceps to remove immature anthers will be needed and must be carried out under strict conditions to avoid any pollen from landing on stigmas. While in species from *Potentilla* or *Ranunculus* flowers are emasculated prior to opening (Asker, 1970a, 1970b; Nogler, 1984), in grasses, anthers are more easily removed during blooming, and hence anther dehiscence must be prevented by using a fog chamber (Burson, 1997). Bagging of flowers without emasculation can only test for self-compatibility as results of this test do not discriminate between pseudogamous apomixis in apomicts and selfing in sexual plants (Hörandl et al., 2008), and embryological techniques (see above) are required for assessing pseudogamy. For bagging, the use of 3D water-repellent paper bags, even in the case of allogamous, insect-pollinated species, is recommended. Vapour-permeable bags help prevent unwanted condensation and the growth of mildew, fungus and bacteria which may deter seed formation. The use of micro-mesh or micro-perforated fabric impermeable to pollen is an expensive option, and cheaper alternatives exist depending on pore size, but will have varying degrees of permeability to wind-borne pollen (Neal and Anderson,

2004) and might not prevent accidental landing of pollen on stigmas (details on types and characteristics of pollination bags can be found at https://en.wikipedia.org/wiki/Pollination_Cags).

3.2.2.2 Castration Experiments

Apomictic reproduction can be detected by using Raunkiaer's castration method (Ostenfeld and Raunkiaer, 1903). Besides those experiments in which different naturalists used dioecious or monoecious plants to test the formation of seeds without the influence of pollen (Braun, 1857, pp. 315–317), or Mendel's accidental progeny test experiments using apomictic *Hieracium* species to confirm his findings on *Pisum* L. (Mendel, 1870), Raunkiaer's castration technique was the first experiment in which reproductive organs were intentionally manipulated to test apomixis (see also Murbeck's experiments in Chapter 6). This quite simple and efficient treatment relies on excluding pollination and fertilization by removing the entire stigmas and upper parts of styles from the capitulum – the inflorescence in Asteraceae – just before opening. By cutting off the female reproductive structures (usually in the upper half of the capitulum), the formation of ripe achenes is possible only through parthenogenetic and autonomous development of seeds, and thus, it represents fairly clear evidence for apomixis. After using this method in *Hieracium* spp., Murbeck (1904) confirmed similar results in *Taraxacum* spp., in which he observed regular embryo development in ca. 85% of the ovules post-castration. Since then, this method of castration has been used in studies for testing apomictic reproduction in experimental hybrids (e.g., van Dijk et al., 1999) and in many taxonomically oriented studies involving related taxa (Gadella, 1984; Richards, 1986; Mráz and Zdvořák, 2019). In grasses, castration or emasculation is used mainly to avoid selfing during experimental crossings. Since most grasses are pseudogamous, emasculation may only be used to search for rare cases of autonomous seed development.

The castration method is simple and effective when testing for autonomous apomictic development of seeds but has three main disadvantages. Perhaps the main limitation of this method is that it is an invasive treatment and might cause ovule abortion or even the abortion of the whole capitulum (Richards, 1997). The second limitation is the little information it provides. For instance, rare occurrence of haploid parthenogenesis will remain undetected, or, in facultative apomicts species, sexuality will not be determined, nor apomixis in species with pseudogamous seed development. A third limitation is that parthenocarpy (i.e., the formation of seedless fruits) can occur simultaneously with apomixis (Picarella and Mazzucato, 2019) and thus challenges the correct identification of apomixis. An advantage might be that is a fast method to select clonal progeny, whereby meiotic female gametes will not be fertilized, and sexual progeny will not develop. In any case, this method requires a molecular confirmation for clonality.

3.3 DETECTING APOMIXIS IN SEEDS

Seeds are the final product of complex development programmes happening in all plant ovules. Different characteristics of the seed and seed tissues retain relevant information which, in most cases, provides clues to understand part of the mechanisms followed during their formation.

3.3.1 Germination Tests

The emergence of multiple seedlings from one single seed may indicate sporophytic apomixis, or – less probable – gametophytic apomixis (Asker and Jerling, 1992; Naumova, 1993). Multiple seedlings *per* seed occur because of the development of more than one embryo, which can be due to (1) the development of adventitious embryos or embryos derived from other cells in the gametophyte than the egg-cell (seedling may be morphologically different depending upon reproductive origin, relative size, initial access to resources); or more rarely due to (2) embryos arising from the simultaneous development of sexual and apomictic female gametophytes in one ovule (concurrent development of aposporous and sexual gametophytes is common in many apomictic species, mostly grasses). Occasionally, the emergence of two embryos can be the result of twin sexual female gametophytes in an ovule.

3.3.2 Flow Cytometric Seed Screen

Flow cytometry (FC) is a complex instrument combining knowledge on fluidics, optics, electronics and informatics which was developed in the twentieth century for cell analysis (Shapiro, 2007). In botany, FC is mainly used to determine the size of plant genomes by measuring the nuclear DNA content. The first application of FC in plant cells is the one by Heller (1973) to analyse the amount of DNA in nuclei of *Vicia faba* cells. Nowadays, the procedure used to prepare a suspension of intact plant cell nuclei for measurement is relatively simple, and nuclei can be stained with a fluorochrome of choice, usually the intercalating dye propidium iodide (but ethidium bromide is also used) or the base-specific minor grove-binding DAPI (AT-specific) fluorochrome (but base-specific Hoechst – AT-specific – or mithramycin, olivomycin and chromomycin – GC-specific – are also used) (Greilhuber et al., 2007).

A technical advance to quantitative studies on apomixis was the development of an approach to recognize the reproductive origin (i.e., sexual or asexual) of dormant seeds. The first rational development of the flow cytometric seed screen (FCSS) method was done by Matzk et al., (2000) and reviewed by Matzk (2007). The method is based on the estimation of DNA content and ploidy in embryo *versus* endosperm tissues of mature seeds and is therefore applicable only to endospermic (albuminous) seeds. The ratio of embryo to endosperm DNA content provides a relative value and information about the mode of reproduction. The rationale behind FCSS relates to the reconstruction of (1) the structure of the female gametophyte (e.g., either with 8 or 4 nuclei), and (2) the type of male and female gametes (reduced *versus* non-reduced) involved in the formation of a seed. This reproductive reconstruction is used to determine the origin of the embryo (zygotic *versus* parthenogenetic) and the origin of the endosperm (pseudogamous *versus* autonomous) (Table 3.1). Hence, in one measurement, the FCSS allows the collection of information on several reproductive steps. However, without a proper embryological evaluation or previous knowledge on the studied sample, certain events during sporogenesis, gametogenesis and even fertilization can only be assumed.

Thus, FCSS can be used to discriminate between seeds formed through pseudogamy, autonomous apomixis or sexual reproduction (Table 3.1, Figure 3.1), as well as to detect cases of incomplete apomixis such as haploid parthenogenesis (e.g., formation of di- or polyhaploids after parthenogenetic development of a reduced egg cell) and B_{III} hybrids (formation of a seed embryo after fertilization of an unreduced egg cell) during the formation of new offspring (Matzk, 2007).

More importantly, the use of FCSS simplifies the simultaneous analysis and collection of reproductive data of large seed samples. Depending upon the size of the seed and the aims of the study, seeds can be analysed individually or bulked. Analysis of bulk samples is recommended as exploratory when searching for variability of the reproductive pathway or in species with tiny seeds to increase the amount of tissue analysed (e.g., in *Hypericum* L. or *Arabidopsis* Heynh.), although in such species, single seeds can be analysed through FCSS (Matzk, 2007). Quantification of progeny classes in bulk samples of facultative apomictic species is to some extent possible by discrimination of peak positions and relative peak sizes in the seed histogram (Krahulcová and Suda, 2006; Barcaccia et al., 2007; Sartor et al., 2011). However, bulking many seeds does not allow for a precise determination of the number of seeds contributing to each reproductive pathway, and will hidden rare cases of sexual seeds or alternative developments.

The method has already been used on several genera, including apomictic species of *Arnica* L., *Boechera* Á. Löve & D. Löve, *Crataegus* L., *Hieracium* L., *Hypericum* L., *Paspalum* L., *Pilosella* Hill., *Poa* L., *Potentilla* L., *Ranunculus* L., *Taraxacum* F.H. Wigg., *Tripsacum* L. (Mazzucato et al., 1994; Matzk et al., 2000; Bantin et al., 2001; Cáceres et al., 2001; Taşkin et al., 2004; Barcaccia et al., 2006, 2007; Wieners et al., 2006; Voigt et al., 2007; Kao, 2007; Talent and Dickinson, 2007a, 2007b; Aliyu et al., 2010; Sartor et al., 2011; Delgado et al., 2014; Hojsgaard et al., 2013, 2014a, 2014b; Hand et al., 2015; Mráz and Zdvořák, 2019; Karunarathne et al., 2020a, 2020b). However, the method is not informative when the relative DNA content/ploidy between embryo and endosperm tissues is the same in sexual and asexual seeds. This is the case in species with adventitious

TABLE 3.1
Expected Ratios of Embryo to Endosperm DNA Contents in Seeds Under Different Ploidy States and Reproductive Mechanisms. Male Gametes are Assumed to Be Meiotically Reduced (As It Is the Case for Many Apomicts)

Mother Plant and Reproductive Mechanism	Zygote/Embryo	Primary Endosperm	
	Egg (m) + sperm (p) cells	Central (1 nucleus; m) + sperm (p) cells	Central (2 nuclei; m) + sperm (p) cells
Diploid sexual	2Cx (1m+1p)	2Cx (1m+1p)	3Cx (2m+1p)
DoFer (2*n*=2*x*)		*Oenothera*	Most angiosperms
Diploid apomict	2Cx (2+0)	2Cx (2+0)?	4Cx (4+0)
AuEnd (2*n*=2*x*)			*Picris*?
Diploid apomict	2Cx (2+0)	3Cx (2+1)[c]	5Cx (4+1)[b]
PsEnd (2*n*=2*x*)		?	*Boechera*
Triploid sexual	~3Cx (1.5+1.5)[a]	~3Cx (1.5+1.5)[a]	~4.5Cx (3+1.5)[a]
DoFer (2*n*=3*x*)		?	*Onosma, Paspalum*
Triploid apomict	3Cx (3+0)	3Cx (3+0)	6Cx (6+0)
AuEnd (2*n*=3*x*)		*Arnica*?[d]	*Boechera, Erigeron, Hieracium, Pilosella, Taraxacum*
Triploid apomict	3Cx (3+0)	4.5Cx (3+1.5)[b]	7.5Cx (6+1.5)[b]
PsEnd (2*n*=3*x*)		?	*Amelanchier, Boechera, Cotoneaster, Crataegus, Paspalum, Rubus, Sorbus*
Tetraploid sexual	4Cx (2+2)	4Cx (2+2)	6Cx (4+2)
DoFer (2*n*=4*x*)		*Oenothera*	Most angiosperms
Tetraploid apomict	4Cx (4+0)	4Cx (4+0)	8Cx (8+0)
AuEnd (2*n*=4*x*)		*Arnica*? [d]	*Antennaria, Crepis, Erigeron, Hieracium, Hypericum, Ixeris, Pilosella, Taraxacum*
Tetraploid apomict	4Cx (4+0)	6Cx (4+2)[b,c]	10Cx (8+2)[b]
PsEnd (2*n*=4*x*)		*Cenchrus, Pennisetum, Brachiaria, Eragrostis, Panicum*	*Amelanchier, Cotoneaster, Crataegus, Hieracium, Hypericum, Paspalum, Poa, Potentilla, Ranunculus, Rubus, Sorbus, Tripsacum*

Cx-values: monoploid genome sizes (Greilhuber et al., 2005); DoFer: double fertilization; AuEnd: autonomous endosperm; PsEnd: pseudogamous endosperm.In each category, examples of genera displaying such developmental patterns are included.

[a] Mostly non-viable due to unbalanced meiosis and formation of aneuploid gametes, but a few triploid sexual species subsist through asymmetric chromosomal segregation (see Hojsgaard, 2018, Kolarčik et al., 2018).

[b] Estimations based on genome contributions by reduced male gametes (most frequently found in sexuals and apomicts). In some cases, either one unreduced male gamete or two reduced male gametes may fertilize the central cell (see e.g. Rutishauser, 1956; Nogler, 1972; Bantin et al., 2001; Aliyu et al., 2010, Burgess et al., 2014).

[c] This pathway of seed formation cannot be differentiated from the sexual one.

[d] Inferred from FCSS analyses but not corroborated by embryology.

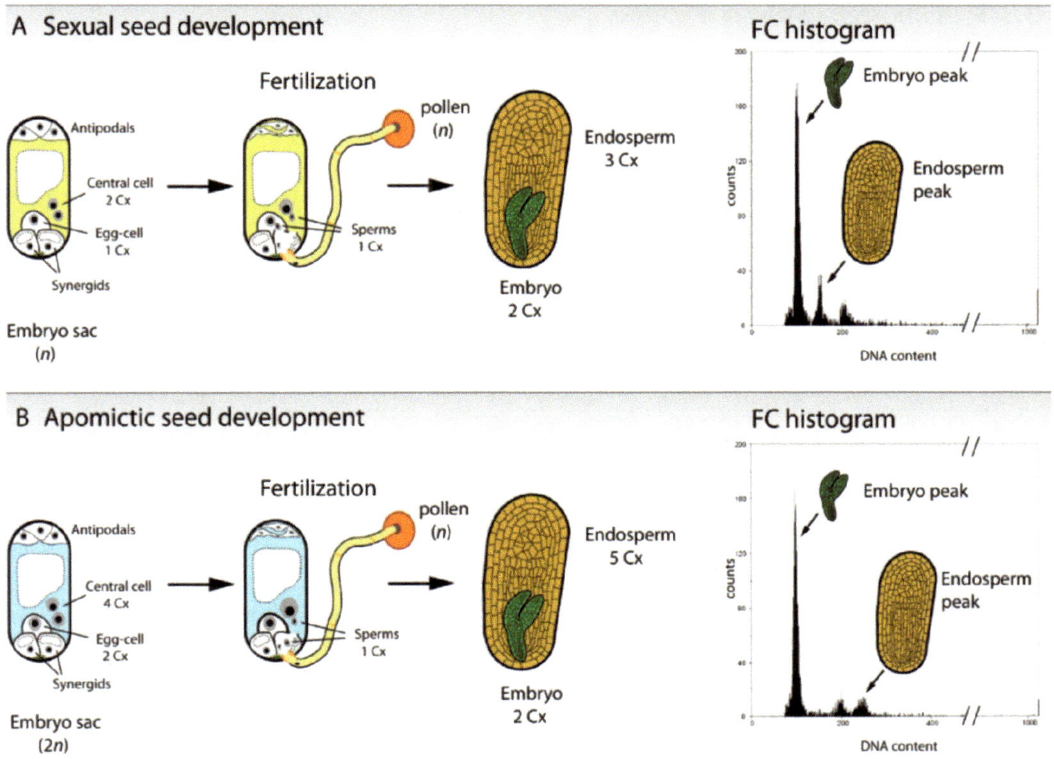

FIGURE 3.1 Schematic representation of the sexual (A) and apomictic (B) developments of seeds in plants with Polygonum type and Hieracium type of female gametophytes, respectively. The expected ratios of Cx-values for the embryo and the endosperm tissues are indicated, together with the relative position of both peaks in Flow Cytometry histograms. DNA content ratios of 2C:3C between the embryo:endosperm tissues are typical of sexual seeds, and 2C:5C of pseudogamous apomictic seeds. Occasional deviations from the normal sexual or apomictic pathways through parthenogenetic development of reduced female gametophytes, or double fertilization in apomictic gametophytes may occur rising seeds with ratios of 1C:2C and 3C:5C, respectively (Matzk et al., 2000). Pollen is represented as reduced as in most pseudogamous apomicts, but some apomicts produce unreduced pollen.

embryony, or in pseudogamous apomictic species exhibiting Panicum type or Eragrostis type of female gametophyte developments, as the apomictic pathway in ovules with four-nucleate gametophytes (which carry one nucleus in the central cell) cannot be differentiated from the standard sexual pathway (Table 3.1). The central cell in four-nucleate Setaria type gametophytes carries two nuclei and hence, sexual and apomictic pathways produce histograms with distinct peaks positions and peak ratios.

Since the amount of embryo or endosperm nuclei is variable according with the seed and endosperm type (farinaceous or non-farinaceous), the relative size of each peak in a FC histogram can vary. Hence, in species where the embryo and the endosperm have the same ploidy, like in some early angiosperm lineages (Williams and Friedman, 2002) or in some species with suspected four-nucleate female gametophytes and autonomous endosperm development (Kao, 2007), FCSS is not applicable or should be complemented with FC analysis of dissected seed tissues and embryological observations of mature female gametophytes.

In taxa with low germination rates or extended seed dormancy in which progeny tests are time-demanding (e.g., in alpine species), FCSS is recommended (Hörandl et al., 2008). Although the number of nuclei within seed tissues or the quality of signals might decrease after storage due to DNA degradation, FCSS analysis of *Hieracium* and *Paspalum* seeds stored for two or more years was carried out successfully (Krahulcová and Suda, 2006; Galdeano et al., 2016; Karunarathne et al., 2020a).

3.3.3 Endosperm Storage Proteins

Another possibility to discriminate between sexual or pseudogamous seed development and autonomous seed development is to analyse the differential accumulation of storage proteins, starch and lipids in the endosperm tissue. Since the endosperm develops asexually (i.e., without male contribution) in autonomous apomicts and sexually (i.e., after the fusion of a sperm to the central cell unit) in sexual and pseudogamous apomicts, variability in storage proteins is expected only in the latter.

Following this reasoning, Agafonov et al., (2004) studied the composition of endosperm storage proteins in *Poa pratensis*. They collected self- and open-pollinated caryopses and used sodium dodecyl sulphate polyacrylamide gel electrophoresis (SDS-PAGE) to assess storage proteins present in the endosperm of different *P. pratensis* accessions. The analysis revealed high levels of similarity in the relative mobility coefficient of storage proteins (73–100%) and the authors concluded that most of the seeds were developed autonomously. Using a similar rationale, El'konin et al. (2012) studied diploid *Sorghum* lines producing aposporous female gametophytes and rarely parthenogenetic proembryos and inferred the genetic composition of the endosperm in maternal-type seedlings by evaluating electrophoretic patterns of kafirins, a class of storage proteins found in *Sorghum* seeds. The electrophoretic patterns of the kafirins of the line producing aposporous female gametophytes were like those of the maternal-type seedlings, but differed from the patterns found in the hybrid seedlings and, hence, they concluded that the endosperm developed autonomously in seeds of the apomictic line of *Sorghum*.

However, caution should be taken when inferring autonomous endosperm development in seeds using storage proteins. Genes related to seed storage proteins were found differentially expressed in apomictic ovules compared to sexual ones (e.g., Polegri et al., 2010), which may suggest an effect of the reproduction mode on expression of storage proteins and their subsequent differential accumulation in cells, irrespective of a pseudogamous or autonomous development of the endosperm. On the other hand, the observation of identical patterns of storage proteins among seed progenies might respond to other genetic effects or regulatory mechanisms. Since apomicts usually display high heterozygosity (see e.g. Paun et al., 2006a, 2006b), inbreeding can be discarded as a possible cause for the accumulation of the same protein groups in endosperm cells. However, parent-of-origin allelic expression in the endosperm during seed development is a widely known mechanism caused by different epigenetic signals on parental genes which determine the sole expression of alleles from one parent (Scott et al., 1998; Hornslien et al., 2019). Such genomic imprinting can cause the unilateral expression of maternal genes during seed development and the accumulation of maternally coded storage proteins, excluding the paternal ones in pseudogamous apomictic species.

3.4 DETECTING APOMIXIS IN PLANTS AND PROGENIES

Beside the above-mentioned techniques, apomixis can also be detected in adult plants by evaluation of morphological and molecular variability. Since one of the consequences of apomixis is the formation of maternal (clonal) offspring (i.e., siblings genetically identical to each other and to the mother), its occurrence in natural or experimental populations may be indicative of apomixis. Thus, several techniques have been developed to identify apomixis in genealogically related individuals.

3.4.1 Progeny Tests

This selection tool is used mainly in animal breeding but also in plant breeding, to determine the breeding value of an individual or genotype, before being used extensively to propagate the best germplasm. Progeny tests are performed by mating a target male (or female) with a number of females (or males) and then evaluating the performance of the offspring to give a breeding value to the male (or female) genotype.

In apomixis research, progeny testing is a useful method when testing for apomixis. The occurrence of maternal individuals in the offspring of experimental crossings indicates apomixis in the female parent. Similarly, the prevalence of individuals with identical features in natural populations pointing to a clonal origin could suggest the existence of apomixis. However, if morphological or genetic homogeneity is detected between the mother and the offspring or among siblings, none of the progeny testing methods can unequivocally establish the reproductive mechanism generating such uniformity and these approaches should be complemented by other analyses, preferentially by embryological observations.

3.4.1.1 Morphological Evaluation

Morphologically uniform progenies from open-pollinated or cross-pollinated heterozygous parentals displaying the maternal phenotype were traditionally used as an indicator for apomictic reproduction (e.g., Tinney and Aamodt, 1940; de Wet et al., 1963; Cameron and Garber, 1968; Burton et al., 1973) before the development of molecular techniques, like protein electrophoresis or genetic markers (see below). In more recent studies, the observation of morphologically uniform offspring is combined with other associated traits (e.g., polyploidy) to infer apomixis (e.g., Kirschner et al., 2006). This approach, however, relies on the capacity of the researcher to differentiate segregating phenotypes, sometimes a tricky task without the presence of heterozygous morphological markers between the parents, especially among grasses. Moreover, maternal phenotypes can occur from selfing in highly homozygous sexual plants (Asker and Jerling, 1992).

Conventional morphological, agronomic or physiological traits are usually not linked to apomixis itself (i.e., to the genetic factors controlling the trait), but to the genotype of the plant that reproduces by apomixis. Monogenically inherited traits associated with the apomictic male parent in homozygous dominant or heterozygous condition have been used as phenotypic markers to reveal F_1 hybrids (e.g., Sampson, 1969; Dujardin and Hanna, 1989; van Dijk et al., 2009; Hojsgaard et al., 2011). When the maternal parent is homozygous recessive for that trait, uniform offspring for the maternal marker would suggest maternal inheritance (Sherwood, 2001).

Nowadays, identification of maternal phenotypes may require detailed measurements of multiple morphological characteristics and the use of complementary techniques, usually genetic markers (e.g., Barcaccia et al., 1997; Hojsgaard et al., 2011; Sailer et al., 2016), to confirm apomixis beyond doubt.

3.4.1.2 Cytological Evaluation

Cytological markers (like multiple female gametophytes, or ploidy levels) can be used to evaluate the occurrence of apomixis. For instance, crossing parents with different ploidy would allow the discrimination between sexual (n+n or B_{II}) plants formed from the union of reduced female and male gametes, sexual (2n+n or B_{III}) plants originating from union between one unreduced and one reduced gamete, or pseudogamous apomictic (2n+0 or M_{II}) plants originating from an unreduced gamete without male contribution. Using chromosome counting or FC to determine ploidy levels in such interploidy crosses can help identifying the occurrence of apomixis among the offspring (e.g., Crane and Thomas, 1939; Einset, 1951; Olden, 1953). While B_{II} plants would have intermediate ploidy (2C value) from both parents, B_{III} plants would show (in most cases) a ploidy higher than that of the parents, and M_{II} plants would be recognized by a 2C value equal to that of the mother (Matzk, 2007). However, this approach relies on the crossability of suitable parents and the development of viable seeds, and in most cases are used in combination with molecular markers (e.g., in *Taraxacum* spp., van Dijk et al., 1999; in *Hypericum* spp., Mayo and Langridge, 2003).

3.4.1.3 Molecular Evaluation

Using the same rationale, biochemical (i.e., proteins, usually enzymes) and genetic (i.e., fragments of DNA) markers are well suited to evaluate polymorphisms between offspring and parents, and to identify recombinant and clonal individuals. Though isozymes and allozymes have been widely used to study genetic variation in natural populations, and specifically for evaluating apomictic systems (Roy and Rieseberg, 1989; Barcaccia et al., 1994; Mazzucato et al., 1995; Poverene and Voigt, 1995; Roy, 1995; Gustine et al., 1996; Schmelzer and Renno, 1999), they have now been replaced by more informative DNA-based markers.

There is a wide range of molecular marker techniques that are used for various applications (Grover and Sharma, 2016). The most commonly used markers in genetic variability assessments under different pollination conditions and in test crosses to identify apomixis in different species are random amplification of polymorphic DNA (RAPD), restriction fragment length polymorphism (RFLP), microsatellites (or short sequence repeats, SSR) and amplified fragment length polymorphism (AFLP) (e.g., in *Allium* spp., Yamashita et al., 2012; in *Brachiaria* spp., Pessino et al., 1997, 1998; in *Cenchrus/Pennisetum* spp., Gustine et al., 1997; Ozias-Akins et al., 1998; Goel et al., 2006; in *Citrus* spp., Andrade-Rodríguez et al., 2004; Ruiz et al., 2000; Kashyap et al., 2018; Woo et al., 2019; in *Erigeron* spp., Noyes and Rieseberg, 2000; in *Hieracium* spp., Shirasawa et al., 2015; in *Hypericum* spp., Barcaccia et al., 2007; in *Panicum* spp., Ebina et al., 2005; in *Paspalum* spp., Ortiz et al., 1997, Pupilli et al., 2001, 2004, Stein et al., 2004, Hojsgaard et al., 2011; in *Poa* spp., Barcaccia et al., 1998, Albertini et al., 2004; in *Poncirus* spp., Kepiro and Roose, 2010; in *Potentilla* spp., Paule, 2010, Paule et al., 2011, Nardi et al., 2018; in *Taraxacum* spp., Vijverberg et al., 2004, 2010; in *Tripsacum* spp., Leblanc et al., 1995b, Grimanelli et al., 1998, Blakey et al., 2001; in *Ranunculus* spp., Paun et al., 2006a, 2006b). When classifying offspring, a certain number of polymorphic fragments (usually ≤5%) among siblings and mother are accepted to consider two individuals as clone mates. A low level of variation between natural and experimental clonal siblings is to be expected as a consequence of technical artefacts or natural mechanisms like mutations arising during replication (e.g., SSR loci tend to show strong mutational dynamics).

Molecular markers can identify dominance and co-dominance within the genome. Dominant markers (like RAPD or AFLP) cannot discriminate heterozygous states and thus, only one allele is visualized. Codominant markers (like isozymes or SSR) can distinguish homozygous and heterozygous states, and thus, more than one allele can be tracked. Overall, since AFLP uses universal primers, a single assay allows analysis of multiple loci randomly distributed in the genome and efficient genotyping of individuals, making it recommended for progeny testing analysis, varietal discrimination, genetic diversity studies and gene flow analysis (Grover and Sharma, 2016; Teixeira-Caixeta et al., 2014).

SSR are highly informative markers and locus specific, and multiple loci can be analysed by multiplexing (i.e., the simultaneous analysis of multiple markers in a single PCR reaction). However, sequence-level information is required for the design of species-specific primers, and due to possible undesired interactions between primers (e.g., primer pairing), the design of a robust multiplex test is not simple (Grover and Sharma, 2016). A pseudo-multiplex strategy in which PCR reactions are run separately and reaction products are then mixed for electrophoresis is an alternative.

Sequencing-based markers, either by conventional Sanger or by next generation sequencing (NGS) methods, provide more detailed information and an enormous amount of single-nucleotide polymorphism (SNP) data, making possible exhaustive genetic diversity assessments and genome-wide association studies (GWAS). Even though NGS methods are revolutionizing research in many fields of plant sciences, including apomixis research, they are not yet used for testing the occurrence of apomixis but rather to characterize the molecular genetics behind apomixis (Hojsgaard, 2021). Even though population genomics is a useful tool for detecting excess clonality and inferring apomixis (e.g., Niissalo et al., 2020), other direct, cheaper and faster methods to identify apomixis (e.g., FCSS) are available.

3.4.2 Marker-Assisted Selection to Screen for Apomixis in Plants

The use of marker-assisted selection (MAS) has proven to be an effective technique for rapid and efficient screening of phenotypes in large samples used in different breeding programmes, helping researchers minimize time and labour (Ribaut and Hoisington, 1998). However, the application of a MAS approach requires preliminary studies verifying that (i) the trait under selection has (preferably) simple genetic inheritance and Mendelian segregation, and that (ii) there are markers closely linked to the trait.

In apomictic plants, MAS has been used to detect apomixis or apomixis elements in population screenings of different species, and in attempts to isolate apomixis genes. Independently of the developmental type, apomixis shows Mendelian segregation (Ozias-Akins and van Dijk, 2007). Depending upon the sort of genetic control regulating the components of apomixis (i.e., a simple locus or separate loci), isozyme and genetic markers linked to apomixis as a unit or to the separation of apomeiosis, parthenogenesis and endosperm development have been identified in many apomictic species. In doing so, researchers used different techniques available and compared the data collected from sexual versus apomictic states. One of the most used approaches involves the generation of segregating progenies or bulk segregant analysis of reproductively phenotyped individuals screened for DNA fragments or sequences present exclusively in apomictic (or sexual) offspring. Here again, the most frequently used molecular markers in this approach are RFLP, RAPD, AFLP and SSR. Markers partially or fully linked to apomixis were found in all the plant systems analysed, including species and hybrids from *Allium* spp., *Brachiaria* spp., *Cenchrus* spp., *Citrus* spp., *Erigeron* spp., *Hieracium* spp., *Paspalum* spp., *Pennisetum* spp., *Poa* spp., *Poncirus* spp., *Taraxacum* spp., *Tripsacum* spp. (Leblanc et al., 1995b; Gustine et al., 1997; Pessino et al., 1997; Barcaccia et al., 1998; Ozias-Akins et al., 1998; Noyes and Rieseberg, 2000; Leblanc and Mazzucato, 2001; Pupilli et al., 2001, 2004; Labombarda et al., 2002; Stein et al., 2004; Nakano et al., 2008a, 2008b, 2012; Kepiro and Roose, 2010; Vijverberg et al., 2010; Hojsgaard et al., 2011; Yamashita et al., 2012; Shirasawa et al., 2015). Genetic studies using segregating progenies have been useful for genetic mapping of apomictic factors and for studying the inheritance and molecular nature of apomixis. Furthermore, fragments that segregate exclusively with apomictic offspring can be isolated, cloned, sequenced and transformed into codominant, locus-specific PCR markers following the design of primers flanking the region of interest. Such markers are then called sequence characterized amplified region (SCAR), and by a simple PCR reaction provide an easily detectable, highly specific, relatively inexpensive and reliable genetic marker that can be used for reproductive phenotyping of individuals, to characterize progeny classes and to aid in transferring apomixis to crop relatives (e.g., Ozias-Akins et al., 1998; Albertini et al., 2001a, 2001b; Labombarda et al., 2002; Roche et al., 2002; Pupilli et al., 2004; Catanach et al., 2006; Goel et al., 2006; Dwivedi et al., 2007, 2015; Schallau et al., 2010; Hojsgaard et al., 2011; Yamashita et al., 2012).

Another possibility is to find markers shared by distinct apomictic individuals, either intraspecific or interspecific within a genus (e.g., Lubbers et al., 1994), or markers associated with a mutant phenotype, either in natural or experimentally created mutants, as compared to a wild-type phenotype (e.g., Catanach et al., 2006). The identified genetic markers can be isolated, sequenced and converted into SCAR markers or used to cytologically and functionally characterize apomixis (Koltunow et al., 2011; Okada et al., 2011). Validation of these markers using a segregating population is important to verify their presence/absence pattern in plant-specific mutants.

A more recent approach involves comparative analysis of RNA populations and relative gene expressions levels in sexual versus apomictic ovules at different developmental stages. This has allowed the identification of alleles exclusively present and/or highly expressed in apomicts of *Boechera* spp., *Hieracium* spp., *Paspalum* spp., *Pennisetum* spp., *Hypericum* spp., *Ranunculus* spp. (e.g., Schallau et al., 2010; Corral et al., 2013; Mau et al., 2013; Conner et al., 2015; Shirasawa et al., 2015; Ortiz et al., 2017; Pellino et al., 2020). Such apomixis-linked sequences can also be transformed into PCR markers, thus offering a simple method for phenotyping apomixis and breeding-related traits in large samples (Mau et al., 2015; Hand et al., 2015).

3.5 STUDYING APOMIXIS-ASSOCIATED FEATURES

In addition to the methods mentioned above to identify apomixis in a species or genus, a wider variety of techniques are used to assess and characterize different developmental and molecular features of apomixis at the individual and population level. These methods include molecular cytogenetics, 3-D reconstructions of ovules, experimental crosses and genetic mapping, deletion mapping, laser-assisted microdissection (LAM), RNA-sequencing, RAD-sequencing, Genotyping-by-Sequencing (GBS), Bisulphite sequencing, etc.

Overall, such studies allowed the identification of specific features associated with the emergence of apomixis in different apomictic species (see also Chapter 9). For example, in several species, apomictic factors are associated with a large genomic region that lacks recombination, and such factors often occur in hemizygosity (e.g., Barcaccia et al., 1998; Grimanelli et al., 1998; Ozias-Akins et al., 1998; Jessup et al., 2002; Labombarda et al., 2002; Akiyama et al., 2004; Pupilli et al., 2004; Calderini et al., 2006; Goel et al., 2009; Sapkota et al., 2016). In different species, apomeiosis and parthenogenesis are inherited either as a single unit or independently (e.g., Noyes and Rieseberg, 2000; Catanach et al., 2006; Schallau et al., 2010; Vijverberg et al., 2010; Koltunow et al., 2011; Ogawa et al., 2013; van Dijk et al., 2020). In all species studied so far, a global dysregulation of gene expression has been detected in apomictic ovules compared to sexual ones (e.g., Sharbel et al., 2010; Polegri et al., 2010; Ortiz et al., 2017; Galla et al., 2019; Pellino et al., 2020). These particularities are discussed elsewhere and, for details about the methodological procedures, the reader should refer to the original studies.

4 Apomictic Grasses

4.1 INTRODUCTION

Compared to other plant groups, apomixis in grasses was discovered later. The credit for the first record of apomixis in Poaceae Barnhart is usually given to Müntzing (1933) for his work on *Poa pratensis* L. and *P. alpina* L., although the occurrence of apomixis in *Poa alpina* was already suggested by Zollikofer (1930), and by Stenar's paper on parthenogenesis in the genus *Calamagrostis* Adans. (Stenar, 1932). Andersen (1927) observed frequent formation of polyembryos in ovules of *Poa pratensis* and *P. compressa* L., but he concluded they were originated from multiple gametophytes developed from different megaspores. Müntzing's work based on cytogenetic analysis and (morphological) progeny tests on different *Poa* biotypes, established apomixis in the genus as an explanation to the constancy in chromosome numbers and morphology observed in certain biotypes compared to sexual ones. This work gave an impetus for further embryological evaluations and research on apomixis in grasses (e.g., Rancken, 1934; Armstrong, 1937; Tinney, 1940). Later, Narayan (1951) and Warmke (1952, 1954) reported apospory in three *Pennisetum* species and in *Panicum maximum* Jacq., respectively, and the occurrence of a special type of female gametophyte that carries only four nuclei at maturity (see Chapter 2). Some years later, Brown and Emery (1958) surveyed 164 grass species belonging to 64 genera and reported that 54 of those species – today belonging to the subfamilies Panicoideae A. Braun and Chloridoideae Kunth ex Beilschm. – were apomictic based on the presence of four-nucleate female gametophytes.

Since those early efforts, different studies have increased the list of apomictic grasses up to more than 240 species (Table 4.1). The present chapter gathers current information on species-level apomixis in the grass family.

4.2 CONSIDERATIONS ABOUT THE OCCURRENCE AND TYPE OF APOMIXIS IN GRASSES

Collected records are based on original reports and previous lists of apomictic species (Stebbins, 1941; Nygren, 1954a, 1967; Brown and Emery, 1958; Carman, 1995; Asker and Jerling, 1992; Naumova, 1993) and regional floras (e.g., Gregor, 2013; Brožová et al., 2019). Whenever possible, studies on the occurrence of apomixis in different species were critically re-evaluated. Species' taxonomy and nomenclature had been revised following Tropicos (https://www.tropicos.org/), the World Checklist of Selected Plant Families (WCSP; https://wcsp.science.kew.org/), The Plant List (http://www.theplantlist.org/) and the World Flora Online (WFO; http://www.worldfloraonline.org/).

We have included basic information about species names, ploidy of studied materials, and the type of apomictic reproduction in Tables 4.1 and 4.2. A substantial number of species have been subjected to changes in their nomenclature. Such changes are indicated in Tables 4.1 and 4.2. Some cases involve the transfer of individual species to related genera [e.g., *Elymus rectisetus* (Nees) Á.Löve & Connor is now *Anthosachne rectiseta* (Nees) Barkworth & S.W.L.Jacobs; or *Eremopogon foveolatus* (Delile) Stapf is now *Dichanthium foveolatum* (Delile) Roberty], the establishment of new genera (e.g., *Nicoraepoa* Soreng & L.J. Gillespie) or the merging of former genera (e.g., *Pennisetum* Rich. with *Cenchrus* L., or *Thrasya* Kunth with *Paspalum* L.).

Illegitimate species names [e.g., *Pennisetum dubium* belong to an undescribed species (see Gildenhuys and Brix, 1959) that correspond to *Pennisetum sphacelatum* (Nees) T. Durand & Schinz] were replaced whenever possible. A few (apparently undescribed) species were cited as apomictic elsewhere but are not found in the online plant databases consulted [e.g., *Bothriochloa acidula*

DOI: 10.1201/9781003088561-4

(Stapf) de Wet, *Capillipedium caucasicum* (Trin.) de Wet; de Wet et al., 1963] and are therefore not included in our lists. Accepted hybrid names [like *Poa* × *herjedalica* Harry Sm., a putative hybrid between *P. alpina* × *P. pratensis*; or *Poa* × *jemtlandica* (Almq.) K. Richt., a hybrid between *P. alpina* × *P. flexuosa*; Nygren, 1954a] are considered as natural apomicts.

Because the "species" category is heterogeneous and might involve more than one species' concepts from the variety of accepted species definitions, species holding apomictic cytotypes are often taxonomically challenging (see e.g., Kellog, 1990; Haveman, 2013; Majeský et al., 2017; Hörandl, 2018). Each apomictic complex or species aggregate should be analysed separately to properly define what it is a species in that group. Examples of this are the species of the *Poa secunda* complex (Kellog, 1987) in this chapter, the *Taraxacum officinale* aggregate (van Baarlen et al., 2002; Majeský et al., 2012) in Chapter 5, the *Crataegus douglasii* complex (Lo et al., 2009) in Chapter 6, or the *Ranunculus auricomus* aggregate (Hörandl et al., 2009; Karbstein, 2021) in Chapter 7. In the *Poa secunda* complex, for instance, some changes are not yet accepted by the whole scientific community [e.g., *Poa juncifolia* Scribn. has been synonymized under *Poa secunda* subsp. *juncifolia* (Scribn.) Soreng] and thus, it might be subjected to future changes. In such cases, we followed the newest taxonomic treatment and left subspecies and varieties individualized in the table for practical purposes.

Studies on species showing multiple embryos arising in a single seed are not considered evidence of sporophytic apomixis unless additional evidence is collected on their somatic development. This is to avoid false classifications as different grass species had shown the occurrence of cleavage polyembryony (e.g., in *Poa pratensis*, Nishimura, 1922), or embryos arising from synergids of the same gametophyte or egg cells from extra gametophytes [e.g., in *Tripsacum dactyloides* (L.) L., Farquharson, 1955; in *Anthosachne scabra* (R. Br.) Nevski sub *Agropyron scabrum* (R. Br.) P. Beauv., Hair, 1956; or in *Cortaderia jubata* (Lemoine ex Carrière) Stapf, Philipson, 1978] which do not correspond to sporophytic apomixis *per se*. Similarly, evidence of adventitious embryony based on experimental hybrids [e.g., in *Elymus rectisetus* (Nees) Á. Löve & Connor, Liu et al., 1995] or infrequent cases of polyembryony in natural or cultivated species (e.g., in *Oryza sativa* L., Wu et al., 1991; in *Zea mays* L., Michel et al., 2018) is not included here. The same goes for cases of gametophytic apomixis in experimental hybrids or cultivated varieties (e.g., in *Sorghum* spp., Rao et al., 1978; Reddy et al., 1979; Wu et al., 1994).

Studies based on the observation of unreduced four-nucleate female gametophytes (e.g., Brown and Emery, 1958) but lacking information about the earlier stages of female gametophyte development (i.e., through apospory or diplospory) are categorized just as holding gametophytic apomixis. This is because four-nucleate gametophytes can be originated in different ways if apospory or diplospory (and details of its anatomy) are not specified (see Chapter 2, e.g., the Panicum type, the Setaria type, the Eragrostis type). Cases in which specific details about apomictic development are partially proven or suggested, although they must be corroborated, were accepted. As an example, Burson and Bennett (1971a, 1971b) suggested autonomous endosperm development in *Paspalum mandiocanum* Trin. based on the observation of initial endosperm development before pollination. However, seed formation without pollen contribution remains to be verified.

In hemisexual species (i.e., species like *Andropogon ternatus* Nees, *Onosma arenaria* Waldst. & Kit., or the *Rosa canina* complex that reproduce through asymmetric meiosis involving asexually inherited univalent-forming and sexually inherited bivalent-forming chromosome sets), testing for apomixis can be tricky (see e.g. Kolarčik et al., 2018) and critical evaluations using embryology, FCSS and progeny tests are required to properly identify the existence of apomixis. Cases like the one of *Bouteloua curtipendula* (Michx.) Torr. which has been classified as apomictic due to an asymmetric first division during meiosis (Harlan, 1949) and later characterized as aposporous (Mohamed and Gould, 1966) remain uncertain. Therefore, we temporarily maintain the classification of apomixis in *B. curtipendula* as aposporous apomixis in Table 4.1, but the mechanism of apomixis itself in this species will have to be re-evaluated.

Apomictic reproduction had been characterized for the first time in different species using FCSS (e.g., Kelley et al., 2009; Galdeano et al., 2016). Thus, in these cases it is only possible to

TABLE 4.1
Table of Apomictic Species (A) within the Family Poaceae Barnhart, Sorted Alphabetically According to Subfamilies

Species	Ploidy	Type of Apomixis[a]	References
Aristidoideae Caro			
Stipagrostis ciliata (Desf.) De Winter		A	Inamuddin and Maghboub (1990)
Chloridoideae Kunth ex Beilschm.			
Bouteloua curtipendula (Michx.) Torr.	*4x*	A; Ap (Hi); Ps	Bryant (1952), Mohamed and Gould (1966)
Buchloe dactyloides (Nutt.) Engelm.		A; Ap	Brown and Emery (1958), Carman (1997)
Chloris andropogonoides E. Fourn.	*4x*	A; Ap	Brown and Emery (1958), Nygren (1967)
Chloris cucullata Bisch.	*4x*	A; Ap	Brown and Emery (1958), Nygren (1967)
Chloris gayana Kunth	*4x*	A; Ap (Pa)	Brown and Emery (1958), Nygren (1967)
Chloris verticillata Nutt.	*4x*	A; Ap	Carman (1995)
Eragrostis caesia Stapf		A; Di (Er); Ps	Voigt and Bashaw (1972, 1976), Burson and Voigt (1996)
Eleusine coracana (L.) Gaertn.	*4x*	A; Ap (Pa); Ps	Rathi et al. (2020)
Eragrostis curvula (Schrad.) Nees (=*Eragrostis chloromelas* Steud.)	*4x, 6x, 7x*	A; Di (Er); Ps	Brown and Emery (1958), Streetman (1963)
Eragrostis heteromera Stapf		A; Di (Er); Ps	Brown and Emery (1958), Streetman (1963)
Eragrostis lehmanniana Nees	*4x*	A; Di (Er); Ps	Voigt (1971), Voigt and Bashaw (1972, 1976), Brix (1974), Vorster and Liebenberg (1984), Voigt et al. (2007)
Eragrostiella bifaria (Vahl) Bor		A; N	Seshavatharam (1983), Sonpipare (1984)
Eustachys paspaloides (Vahl) Lanza & Mattei	*4x*	A; Ap (Hi)	Strydom and Spies (1994)
Fingerhuthia africana Lehm		A; Di (Er)	Brown and Emery (1958)
Harpochloa falx (L.f.) Kuntze	*4x, 6x*	A; Ap (Hi)	Strydom and Spies (1994)
Microchloa altera (Rendle) Stapf [≡*Rendlia altera* (Rendle) Chiov.]	*4x*	A; Ap (Hi)	Strydom and Spies (1994)
Perotis hordeiformis Nees ex Hook. & Arn.		A; N	Venkateswarlu and Devi (1964)
Danthonioideae H.P. Linder & N.P. Barker			
Cortaderia jubata (Lem.) Stapf	*12x*	A; Ap (Hi); Au	Philipson (1978), Astegiano et al. (1995)
Cortaderia rudiuscula Stapf	*12x*	A; Ap (Hi); Au	Philipson (1978), Astegiano et al. (1995)
Danthonia spicata (L.) Roem. & Schult. [sub *Danthonia spicata* (L.) Beauv.]		A; Ap (Hi); Ps	Philipson (1986)
Pentameris eriostoma (Nees) Steud. (sub *Pentaschistis eriostoma*)	$2x_2$	A; Ap (Pa)	Spies et al. (1999)
Panicoideae A. Braun			
Anthephora pubescens Nees		A; Ap	Brown and Emery (1958), Carman (1997)
Apluda mutica L. [sub *Apluda mutica var. aristata* (L.) Pilger]	*6x, 7x*	A; Ap (Se, Pa); Ps	Murty (1973)
Bothriochloa bladhii (Retz.) S.T. Blake	*4x, 5x, 6x, 8x*	A; Ap (Pa, Se); Ps	Celarier and Harlan (1957), Harlan et al. (1964), Choda and Bhanwra (1977), Bhanwra (1988), Saran and de Wet (1970)

(Continued)

TABLE 4.1 (Continued)
Table of Apomictic Species (A) within the Family Poaceae Barnhart, Sorted Alphabetically According to Subfamilies

Species	Ploidy	Type of Apomixis[a]	References
Bothriochloa decipiens (Hack.) C.E. Hubb.		A; Ap (Pa); Ps	Brown and Emery (1958), de Wet and Harlan (1970)
Bothriochloa ewartiana (Domin.) C.E. Hubb.		A; Ap (Pa); Ps	Carman (1995)
Bothriochloa grahamii (Haines) Bor		A; AE; Ap; Ps	Harlan et al. (1964)
Bothriochloa insculpta (A. Rich.) A. Camus		A; Ap (Pa); Ps	Celarier and Harlan (1957), Harlan et al. (1964)
Bothriochloa ischaemum (L.) Keng.	*4x, 5x, 6x*	A; N; Ap (Pa, Se); Ps	Brown and Emery (1957), Celarier and Harlan (1957), Harlan et al. (1964), Christoff and Moskova (1972), Moskova and Yakovlev (1974), Moskova (1975)
Bothriochloa kuntzeana (Hack.) Henrard		A; Ps	de Wet et al. (1963)
Bothriochloa pertusa (L.) A. Camus	*4x*	A; Ap (Pa, Se); Ps	Celarier and Harlan (1957), Harlan et al. (1964), Gupta (1968, 1969 , 1970)
Bothriochloa pseudoischaemum (Nees ex Steud.) Henrard [=*Bothriochloa oryzetorum* (Hack.) Bor]		A; Ps	de Wet et al. (1963)
Bothriochloa radicans (Lehm.) A. Camus		A; Ap (Pa); Ps	Brown and Emery (1958)
Bothriochloa woodrovii (Hook.f.) A. Camus	*5x*	A; Ps	de Wet et al. (1963)
Hemisorghum venustum (Thwaites) Clayton [≡*Bothriochloa venusta* (Thw.) A. Camus]		A; Ap (Pa, Se)	Celarier and Harlan (1957)
Brachiaria bovonei (Chiov.) Robyns		A; Ap (Pa); Ps	do Valle and Glienke (1991), do Valle and Miles (2001)
Brachiaria brizantha (A. Rich.) Stapf	*4x*	A; Ap (Pa); Ps	Brown and Emery (1958), Lutts et al. (1991, 1994), Alves et al. (2001), Araujo et al. (2000, 2005)
Brachiaria decumbens Stapf	*4x*	A; Ap (Pa); Ps	Pritchard (1967), Lutts et al. (1991, 1994), Pessino et al. (1997), Dusi (2001)
Brachiaria jubata (Fig. & De Not) Stapf		A; Ap (Pa); Ps	do Valle and Glienke (1991)
Brachiaria nigropedata (Ficalho & Hiern) Stapf		A; Ap (Pa); Ps	do Valle and Glienke (1991)
Brachiaria platynota (K. Schum.) Robyns		A; Ap (Pa); Ps	do Valle and Glienke (1991)
Brachiaria serrata (Thanb.) Stapf		A; Ap (Pa); Ps	Brown and Emery (1958)
Brachiaria subulifolia (Mez) Clayton		A; Ap (Pa); Ps	do Valle and Glienke (1991), do Valle and Miles (2001)
Capillipedium kwashotense (Hayata) C.C. Hsu [sub *Capillipedium kwashotense* (Hyata) de Wet]		A	de Wet et al. (1963)
Capillipedium huegelli (Hack.) A. Camus	*4x*	A; Ap (Pa, Se); Ps	Choda and Bhanwra (1980), Bhanwra (1988)
Capillipedium parviflorum (R.Br.) Stapf	*4x, 5x*	A; Ap (Pa); Ps	Celarier and Harlan (1957), Bhanwra et al. (1982), Bhanwra (1988)

TABLE 4.1 (Continued)

Species	Ploidy	Type of Apomixis[a]	References
Capillipedium spicigerum S.T. Blake	*4x*	A; Ap (Pa); Ps	Celarier and Harlan (1957)
Cenchrus ciliaris L. [≡*Pennisetum ciliare* (L.) Link; =*Cenchrus glaucus* C.R. Mudaliar & Sundararaj]	*4x, 6x*	A; N; Ap (Pa, Se); Ps	Fisher et al. (1954), Snyder et al. (1955), Simpson and Bashaw (1969), Gupta and Yashvir (1971), Shanthamma and Narayan (1976, 1977), Bray (1978), Shanthamma (1982), Febulaus and Pullaiah (1995), Visser et al. (1999)
Cenchrus caudatus (Schrad.) Kuntze (=*Pennisetum macrourum* Trin.)	*4x*	A; Ap	Dujardin and Hanna (1984a)
Cenchrus clandestinus (Hochst. ex Chiov.) Morrone (≡*Pennisetum clandestinum* Hochst ex Chiov.)	*4x*	A; Ap (Pa); Ps	Narayan (1955)
Cenchrus elegans (Hassk.) Veldkamp [=*Pennisetum macrostachyum* (Brongn.) Trin.]	*6x*	A; Ap	Dujardin and Hanna (1984a)
Cenchrus flaccidus (Griseb.) Morrone (=*Penisetum flaccicum* Griseb.)	*4x, 5x*	A; Ap (Pa)	Bashaw et al. (1992), Chatterji and Timothy (1969)
Cenchrus hohenackeri (Hochst. ex Steud.) Morrone (=*Pennisetum hohenackeri* Hochst. Ex Steud.)		A; Ap (Pa); Ps	Kurup (1973), Seshavatharam (1983)
Cenchrus hordeoides (Lam.) Morrone [≡*Pennisetum hordeoides* (Lam.) Steud.]	*4x, 5x*	A	Schmelzer and Renno (1999), Ozias-Akins et al. (2003)
Cenchrus latifolius (Spreng.) Morrone (≡*Pennisetum latifolium* Spreng.)	*4x*	A	Ozias-Akins et al. (2003)
Cenchrus longisetus M.C. Johnst. (=*Pennisetum villosum* Fresen)	*3x-6x*	A; Ap (Pa, Hi); Ps	Narayan (1951), Fisher et al. (1954), Snyder et al. (1955), Gildenhuys and Brix (1959)
Cenchrus massaicus (Stapf) Morrone (≡*Pennisetum massaicum* Stapf)	*4x*	A	Saran and Mishra (1986), D'Cruz and Reddy (1968)
Cenchrus mezianus (Leeke) Morrone (≡*Pennisetum mezianum* Leeke)	*4x*	A; N; Ap	Shanthamma (1973), Shanthamma and Narayan (1976), Bashaw et al. (1992)
Cenchrus orientalis (Rich.) Morrone (≡*Pennisetum orientale* Rich.)	*3x-6x*	A; Ap (Pa); Ps, Au	Narayan (1951), Chatterji (1969), Choda and Sharma (1981), Emery (1957), Bhanwra (1988)
Cenchrus pedicellatus (Trin.) Morrone (≡*Pennisetum pedicellatum* Trin.)	*4x-6x*	A; Ap (Pa); Au	Chatterji and Pillai (1970), Shobha (1988), Schmelzer and Renno (1999)
Cenchrus pennisetiformis Steud. (≡*Cenchrus pennisetiformis* Hochst.)	*6x*	A; Ap (Pa)	Islam and Das (1970), Das and Islam (1977), Kumar et al. (2017a, 2017b)
Cenchrus pilcomayensis (Mez) Morrone (=*Pennisetum frutescens* Leeke)	*7x*	A	Ozias-Akins et al. (2003)
Cenchrus purpureus (Schumach.) Morrone (≡*Pennisetum purpureum* Schumach.)	*4x*	A; Ap (Pa)	Brown and Emery (1958)
Cenchrus ramosus (Hochst.) Morrone [≡*Pennisetum ramosum* (Hochst.) Schweinf. & Asch.]	*4x*	A	Narayan (1962)
Cenchrus setaceus (Forssk.) Morrone [≡*Pennisetum setaceum* (Forssk.) Chiov.]	*4x*	A; Ap (Pa); Au	Narayan (1951), Emery (1957), Simpson and Bashaw (1969)

(Continued)

TABLE 4.1 (Continued)
Table of Apomictic Species (A) within the Family Poaceae Barnhart, Sorted Alphabetically According to Subfamilies

Species	Ploidy	Type of Apomixis[a]	References
Cenchrus setigerus Vahl	*4x, 6x*	A; Ap (Pa)	Fisher et al. (1954), Das and Islam (1977), Bhanwra et al. (1991)
Cenchrus setosus var. setosus [=*Pennisetum subangustum* (Schumach.) Stapf & C.E. Hubb.]	*4x, 6x*	A	Schmelzer and Renno (1999)
Cenchrus sphacelatus (Nees) Morrone [=*Pennisetum dubium*, undescribed species (see Gildenhuys and Brix, 1959) for *Pennisetum sphacelatum* (Nees) T. Durand & Schinz]	*7x*	A; Ap (Pa, Se); Ps	Gildenhuys and Brix (1959)
Cenchrus squamulatus (Fresen.) Morrone (≡*Pennisetum squamulatum* Fresen.)	*6x*	A; Ap (Pa, Se); Au	Sindhe (1976), Dujardin and Hanna (1984a), Wen et al. (1998)
Chionachne gigantea (J. Koenig) Veldkamp [≡*Coix gigantea* J. Koenig; =*Chionachne koenigii* (Spreng.) Thwaites]		A; Ap (Pa); Ps	Narayan (1967, 1968), Seshavatharam (1983)
Chrysopogon aciculatus (Retz.) Trin.		A	Seshavatharam (1983)
Chrysopogon serrulatus Trin.		A; Di	Faruqi et al. (1975)
Coix aquatica Roxb.		A; Ap (Pa)	Koul (1970), Venkateswarlu and Rao (1966), Venkateswarlu and Rao (1975a, 1975b)
Coix lachryma-jobi L.		A	Seshavatharam (1983), Mello et al. (1995)
Cymbopogon commutatus (Steud.) Stapf (=*Cymbopogon parkeri* Stapf)	*4x-6x*	A	Faruqi et al. (1975)
Cymbopogon winterianus Jowitt ex Bor		A	Khan (1990)
Dichanthium annulatum (Forssk.) Stapf [=*Dichanthium nodosum* Willemet; =*Dichanthium papillosum* (Hochst. ex A.Rich.) Stapf]	*4x, 6x*	A; Ap (Pa, Se, Hi); Ps	Celarier and Harlan (1957), de Wet and Harlan (1968), Gupta (1968), Reddy and D'Cruz (1969), Gupta et al. (1969), Carman (1995)
Dichanthium aristatum (Poir.) C.E. Hubb.	*4x, 6x*	A; Ap (Pa); Ps	Celarier and Harlan (1957), de Wet and Harlan (1968)
Dichanthium caricosum (L.) A. Camus	*4x*	A; Ap (Pa); Ps	Celarier and Harlan (1957), de Wet and Harlan (1968)
Dichanthium concanense (Hook. f.) S.K. Jain & Deshp. [≡*Bothriochloa concanensis* (Hook.f.) Henrard]	*5x*	A	de Wet and Harlan (1970)
Dichanthium fecundum S.T. Blake		A	de Wet et al. (1963)
Dichanthium foulkesii (Hook.f.) S.K. Jain & Deshp.		A	de Wet et al. (1963)
Dichanthium foveolatum (Delile) Roberty [≡*Eremopogon foveolatus* (Delile) Stapf]		A; Ap (Pa)	Bhanwra and Choda (1981), Satyamurthy and Seshavathram (1983)
Dichanthium tenue (R.Br.) A. Camus		A	de Wet et al. (1963)
Echinochloa frumentacea Link	*6x*	A; Ap	Narayanaswami (1955)
Echinochloa stagnina (Retz.) P. Beauv.	*4x, 6x, 8x, 12x, 14x*	A; Ap (Pa)	Muniyamma (1978)

TABLE 4.1 (Continued)

Species	Ploidy	Type of Apomixis[a]	References
Eriochloa stapfiana Clayton (=*Eriochloa borumensis* Hack.)		A; Ap	Brown and Emery (1958)
Eriochloa procera (Retz.) C.E. Hubb.		A; N; Ap (Pa)	Shanthamma and Narayan (1976, 1977), Seshavatharam (1983)
Eriochloa sericea (Scheele) Munro ex Vasey		A; Ap	Brown and Emery (1958)
Eulaliopsis binata (Retz.) C.E. Hubb.	*4x*	A; N; Ap (Pa); Au	Yao et al. (2004, 2007)
Heteropogon contortus (L.) P. Beauv. ex Roem. & Schult.		A; Ap (Pa)	Gupta (1968), Tothill (1968), Emery and Brown (1958)
Hyparrhenia diplandra (Hack.) Stapf	*4x*	A; Ps	Durand et al. (2000)
Hyparrhenia hirta (L.) Stapf	*3x-6x*	A; Ap (Pa); Ps	Brown and Emery (1958), McWilliam et al. (1970)
Hyparrhenia rufa (Nees) Stapf		A; Ap (Pa)	Brown and Emery (1958)
Iseilema anthephoroides Hack.		A; N	Venkateswarlu and Devi (1964)
Melinis repens subsp. *grandiflora* (Hochst.) Zizka [=*Rhynchelytrum villosum* (Parl.) Chiov.]	*4x*	A; Ap	Shanthamma and Narayan (1976, 1977); Inamuddin et al. (2009)
Panicum antidotale Retz.		A; Ap (Hi)	Shamakumari (1960), Narayan (1962)
Panicum deustum Thub.	*4x*	A; Ap	Brown and Emery (1958)
Panicum maximum Jacq.	*4x*	A; Ap (Hi; Pa); Ps	Warmke (1954), Brown and Emery (1958)
Panicum obtusum Kunth	*4x*	A; Ap (Pa)	Brown and Emery (1958), Carman (1995)
Panicum repens L. (=*Panicum notatum* Retz.)		A; Ap (Hi)	Febulaus (1992)
Panicum virgatum L.	*4x, 6x, 8x, 9x, 10x*	A; Ap (Pa)	Brown and Emery (1958)
Paspalum almum Chase	*4x*	A; Ap (Pp); Ps	Burson (1975), Quarin and Hanna (1980)
Paspalum arundinellum Mez	*4x*	A; Ap (Pp); Ps	Bashaw et al. (1970)
Paspalum atratum Swallen	*4x*	A; Ap (Pp); Ps	Quarin et al. (1997)
Paspalum buckleyanum Vasey (=*Paspalum alcalinum* Mez)	*4x, 5x, 6x*	A; Ap (Pp); Ps	Burson (1997), Sartor et al. (2011)
Paspalum chaseanum Parodi	*4x*	A; Ap (Pp); Ps	Novo et al. (2014)
Paspalum compressifolium Swallen	*4x, 6x*	A; Ap (Pp); Ps	Quarin et al. (1996)
Paspalum commune Lillo	*4x*	A; Ps	Galdeano et al. (2016)
Paspalum conjugatum P.J. Bergius	*4x, 6x*	A; Ap (Pp); Di (Ta); Ps, Au	Chao (1980), Ma et al. (2009), Galdeano et al. (2016)
Paspalum coryphaeum Trin.	*4x*	A; Ap (Pp); Ps	Burson (1975), Quarin and Urbani (1990)
Paspalum cromyorhizon Trin. ex Döll (sub *Paspalum guaraniticum* Parodi)	*3x, 4x*	A; Ap (Pp); Ps	Bashaw et al. (1970), Burson and Bennett (1971a, 1971b), Quarin et al. (1982), Martínez et al. (1999)
Paspalum dedeccae Quarin	*4x*	A; Ap (Pp); Ps	Quarin and Burson (1991)
Paspalum denticulatum Trin.	*3x, 4x*	A; Ap (Pp); Ps	Quarin and Burson (1991), Sartor et al. (2011)
Paspalum dilatatum Poir.	*5x, 6x*	A; Ap (Pp); Ps	Bennett (1944), Smith (1948), Hayman (1956), Bashaw and Holt (1958), Brown and Emery (1958), Burson (1985), Burson et al. (1991)
Paspalum distichum L.	*4x, 6x*	A; Ap (Pp, Hi); Ps	Bashaw et al. (1970), Choda and Bhanwra (1977), Burson and Bennett (1971a), Bhanwra (1988), Quarin and Burson (1991), Srivastava (1982), Sherman et al. (1991)

(Continued)

TABLE 4.1 (Continued)
Table of Apomictic Species (A) within the Family Poaceae Barnhart, Sorted Alphabetically According to Subfamilies

Species	Ploidy	Type of Apomixis[a]	References
Paspalum durifolium Mez	*5x, 6x*	A; Ap (Pp); Ps	Burson (1985), Galdeano et al. (2016)
Paspalum erianthum Nees ex Trin.	*8x*	A; Ps	Galdeano et al. (2016)
Paspalum exaltatum J. Presl (=*Paspalum arechavaletae* Hack. ex Arechav.)	*4x*	A; Ap (Pp); Ps	Burson and Bennett (1971a)
Paspalum falcatum Nees ex Steud.	*4x*	A; Ap (Pp); Ps	Burson (1997)
Paspalum foliiforme S. Denham [=*Thrasya petrosa* (Trin.) Chase]	*4x*	A; Ap (Pp); Ps	Acuña et al. (2005)
Paspalum guenoarum Arechav. (=*Paspalum kempffii* Killeen; =*Paspalum macedoi* Swallen; =*Paspalum rojasii* Hack.)	*4x*	A; Ap (Pp); Ps	Bashaw et al. (1970), Pritchard (1970), Burson and Bennett (1971a, 1971b), Espinoza et al. (2001)
Paspalum hartwegianum E. Fourn.	*4x*	A; Ap (Pp);	Brown and Emery (1958)
Paspalum haumanii Parodi	*4x*	A; Ap (Pp); Ps	Burson (1975), Norrmann et al. (1989)
Paspalum humboldtianum Flüggé	*8x*	A; Au	Galdeano et al. (2016)
Paspalum intermedium Munro ex Morong	*4x*	A; Ap (Pp); Ps	Norrmann et al. (1989), Karunarathne et al. (2018)
Paspalum ionanthum Chase (=*Paspalum guaraniticum* Parodi)	*8x*	A; Ap (Pp); Ps	Burson and Bennett (1971b)
Paspalum lenticolare Kunth (=*Paspalum limbatum* Henrard)	*4x, 6x*	A; Ap (Pp); Ps	Espinoza et al. (2001), Galdeano et al. (2016)
Paspalum lividum Trin.	*4x*	A; Ap (Pp); Ps	Burson and Bennett (1971a); Sartor et al. (2011)
Paspalum longifolium Roxb.	*4x*	A; Di (Ta); Ps	Chao (1974)
Paspalum maculosum Trin.	*4x*	A; Ap (Pp); Ps	Norrmann et al. (1989)
Paspalum malacophyllum Trin.	*4x*	A; Ap (Pp); Ps	Brown and Emery (1958), Burson and Hussey (1998), Hojsgaard et al. (2008)
Paspalum mandiocanum Trin.	*5x, 6x*	A; Ap (Pp); Au	Burson and Bennett (1971a); Galdeano et al. (2016)
Paspalum minus L.	*5x*	A; Ap (Pp); Di (Ta); Ps	Bonilla and Quarin (1997)
Paspalum modestum Mez	*4x*	A; Ap (Pp); Ps	Karunarathne et al. (2020a)
Paspalum nicorae Parodi	*4x*	A; Ap (Pp); Ps	Bashaw (1970), Burson and Bennett (1971a), Sartor et al. (2011)
Paspalum notatum Flüggé	*4x*	A; Ap (Pp); Ps	Burton (1948), Bashaw et al. (1970)
Paspalum orbiculare G. Forst	*4x-6x*	A; Di (Ta)	Chao (1964)
Paspalum oteroi Swallen	*4x*	A; Ps	Galdeano et al. (2016)
Paspalum ovale Nees ex Steud.	*8x*	A; Ps	Galdeano et al. (2016)
Paspalum palustre Mez	*4x*	A; Ps	Galdeano et al. (2016)
Paspalum paucifolium Swallen	*4x*	A; Ap (Pp); Ps	Burson (1997)
Paspalum pauciciliatum (Parodi) Herter	*4x*	A; Ap (Pp); Ps	Bennett and Bashaw (1966), Bashaw et al. (1970)
Paspalum plicatulum Michx.	*4x*	A; Ap (Pp); Ps	Bashaw et al. (1970), Pritchard (1970), Burson and Bennett (1971b)
Paspalum polyphyllum Nees ex Trin.	*4x*	A; Ap (Pp); Ps	Burson (1997)
Paspalum procurrens Quarin	*4x*	A; Ap (Pp); Ps	Hojsgaard et al. (2008)
Paspalum proliferum Arechav.	*4x, 6x*	A; Ap (Pp); Ps	Quarin et al. (1982), Burson (1975)

TABLE 4.1 (Continued)

Species	Ploidy	Type of Apomixis[a]	References
Paspalum quadrifarium Lam.	*3x, 4x*	A; Ap (Pp); Ps	Bashaw et al. (1970), Norrmann et al. (1989), Quarin and Lombardo (1986)
Paspalum quarini Morrone & Zuloaga (=*Paspalum brunneum* Mez)	*4x*	A; Ap (Pp); Ps	Burson (1975), Norrmann et al. (1989)
Paspalum ramboi I.L. Barreto	*6x*	A; Ap (Pp); Ps	Quarin and Burson (1991)
Paspalum rufum Nees ex Steud.	*4x*	A; Ap (Pp); Ps	Burson (1975), Norrmann et al. (1989), Quarin et al. (1998), Siena et al. (2008)
Paspalum scrobiculatum L. (=*P. commersonii* Lam.)	*6x, 10x,*	A; Ap; Di (Ta); Ps; Au	Chao (1974), Ma et al. (2009)
Paspalum secans Itchc. & Chase	*4x*	A; Ap (Hi); Ps;	Snyder (1957), Bashaw et al. (1970)
Paspalum simplex Morong	*3x, 4x*	A; Ap (Hi; Pp); Ps	Caponio and Quarin (1987), Cáceres et al. (2001), Urbani et al. (2002)
Paspalum thunbergii Kunth ex Steud.	*4x*	A; Ap; Ps	Chung (1974), Ma et al. (2004)
Paspalum unispicatum (Scribn. & Merr.) Nash	*3x, 4x*	A; Ap (Pp); Ps	Burson (1997), Galdeano et al. (2016)
Paspalum wrightii Hitchc. & Chase (=*Paspalum hydrophilum* Henrard)	*4x*	A; Ap (Pp); Ps	Norrmann (1981)
Polytrias indica (Houtt.) Veldkamp [≡*Ischaemum indicum* (Houtt.) Merr.		A; Ap (Pa); Ps	Seshavatharam (1983)
Schizachyrium sanguineum (Retz.) Alston	*4x, 10x*	A; Ap	Carman and Hatch (1982)
Sehima ischaemoides Forssk.		A; Ap (Pa); Ps	Seshavatharam (1983)
Sehima nervosum (Rottler) Stapf (=*Ischaemum laxum* R.Br.)		A; Ap (Hi)	Seshavatharam (1983)
Setaria leucophila (Scribn. & Merr.) K. Schum.	*6x*	A; Ap (Pa); Ps	Emery (1957)
Setaria parviflora (Poir.) Kerguélen [=*Pennisetum polystachyon* (L.) Schult.; =*Pennisetum setosum* (Swartz.) L.Rich.]	*4x, 6x, 7x*	A; Ap (Pa); Ps	Birari (1981), Dujardin and Hanna (1984a); Saran and Mishra (1986), Schmelzer and Renno (1999), Ozias-Akins et al. (2003)
Setaria villosissima (Scribn. & Merr.) K. Schum.	*6x*	A; Ap (Pa); Ps	Emery (1957)
Sorghum bicolor (L.) Moench	*2x, 4x*	A; N; Ap (Hi); Ps	Hanna et al. (1970)
Sorghum halepense (L.) Pers.		A; Ap (Pa); Ps	Seshavatharam (1983)
Themeda arundinacea (Roxb.) A. Camus	*6x*	A; Ap (Pa)	Birari (1980)
Themeda dacruzii Birari	*6x*	A; Ap (Pa)	Birari (1980)
Themeda longispatha (Hackel.) Raizada & Jain.	*5x*	A; Ap (Pa)	Birari (1980)
Themeda quadrivalvis (L.) Kuntze	*4x*	A; Ap (Pa)	Brown and Emery (1958), Evans and Knox (1969), Birari (1980), Liebenberg (1990)
Themeda triandra Forssk. [=*Themeda australis* (R.Br.) Stapf]	*3x-6x, 8x, 9x, 11x*	A; Ap (Pa, Se); Ps	Brown and Emery (1957), Evans and Knox (1969), Liebenberg (1990)
Trachys muricata (L.) Pers.		A; Ap (Pa); Ps	Seshavatharam (1983)
Tricholaena monachne (Trin.) Stapf & C.E. Hubb.	*4x*	A; Ap	Brown and Emery (1958), Carman (1995)

(Continued)

TABLE 4.1 (Continued)
Table of Apomictic Species (A) within the Family Poaceae Barnhart, Sorted Alphabetically According to Subfamilies

Species	Ploidy	Type of Apomixis[a]	References
Tripsacum dactyloides (L.) L.	*4x-6x, 8x, 10x*	A; Di (Ta; An); Ps	Farquharson (1955), Burson et al. (1990), Kindiger and Dewald (1994), Leblanc et al. (1995a, 1995b)
Tripsacum intermedium De Wet & J.R.Harlan	*3x-5x*	A; Di (An); Ps	Leblanc et al. (1995a, 1995b)
Tripsacum lanceolatum E. Fourn.	*4x*	A; Di (An); Ps	Leblanc et al. (1995a, 1995b)
Tripsacum maizar Hern.-Xol. & Randolph	*3x, 4x*	A; Di (An); Ps	Leblanc et al. (1995a, 1995b)
Tripsacum pilosum Scribn. & Merr.	*4x*	A; Di (An); Ps	Leblanc et al. (1995a, 1995b)
Tripsacum zopolitense Hern.-Xol. & Randolph	*3x, 4x*	A; Di (Ta; An); Ps	Leblanc and Savidan (1994), Leblanc et al. (1995a, 1995b)
*Urochloa bolbodes (*Steud.) Stapf	*4x*	A; Ap (Pa)	Brown and Emery (1958)
Urochloa dictyoneura (Fig. & De Not.) Veldkamp [≡*Brachiaria dictyoneura* (Fig. & De Not) Stapf]		A; Ap (Pa); Ps	do Valle and Glienke (1991)
Urochloa humidicola (Rendle) Morrone & Zuloaga [≡*Brachiaria humidicola* (Rendle) Schweick]	*4x*	A; Ap; Ps	do Valle and Glienke (1991)
Urochloa panicoides P. Beauv.		A; Ap	Khan (1990)
Urochloa setigera (Retz.) Stapf [≡*Brachiaria setigera* (Retz.) C.E. Hubb.]	*4x*	A; Ps	Muniyamma (1977)
Urochloa trichopus (Hoschst.) Stapf [=*Urochloa mossambicensis* (Hack.) Dandy; =*Urochloa pullulans* Stafp.]	*4x*	A; Ap (Pa)	Brown and Emery (1958), Pritchard (1970)
Zea mexicana (Schrad.) Kuntze		A	Tantravahi (1965)
Pooideae Benth.			
Anthosachne rectiseta (Nees) Barkworth & S.W.L. Jacobs [≡*Elymus rectisetus* (Nees) Á. Löve & Connor]		A; Di (Ta)	Hair (1956), Crane and Carman (1987)
Anthosachne scabra (R. Br.) Nevski [≡*Elymus scabrus* (R.Br.) Á. Löve]		A; Ap; Ps	Hair (1956)
Anthoxanthum australe (Schrad.) Veldkamp [≡*Hierochloe australis* (Schrad.) Roem. & Schult.]		A; Ap (Hi)	Weimarck (1967a)
Anthoxanthum monticola (Bigelow) Veldkamp [≡*Hierochloe monticola* (Bigel.) Love & Love; =*Hierochloe alpina* (Sw.) Roem. & Schult.]	*8x, 9x*	A; Ap (Hi); Au	Weimarck (1967b, 1970)
Anthoxanthum nitens (Weber) Y. Schouten & Veldkamp [=*Hierochloe odorata* (L.) P. Beauv.]	*8x*	A; Ap (Pa); Ps	Norstog (1963), Weimarck (1967a)
Anthoxanthum odoratum L.		A; Ap	Shanthamma and Narayan (1976)
Anthoxanthum wendelboi (G.Weim.) Veldkamp (≡*Hierochloe wendelboi* G. Weim.)	*4x*	A; Ap (Hi)	Weimarck (1977)
Bromus inermis Leyss.		A; N	Naumova (1993)

TABLE 4.1 (Continued)

Species	Ploidy	Type of Apomixis[a]	References
Calamagrostis arundinacea (L.) Roth (=*Calamagrostis brachytricha* Steud.)	*7x, 8x*	A	Tateoka (1969)
Calamagrostis canadensis (Michx.) P. Beauv.	*7x-9x*	A; Di	Nygren (1954b), Jørgensen et al. (1958)
Calamagrostis chalybaea (Laest.) Fr.	*6x*	A; Di (An); Au	Stenar (1932), Nygren (1946)
Calamagrostis crassiglumis Thurb.	*20x*	A; Di (An); Au	Nygren (1954b), Jørgensen et al. (1958)
Calamagrostis inexpansa A. Gray		A; Di (Ta); An)	Greene (1984)
Calamagrostis lapponica (Wahlenb.) Hartm.	*6x-16x*	A; Di (An); Au	Nygren (1946), Jørgensen et al. (1958)
Calamagrostis purpurascens R.Br.	*6x, 8x*	A; Di (An)	Nygren (1954b)
Calamagrostis purpurea (Trin.) Trin.	*8x-12x*	A; Di (An); Au	Nygren (1954b)
Calamagrostis stricta (Timm) Koeler		A; Di (An); Au	Greene (1984)
Festuca drymeja Mert. & W.D.J. Koch		A	Crane and Carman (1987), Shishkinskaya and Borodko (1987)
Festuca olgae (Regel) Krivot.		A; Ps	Brožová et al. (2019)
Nardus stricta L.	*4x*	A; Di; Au	Rychlewski (1961), Coulon (1923), Kissling et al. (2006)
Nicoraepoa erinacea (Speg.) Soreng & L.J. Gillespie (≡*Poa erinacea* Speg.)		A	Kelley et al. (2009)
Poa alpina L.	*4x-10x*	A; Di (An); Ps	Müntzing (1933), Håkansson (1943, 1944), Müntzing (1966), Nygren (1967), Kelley et al. (2009)
Poa araratica subsp. *psilolepis* (Keng) Olonova & G.H. Zhu (≡*Poa psilolepis* Keng)		A	Kelley et al. (2009)
Poa arctica R.Br.	*8x, 10x*	A; Ap (Hi); Ps	Engelbert (1941), Nygren (1950b), Soreng et al. (2017)
Poa arida Vasey	*9x*	A; Ap	Conner (1979)
Poa attenuata Trin.	*3x*	A; Ps	Kelley et al. (2009), Brožová et al. (2019)
Poa bactriana Roshev.		A; Au	Kelley et al. (2009)
Poa binata Nees		A	Kelley et al. (2009)
Poa bulbosa L.	*4x, 5x, 10-14x*	A; Au	Stebbins (1949), Kelley et al. (2009)
Poa chaixii Vill.		A	Kelley et al. (2009)
Poa compressa L.	*5x-8x*	A; Ap (Hi)	Armstrong (1937), Kelley et al. (2009)
Poa curtifolia Scribn.	*6x*	A; Ap	Kellog (1987)
Poa densa Troitsky		A	Kelley et al. (2009)
Poa glauca Vahl	*6x-10x*	A; Ap; Di (An)	Åkerberg (1941), Nygren (1950a, 1950b), Kelley et al. (2009)
Poa granitica Braun-Blanq.	*9x-13x*	A; Ap	Skalińska (1959)
Poa × *herjedalica* Harry Sm.	*6x-11x*	A; Ap; Di	Nannfeldt (1937), Nygren (1950a, 1950b)
Poa × *jemtlandica* (Almq.) K. Richt.	*5x*	A; Ap; Di	Nannfeldt (1937), Nygren (1950a, 1950b, 1955)
Poa humilis Ehrh. ex Hoffm. (=*Poa irrigata* Lindm.)		A; Ap; Ps	Löve (1952)
Poa interior Rydb.	*4x*	A	Kelley et al. (2009)
Poa nemoralis L.	*4x, 6x, 8x*	A; Ap; Di (An); Ps	Nygren (1950a, 1950b), Naumova et al. (1999), Kelley et al. (2009)

(*Continued*)

TABLE 4.1 (Continued)
Table of Apomictic Species (A) within the Family Poaceae Barnhart, Sorted Alphabetically According to Subfamilies

Species	Ploidy	Type of Apomixis[a]	References
Poa nervosa (Hook.) Vasey	*4x, 8x, 9x, 12x*	A; Ap (Hi); Di (An); Au	Grun (1955), Kelley et al. (2009)
Poa palustris L.	*4x, 6x*	A; Ap; Di (An); Ps	Kiellander (1935, 1937), Wieners et al. (2006), Kelley et al. (2009)
Poa pratensis L.	*7x-12x*	A; N; Ap; Di; Ps	Tinney (1940), Kiellander (1941), Åkerberg (1941), Nielsen (1946), Juhl-Noodt (1955), Grazi et al. (1961), Batygina and Freiberg (1979)
Poa riphaea (Asch. & Graebn.) Fritsch	*8x*	A	Hoták et al. (2013)
Poa secunda J. Presl	*8x-14x*	A; Ap; Ps	Kellog (1987), Kelley et al. (2009)
Poa secunda var. *ampla* (Merr.) Soreng (=*Poa ampla* Merr.)	*8x, 9x*	A; Ap	Kellog (1987)
Poa secunda var. *gracillima* (Vasey) Soreng (=*Poa gracillima* Vasey)	*7x*	A; Ap; Ps	Kellog (1987)
Poa secunda subsp. *juncifolia* (Scribn.) Soreng (=*Poa juncifolia* Scribn.)	*8x*	A; Ap; Ps	Kellog (1987)
Poa secunda var. *nevadensis* (Vasey) Soreng (=*Poa nevadensis* Vasey)		A; Ap; Ps	Kellog (1987)
Poa secunda subsp. *secunda*. [=*Poa canbyi* (Scribn.) Howell; =*Poa sandbergii* Vasey]	*10x*	A; Ap; Ps	Kellog (1987)
Poa secunda var. *scabrella* (Thurb.) Soreng [=*Poa scabrella* (Thurb.) Benth. ex Vasey]	*12x*	A; Ap; Ps	Kellog (1987)
Poa sibirica Roshev.		A	Kelley et al. (2009)
Poa sterilis M. Bieb.		A; Ps	Kelley et al. (2009), Brožová et al. (2019)
Poa versicolor Besser		A	Kelley et al. (2009)
Stipa splendens Trin. [≡*Achnatherum splendens* (Trin.) Nevski]		A; Ps	Brožová et al. (2019)

[a]For species with sporophytic apomixis, AE: adventitious embryony; N: nucellar embryony. For species with gametophytic apomixis, Ap ( ): apospory (type of female gametophyte development); Di ( ): diplospory (type of female gametophyte development); Ps: pseudogamous endosperm development; Au: autonomous endosperm development.

Hi: Hieracium type; Pa: Panicum type; Se: Setaria type; Pp: Paspalum type; Ta: Taraxacum type; An: Antennaria type; Er: Eragrostis type.

hypothesize about the type of female gametophyte development. For instance, in the case of *Poa* L., a genus containing species that reproduce through apospory, diplospory, or using both mechanisms, making conclusions on the type of female gametophyte development in any of the ten new apomictic species described by Kelley et al. (2009) would be speculative. The same is true for *Paspalum* L., even though species of this genus reproduce almost exclusively by apospory, they show the Paspalum type of female gametophyte development, and there is a high probability that the new six polyploid apomictic species characterized by FCSS (Galdeano et al., 2016) will show a similar type of development. Hence, in all these cases no extrapolations were made, and specific analyses are required to reveal the types of apomictic developments.

We apologize for any missing data or errors resulting from lack of access to certain works or from our language limitations (e.g., Miroshnichenko, 1970; Shishkinskaya and Borodko, 1987; Yudakova and Shishkinskaya, 2008).

4.2.1 Doubtful Cases of Apomixis in Grasses

In keeping with current times, doubtful cases are placed on a quarantine list (Table 4.2). Although circumstantial evidence points to the occurrence of apomixis in these species, the information collected appears to be insufficient to conclude the existence of apomixis in these taxa.

For instance, in the South African species of *Schmidtia* Steud. ex J.A. Schmidt [*Schmidtia pappophoroides* Steud. ex J.A. Schmidt (=*S. bulbosa* Stapf) and *S. kalahariensis* Stent], Brown and Emery (1958) did not observe antipodal cells. The lack of antipodal cells in grasses is linked to (aposporous) apomixis. However, the authors were unable to reach any conclusions concerning their embryology, since samples at the correct stages of development were limited, and the development of the female gametophyte in the available material appeared abnormal.

Phleum L. was categorized as diplosporous by Czapik's review on apomixis in monocotyledons (Czapik, 2000). However, we have not been able to find a specific reference for this record or the name of the taxon for which diplospory is claimed.

In *Arundinella mesophylla* Nees ex Steud., embryological evaluation of a diploid cytotype ($2n=2x=20$) showed the occurrence of around 10% of aposporous female gametophytes with no apparent organization (either four-nucleate or eight-nucleate gametophytes) and no evidence of parthenogenetic development (Basappa and Muniyamma, 1981). Since the development of twin female gametophytes from two archesporial cells was detected in a related species (*A. purpurea* Hochst. ex Steud., Basappa and Muniyamma, 1981), the evidence points to *A. mesophylla* being a diploid sexual cytotype with spontaneous occurrence of apospory but without functional apomixis, as it happens in many other apomictic species holding diploid sexual cytotypes and polyploid apomicts (see e.g. in *Paspalum* spp., Ortiz et al., 2013). Although there is a possibility of finding polyploid cytotypes of *Arundinella mesophylla* with apomixis, until then it is appropriate to include this species in the quarantine list (Table 4.2).

A similar case is found among species of *Tribolium* Desv. Visser and Spies (1994) found embryological evidence of aposporous female gametophytes in diploid cytotypes of *T. utriculosum* (Nees) Renvoize, *T. echinatum* (Thunb.) Renvoize and *T. uniolae* (L.f.) Renvoize. However, only images of the latter two species have been published, in which it is possible to see smaller (immature) aposporous female gametophytes, with a lateral position relative to the normal sexual gametophyte in the ovule (Visser and Spies, 1994, pp. 25, Fig. 2D-I) and without apparent signs of vacuolization (like that observed in meiotic female gametophytes (Visser and Spies, 1994, Fig. 1 and Fig. 2D-F), which would suggest active growth and differentiation. Although the authors studied 20 ovules ("embryo sacs") *per* accession ("specimen") and a total of 64 accessions (Visser and Spies, 1994, Table 1, pp. 23), no information was presented on the frequency of occurrence of aposporous female gametophytes. Unfortunately, a description of the development of the aposporous pathway is missing, and Hieracium type female gametophytes are only mentioned in the discussion and abstract sections. While mature female gametophytes of the Polygonum type were portrayed (Visser and Spies, 1994, pp. 24, Fig. 1A-F), mature female gametophytes of the Hieracium type were not. Furthermore, following the accession numbers of the material studied (Visser and Spies, 1994, pp. 22, 23 and Table 1), all accessions from *T. uniolae* ($n=34$) were either tetraploid or hexaploid, except accession Spies 4585 which included flowers with diploid and tetraploid chromosome numbers (Visser and Spies, 1994, Table 1). Yet, there is no mention of female gametophyte development in the polyploids. Therefore, we consider that this case requires further clarification and, therefore, we place the species of *Tribolium* in Table 4.2.

In *Calamagrostis* Adans., based on examination of herbarium material of *Calamagrostis inexpansa* A. Gray (=*Calamagrostis hyperborea* Lange) showing defective and shrivelled pollen, Jørgensen et al.

TABLE 4.2
Table of Doubtful Apomictic Species (A) within the Family Poaceae Barnhart, Sorted Alphabetically According to Subfamilies

Species	Ploidy	Type of Apomixis[a]	References
Chloridoideae Kunth ex Beilschm.			
Schmidtia pappophoroides Steud. ex J.A. Schmidt (=*Schmidtia bulbosa* Stapf)			Brown and Emery (1958)
Schmidtia kalahariensis Stent			Brown and Emery (1958)
Danthonioideae H.P. Linder & N.P. Barker			
Cortaderia speciosa Stapf		A	Conner (1979), Astegiano et al. (1995)
Cortaderia hieronymi (Kuntze) N.P. Barker & H.P. Linder [≡*Lamprothyrsus hieronymi* (O. Kuntze) Pilger]	*ca. 16x*	A	Connor and Dawson (1993)
Cortaderia peruviana (Hitchc.) N.P. Barker & H.P. Linder (≡*Lamprothyrsus peruvianus* Hitchc.)		A	Connor and Dawson (1993)
Capeochloa arundinacea (P.J. Bergius) N.P. Barker & H.P. Linder [≡*Merxmuellera arundinacea* (P.J. Bergius) Conert]		A	Verboom et al. (1994)
Geochloa rufa (Nees) N.P. Barker & H.P. Linder [≡*Merxmuellera rufa* (Nees) Conert]		A; Ap	Verboom et al. (1994)
Rytidosperma pallidum (R.Br.) A.M. Humphreys & H.P. Linder (≡*Danthonia pallida* R.Br.)		A	Verboom et al. (1994)
Tribolium utriculosum (Nees) Renvoize	*2x*	A; Ap	Visser and Spies (1994)
Tribolium echinatum (Thunb.) Renvoize	*2x*	A; Ap	Visser and Spies (1994)
Tribolium uniolae (L.f.) Renvoize	*2x, 4x, 6x*	A; Ap	Visser and Spies (1994)
Oryzoideae Kunth ex Beilschm.			
Potamophila parviflora R. Br.	*2x-4x*	A; Di	Wheeler et al. (2001)
Panicoideae A. Braun			
Arundinella mesophylla Nees ex Steud.	*2x*	A; Ap (Pa)	Basappa and Muniyamma (1981)
Saccharum officinarum L.	*8x*	A; Di?	Bremer (1923), Narayanaswami (1940)
Saccharum spontaneum L.	*6x-14x*	A; Di?	Narayanaswami (1940)
Pooideae Benth.			
Phleum sp.		A; Di	Czapik (2000)

[a]For species with gametophytic apomixis, Ap ( ): apospory (type of female gametophyte development); Di ( ): diplospory (type of female gametophyte development); Ps: pseudogamous endosperm development; Au: autonomous endosperm development.

Hi: Hieracium type; Pa: Panicum type; Se: Setaria type; Pp: Paspalum type; Ta: Taraxacum type; An: Antennaria type; Er: Eragrostis type.

(1958) indicated that "in all probability [the species is] apomictic". Greene (1984) classified the species as possibly apomict. Although pollen sterility has been observed in other apomictic species of *Calamagrostis*, here we decided not to include this and other similar cases (such as *C. hakonensis* Franch. & Sav. and *C. sachalinensis* F. Schmidt, Tateoka, 1968, 1973, 1984) in our lists, mainly because pollen sterility does not strictly correlate with apomixis in plants. Similarly, species for which there is some indication of apomixis, but which have not been analysed, are excluded from our lists. An example of this is *Thyrsostachys siamensis* Gamble, an interesting species of the Bambusoideae Luerss. for which signs of aposporic development had been detected in embryological analysis (Bhanwra et al., 1991), but apomixis *per se* was not proven and the observed development could simply pertain to natural variation in developmental programmes observed in plants.

Saccharum L. is a genus whose species present a complicated embryology, showing a combination of parthenogenesis and endopolyploidy as observed in *Rudbeckia* L. (Nygren, 1954a). Different studies have shown the formation of unreduced gametes and B_{III} hybrids between *S. officinarum* L. and *S. spontaneum* L. (Bremer, 1923, 1946). Such unreduced gametes are formed through asymmetric meiotic division followed by fusion of consecutive megaspores (Narayanaswami, 1940; Raghavan, 1951) or endomitosis in the chalazal megaspore (Bremer, 1946) to form unreduced female gametophytes. Subramanian (1946) revealed that after asymmetric division and female gametophyte formation, the egg cell divides early without formation of a cell wall, thus doubling the number of chromosomes. In his critical review of apomixis in sugarcane, Price (1959) concludes that, although some forms of apomixis may occasionally occur in *Saccharum* spp., they must do so very infrequently, and yet it must be shown that progeny derived from apomixis are produced. *Saccharum* spp. is therefore included in Table 4.2.

4.3 PATHWAYS TO ASEXUAL SEEDS IN GRASSES

A total of 242 grass species have at least one record of apomixis, either sporophytic or gametophytic apomixis (see Chapter 2).

The pathways of apomictic development are known for 81.0% of species (*n*=196), while partial or no information on reproductive development was obtained for 46 taxa. This includes most of the new species categorized as apomictic following FCSS analyses (see below) which are included under gametophytic apomixis, and 6 species for which apomixis has been stablished only through progeny tests or molecular analyses [e.g., *Cenchrus pilcomayensis* (Mez) Morrone (=*Pennisetum frutescens* Leeke), Ozias-Akins et al., 2003; *Hyparrhenia diplandra* (Hack.) Stapf, Durand et al., 2000], and no further inferences can be made about the reproductive development. Among species with gametophytic apomixis with embryological details, ten taxa show both apospory and diplospory (Table 4.1). Sporophytic apomixis is recorded for only 11 taxa, all cases of nucellar embryony (Table 4.1). Seven taxa show both sporophytic and gametophytic apomixis, mostly apospory (Table 4.1).

Information about the development of the endosperm is collected for about 63.2% of all apomictic grasses (*n*=153; Table 4.1). Most recorded apomicts are pseudogamous (*n*=133), and only 9.1% of species (*n*=22) show autonomous endosperm development.

Ploidy is known for 60.3% of apomictic taxa (*n*=146) and 53 cases have two or more ploidy records (Table 4.1). All species but one are polyploid, ranging from triploids to icosaploid (*x*=20, in *Calamagrostis crassiglumis* Thurb., Nygren, 1954b; Jørgensen et al., 1958). The most frequent ploidy is tetraploid (*n*=95), followed by hexaploid (*n*=40), octoploid (*n*=26), and among odd ploidies, pentaploid (*n*=18) and nonaploid (*n*=13). Other ploidies found among apomictic grasses are decaploid (*n*=13), triploid (*n*=11), and heptaploid (*n*=11), and a few more cases with higher ploidy (Table 4.1).

The only diploid detected is *Pentaschistis eriostoma* (Nees) Stapf (Spies et al., 1999), a species expected to be a secondary diploid. The species has $2n=2x=26$ chromosomes, likely correspond to a polyploid hybrid species with a secondary basic chromosome number ($x=6+7$ or $x=7+7-1$, Du Plessis and Spies, 1992). Most *Pentaschistis* species exhibit a basic chromosome number of $x=7$, while only two species show $x=13$ and *P. eriostoma* genomes (Klopper et al., 1998). Other cases of apomixis occurring in diploids represent sporadic events of apomictic developments rather than established asexual diploid lineages (e.g., in *Paspalum* spp., Siena et al., 2008; Delgado et al., 2014).

A wide diversity of unreduced female gametophyte developments is observed among aposporous and diplosporous grasses, whereby monopolar developments including the Panicum type, the

Setaria type, the Paspalum type, or the Eragrostis type, or bipolar developments like the Hieracium type, the Antennaria type and the Taraxacum type can be found among species and genera.

4.3.1 Sporophytic Apomixis in Grasses

Adventitious embryos have been found in eleven species of grasses from three subfamilies, in all cases arising from the nucellar cells (Table 4.1). In the subfamily Panicoideae, seven species belonging to five genera show sporophytic apomixis, including *Bothriochloa* spp. (1 species; Moskova, 1975), *Cenchrus* spp. (2 species; Fisher et al., 1954; Shanthamma and Narayan, 1976), *Eulaliopsis* spp. (1 species; Yao et al., 2004, 2007), *Eriochloa* spp. (1 species; Venkateswarlu and Devi, 1964), *Iseilema* spp. (1 species; Venkateswarlu and Devi, 1964), and *Sorghum* spp. (1 species; Hanna et al., 1970). Within Pooideae Benth., the genera *Bromus* spp. (1 species; Naumova, 1993) and *Poa* spp. (1 species; Batygina and Freiberg, 1979; Batygina, 1991) show nucellar embryony, as well as the genera *Eragrostiella* spp. (1 species; Seshavatharam, 1983; Sonpipare, 1984) and *Perotis* spp. (1 species; Venkateswarlu and Devi, 1964) of the subfamily Chloridoideae.

4.3.2 Gametophytic Apomixis in Grasses

In total, 238 species of grasses are recorded as apomictic by the formation of unreduced female gametophytes, including seven out of eleven species showing sporophytic apomixis (Table 4.1). The pathways of apomictic development are known for 83.1% of these species; the remaining 16.9% of cases have been determined without details on the type of apomixis (for example by the presence of four-nucleate unreduced female gametophytes without information on megasporogenesis or by FCSS).

4.3.2.1 The Aposporous Development

Apospory is documented in 70% of the species with gametophytic apomixis (n=166; Table 4.1). The development of an unreduced female gametophyte through apospory starts with the enlargement and acquisition of a gametogenetic fate by one or more nucellar cells near the megaspore mother cell, the meiotic megaspores or the developing meiotic gametophyte (Asker and Jerling, 1992). The germline in the ovules of these apomicts often degenerates before or during the meiotic division. In some cases, the developing meiotic gametophyte may degenerate or develop to maturity and produce a sexual seed.

4.3.2.1.1 The Hieracium Type

This type of unreduced female gametophyte development is present in about 24 species of different subfamilies (Table 4.1). Within Panicoideae, the Hieracium type occur in species of *Cenchrus* spp. (2 species; Narayan, 1951, 1955), in *Dichanthium* spp. (1 species; Celarier and Harlan, 1957), in *Paspalum* spp. (3 species; Snyder, 1957; Bashaw et al., 1970; Caponio and Quarin, 1987), in *Panicum* spp. (3 species; Warmke, 1954; Narayan, 1962), in *Sehima* spp. (1 species; Seshavatharam, 1983), and in *Sorghum* spp. (1 species; Hanna et al., 1970). In the subfamily Pooideae, it occurs in species of *Anthoxanthum* spp. (3 species; Weimarck 1967a, 1967b, 1970) and in *Poa* spp. (3 species; Armstrong, 1937; Nygren, 1950a, 1950b). Among Chloridoideae, gametophytes of Hieracium type had been recorded in *Bouteloua* spp. (1 species; Bryant, 1952; Mohamed and Gould, 1966), and in *Eustachys* spp. (1 species), in *Harpochloa* spp. (1 species), and in *Microchloa* spp. (1 species) by Strydom and Spies (1994). In Danthonioideae H.P. Linder & N.P. Barker, this type of development occurs in the genus *Cortaderia* spp. (2 species; Philipson, 1978; Astegiano et al., 1995) and in *Danthonia* spp. (1 species; Philipson, 1986).

4.3.2.1.2 The Panicum Type

The Panicum type of aposporous development is the most common type of monopolar aposporous development and has been observed in 64 species of grasses with gametophytic apomixis (Table 4.1).

This type is widely distributed among the genera of the subfamily Panicoideae but is also present in the subfamily Pooideae (*Anthoxanthum* spp.), in Chloridoideae (*Chloris* spp., *Eleusine* spp.), and in Danthonioideae (*Pentameris* spp.). Within Panicoideae, it occurs in 21 genera (Table 4.1) including *Bothriochloa* spp. (7 species; Brown and Emery, 1958; Celarier and Harlan, 1957), *Capillipedium* spp. (3 species; Celarier and Harlan, 1957), *Dichanthium* spp. (4 species; Celarier and Harlan, 1957; de Wet and Harlan, 1970), *Brachiaria* spp. (8 species; Brown and Emery, 1958; do Valle and Glienke, 1991), *Panicum* spp. (3 species; Warmke, 1954; Brown and Emery, 1958; Febulaus, 1992), *Cenchrus/Pennisetum* spp. (13 species; e.g., Narayan, 1951; Brown and Emery, 1958; Dujardin and Hanna, 1984a, 1984b) and *Themeda* spp. (5 species; Brown and Emery, 1958; Birari, 1980).

4.3.2.1.3 The Setaria Type

The Setaria type is found only in genera of the subfamily Panicoideae (Table 4.1), in some species of *Setaria* spp. (3 species; Emery, 1957; Birari, 1981) but also in *Bothriochloa* spp. (3 species; Celarier and Harlan, 1957), *Cenchrus* spp. (3 species; Fisher et al., 1954; Dujardin and Hanna, 1984a, 1984b), *Dichanthium* spp. (2 species; Celarier and Harlan, 1957; Choda and Bhanwra, 1980), *Themeda* spp. (1 species; Brown and Emery, 1957), *Heteropogon* spp. (1 species; Brown and Emery, 1958), *Hemisorghum* spp. (1 species; Celarier and Harlan, 1957) and *Apluda* spp. (1 species; Murty, 1973). However, since in most cases this type of development has been found co-occurring with female gametophytes of the Panicum type, it is likely to occur more frequently among aposporous grass species.

4.3.2.1.4 The Paspalum Type

Although this type of female gametophyte development occurs only in *Paspalum* spp., it is consistently found in 44 species (Table 4.1). The occasional occurrence of 2-celled gametophytes has been mentioned for *Cenchrus sphacelatus* (Nees) Morrone, a species in which Gildenhuys and Brix (1959) observed a few female gametophytes with an egg cell and a central cell with three nuclei.

4.3.2.2 The Diplosporous Development

Diplospory is recorded in 15% of all species with gametophytic apomixis (n=35; Table 4.1). The development of an unreduced female gametophyte through diplospory begins with the interruption of the steps of meiosis (either entry into meiosis or its progression) with the subsequent development of an unreduced megaspore that develops into a mature female gametophyte (Asker and Jerling, 1992). In occasional cases, the germline may progress normally through meiosis and produce reduced megaspores, a female gametophyte and a sexual seed after fertilization.

4.3.2.2.1 The Antennaria Type

The Antennaria type of female gametophyte development is found in a total of 20 species of the subfamilies Panicoideae and Pooideae (Table 4.1). Species from the genera *Calamagrostis* spp. (8 species; Nygren, 1954b; Greene, 1984), *Poa* spp. (5 species; Müntzing, 1933; Kiellander, 1935; Åkerberg, 1941; Nygren, 1950a, 1950b; Grun, 1955), *Tripsacum* spp. (6 species; Farquharson, 1955; Leblanc et al., 1995a, 1995b, 1995c) and *Nardus* spp. (1 species; Coulon, 1923; Rychlewski, 1961) show this type of gametophyte development.

4.3.2.2.2 The Taraxacum Type

This type of development is documented for nine species (Table 4.1). Within the subfamily Panicoideae it has been found in *Paspalum* spp. (5 species; Chao, 1964, 1974, 1980; Bonilla and Quarin, 1997) and in *Tripsacum* spp. (2 species; Farquharson, 1955; Leblanc et al., 1995a, 1995b, 1995c). Within Pooideae, it occurs in *Anthosachne* spp. (1 species; Hair, 1956; Crane and Carman, 1987) and in *Calamagrostis* spp. (1 species; Greene, 1984).

4.3.2.2.3 The Eragrostis Type

This is the least frequently observed type of female gametophyte development among grasses (apart from the Ixeris type) and is found only in species of the subfamily Chloridoideae, in the genera *Eragrostis* spp. (4 species; Brown and Emery, 1958; Streetman, 1963; Voigt and Bashaw, 1972) and *Fingerhuthia* spp. (1 species; Brown and Emery, 1958).

4.3.2.3 Variations in Female Gametophyte Ontogeny and Anatomy

In many species, variations on the type of female gametophyte development had been recorded. Except for those *Setaria* species in which only the Setaria type of female gametophyte development has been observed, in all other species such as *Cenchrus ciliaris* L., *Bothriochloa ischaemum* (L.) Keng, *Capillipedium huegelli* (Hack.) A. Camus, *Themeda triandra* Forssk. or *Dichanthium foveolatum* (Delile) Roberty, gametophyte developments of the Panicum type and the Setaria type have been observed simultaneously, in each case the Panicum type being the most frequent (Table 4.1; Brown and Emery, 1957; Gildenhuys and Brix, 1959; Choda and Bhanwra, 1980; Sathyamurthy and Seshavatharam, 1983; Dujardin and Hanna, 1984a, 1984b; Ozias-Akins et al., 2003). In other species, including *Cenchrus ciliaris* L. (≡*Pennisetum ciliare* (L.) Link, Fisher et al., 1954; Snyder et al., 1955), *Cenchrus longisetus* M.C.Johnst. (=*Pennisetum villosum* Fresen, Narayan, 1951), *Dichanthium annulatum* (Forssk.) Stapf (Celarier and Harlan, 1957), *Panicum maximum* Jacq. (Warmke, 1954) and *Paspalum simplex* Morong (Caponio and Quarin, 1987), the simultaneous occurrence of the Panicum type or Paspalum type together with the Hieracium type and the Polygonum type of female gametophyte developments has also been recorded.

A similar situation can be found among diplosporous species such as *Tripsacum dactyloides* (L.) L. and *T. zopolitense* Hern.-Xol. & Randolph (Leblanc et al., 1995a, 1995b, 1995c) or *Calamagrostis inexpansa* A. Gray (Greene, 1984) whereby most ovules show the Antennaria type of female gametophyte development and some ovules develop by the Taraxacum type. The co-occurrence of aposporous and diplosporous pathways is also part of the developmental variation observed among apomictic grasses. In *Paspalum conjugatum* P.J. Bergius (Chao, 1980) and *P. minus* L. (Bonilla and Quarin, 1997), female gametophytes can be formed by either Paspalum type or Taraxacum type development.

Likewise, multiple, unorganized, multinucleate and non-cellularized female gametophytes at anthesis are not uncommon in many grass species (e.g., Muniyamma, 1978; Battaglia, 1991; Bonilla and Quarin, 1997; Yudakova, 2009). Female gametophytes with an abnormal number of nuclei may result from lack of polarization and desynchronization of mitosis between nuclei (Yudakova, 2009) and could result in mature female gametophytes lacking specific cells or cell types. These deviations are not uncommon in grasses.

In general, variations in the pattern of polarization, nuclear divisions and cellularization of coenocytic female gametophytes that produce the observed variability, including 2-celled, 3-celled, 4-celled or 7-celled gametophytes with one or two nuclei in the central cell, likely respond to the genetically and metabolically altered gametogenesis developmental programmes observed in apomictic species (Schmidt, 2020; Hojsgaard, 2020).

5 Apomictic Daisies

5.1 INTRODUCTION

Early cytological studies in daisies pioneered the description of the main patterns of apomictic development in plants. Thus, most of the current names used to describe alternative apomictic pathways belong to genera of the family Asteraceae Giseke. The alternative diplosporous pathways for the formation of unreduced female gametophytes were first described by Juel (1898, 1900, 1904) in *Antennaria alpina* (L.) Gaertn. (the Antennaria type) and in *Taraxacum officinale* F.H. Wigg. (the Taraxacum type), and by Okabe (1932) in *Ixeridium dentatum* (Thunb.) Tzvelev [≡*Ixeris dentata* (Thunb.) Nakai; the Ixeris type]. Later studies revealed the occurrence of the Antennaria type not only in other species of *Antennaria* Gaertn. (e.g., Stebbins, 1932) but also in many other species of different genera, including *Brachyscome* Cass. (Davis, 1963, 1964b), *Erigeron* L. (Tahara, 1915; Holmgren, 1919; McDonald, 1927), *Eupatorium* L. (e.g., Sullivan, 1976), *Hieracium* L. (e.g., Gentscheff, 1937; Gentscheff and Gustafsson, 1940; Bergman, 1941; Skalińska, 1967) and *Parthenium* L. (Esau, 1944, 1946; Powers and Rollins, 1945; Gerstel et al., 1953), among others. In turn, the Taraxacum type of development has been mainly found in species from *Chondrilla* L. (Rosenberg, 1912; Poddubnaja-Arnoldi, 1933, 1964) and *Hieracium* L. (Gentscheff, 1938; Gentscheff and Gustafsson, 1940; Bergman, 1941), along with a significant number of *Taraxacum* species (e.g., Ikeno, 1910; Schkorbatow, 1912; Osawa, 1913; Sears, 1917, 1922; Stork, 1920; Gustafsson, 1935; Battaglia, 1948; Małecka, 1967). The Ixeris type is currently known in species of *Townsendia* Hook. (Beaman, 1957), *Rudbeckia* L. (Battaglia, 1946a, 1946b, 1946c, 1946d, 1952, 1955; Fagerlind, 1946) and *Erigeron* L. (Holmgren, 1919; Bergman, 1944; Fagerlind, 1947; Battaglia, 1950).

Similarly, the aposporous pathway for the formation of eight-nucleate unreduced female gametophytes (the Hieracium type) was first described by Rosenberg (1907) in *Hieracium excellens* Błocki [=*Pilosella densiflora* (Tausch) Soják], *H. flagellare* Willd. [≡*Pilosella flagellaris* (Willd.) Arv.-Touv.] and *H. aurantiacum* L. [≡*Pilosella aurantiaca* (L.) F.W.Schultz & Sch.Bip.]. Later studies described the same type of apomictic female gametophyte development in species from varied genera, including species of *Crepis* L. (Babcock and Stebbins, 1938; Stebbins and Jenkins, 1939), *Calotis lappulacea* Benth. (Davis, 1968), *Campovassouria cruciata* (Vell.) R.M. King & H. Rob. (Coleman and Coleman, 1984), *Minuria integerrima* (DC.) Benth. (Davis, 1964a), in a few species of *Hieracium* L. (Gentscheff, 1937; Christoff and Christoff, 1948), as well as in different species of *Pilosella* Hill (Haberlandt, 1921a, 1921b; Rosenberg, 1930; Christoff, 1942; Skalińska, 1973, 1976).

Since those early studies many more assessments of reproductive biology in daisies have expanded the list of apomictic taxa to more than 350 species and subspecies (Table 5.1). The present chapter brings together current information on apomixis at the species level in the family Asteraceae.

5.2 CONSIDERATIONS ABOUT THE OCCURRENCE AND TYPE OF APOMIXIS IN DAISIES

Collected records are based on original reports and previous lists of apomictic species (Stebbins, 1941; Nygren, 1954a, 1954b, 1967; Davis, 1967; Pullaiah, 1984; Carman, 1995, 1997; Asker and Jerling, 1992; Naumova, 1993; Noyes, 2007) and regional floras (e.g., Richards, 2003; Gregor, 2013; Pegoraro et al., 2020). Several particularly old studies were accessed through the Biodiversity Heritage Library (BHL; https://www.biodiversitylibrary.org/) and the Academic Archive Online (DiVA, http://www.diva-portal.org/). Whenever possible, studies on the occurrence of apomixis in different species were critically re-evaluated. Species' taxonomy and nomenclature had been revised following the International Plant Names Index (IPNI; https://www.ipni.org/), Tropicos (https://www.tropicos.org/), the Plants of the World Online (http://www.plantsoftheworldonline.org/), The Plant List (http://www.theplantlist.org/) and the World Flora Online (WFO; http://www.worldflora-online.org/), the Global Compositae Checklist (https://compositae.landcareresearch.co.nz/Default.aspx) and the Global Compositae Database (https://compositae.org/gcd.php).

We have included basic information about species names, ploidy of studied materials, and the type of apomictic reproduction in Tables 5.1 and 5.2. As with apomictic grasses, a number of daisy species had been subjected to changes in their nomenclature. Such changes are indicated in Tables 5.1 and 5.2. Most cases involve synonyms, either as new species or subspecies [e.g., *Erigeron mucronatus* DC. is now placed under the name *Erigeron karvinskianus* DC., or *Arnica amplexicaulis* Nutt. is now *Arnica lanceolata* subsp. *prima* (Maguire) Strother & S.J. Wolf] or placed under new genera [e.g., *Ixeris dentata* (Thunb.) Nakai is now *Ixeridium dentatum* (Thunb.) Tzvelev, or *Hieracium pilosella* L. is considered here under the name *Pilosella officinarum* Vaill.]. A few species are moved into newly established genera [e.g., *Hieracium intybaceum* All. is now under the name *Schlagintweitia intybacea* (All.) Griseb.].

A few cases required specific considerations. For example, *Hieracium barrelieri* and *H. brachytrichellum* cited as apomicts by Mráz and Zdvořák (2019) do not appear as species (nor subspecies) in databases. For *Hieracium bauhinii* Besser (Krahulcová et al., 2001) we followed the name *Hieracium bauhini* Schult. since it was Schultz who published the description of the species in the year 1809, and therefore, this taxon is listed as *Pilosella piloselloides* subsp. *bauhinii* (Schult.) S. Bräut. & Greuter (=*Hieracium bauhini* Schult.) in Table 5.1. Other cases like *Hieracium boreale* for which the original work could not be accessed (Zahn, 1904; Haberlandt, 1921a, 1921b; Rosenberg, 1927; Bergman, 1941), or when author citation was not provided (Rosenberg, 1930; Mráz and Zdvořák, 2019) and different authors are valid (e.g., *Hieracium sudeticum* Sternb. and *Hieracium sudeticum* Tausch ex Link), priority given by IPNI was followed. Triploid *Chondrilla pauciflora* is cited as apomict by Poddubnaja-Arnoldi (1933). Since this work could not be accessed, the name *Chondrilla pauciflora* Ledeb. was chosen and listed in Table 5.1 despite *Chondrilla pauciflora* D. Don [≡*Pyrrhopappus pauciflorus* (D. Don) DC.] is described first according to IPNI.

Illegitimate species names [e.g., *Eupatorium glandulosum* H.B.K. correspond to *Ageratina adenophora* (Spreng.) R.M. King & H. Rob.] are replaced whenever possible. Nine unresolved nomenclatural names or unpublished names cited elsewhere as apomicts are not included in the table of apomictic species [i.e., *Antennaria nordhagiana* Rune & Rønning; *Hieracium anodon* Brenner; *Hieracium colophyllum* N. & P. subsp. *leiopogon* Gren.; *Hieracium longifolium* Schleich.; *Hieracium membrosum* Johansson & Samuelsson; *Hieracium meticeps* Almq ex Dahlst.; *Hieracium serratifrons* Almqv. subsp. *crispatum* Dahlst. ined. (Murbeck, 1904); *Taraxacum alpinum* K. Koch; and *Taraxacum vulgare* (Lam.) Raunk.].

In the Asteraceae, the taxonomic treatment of apomictic taxa is seemly more complex than in the Poaceae (see e.g. Tyler and Jönsson, 2013). In the rich genera *Hieracium* L. and *Pilosella* Hill, alternative schools using slightly different species' circumscription systems coexist (Schuhwerk, 2002; Majeský et al., 2017). The view that considers a formal hierarchy of ancestral (mostly sexual, non-hybrid) "basic species" or "Hauparten" (following the original German word) with unique characters and (hybrid) "intermediate species" or "Zwischenarten" combining characters from basic species is the most widely accepted. Yet, molecular studies show that species in both groups frequently are of hybrid origin (e.g., Fehrer et al., 2007), and the alternative use of informal "species aggregates" based on multivariate morphometry (Tyler, 2006) might work as a convenient choice to bring both taxonomic schools together and simplify comparative studies.

Here, we agreed with the different schools, accept the concept of agamospecies, i.e., a group of (morphologically, cytologically and phylogenetically) related asexually reproducing biotypes (Gustafsson, 1947a, 1947b; Asker and Jerling, 1992), and consider microspecies and infraspecific delimitations as valid taxonomic subdivisions of apomictic complexes. Even when we are aware of the use of different taxonomic treatments better suiting the biosystematics of each agamic complex or species aggregate might introduce some bias in the total number of taxa, the taxonomic debate on how to treat apomictic taxa is far from being resolved (see Majeský et al., 2017; Hörandl, 2018) and is out of the scope of the present work (but see comments in Chapter 6). Thus, for each genus, we have basically followed the taxonomic concepts accepted by specialists (i.e., the authors of the article on each case). Agamospecies are considered as categorized (notho)taxa following current taxonomic evaluations of the sources consulted. Accepted subspecies and varieties follow the most recent taxonomic treatments.

For example, the species-rich genus *Antennaria* Gaertn. is taxonomically complicated, but it is a remarkably interesting genus considering the biology and evolution of its species and agamospecies, including different apomictic forms. The agamic complex of *A. neodioica* s.l. (containing four subspecies, Bayer and Stebbins, 1987) has been renamed as *A. howellii* subsp. *canadensis* (Greene) R.J. Bayer, *A. howellii* subsp. *neodioica* (Greene) R.J. Bayer and *A. howellii* subsp. *petaloidea* (Fernald) R.J. Bayer (Table 5.1; Flora of North America Editorial Committee, 2006). The other most known agamic complex of *A. parlinii* s.l. has been maintained with two subspecies, *A. parlinii* Fernald subsp. *parlinii* and *A. parlinii* subsp. *fallax* (Greene) R.J. Bayer & Stebbins (Flora of North America Editorial Committee, 2006). Other examples of taxonomic changes in *Antennaria* spp. involve *A. rosea* subsp. *pulvinata* (Greene) R.J. Bayer (=*A. fusca* E.E. Nelson; Bayer and Stebbins, 1987), *A. chilensis* J. Rémy var. *chilensis* and *A. chilensis* var. *magellanica* (Sch.Bip.) Reiche (Zuloaga et al., 2008).

Whenever possible, ploidy levels from the studied materials are included (see Tables 5.1 and 5.2). However, in a few cases where ploidy was not evaluated or was not provided in the original studies, the ploidy level is added for those species for which a single ploidy record is available from additional studies (without providing references).

We tried to mention the first study reporting on apomixis, but there is a chance that we may have overlooked the original records in some cases, especially in genus *Hieracium* L. and *Taraxacum* F.H. Wigg. whereby many studies on apomixis are based on repeated castration experiments. Seed set analyses using emasculation/castration experiments on taxa with odd ploidies are listed in Table 5.1 but would require embryological confirmation. Cases of odd-ploidy taxa displaying high seed set but lacking experimental pollen exclusion tests are not listed [e.g., *Amauriopsis dissecta* (A. Gray) Rydb., Keil et al., 1988 in Noyes, 2008]. Species categorized as autonomous apomicts based on lack of male plants in field explorations are accepted only if apomixis was confirmed through other analyses (e.g., in a few species of *Antennaria* Gaertn., Bayer and Stebbins, 1987; Table 5.1). Embryological and cytological analyses are recommended to properly characterize the type of apomictic development in those taxa.

Evidence of experimental apomixis obtained by subjecting plants either to different laboratory conditions or to artificial crossings is not considered here. For example, Voronova (2013) found ovules showing aposporous development and integumentary embryony after interspecific crosses between sunflower (*Helianthus annuus* L.) lines and wild perennial sunflower species.

In different genera, apomixis has been characterized both through castration experiments and embryologically (e.g., Richards, 1970). Here, reports of spontaneous (infrequent) developments are not considered for classifying the reproductive biology of the species. For example, *Parthenium argentatum* A. Gray and *P. incanum* Kunth were both considered as diplosporous in Table 5.1 despite aposporous female gametophytes were rarely observed (Esau, 1946; Noyes, 2007). The occurrence of low frequencies of supernumerary aposporous female gametophytes was observed not only in the polyploid apomicts but also in the sexual diploid materials.

Embryological evaluation of *Erigeron* species has led to different authors to conclude their type of development being Antennaria or Ixeris (Table 5.1). This basically depends upon the identification of the first mitosis during the development of the female gametophyte (either the second meiotic division occurs or it corresponds to the first mitosis of the gametogenesis), which consequently determine the initial state of the female gametophyte (one or two nuclei). The occurrence of species exhibiting bispory and tetraspory with alternative deviations from these developmental patterns (Pagni, 1954; Ferri, 1961) makes this genus particularly interesting for understanding the origin, genetics and evolution of apomixis.

TABLE 5.1
Table of Apomictic Species (A) within the Family Asteraceae Giseke, Sorted Alphabetically According to Subfamilies

Species	Ploidy	Type of Apomixis[a]	References
Asteroideae Lindl.			
Ageratina adenophora (Spreng.) R.M. King & H. Rob. (sub *Eupatorium glandulosum* H.B.K. illeg.)	*3x*	A; Di (An); Au	Holmgren (1916, 1919).
Ageratina riparia (Regel) R.M. King & H. Rob. (=*Eupatorium riparium* Regel)		A	Sparvoli (1960)
Antennaria alpina (L.) Gaertn. [=*Antennaria cannescens* (Lange) Malte; =*Antennaria glabrata* (J. Vahl) Greene; =*Antennaria intermedia* (Rosenv.) Porsild; =*Antennaria porsildii* E. Ekman]	*4x, 6x, 8x?*	A; Di (An); Au	Juel (1898, 1900), Stebbins (1932), Bergman (1935c)
Antennaria aromatica Evert	*4x*	A; Au	Bayer and Stebbins (1987)
Antennaria atriceps Fern.	*4x*	A; Au	Bayer and Stebbins (1987)
Antennaria chilensis J. Rémy var. *chilensis*	*4x*	A; Di; Au	Bergman (1937), Bayer and Stebbins (1987)
Antennaria chilensis var. *magellanica* (Sch. Bip.) Reiche	*4x*	A; Di (An); Au	Bergman (1937), Bayer and Stebbins (1987)
Antennaria friesiana (Trautv.) Ekman [sub *Antennaria friesiana* (Trautv.) Ekman subsp. *friesiana*]	*4x, 8x?*	A; Au	Bayer and Stebbins (1987), Bayer (1991)
Antennaria howellii subsp. *canadensis* (Greene) R.J. Bayer (=*Antennaria canadensis* Greene)	*6x*	A; Di (An); Au	Stebbins (1932)
Antennaria howellii subsp. *neodioica* (Greene) R.J. Bayer (=*Antennaria neodioica* Greene)	*4x, 6x, 10x*	A; Di (An); Au	Stebbins (1932), Bayer and Stebbins (1987)
Antennaria howellii subsp. *petaloidea* (Fernald) R.J. Bayer [=*Antennaria petaloidea* (Fernald) Fernald]	*6x*	A; Di (An); Au	Stebbins (1932), Bayer and Stebbins (1987)
Antennaria marginata Greene	*4x, 6x, 8x, 10x*	A; Au	Bayer and Stebbins (1987)
Antennaria media Greene	*4x, 7x, 8x*	A; Au	Bayer et al. (1990)
Antennaria microphylla Rydb. (=*Antennaria bracteosa* Rydb.)	*3x, 4x*	A; Au	Bayer and Stebbins (1987)
Antennaria monocephala DC.	*4x, 5x*	A; Au	Bayer (1991)
Antennaria parlinii Fernald (sub *Antennaria parlinii* Fernald subsp. *parlinii*)	*4x, 5x, 6x, 8x*	A; Di (An); Au	Stebbins (1932), Bayer and Stebbins (1983), Bierzychudek (1985)
Antennaria parlinii subsp. *fallax* (Greene) R.J. Bayer & Stebbins (=*Antennaria brainerdi* Fernald; =*Antennaria occidentalis* Greene)	*3x, 6x*	A; Di (An); Au	Stebbins (1932)
Antennaria parvifolia Nutt.	*4x, 6x, 8x, 10x*	A; Au	Bayer and Stebbins (1987)

TABLE 5.1 (Continued)

Species	Ploidy	Type of Apomixis[a]	References
Antennaria rosea (D.C. Eaton) Greene [=*Antennaria rosea* Greene subsp. *rosea*; =*Antennaria alborosea* A.E. Porsild ex Porsild; =*Antennaria imbricata* E.E. Nelson, =*Antennaria oxyphylla* Greene]	*3x-5x*	A; Au	Bayer (1997)
Antennaria rosea subsp. *arida* (E.E. Nelson) R.J. Bayer (=*Antennaria arida* E.E. Nelson)		A; Au	Bayer and Stebbins (1987)
Antennaria rosea subsp. *confinis* (Greene) R.J. Bayer (=*Antennaria aprica* E.E. Nelson, =*Antennaria concinna* E.E. Nelson, =*Antennaria incarnata* A.E. Porsild, =*Antennaria sedoides* Greene, =*Antennaria subviscosa* Fernald)	*4x, 6x*	A; Au	Bayer and Stebbins (1987)
Antennaria rosea subsp. *pulvinata* (Greene) R.J. Bayer [=*Antennaria fusca* E.E. Nelson; =*Antennaria gaspensis* (Fernald) Fernald]		A; Au	Bayer and Stebbins (1987)
Antennaria soliceps S.F. Blake	*12x*	A	Bayer and Minish (1993)
Arnica alpina Salisb.	*3x*	A; Di (An); Au	Afzelius (1936), Engell (1970), Bierzychudek (1985), Kao (2007)
Arnica lanceolata subsp. *prima* (Maguire) Strother & S.J. Wolf (=*Arnica amplexicaulis* Nutt.)	*3x, 4x*	A; Au	Bierzychudek (1985)
Arnica angustifolia Vahl	*3x, 4x*	A; Au	Bierzychudek (1985)
Arnica chamissonis Less.	*3x*	A; Di (An); Au	Afzelius (1936), Bierzychudek (1985)
Arnica cordifolia Hook.	*3x, 4x, 5x*	A; Ps; Au	Wolf (1987), Kao (2007, 2008)
Arnica ovata Greene (=*Arnica diversifolia* Greene)	*3x, 4x*	A; Di (An); Au	Afzelius (1936)
Arnica lessingii (Torr. & A. Gray) Greene	*4x*	A; Au	Bierzychudek (1985)
Arnica lonchophylla Greene	*3x, 4x*	A; Au	Bierzychudek (1985)
Arnica longifolia D.C. Eaton	*3x, 4x*	A; Au	Bierzychudek (1985)
Arnica louiseana Farr	*4x, 5x*	A; Au	Bierzychudek (1985)
Arnica mollis Hook.	*3x, 4x, 5x, 6x, 7x, 8x*	A; Au	Bierzychudek (1985)
Arnica nevadensis A. Gray	*4x*	A; Au	Bierzychudek (1985)
Arnica parryi A. Gray	*4x*	A; Au	Bierzychudek (1985)
Arnica rydbergii Greene	*4x*	A; Au	Bierzychudek (1985)
Ayapana triplinervis (Vahl) R.M. King & H. Rob. (=*Eupatorium triplinerve* Vahl)	*4x*	A; Di (An); Au	Mukherjee and Desai (1980)
Blumea eriantha DC.	*3x*	A; Di (Ta); Au	Chennaveeraiah and Patil (1971)

(Continued)

TABLE 5.1 (Continued)
Table of Apomictic Species (A) within the Family Asteraceae Giseke, Sorted Alphabetically According to Subfamilies

Species	Ploidy	Type of Apomixis[a]	References
Blumea oxydonta DC.	*3x*	A; Di; Au	Chennaveeraiah and Patil (1971)
Brachyscome ciliaris Less.	*3x, 4x*	A; Di (An); Au	Davis (1964b)
Brachyscome ciliaris var. *lanuginosa* (Steetz) Benth.		A; Di (An); Au	Davis (1963)
Calea triantha (Vell.) Pruski [=*Calea hispida* (DC.) Baker]		A; Au	Werpachowski et al. (2004)
Calotis lappulacea Benth.	*4x*	A; Ap (Hi); Au	Davis (1968)
Campovassouria cruciata (Vell.) R.M. King & H. Rob. (=*Eupatorium bupleurifolium* DC.)	*3x*	A; Ap (Hi)	Coleman and Coleman (1984)
Campuloclinium macrocephalum (Less.) DC.	*3x*	A; Ap (Hi)	Farco et al. (2012)
Chromolaena callilepis (Sch.Bip. ex Baker) R.M. King & H. Rob. (=*Eupatorium callilepis* Sch. Bip. ex Baker)	*3x*	A; Di	Coleman and Coleman (1984)
Chromolaena laevigata (Lam.) R.M. King & H. Rob. (=*Eupatorium laevigatum* Lam.)	*6x*	A; Di (An); Au	Bertasso-Borges and Coleman (2005)
Chromolaena odorata (L.) R.M. King & H. Rob. (=*Eupatorium odoratum* L.)	*6x*	A; Di (An); Au	Pullaiah (1976, 1982), Maheswari Devi and Pullaiah (1977a); Coleman (1989)
Chromolaena squalida (DC.) R.M. King & H. Rob. (=*Eupatorium squalidum* DC.)	*3x*	A	Coleman and Coleman (1988)
Erigeron annuus (L.) Pers.	*3x*	A; Di (An, Ix); Au	Tahara (1915), Holmgren (1919), McDonald (1927)
Erigeron compositus Pursh	*4x, 6x, 7x*	A; Au	Noyes and Soltis (1996)
Erigeron divergens Torr. & A. Gray	*3x, 4x*	A; Au	Noyes (2000)
Erigeron grandiflorus Hook.	*4x*	A; Au	Spongberg (1971), Noyes (2000)
Erigeron lackschewitzii G.L. Nesom & W.A. Weber	*4x*	A; Au	Kerstetter (1994)
Erigeron lanatus Hook.	*4x*	A; Au	Spongberg (1971)
Erigeron karvinskianus DC. (=*Erigeron mucronatus* DC.)	*3x, 4x*	A; Di (Ix); Au	Carano (1921, 1924), Fagerlind (1947), Battaglia (1950), Beals (1970), Mehra and Gill (1972), Pegoraro et al. (2020)
Erigeron strigosus Muhl. ex Willd. [=*Erigeron ramosus* (Walter) Britton, Sterns & Poggenb.]	*3x, 4x*	A; Di (An)	Holmgren (1919)
Erigeron trifidus Hook.	*5x*	A; Au	Burke and Bain (2008)
Erigeron altissimum L.	*3x, 4x*	A; Di (An); Au	Sullivan (1976)
Erigeron cuneifolium Willd.	*3x, 4x*	A; Au	Sullivan (1976)
Erigeron hyssopifolium L. (=*Eupatorium lecheifolium* Greene)	*4x*	A; Au	Sullivan (1976)
Erigeron leucolepis (DC.) Torr. & A. Gray		A; Au	Sullivan (1976)

TABLE 5.1 (Continued)

Species	Ploidy	Type of Apomixis[a]	References
Erigeron pilosum Walter	*3x, 4x*	A; Di (An); Au	Sullivan (1976)
Erigeron rotundifolium L.	*3x*	A; Au	Sullivan (1976)
Erigeron sessiliflorum L.	*3x*	A; Au	Sullivan (1976)
Erigeron tanacetifolium Gillies ex Hook. & Arn. (=*Gyptis pinnatifida* Cass.)	*3x*	A; Di (An); Au	Rozenblum et al. (1988)
Grazielia intermedia (DC.) R.M. King & H. Rob. (=*Eupatorium intermedium* DC.)		A; Au	Valentin-Silva et al. (2016).
Leontopodium brachyactis Gand. (=*Leontopodium alpinum* Cass.)		A; Di (An); Ps	Sokolowska-Kulzyca (1959), Maugini (1962)
Leucanthemopsis alpina (L.) Heywood	*3x*	A;	Pegoraro et al. (2020)
Melampodium divaricatum (Rich. ex Rich.) DC.	*2x*	A; I	Maheswari Devi and Pullaiah (1976b)
Minuria integerrima (DC.) Benth.	*4x*	A; Ap (Hi); Au	Davis (1964a); Watanabe et al. (1996)
Parthenium argentatum A. Gray	*3x, 4x, 5x, 6x, 7x, 8x*	A; Ap (Hi); Di (An); Ps	Esau (1944, 1946), Powers and Rollins (1945), Gerstel et al. (1953), Cruz et al. (2017)
Parthenium incanum Kunth	*3x*	A; Ap (Hi); Di (An); Ps	Esau (1946), Powers and Rollins (1945), Gerstel et al. (1953), Cruz et al. (2017)
Praxelis pauciflora (Kunth) R.M. King & H. Rob. (=*Eupatorium pauciflorum* Kunth)	*3x*	A; Di; Au	Bertasso-Borges and Coleman (1998)
Rudbeckia californica A. Gray	*4x*	A; Di	Nygren (1954a, 1954b)
Rudbeckia columnaris Sims		A; Di (Ix)	Battaglia (1952)
Rudbeckia fulgida var. *deamii* (S.F. Blake) Perdue (=*Rudbeckia deami* S.F. Blake)	*4x*	A; Di (Ix)	Battaglia (1952)
Rudbeckia fulgida var. *speciosa* (Wender.) Perdue (=*Rudbeckia speciosa* Wender.)	*4x*	A; Di (Ix); Ps	Battaglia (1946a, 1946b, 1946c, 1946d)
Rudbeckia fulgida var. *sullivantii* (C.L. Boynton & Beadle) Cronquist (=*Rudbeckia sullivantii* C.L. Boynton & Beadle)	*4x*	A; Di (Ix); Ps	Battaglia (1955)
Rudbeckia laciniata L.	*4x*	A; Di (Ix); Ps	Battaglia (1945, 1947a, 1952), Solntzeva (1973b), Movsesyan and Kasparova (1977a, 1977b), Kościńska-Pająk et al. (2014)
Rudbeckia triloba L.	*3x, 4x, 5x, 6x, 7x, 8x*	A	Battaglia (1952), McCrea (1981)
Thymophylla tenuiloba (DC.) Small	*3x*	A; Au	Strother (1989)
Townsendia condensata Parry	*4x*	A; Di (Ix); Au	Beaman (1957)
Townsendia exscapa (Richardson) Porter	*3x, 4x*	A; Au	Beaman (1957)
Townsendia grandiflora Nutt.	*3x, 4x*	A; Au	Beaman (1957)
Townsendia hookeri Beaman	*3x, 4x*	A; Di (Ix); Au	Beaman (1957), Thompson and Whitton (2006)

(*Continued*)

TABLE 5.1 (Continued)
Table of Apomictic Species (A) within the Family Asteraceae Giseke, Sorted Alphabetically According to Subfamilies

Species	Ploidy	Type of Apomixis[a]	References
Townsendia incana Nutt.	*3x,4x*	A; Di (Ix); Au	Beaman (1957)
Townsendia leptotes (A. Gray) Osterh.	*3x, 4x*	A; Au	Beaman (1957)
Townsendia minima Eastw. [=*Townsendia montana* M.E. Jones var. *minima* (Eastwood) Beaman]	*3x*	A; Au	Beaman (1957)
Townsendia montana M.E. Jones (=*Townsendia montana* M.E. Jones var. *montana*)	*4x*	A; Au	Beaman (1957)
Townsendia parryi D.C. Eaton	*4x*	A; Di (Ix); Au	Beaman (1957)
Townsendia rothrockii A. Gray ex Rothrock.	*4x*	A; Di (Ix); Au	Beaman (1957)
Townsendia scapigera D.C. Eaton		A; Au	Beaman (1957)
Townsendia spathulata Nutt.		A; Au	Beaman (1957)
Townsendia strigosa Nutt.		A; Au	Beaman (1957)
Carduoideae Cass. ex Sweet			
Carthamus tinctorius L.		A; I	Maheswari Devi and Pullaiah (1976a)
Centaurea scabiosa subsp. *alpestris* (Hegetschw.) Nyman	*4x*	A	Pegoraro et al. (2020)
Cichorioideae Chevall.			
Chondrilla acanthalepis Boiss.	*3x*	A; Di (Ta); Au	Poddubnaja-Arnoldi (1933, 1964)
Chondrilla brevirostris Fisch. & C.A. Mey.	*3x*	A; Di (Ta)	Musiał and Kościńska-Pająk (2019)
Chondrilla graminea M. Bieb.	*4x*	A; Di (Ta); Au	Poddubnaja-Arnoldi (1933)
Chondrilla juncea L.	*3x*	A; Di (Ta); Au	Rosenberg (1912), Poddubnaja-Arnoldi (1933), Bergman (1944, 1950), Battaglia (1949), Chaboudez (1994), van Dijk (2003), Pegoraro et al. (2020)
Chondrilla kusnezovii Iljin	*3x*	A; Au	Poddubnaja-Arnoldi (1933)
Chondrilla lejosperma Kar. & Kir. (=*Chondrilla coronifera* Iljin)	*3x*	A; Au	Poddubnaja-Arnoldi (1933)
Chondrilla ornata Iljin	*4x*	A; Au	Poddubnaja-Arnoldi (1933)
Chondrilla pauciflora Ledeb.	*3x*	A; Di (Ta); Au	Poddubnaja-Arnoldi (1933)
Crepis acuminata Nutt.	*3x, 4x*	A; Ap (Hi); Au	Babcock and Stebbins (1938), Stebbins and Jenkins (1939), Bierzychudek (1985)
Crepis atribarva A. Heller (=*Crepis exilis* Osterh.)	*4x, 5x*	A; Ap (Hi); Au	Babcock and Stebbins (1938), Stebbins and Jenkins (1939)
Crepis bakeri Greene	*4x*	A; Au	Babcock and Stebbins (1938), Stebbins and Jenkins (1939)
Crepis barbigera Leiberg ex Coville	*4x, 7x, 8x*	A; Ap (Hi); Au	Babcock and Stebbins (1938), Stebbins and Jenkins (1939), Hersh et al. (2016)

TABLE 5.1 (Continued)

Species	Ploidy	Type of Apomixis[a]	References
Crepis intermedia A. Gray	*3x, 4x, 5x, 8x*	A; Ap (Hi); Au	Babcock and Stebbins (1938), Stebbins and Jenkins (1939)
Crepis modocensis Greene	*4x*	A; Ap (Hi); Au	Babcock and Stebbins (1938), Stebbins and Jenkins (1939)
Crepis monticola Coville	*4x,5x*	A; Ap (Hi); Au	Babcock and Stebbins (1938), Stebbins and Jenkins (1939)
Crepis occidentalis Nutt.	*3x,4x,7x*	A; Ap (Hi); Au	Babcock and Stebbins (1938), Stebbins and Jenkins (1939), Gustafsson (1946, 1947a, 1947b)
Crepis pleurocarpa A. Gray	*5x*	A; Ap (Hi); Au	Babcock and Stebbins (1938), Stebbins and Jenkins (1939)
Crepis versicaria L.	*2x*	A	Pegoraro et al. (2020)
Hieracium alpinum L.	*3x*	A; I; Di (An); Au	Skawinska (1962), Naumova (1993)
Hieracium amplexicaule L.	*3x*	A; Di (An); Au	Gentscheff and Gustafsson (1940), Mráz et al. (2009), Hand et al. (2015), Pegoraro et al. (2020)
Hieracium atratum Fr.	*3x, 4x*	A; Au	Mráz and Zdvořák (2019)
Hieracium barrelieri Gottschl., Raimondo, Greuter & Di Grist.	*4x*	A; Au	Mráz and Zdvořák (2019)
Hieracium bifidum Kit. ex Hornem.	*3x, 4x*	A; Di (An); Au	Hand et al. (2015), Pegoraro et al. (2020)
Hieracium bifidum subsp. *maculosum* (Dahlst.) Zahn (=*Hieracium maculosum* Dahlst.)		A; Di (An)	Bergman (1941)
Hieracium brachiatum Bert. ex DC.	*3x, 4x, 5x, 7x, 8x*	A; Au	Rotreklová et al. (2002, 2005)
Hieracium bupleuroides C.C. Gmel.	*3x,4x*	A; Au	Mráz and Zdvořák (2019)
Hieracium caesium Fr.	*3x*	A; Di (Ta); Au	Ostenfeld (1906), Bergman (1941), Hand et al. (2015), Mráz and Zdvořák (2019)
Hieracium caesium subsp. *galbanum* Dahlst. (=*Hieracium ravidum* Brenner)	*4x*	A; Au	Tyler and Jönsson (2013)
Hieracium calenduliflorum Backh.*f.*	*4x*	A; Au	Shi et al. (1996)
Hieracium carpaticum Griseb.	*4x*	A; Au	Mráz and Zdvořák (2019)
Hieracium chlorocephalum subsp. *stygium* (R. Uechtr.) Zahn (=*Hieracium stygium* Uechtr.)	*4x*	A; Au	Mráz and Zdvořák (2019)
Hieracium chloropsis Gren. & Godr.	*4x*	A; Au	Mráz and Zdvořák (2019)
Hieracium christianense Dahlst. ex Stenstr. (=*Hieracium maculosum* Dahlst. ex Stenstr.)		A; Di (Ta); Au	Bergman (1941)
Hieracium chrysostyloides (Zahn) Chrtek f.	*5x*	A; Au	Chrtek et al. (2007)

(*Continued*)

TABLE 5.1 (Continued)
Table of Apomictic Species (A) within the Family Asteraceae Giseke, Sorted Alphabetically According to Subfamilies

Species	Ploidy	Type of Apomixis[a]	References
Hieracium crassipedipilum (Pawł. & Zahn) Chrtek	*3x, 4x*	A; Au	Chrtek and Marhold (1998), Štorchová et al. (2002)
Hieracium cyathis (Ley) W.R. Linton	*3x*	A; Au	Rich et al. (2008)
Hieracium cydoniifolium Vill.	*3x*	A;	Pegoraro et al. (2020)
Hieracium danicum Dahlst.		A; Au	Ostenfeld (1906)
Hieracium dovrense Fr.		A; Au	Ostenfeld (1906)
Hieracium expallideforme Dahlst.		A; Di (Ta); Au	Bergman (1941)
Hieracium froelichianum H. Buek	*3x*	A; Au	Mráz and Zdvořák (2019), Pegoraro et al. (2020)
Hieracium froelichianum subsp. *macilentum* (Fr.) Gottschl. & Greuter (=*Hieracium macilentum* Fr.)	*3x*	A; Au	Mráz and Zdvořák (2019)
Hieracium fuscoatrum Nägeli & Peter	*4x*	A; Au	Krahulcová et al. (2001), Krahulec et al. (2004)
Hieracium glaucinum subsp. *jaubertianum* (Timb.-Lagr. & Loret) O. Bolòs & Vigo (=*Hieracium jaubertianum* Timb.-Lagr. & Loret.)	*3x, 4x*	A; Au	Mráz and Zdvořák (2019)
Hieracium glaucinum subsp. *petiolare* (Jord.) Greuter (=*Hieracium petiolare* Jord.)	*3x*	A; Au	Mráz and Zdvořák (2019)
Hieracium glaucopsis Gren. & Godr.	*3x*	A	Pegoraro et al. (2020)
Hieracium glaucum All.	*3x, 4x*	A; Au	Ostenfeld (1906), Mráz and Zdvořák (2019), Pegoraro et al. (2020)
Hieracium grovesianum subsp. *rigoanum* (Zahn) Zahn	*4x*	A; Au	Mráz and Zdvořák (2019)
Hieracium halleri Vill.	*3x*	A; Au	Mráz and Zdvořák (2019)
Hieracium holosericeum Backh.	*3x, 4x*	A; Au	Shi et al. (1996)
Hieracium hoppeanum Schult.	*5x*	A; Ap (Hi); Au	Christoff and Christoff (1948)
Hieracium humile Jacq.	*3x, 4x*	A; Di (An); Au	Hand et al. (2015), Pegoraro et al. (2020)
Hieracium hypochoeroides subsp. *wiesbaurianum* (R. Uechtr.) Greuter (=*Hieracium wiesbaurianum* R.Uechtr.)		A	Reisch et al. (2015)
Hieracium jankae R. Uechtr.	*3x*	A; Au	Mráz and Szelag (2004)
Hieracium jarzabczynum (Pawł. & Zahn) Mráz & Chrtek f.	*4x*	A; Au	Chrtek et al. (2007)
Hieracium jurassicum Griseb.	*3x*	A; Au	Mráz and Zdvořák (2019)
Hieracium krivanense (Woł. & Zahn) Schljakov	*4x*	A; Au	Štorchová et al. (2002)
Hieracium lachenalii subsp. *cruentifolium* (Dahlst. & Lübeck) Zahn (=*Hieracium vulgatum* Fr.)	*3x*	A; Ap; Di (Ta, An); Au	Gentscheff (1937), Bergman (1941)

TABLE 5.1 (Continued)

Species	Ploidy	Type of Apomixis[a]	References
Hieracium lachenalii subsp. *cruentifolium* (Dahlst. & Lübeck) Zahn (=*Hieracium cruentifolium* Dahlst. & Lübeck. ex Dahlst. & Lübeck)	*3x*	A; Au	Tyler and Jönsson (2013)
Hieracium lachenalii Suter	*3x, 4x*	A; Au	Mráz and Szelag (2004), Mráz and Zdvořák (2019)
Hieracium laevigatum subsp. *rigidum* (Hartm.) Čelak. (=*Hieracium rigidum* Hartm.)		A; Di (Ta); Au	Bergman (1941)
Hieracium laevigatum Willd.	*3x*	A; Di (An); Au	Hand et al. (2015), Mráz and Zdvořák (2019)
Hieracium lawsonii Vill.	*3x*	A	Pegoraro et al. (2020)
Hieracium lindebergii (Nyman) Dahlst.	*3x*	A; Au	Tyler and Jönsson (2013)
Hieracium marmoreum Pančić & Vis.	*3x*	A; Di (An); Au	Gentscheff (1937)
Hieracium murorum L. [=*Hieracium hyparcticum* (Almq.) Elfstr.]	*3x, 4x*	A; Di (An); Au	Ostenfeld (1906), Hand et al. (2015), Mráz and Zdvořák (2019), Pegoraro et al. (2020)
Hieracium murorum subsp. *sylvularum* (Boreau) Zahn (=*Hieracium grandidens* Dahlst.)	*3x*	A; Di (An); Au	Murbeck (1904), Bergman (1941)
Hieracium neocerinthe Fr.		A; Au	Ostenfeld (1906)
Hieracium nigrescens subsp. *decipiens* (Tausch) Zahn (=*Hieracium decipiens* Tausch)	*4x*	A; Au	Chrtek et al. (2007)
Hieracium nigrescens Willd.	*3x, 4x*	A; Di (An); Au	Hand et al. (2015), Mráz and Zdvořák (2019)
Hieracium olympicum Boiss.	*3x*	A; Di (An); Au	Hand et al. (2015)
Hieracium pannosum subsp. *bornmuelleri* (Freyn) Murr & Zahn (=*Hieracium bornmuelleri* Freyn)		A; Au	Ostenfeld (1906)
Hieracium picroides Vill.	*3x*	A; Au	Mráz and Zdvořák (2019)
Hieracium piliferum Hoppe	*3x, 4x*	A; Au	Mráz and Zdvořák (2019), Pegoraro et al. (2020)
Hieracium piliferum subsp. *glanduliferum* (Hoppe) Zahn (=*Hieracium glanduliferum* Hoppe)	*3x*	A; Au	Mráz and Zdvořák (2019)
Hieracium pilosum Froel.	*3x*	A; Au	Mráz and Zdvořák (2019)
Hieracium pinetophilum (Degen & Zahn) Chrtek	*3x*	A; Au	Štorchová et al. (2002)
Hieracium plumbeum Blytt & Fr.	*4x*	A; Au	Tyler and Jönsson (2013)
Hieracium praecurrens Vuk.	*3x*	A; Au	Mráz and Zdvořák (2019)
Hieracium prenanthoides Vill.	*3x*	A; Au	Mráz and Zdvořák (2019)
Hieracium racemosum subsp. *crinitum* (Sm.) Rouy (sub *Hieracium crinitum*)	*3x*	A; Au	Mráz and Zdvořák (2019)
Hieracium racemosum Waldst. & Kit. ex Willd.	*3x*	A; Au	Mráz and Zdvořák (2019)

(Continued)

TABLE 5.1 (Continued)
Table of Apomictic Species (A) within the Family Asteraceae Giseke, Sorted Alphabetically According to Subfamilies

Species	Ploidy	Type of Apomixis[a]	References
Hieracium ramosissimum subsp. *lactucifolium* (Arv.-Touv.) Rouy	*3x*	A	Pegoraro et al. (2020)
Hieracium ramosum Waldst. & Kit. ex Willd.		A; Ap (Hi); Di; Au	Gentscheff (1937), Battaglia (1947d)
Hieracium rapunculoides Arv.-Touv.	*4x*	A; Au	Mráz and Zdvořák (2019)
Hieracium rigidum Fr.	*3x*	A; Di (An); Au	Ostenfeld (1906), Bergman (1941)
Hieracium robustum Fr.	*3x, 4x*	A; Di (Ta); Au	Gentscheff and Gustafsson (1940)
Hieracium rohacsense Kit. ex Kit.	*3x*	A; Au	Mráz et al. (2001)
Hieracium sabaudum L.	*3x, 4x*	A; Di (An); Au	Hand et al. (2015), Mráz and Zdvořák (2019)
Hieracium sabaudum subsp. *boreale* (Fr.) Hayek (=*Hieracium boreale* Fr.)		A; Di (Ta, An); Au	Zahn (1904, in Ostenfeld (1906, p. 234), Rosenberg (1927), Bergman (1941)
Hieracium saxifragum Fr.	*4x*	A; Di (An); Au	Hand et al. (2015)
Hieracium scitulum Woł.	*3x, 4x*	A; Au	Mráz and Zdvořák (2019)
Hieracium slovacum Chrtek	*4x*	A; Au	Štorchová et al. (2002)
Hieracium sparsidens Dahlst.		A; Di (An); Au	Bergman (1941)
Hieracium sparsum subsp. *kotschyanum* (Heuff.) Zahn (=*Hieracium kotschyanum* Heuff.)	*3x*	A; Au	Mráz and Szelag (2004)
Hieracium sparsum subsp. *lubricicaule* Zahn [=*Hieracium lubricicaule* (Zahn) Borza]	*3x*	A; Au	Mráz and Szelag (2004)
Hieracium sparsum subsp. *magocsyanum* (Jáv.) Zahn (=*Hieracium magocsyanum* Jáv.)	*3x*	A; Au	Mráz and Szelag (2004)
Hieracium sparsum subsp. *silesiacum* (E. Krause) Zahn (=*Hieracium silesiacum* E. Krause)	*4x*	A; Au	Mráz and Zdvořák (2019)
Hieracium sparsum subsp. *sparsiforme* (Degen & Zahn) Greuter (=*Hieracium ostii-bucurae* Nyár. ex Szeląg)	*3x*	A; Au	Mráz and Szelag (2004)
Hieracium sparsum subsp. *telekianum* (Boros & Lengyel) Greuter (=*Hieracium telekianum* Boros & Lengyel)	*3x*	A; Au	Mráz and Szelag (2004), Mráz and Zdvořák (2019)
Hieracium sparsum subsp. *tomiasae* Nyár. & Zahn (=*Hieracium tomiasae* (Nyár. & Zahn) Nyár.)	*3x*	A; Au	Mráz and Szelag (2004)
Hieracium sparsum subsp. *tubulatum* (Zahn) Greuter (=*Hieracium tubulare* Nyár.)	*3x*	A; Au	Mráz and Szelag (2004)
Hieracium studeticum Sternb.	*4x*	A; Au	Mráz and Zdvořák (2019)
Hieracium substoloniferum (Nägeli & Peter) Üksip		A; Au	Ostenfeld (1906)
Hieracium sylvaticum (L.) Gouan		A; Au	Ostenfeld (1906)
Hieracium tenuifrons P.D. Sell & C. West	*3x, 4x*	A; Au	Shi et al. (1996)

TABLE 5.1 (Continued)

Species	Ploidy	Type of Apomixis[a]	References
Hieracium tomentosum L.	*3x*	A	Pegoraro et al. (2020)
Hieracium umbellatum L. (=*Hieracium canadense* Michx.)	*3x*	A; Ap; Di (Ta, An); Au	Gentscheff (1937), Bergman (1941)
Hieracium vagneri Pax	*4x*	A; Au	Mráz and Zdvořák (2019)
Hieracium valdepilosum Vill.	*3x, 4x*	A; Au	Mráz and Zdvořák (2019), Pegoraro et al. (2020)
Hieracium villosum Jacq.	*3x, 4x*	A; Di (An); Au	Hand et al. (2015), Mráz and Zdvořák (2019), Pegoraro et al. (2020)
Hieracium virosum Pall.	*3x, 4x*	A; Di (An); Au	Ostenfeld (1906), Hand et al. (2015)
Hieracium vapenicanum (Lengyel & Zahn) Chrtek f. & Mráz	*3x*	A; Au	Chrtek et al. (2007)
Ixeridium dentatum (Thunb.) Tzvelev [=*Ixeris dentata* (Thunb.) Nakai]	*3x*	A; Di (Ix); Au	Okabe (1932)
Picris hieracioides L.	*2x*	A; Ap	Bergman (1932), Pegoraro et al. (2020)
Pilosella alpicola (Steud. & Hochst.) F.W. Schultz & Sch.Bip.	*3x, 4x, 5x*	A; Ap (Hi); Au	Šingliarová et al. (2011)
Pilosella aurantiaca (L.) F.W. Schultz & Sch.Bip. (≡*Hieracium aurantiacum* L.)	*3x, 4x, 5x, 6x, 7x*	A; Ap (Hi); Au	Christoff (1942), Skalińska (1973, 1976), Bicknell (1994, 1997), Krahulcová and Krahulec (2000), Krahulcová et al. (2001, 2013), Pegoraro et al. (2020)
Pilosella blyttiana (Fr.) F.W. Schultz & Sch. Bip. (≡*Hieracium blyttianum* Fr.)	*3x, 4x, 5x*	A; Au	Krahulec et al. (2004), Krahulcová et al. (2001)
Pilosella caespitosa (Dumort.) P.D. Sell & C. West (≡*Hieracium caespitosum* Dumort.; =*Hieracium pratense* Tausch)	*4x, 5x*	A; Ap (Hi); Au	Rosenberg (1906, 1907, 1930), Gustafsson (1946, 1947a, 1947b), Skalińska (1967), Turesson and Turesson (1960), Skalińska and Kubien (1972), Krahulcová and Krahulec (2000), Krahulcová et al. (2001, 2013), Okada et al. (2011)
Pilosella cymosa subsp. *vaillantii* (Tausch) S. Bräut. & Greuter [=*Hieracium cymosum* subsp. *cymigerum* (Reichenb.) Peter]	*5x*	A; Au	Krahulec et al. (2004)
Pilosella densiflora (Tausch) Soják (=*Hieracium excellens* Błocki)		A; Ap (Hi); Au	Rosenberg (1930)
Pilosella echioides (Lumn.) F.W. Schultz & Sch.Bip. (≡*Hieracium echioides* Lumn.)	*4x*	A; Au	Peckert et al. (2005), Chrtek et al. (2014)
Pilosella flagellaris (Willd.) Arv.-Touv. (≡*Hieracium flagellare* Willd.)		A; Ap (Hi); Au	Haberlandt (1921a, 1921b), Rosenberg (1930)

(Continued)

TABLE 5.1 (Continued)
Table of Apomictic Species (A) within the Family Asteraceae Giseke, Sorted Alphabetically According to Subfamilies

Species	Ploidy	Type of Apomixis[a]	References
Pilosella floribunda (Wimm. & Grab.) Fr. (≡*Hieracium floribundum* Wimm. & Grab.)	*3x, 4x*	A; Au	Skalińska (1967), Krahulcová and Krahulec (1999), Krahulcová et al. (2001), Krahulec et al. (2004)
Pilosella glomerata (Froel.) Fr. (≡*Hieracium glomeratum* Froel.)	*4x, 5x*	A; Au	Turesson and Turesson (1960), Krahulcová and Krahulec (1999, 2000), Krahulcová et al. (2001, 2013), Krahulec et al. (2004)
Pilosella iserana (R. Uechtr.) Soják (≡*Hieracium iseranum* R. Uechtr.)	*4x, 5x*	A; Au	Krahulcová and Krahulec (1999), Krahulcová et al. (2001, 2013), Krahulec et al. (2004)
Pilosella leptophyton (Nägeli & Peter) P. D. Sell & C. West (≡*Hieracium leptophyton* Nägeli & Peter)	*5x*	A; Au	Křišťálová et al. (2010)
Pilosella officinarum Vaill. (=*Hieracium pilosella* L.)	*5x, 6x, 7x, 9x, 10x*	A; Ap (Hi); Au	Gadella (1982), Fehrer et al. (2007)
Pilosella piloselliflora (Nägeli & Peter) Soják (≡*Hieracium piloselliflorum* Nägeli & Peter; =*Hieracium apatelium* Nägeli & Peter)	*4x, 5x, 6x*	A; Ap (Hi); Au	Krahulcová and Krahulec (1999, 2000), Krahulcová et al. (2001), Krahulec et al. (2004), Fehrer et al. (2007)
Pilosella piloselloides (Vill.) Soják (≡*Hieracium piloselloides* Vill.)	*3x, 4x, 5x, 6x*	A; Ap (Hi); Au	Okada et al. (2013)
Pilosella piloselloides subsp. *bauhinii* (Schult.) S.Bräut. & Greuter (=*Hieracium bauhini* Schult.)	*4x, 5x, 6x*	A; Ap (Hi); Au	Krahulcová et al. (2001, 2013), Rotreklová et al. (2002), Křišťálová et al. (2010), Fehrer et al. (2007)
Pilosella piloselloides subsp. *praealta* (Gochnat) S. Bräut. & Greuter (=*Hieracium praealtum* Vill. ex Gochnat)	*4x, 5x*	A; Ap (Hi); Au	Křišťálová et al. (2010), Okada et al. (2011, 2013), Kotani et al. (2014)
Pilosella rothiana (Wallr.) F.W. Schultz & Sch.Bip.	*4x*	A; Au	Rotreklová et al. (2002), Křišťálová et al. (2010), Krahulcová et al. (2013)
Pilosella rubra (Peter) Soják (≡*Hieracium rubrum* Peter)	*6x*	A; Au	Krahulcová and Krahulec (1999, 2000), Krahulcová et al. (2001, 2004), Krahulec et al. (2004), Fehrer et al. (2007)
Pilosella scandinavica (Dahlst.) Schljakov	*4x*	A; Ap (Hi); Au	Bergman (1941), Krahulec et al. (2004), Krahulcová et al. (2013)
Pilosella schultesii (F.W. Schultz) F.W. Schultz & Sch.Bip. (≡*Hieracium schultesii* F.W. Schultz)	*4x, 5x*	A; Au	Krahulcová and Krahulec (1999), Krahulec et al. (2004)
Pilosella setigera Fr. (=*Hieracium fallax* Willd.)		A; Au	Křišťálová et al. (2010)

TABLE 5.1 (Continued)

Species	Ploidy	Type of Apomixis[a]	References
Pilosella stoloniflora (Waldst. & Kit.) F.W. Schultz & Sch.Bip. (≡*Hieracium stoloniflorum* Waldst. & Kit.)	*6x*	A; Au	Krahulcová and Krahulec (1999), Krahulec et al. (2004)
Pilosella tubulascens Norrl. [≡*Hieracium tubulascens* (Norrl.) Norrl.]	*4x*	A; Au	Krahulcová et al. (2001), Krahulec et al. (2004)
Pilosella visianii F.W. Schultz & Sch.Bip.	*4x*	A; Au	Křišťálová et al. (2010)
Pilosella ziziana (Tausch) F.W. Schultz & Sch.Bip.	*3x, 4x*	A; Au	Rotreklová et al. (2005), Křišťálová et al. (2010)
Schlagintweitia huteri (Bamb.) Gottschl. & Greuter (=*Hieracium pallidiflorum* Hausm.)	*3x*	A; Au	Mráz and Zdvořák (2019)
Schlagintweitia intybacea (All.) Griseb. (=*Hieracium intybaceum* All.; =*Hieracium albidum* Vill.)	*4x*	A; Au	Ostenfeld (1906), Mráz and Zdvořák (2019)
Taraxacum alatum H. Lindb.	*3x*	A	Kirschner et al. (2016)
Taraxacum albidum Dahlst.	*4x,5x*	A; Di (Ta); Au	Ikeno (1910), Osawa (1913), Menken and Morita (1989)
Taraxacum alpestre (Tausch) DC.		A; Au	Kirschner and Štěpánek (1994)
Taraxacum amborum G.E. Haglund	*3x*	A; Di (Ta)	Gustafsson (1935)
Taraxacum amplum Markl. (=*Taraxacum robustum* Puol.; sub *Taraxacum robustum* Schischk.)	*3x*	A; Di (Ta); Au	Poddubnaja-Arnoldi and Dianova (1964)
Taraxacum ancoriferum Hudziok	*3x*	A; Au	Battjes et al. (1992)
Taraxacum arcticum (Trautv.) Dahlst.	*5x*	A; Au	Kirschner et al. (2015)
Taraxacum atactum Sahlin & Soest	*3x*	A; Au	Mogie (1985)
Taraxacum atricapillum Sonck	*3x*	A; Di (Ta)	Musiał et al. (2015)
Taraxacum balticum Dahlst.	*4x*	A; Di (Ta)	Gustafsson (1935)
Taraxacum baltistanicum Soest		A; Au	Kirschner et al. (2017)
Taraxacum banhyhalense Soest	*4x*	A; Di (Ta); Au	Kaul and Singh (1980)
Taraxacum boekmanii Borgv.		A; Au	Mogie (1985)
Taraxacum brunneum Soest	*3x*	A; Au	Richards (1970)
Taraxacum bellicum Sonck (=*Taraxacum prunicolor* Schmid, Vašut & Oosterv.)	*3x*	A	Vašut et al. (2005), Majeský et al. (2015)
Taraxacum belorussicum Val.N. Tikhom.	*3x*	A; Di (Ta)	Janas et al. (2016)
Taraxacum brachyglossum Dahlst.	*3x*	A; Ps; Au	Richards (1970), Hughes (1987)
Taraxacum campylodes G.E. Haglund [=*Taraxacum vulgare* (Lam.) Schrank]	*3x*	A; Di (Ta); Au	Raunkiaer (1903), Juel (1904), Sears (1917)
Taraxacum canum Soest		A; Au	Kirschner et al. (2017)
Taraxacum ceratophorum (Ledeb.) DC.	*3x, 4x*	A; Di (Ta); Au	Gustafsson (1935)
Taraxacum cordatum Palmgr.	*3x*	A; Au	Richards (1970)
Taraxacum cristatum Kirschner, Štěpánek & Vašut	*3x*	A; Au	Vašut et al. (2005), Mártonfiová et al. (2010), Majeský et al. (2015)
Taraxacum croceum Dahlst.	*4x*	A; Di (Ta)	Gustafsson (1935)
Taraxacum cyprium H. Lindb.	*3x, 4x*	A; Au	Haran (1952)
Taraxacum danubium A.J. Richards	*3x*	A; Au	Richards (1970), Mártonfiová et al. (2010)

(Continued)

TABLE 5.1 (Continued)
Table of Apomictic Species (A) within the Family Asteraceae Giseke, Sorted Alphabetically According to Subfamilies

Species	Ploidy	Type of Apomixis[a]	References
Taraxacum dilutisquameum M.P. Christ.		A	Mes et al. (2002)
Taraxacum ekmanii Dahlst.	*3x*	A	Kirschner et al. (2016)
Taraxacum erythrospermum Andrz. ex Besser [=*Taraxacum laevigatum* (Willd.) DC.]	*3x*	A; Di (Ta); Au	Sears (1917), Poddubnaja-Arnoldi and Dianova (1964)
Taraxacum expallidiforme Dahlst. (=*Taraxacum subcyanolepis* M.P. Christ.)	*3x*	A; Di (An); Au	Bergman (1941), Richards (1970)
Taraxacum flavum Soest	*3x*	A; Di (Ta); Au	Kaul and Singh (1980)
Taraxacum fontanum Hand.-Mazz.	*4x*	A; Di (Ta)	Gustafsson (1935)
Taraxacum fulviforme Dahlst.	*4x*	A; Au	Ford and Richards (1985)
Taraxacum fulvum Raunk.	*4x*	A; Di (Ta)	Gustafsson (1935)
Taraxacum gulmargense Soest	*3x*	A; Di (Ta); Au	Kaul and Singh (1980)
Taraxacum haematopus H. Lindb. (sub *Taraxacum haematopus* Dahlst.)	*3x*	A; Di (Ta)	Gustafsson (1935)
Taraxacum hamatiforme Dahlst.	*3x*	A; Au	Mogie (1985)
Taraxacum hamatum Raunk.	*3x*	A; Au	Richards (1970), Mogie (1985)
Taraxacum hamiferum Dahlst.		A; Au	Mogie (1985)
Taraxacum harbhajan-singhii subsp. *pahalgamense* Soest	*3x*	A; Di (Ta); Au	Kaul and Singh (1980)
Taraxacum hemicyclum G.E. Haglund	*3x*	A	Kirschner et al. (2016)
Taraxacum hepaticum Rail.	*3x*	A	Kirschner et al. (2016)
Taraxacum hollandicum Soest	*3x,4x*	A; Au	Battjes et al. (1992)
Taraxacum hondoense Nakai ex H. Koidz.	*3x,4x,5x*	A; Au	Akhter et al. (1993)
Taraxacum hybernum Steven (sub *Taraxacum hybernum* Dt)	*3x,4x*	A; Di (Ta); Au	Poddubnaja-Arnoldi and Dianova (1964)
Taraxacum interveniens G.E.Haglund	*3x*	A; Au	Kirschner et al. (2016)
Taraxacum isophyllum G.E. Haglund	*3x*	A; Au	Richards (1970)
Taraxacum kernianum Hagend., Soest & Zevenb.	*3x*	A; Au	Mogie (1985)
Taraxacum kok-saghyz L.E. Rodin (=*Taraxacum brevicorniculatum* Korol.)	*3x*	A; Au	Warmke (1943), Kirschner et al. (2013)
Taraxacum lacistophyllum (Dahlst.) Raunk.	*3x*	A; Au	Ford and Richards (1985)
Taraxacum laeticolor Dahlst.	*3x*	A; Di (Ta)	Gustafsson (1935)
Taraxacum leucanthum (Ledeb.) Ledeb.	*3x,4x*	A	Kirschner et al. (2006)
Taraxacum litorale Raunk.	*3x*	A; Di (Ta)	Gustafsson (1935)
Taraxacum luridum G.E. Haglund	*3x*	A; Au	Kirschner et al. (2006, 2015)
Taraxacum macranthoides G.E. Haglund	*3x*	A; Au	Kirschner et al. (2016)
Taraxacum maricum Vašut, Kirschner & Štěpánek	*3x*	A; Au	Mártonfiová et al. (2010)
Taraxacum melleum Soest		A; Au	Kirschner et al. (2017)
Taraxacum microspermum Schischk.	*3x*	A; Di (Ta); Au	Poddubnaja-Arnoldi and Dianova (1964)
Taraxacum mitalii Soest		A	Kirschner et al. (2017)
Taraxacum mongolicum Hand.-Mazz.	*3x*	A; Di (Ta); Au	Battaglia (1948)

TABLE 5.1 (Continued)

Species	Ploidy	Type of Apomixis[a]	References
Taraxacum naevosiforme Dahlst.	*4x*	A; Au	Ford and Richards (1985)
Taraxacum naevosum Dahlst.		A	Mes et al. (2002)
Taraxacum nordstedtii Dahlst.	*6x*	A; Au	Ford and Richards (1985)
Taraxacum obliquum (Fr.) Dahlst.	*3x*	A; Au	van Oostrum et al. (1985)
Taraxacum obtusifrons Markl.	*3x*	A	Kirschner et al. (2016)
Taraxacum officinale F.H. Wigg.	*3x*	A; Di (Ta); Au	Juel (1904), Schkorbatow (1912), Osawa (1913), Stork (1920), Rosenberg (1930), Cooper and Brink (1949), Pegoraro et al. (2020)
Taraxacum ostenfeldii Raunk. (=*Taraxacum duplidens* H. Lindb.)	*3x*	A; Di (Ta); Au	Richards (1970), Kaul and Singh (1980)
Taraxacum paludosum (Scop.) Crépin	*4x*	A; Di (Ta); Au	Záveský et al. (2007)
Taraxacum parnassicum Dahlst.	*3x*	A; Au	Mártonfiová et al. (2010)
Taraxacum parvulum DC. (=*Taraxacum himalaicum* ssp. *himalaicum* Soest)		A; Di (Ta); Au	Kaul and Singh (1980)
Taraxacum paucilobum Hudziok	*3x*	A; Au	Battjes et al. (1992)
Taraxacum perincisum (Murr) Murr (=*Taraxacum oxoniense* Dahlst.)	*4x*	A; Au	Ford and Richards (1985)
Taraxacum persquamulosum Kirschner & Štěpánek		A; Au	Kirschner et al. (2017)
Taraxacum piceatum Dahlst.	*3x*	A; Au	Kirschner et al. (2016)
Taraxacum polyodont Dahlst.	*3x*	A; Di (Ta); Au	Kaul and Singh (1980)
Taraxacum porphyranthum Boiss.	*3x*	A; Di (Ta)	Gustafsson (1935)
Taraxacum praestans H. Lindb.	*3x,4x*	A; Di (Ta); Au	Gustafsson (1935), Ford and Richards (1985)
Taraxacum proximum (Dahlst.) Raunk. (=*Taraxacum proximum* Dahlst.)	*3x,4x*	A; Au	Ford and Richards (1985), Mártonfiová et al. (2010)
Taraxacum pseudohamatum Dahlst.	*3x*	A; Au	Hughes and Richards (1988, 1989)
Taraxacum pseudofulvum H. Lindb. (sub *Taraxacum pseudofulvum* H. Lindb. fil.)		A; Di (Ta)	Gustafsson (1935)
Taraxacum pudicum Vašut & Majeský	*3x*	A; Di (Ta); Au	Vašut and Majeský (2015)
Taraxacum pulchrifolium Markl.	*3x*	A; Au	Kirschner et al. (2016)
Taraxacum quadrans H. Øllg.	*3x*	A; Au	Mogie (1985)
Taraxacum retroflexum H. Lindb.	*3x*	A; Di (Ta); Au	Poddubnaja-Arnoldi and Dianova (1964), Gustafsson (1935)
Taraxacum ruberulum Borgv. & Dahlst.	*3x*	A; Di (Ta)	Gustafsson (1935)
Taraxacum scanicum Dahlst.	*3x*	A; Di (Ta); Au	Małecka (1967), Vašut et al. (2005), Mártonfiová et al. (2010), Majeský et al. (2015)
Taraxacum sellandii Dahlst.	*3x*	A; Au	Richards (1970)
Taraxacum silesiacum Dahlst.	*3x*	A; Au	Richards (1970)
Taraxacum sinicum Kitag. (=*Taraxacum armeriifolium* Soest)	*3x*	A	Kirschner et al. (2006)
Taraxacum slovacum Klášt.		A; Au	Richards (1970)

(Continued)

TABLE 5.1 (Continued)
Table of Apomictic Species (A) within the Family Asteraceae Giseke, Sorted Alphabetically According to Subfamilies

Species	Ploidy	Type of Apomixis[a]	References
Taraxacum sonchoides (D. Don) Sch.Bip. [=*Taraxacum montanum* (C.A. Mey.) DC.]	*5x*	A; Di (Ta); Au	Poddubnaja-Arnoldi and Dianova (1964)
Taraxacum speciosum Raunk.	*3x*	A; Di; Au	Murbeck (1904)
Taraxacum stictophyllum Dahlst.		A; Au	Mes et al. (2002)
Taraxacum subalpinum Hudziok	*3x*	A; Au	Battjes et al. (1992)
Taraxacum sublaeticolor Dahlst.	*3x*	A; Di (Ta)	Gustafsson (1935)
Taraxacum subnaevosum A.J. Richards	*4x*	A; Au	Ford and Richards (1985)
Taraxacum tenebristylum Soest	*4x*	A; Di (Ta); Au	Kaul and Singh (1980)
Taraxacum tenuiculum Kirschner & Štěpánek		A; Au	Kirschner et al. (2017)
Taraxacum tenuilobum (Dahlst.) Dahlst.	*3x*	A; Di (Ta)	Gustafsson (1935)
Taraxacum tortilobum Florstr.	*3x*	A; Au	Reisch (2004)
Taraxacum tricolor Soest		A; Au	Kirschner et al. (2017)
Taraxacum udum Jord.	*3x*	A; Di (Ta); Au	Musiał et al. (2013)
Taraxacum uliginosum Kirschner & Štěpánek	*3x*	A; Au	Battjes et al. (1992)
Taraxacum unguilobum Dahlst.	*4x*	A; Au	Mogie (1985), Ford and Richards (1985), Hughes and Richards (1989)
Taraxacum vindobonense Soest	*4x*	A; Au	Battjes et al. (1992)
Taraxacum virgineum Kirschner, Štěpánek & Klimes	*3x*	A; Au	Kirschner et al. (2006)
Taraxacum vulpinum f. *pallido-carpum* Soest	*3x*	A; Di (Ta); Au	Kaul and Singh (1980)
Taraxacum wendelboanum Soest		A; Au	Kirschner et al. (2017)

[a]For species with sporophytic apomixis, I: integumentary embryony. For species with gametophytic apomixis, Ap ( ): apospory (type of female gametophyte development); Di ( ): diplospory (type of female gametophyte development); Ps: pseudogamous endosperm development; Au: autonomous endosperm development.

Hi: Hieracium type; An: Antennaria type; Ta: Taraxacum type; Ix: Ixeris type.

Despite the frequent observation of abnormal pollen development and asynapsis in apomictic daisies, indications about the occurrence of apomixis based on observations of meiotic abnormalities during the development of the male gametophyte and poor pollen quality [e.g., in *Praxelis kleinioides* (H.B.K.) Schultz-Bip, Robinson et al., 1989; in species of *Eupatorium* L., Kawahara et al., 1989; in *Stevia* Cav., King and Robinson, 1967; Grashoff et al., 1972; Soejima, et al., 2001; or in species of *Taraxacum* F.H. Wigg., Sterk et al., 1982; Uhlemann, 2001] or based on biosystematics studies involving ploidy distributions [e.g., in *Ixeridium chinense* (Thunb.) Tzvelev sub *Ixeris chinensis* (Thunb.) Nakai, Pak and Kawano, 1990; Pak et al., 1997] are considered only when accompanied by embryological evaluation of the female gametophyte development in at least one of the species (e.g., in *Townsendia* Hook., Beaman, 1957). Preliminary occurrence of apomixis in taxa based on above analyses (e.g., Vašut et al., 2005) was accepted only once corroborated by other experimental evidence, like castration experiments, embryology or progeny tests – by morphological, cytological or molecular evaluation (e.g., Štorchová et al., 2002; Majeský et al., 2015). In several cases involving *Taraxacum* spp., due to the taxonomic complexity and study aims (e.g., characterization of genotypic variability in single populations), the research was conducted independently from taxonomic determination (e.g., Menken et al., 1995; van der Hulst et al., 2003; Mártonfiová, 2006). Hence, in those cases and despite molecular evidence of clonality, apomixis could not be assigned to specific taxa.

Molecular evidence for clonality is based on tests using five or more individuals (e.g., Tyler and Jönsson, 2013). Some studies were designed to disclose genotypic variability rather than clonality and, hence, are not considered here.

Triploid *Blumea eriantha* DC. and *Blumea oxydonta* DC. are classified as diplosporous apomicts based on embryology, male sterility, high seed set and morphologically uniform offspring (Chennaveeraiah and Patil, 1971). However, alterations in the sexual developmental programme appear to involve automixis, i.e., the formation and fusion of reduced gametes. Although automixis can be considered as a type of asexuality enabling the plant to produce genetically uniform seeds and offspring, the genetics and developmental mechanisms underlying this type of development are different from those observed in gametophytic or sporophytic apomixis (Hojsgaard and Schartl, 2021).

Different new apomictic daisies have been detected using FCSS (e.g., Mártonfiová et al., 2010; Mráz and Zdvořák, 2019; Pegoraro et al., 2020). In these cases, it is only possible to speculate on the type of female gametophyte development, so embryological analyses are needed. Even though the observed types of female gametophyte developments are uniform within a genus, this may depend on the number of species studied, or the number of ovules and individuals analysed within a species. For instance, in *Hieracium*, a widely studied genus exhibiting an apomictic reproductive mode tightly associated with diplospory (except for a few cases showing apospory), alternative types of diplosporous developments (i.e., Antennaria type and Taraxacum type) have been recorded for different species (Bergman, 1935a, 1941; Battaglia, 1947d; Skawinska, 1962; Hand et al., 2015).

We apologize for any missing data or errors resulting from lack of access to certain works or from our language limitations (e.g., Movsesyan, 1964; Poddubnaja-Arnoldi, 1964, 1976; Chochlov et al., 1978).

5.2.1 Doubtful Cases of Apomixis in Daisies

Like in grasses, the occurrence of apomixis is yet to be confirmed in several species of daisies (n=42; Table 5.2). Despite the circumstantial evidence supporting the existence of apomixis, more specific analysis or experiments are needed in those taxa to corroborate the formation of asexual seeds. In general, we agree with the observations of Noyes (2007), and have added new doubtful cases to our list, or resolved some of the old ones based on new evidence. Most of the doubtful cases listed in Table 5.2 involve diploids (76.3% of the species with ploidy records) and are likely to be based on the spontaneous occurrence of apomixis-like developments lacking functional apomixis. This is a widely known phenomenon that occurs in most apomictic systems that also fits the current hypothesis on the origin and establishment of apomixis in natural populations (Hojsgaard and Hörandl, 2019).

For instance, diploid *Antennaria dioica* (L.) Gaertn. has previously been classified as apomictic (see details in Noyes, 2007) on the basis of observations of aposporous developments (Haberlandt, 1921a, 1921b). Yet, the tendency to apospory observed in this dioecious species does not corroborate functional apomixis (Gustafsson, 1946), and all cases of apomixis tested in *Antennaria* spp. involve diplosporous development (see Table 5.1). Therefore, the frequency and function of aposporous gametophytes should be checked before drawing any conclusions.

For *Bellis perennis* L., Carano (1913) reported apospory and Murthy and Sharma (1977) corroborated apomictic tendencies by suggesting that the diploid cytotype is a casual apomict. However, the observations probably correspond to spontaneous abnormal developments (Noyes, 2007) and functional apomixis should be corroborated for this species. *Artemisia nitida* Bertol. (Chiarugi, 1926), *Eurybia sibirica* (L.) Nesom (≡*Aster sibiricus* L., Poddubnaja-Arnoldi, 1976), *Felicia amelloides* (L.) Voss. (=*Aster capensis* Less., Davis, 1967 based on Dahlgren, 1920) and several species of *Centaurea* L. (Bergman, 1935b; Czapik 1954; Poddubnaja-Arnoldi, 1976) are other examples of diploids in which apomixis is probably non-functional and previous records were based on observations of spontaneous occurrence of apomictic-like developments (for details, see Noyes, 2007). Likewise, *Symphyotrichum undulatum* (L.) G.L. Nesom (≡*Aster undulatus* L.) and *S. novae-angliae*

TABLE 5.2
Table of Doubtful Apomictic Species (A) within the Family Asteraceae Giseke, Sorted Alphabetically According to Subfamilies

Species	Ploidy	Type of Apomixis[a]	References
Asteroideae Lindl.			
Achillea collina (Becker ex Rchb.f.) Heimerl (sub *Achillea collina* Becker ex Rchb.)	4*x*	A; N; Ap	Terziĭski et al. (1995a, 1995b)
Achillea seidlii J. Presl & C. Presl (=*Achillea pannonica* Scheele)		A; N	Terziĭski et al. (1995a, 1995b)
Achillea millefolium L.		A; N	Terziĭski et al. (1995a, 1995b)
Amphiachyris dracunculoides (DC.) Nutt. [=*Gutierrezia dracunculoides* (DC.) S.F. Blake]	2*x*	A; Ap	Davis (1967)
Antennaria dioica (L.) Gaertn.	2*x*	A; Ap	Haberlandt (1921a, 1921b)
Antennaria carpatica (Wahlenb.) Bluff & Fingerh.	3*x*	A; Ap (Hi); Di (Ta)	Bergman (1951)
Arnica montana L.	2*x*	A; Di (Ta)	Yankova-Tsvetkova et al. (2016)
Artemisia nitida Bertol.	3*x*,6*x*	A; Ap; Di	Chiarugi (1926)
Bellis perennis L.	2*x*	A; Ap (Hi)	Carano (1913, 1921), Murthy and Sharma (1977)
Bidens tripartita L.	4*x*	A; Ap	Davis (1967, based on Dahlgren, 1920)
Brachyscome iberidifolia Benth.	2*x*	A; Di	Sharma and Murthy (1977)
Calotis scapigera Hook.	3*x*	A; Ap	Stace (1978)
Calotis suffruticosa Domin	4*x*	A; Ap	Stace (1978)
Coreopsis tinctoria Nutt.	2*x*	A; Ap	Gustafsson (1946, 1947a, 1947b, based on Gelin, 1934)
Erigeron glabellus Nutt.	2*x*	A; Ap (Hi)	Carano (1921)
Erigeron elatus (Hook.) Greene	4*x*	A; Ap (Hi)	Bergman (1942)
Eurybia sibirica (L.) G.L. Nesom (≡*Aster sibiricus* L.)	2*x*	A; Ap	Poddubnaja-Arnoldi (1964)
Felicia amelloides (L.) Voss. (=*Aster capensis* Less.)	2*x*	A; Ap	Davis (1967)
Grindelia ciliata (Nutt.) Sprengel	2*x*	A; Ap	Davis (1967)
Grindelia lanceolata Nutt.	2*x*	A; Ap	Davis (1967)
Helianthus annuus L.	2*x*	A; Ap	Poddubnaja-Arnoldi (1964), Ustinova (1955, 1970)
Helianthus maximiliani Schrad.	2*x*	A; Ap	Davis (1967, based on Goldflus, 1898, 1899)
Laggera alata (D. Don) Sch. Bip. ex Oliv.	2*x*	A; Ap	Pullaiah (1984)
Laggera aurita (L. f.) Benth. ex C.B. Clarke	2*x*	A; Ap	Pullaiah (1984)
Leucanthemum lacustre (Brot.) Samp.		A; Ap	Davis (1967, based on Goldflus, 1898, 1899)
Leucanthemum vulgare Lam. (sub *Chrysanthemum leucanthemum* L.)		A; Ap	Davis (1967, based on Goldflus, 1898, 1899)
Melampodium divaricatum (Rich. ex Rich.) DC.	2*x*	A; Ap	Maheswari Devi and Pullaiah (1976a, 1976b)
Rudbeckia bicolor Nutt.	2*x*	A; Ps	Kościńska-Pająk (1980); Musiał (2012)

(Continued)

TABLE 5.2 (Continued)
Table of Doubtful Apomictic Species (A) within the Family Asteraceae Giseke, Sorted Alphabetically According to Subfamilies

Species	Ploidy	Type of Apomixis[a]	References
Rudbeckia maxima Nutt.	*2x*	A	Movsesyan (1964)
Solidago purpurea L. (illeg.)		A; Ap	Mukherjee and Desai (1990)
Stevia rebaudiana (Bertoni) Bertoni	*2x*	A; N; I	Monteiro (1980), Caponio et al. (2016)
Symphyotrichum undulatum (L.) G.L. Nesom (≡*Aster undulatus* L.)	*2x,4x*	A; Ap	Davis (1967, based on Opperman, 1904)
Symphyotrichum novae-angliae (L.) G.L. Nesom (≡*Aster novae-angliae* L.)	*2x*	A; Ap	Davis (1967)
Tridax trilobata (Cav.) Hemsl.	*2x*	A; Ap	Hjelmqvist (1951)
Cichorioideae Chevall.			
Leontodon hispidus L.	*2x* (?)	A; Ap (Hi); Ps	Bergman (1932)
Krigia wrightii (A. Gray) K.L. Chambers ex K.J. Kim	*3x* (?)	A	Kim and Turner (1992)
Krigia occidentalis Nutt.	*2x*	A	Kim and Turner (1992)
Krigia cespitosa (Raf.) K.L. Chambers	*2x*	A	Kim and Turner (1992)
Carduoideae Cass. ex Sweet			
Centaurea scabiosa L.	*2x*	A; Ap	Bergman (1935a, 1935b, 1935c), Czapik (1954), Poddubnaja-Arnoldi (1964)
Cyanus segetum Hill (=*Centaurea cyanus* L.)	*2x*	A; Ap	Bergman (1935a, 1935b, 1935c), Czapik (1954), Poddubnaja-Arnoldi (1964)

[a]For species with sporophytic apomixis; N: nucellar embryony; I: integumentary embryony. For species with gametophytic apomixis; Ap ( ): apospory (type of female gametophyte development); Di ( ): diplospory (type of female gametophyte development); Ps: pseudogamous endosperm development; Au: autonomous endosperm development.

Hi: Hieracium type; Ta: Taraxacum type.

(L.) G.L. Nesom (≡*Aster novae-angliae* L.) were listed as apomicts by Davis (1967) based on Opperman (1904). Yet, Opperman reported the occurrence of apogamety in antipodal cells of the female gametophyte rather than apomixis (Opperman, 1904, pp. 359–360).

Similar cases in which apomixis elements had been clearly demonstrated but lacked evidence of functional apomixis are removed or placed in our quarantine table (Table 5.2). The use of novel techniques will provide an opportunity to revise such cases. An interesting example is *Antennaria carpatica* (Wahlenb.) Hook. In this dioecious species, Bergman (1935c, 1951) described the development of unreduced gametophytes of the Taraxacum type (diplosporous) and Hieracium type (aposporous) in ovules of female plants. However, the tested plants produced almost no seeds. Whether this is caused by malfunctioning of a parthenogenesis gene (i.e., if autonomous development is expected, as observed in most other *Antennaria* species) or due to a lack of pollen (i.e., if pseudogamous development is the case, and male plants are absent) remains to be tested.

Cases of apomicts categorized as diploids, like *Leontodon hispidus* L. ($2n$=14; Bergman, 1935b), may actually be polyploids (or paleopolyploids) since other species of the genus show varied chromosomal constitutions ($2n$=8, 12, 14, 21), even when no evidence of apomixis has been found among these cytotypes (Rousi, 1973). Yet, of the many plants analysed, Bergman only observed aposporous development in two plants growing in the botanical garden at the University of Stockholm (the Hortus Bergianus), and in those plants apospory occurred in a few flower heads with varying frequency, in certain cases in up to 10-20% of the flowers (Bergman, 1935b, pp. 163) and was reduced to a frequency of <0.2% in the F_1 offspring. The study also provides evidence that the unreduced female gametophytes are functional and can develop into a triploid embryo ($2n$=21) and a heptaploid endosperm ($2n$=49) after fertilization (Bergman, 1935b, pp. 180). According to Bergman (1935b, pp. 182), the chalazal position of the aposporous gametophytes in the ovules of *L. hispidus* is responsible for their lack of functionality. The occurrence of non-functional (or rarely functional), unreduced female gametophytes is common among sexual diploids of conspecific apomictic polyploids, and it is assumed to have a central role in the establishment of polyploid apomictic populations (Hojsgaard, 2018; Hojsgaard and Hörandl, 2019). Regardless of the ploidy level of the studied cytotype of *L. hispidus*, it seems unlikely that it is a regular apomict.

Similarly, in a recent study in *Arnica montana* L., Yankova-Tsvetkova et al. (2016) recorded a low frequency of Taraxacum type development (and possibly parthenogenetic embryo development) in diploid individuals that otherwise showed Polygonum type sexual development. Apomixis is probably not stabilized in the diploid cytotype of the species. Like in other species of *Arnica* L., detailed cytogeographical studies are likely to reveal the existence of polyploid apomictic cytotypes in *A. montana*.

Monteiro (1980) showed the occurrence of integumentary embryony and pseudogamy in diploid material of *Stevia rebaudiana* (Bertoni) Bertoni. However, in a recent analysis using different regional ecotypes, Caponio et al. (2016) demonstrated normal Polygonum type development indicating sexual reproduction. Therefore, the species is placed in our quarantine table for further consideration and evaluation of alternative ecotypes and polyploid cytotypes.

Embryologically complicated genera, such as *Rudbeckia* L. displaying the occurrence of hemigamy and other developmental alterations, require revision by a combination of classic and novel techniques. Recent studies are making progress in this direction (e.g., Musiał et al., 2012). Here we place two *Rudbeckia* species in Table 5.2 and maintain *Rudbeckia californica* A. Gray in Table 5.1 following Nygren's classification of the species as apomictic based on Battaglia's studies (Nygren, 1954a, 1954b). However, not all Battaglia's original studies were accessible (e.g., Battaglia, 1946d, 1947a, 1947b, 1947c).

Two apparently aposporous species of *Erigeron* L. (Carano, 1921; Bergman, 1942) are also included in the quarantine list (Table 5.2) following Noyes' observations on the widespread occurrence of diplospory among *Erigeron* species (Noyes, 2007) and the fact that one of these species is consistently diploid in nature.

The record of apospory for the pentaploid *Hieracium hoppeanum* Schult. (Christoff and Christoff, 1948) is maintained in Table 5.1 but may need to be revised, as this taxon has been found to be diploid in recent surveys (Uhríková and Májovský, 1980; Papanicolaou, 1984; Gadella, 1992; Pavlova, 1999).

5.3 PATHWAYS TO ASEXUAL SEEDS IN DAISIES

A total of 354 taxa, including 326 daisy species and 23 subspecies, have at least one record of apomixis, either sporophytic or gametophytic apomixis (see Chapter 2).

The pathways of apomictic development are known only for 42.9% of these species (n=152); in the remaining 57.1% of the cases apomixis has been determined without details on the developmental pathway (mostly through castration experiments or FCSS). In the family Asteraceae, apomixis is tightly associated with the gametophytic type and all studies using castration experiments to test for apomixis that were subsequently corroborated embryologically show gametophytic apomixis. Therefore, for simplicity, those cases are included under Section 5.3.2. A total of 48 taxa, mostly from the genera *Hieracium* and *Taraxacum*, are categorized as apomictic based on tests using molecular markers [e.g., *Hieracium krivanense* (Woł. & Zahn) Schljakov, Štorchová et al., 2002; *H. caesium* subsp. *galbanum* Dahlst. (=*H. ravidum* Brenner), Tyler and Jönsson, 2013; *H. hypochoeroides* subsp. *wiesbaurianum* (R. Uechtr.) Greuter (=*H. wiesbaurianum* R. Uechtr.), Reisch et al., 2015;

or *Taraxacum fulviforme* Dahlst., Ford and Richards, 1985; *T. paucilobum* Hudziok, Battjes et al., 1992; *T. bellicum* Sonck (=*Taraxacum prunicolor* Schmid, Vašut & Oosterv.), Majeský et al., 2015].

Among species with gametophytic apomixis with embryological details, five taxa show both apospory and diplospory (Table 5.1). Sporophytic apomixis is recorded only for three taxa, all cases of integumentary embryony (Table 5.1). One taxon shows both sporophytic and gametophytic apomixis (Table 5.1).

Information about the development of the endosperm is collected for about 84.7% of all apomictic daisies (*n*=300; Table 5.1). Most of the apomicts recorded show autonomous endosperm development (*n*=294), and only 2.3% of species (*n*=8) show pseudogamous development, including two species showing both types of development.

Ploidy is known for 86.4% of apomictic taxa (*n*=306) and 85 cases have two or more ploidy records (Table 5.1). All species but three are polyploid, ranging from triploids to dodecaploid (*x*=12, in *Antennaria soliceps* S.F. Blake, Bayer and Minish, 1993). The most frequent ploidy is triploid (*n*=196), followed by tetraploid (*n*=150). Other frequent ploidies among apomictic daisies are pentaploid (*n*=37), hexaploid (*n*=24), heptaploid (*n*=10) and octoploid (*n*=12). The last cases with ploidy record are decaploid (*n*=4), nonaploid (*n*=1) and dodecaploid (*n*=1) (Table 5.1).

As in grasses, apomictic daisies are mainly polyploid and often contain diploid conspecific cytotypes showing low levels of apomictic development in embryological studies (e.g., in *Parthenium* spp., Esau, 1946). Yet, all previous reports of apomixis in diploids without appropriate confirmation of the formation of apomictic seeds are here categorized as doubtful and placed in our quarantine table (Table 5.2). They represent 76.3% of the listed species having known ploidy (*n*=38), and likely correspond to sporadic events of apomictic developments but without the expected regular formation of functional apomictic seeds.

Two out of the three apomictic diploids mentioned above were reported together with other 22 polyploid apomicts in a recent study using FC to screen for the development of sexual versus apomictic seeds among 229 species from the Asteraceae family distributed throughout the European Alps (Pegoraro et al., 2020). The two diploid species are *Crepis versicaria* L. and *Picris hieracioides* L. (Table 5.1), the latter partially confirming earlier embryological analyses indicating the occurrence of apospory without parthenogenesis in low frequency in one individual of the same species (Bergman, 1932) but contrasting the results from a reproductive screening in 32 diploid populations of *P. hieracioides* from the same geographic region found to be sexual (Slovak et al., 2007). Surprisingly, in the above two diploid species and in one tetraploid species, i.e., *Centaurea scabiosa* L. subsp. *alpestris* (Hegetschw.) Nyman, Pegoraro et al. (2020) report separate populations being either sexual or apomictic (based on data from five seeds in each case). Likewise, the authors did also report the occurrence of apomixis in a tetraploid population of *Pilosella officinarum* Vaill., a species widely known to have sexual tetraploid cytotypes and higher ploidy apomictic cytotypes, varying from pentaploid to decaploid (Mráz et al., 2008). Special attention should be paid to the above four cases, since if corroborated by more detailed analyses, they will represent exceedingly rare cases of species having homoploid populations displaying divergent reproductive strategies. In species with gametophytic apomixis, apomicts originate from sexual relatives either by polyploidization (autopolyploids) or by hybridization often associated to polyploidization when parents are diploids (in allopolyploids), and therefore, apomicts display higher ploidies than sexual counterparts (Hojsgaard et al., 2014a, 2014b). The spontaneous occurrence of diploid sexual and apomictic individuals can simplify genetic analyses and the understanding of the emergence of apomixis in natural populations.

The third apomictic diploid is *Melampodium divaricatum* (Rich. ex Rich.) DC. and represents a case of integumentary embryony (Maheswari Devi and Pullaiah, 1976b). Sporophytic apomixis is common among species with diploid chromosome numbers (see Chapter 7).

A broad diversity of unreduced female gametophyte developments is observed among aposporous and diplosporous daisies, whereby bipolar developments including the Antennaria type, Taraxacum type, Ixeris type or Hieracium type can be found among species and genera. No monopolar developments (e.g., the Panicum type, or the Eragrostis type) were listed for the Asteraceae family.

5.3.1 Sporophytic Apomixis in Daisies

Adventitious embryo formation is exceedingly rare in the family Asteraceae and has only been observed in three species of three subfamilies, arising in all cases from integumentary tissue (Table 5.1). In the subfamily Asteroideae Lindl., one species from *Melampodium* spp. shows this type of adventitious embryos (Maheswari Devi and Pullaiah, 1976b). Within the Carduoideae Cass. ex Sweet, the genus *Carthamus* spp. (1 species; Maheswari Devi and Pullaiah, 1976a) shows integumentary polyembryony, and the genus *Hieracium* spp. (1 species; Naumova, 1993) for the subfamily Cichorioideae Chevall.

5.3.2 Gametophytic Apomixis in Daisies

In total, 300 species and subspecies of daisies are currently recorded as apomict by direct or indirect evidence of the formation of unreduced gametophytes, including one species also showing sporophytic apomixis (Table 5.1). The pathways of apomictic development are known for only 48.7% of these species; the remaining 51.3% of cases have been determined without details on the type of apomixis (through castration experiments or by FCSS).

5.3.2.1 The Aposporous Development

Apospory has been recorded in around 10.0% of all species with gametophytic apomixis (n=30; Table 5.1). Like in grasses, in daisies the aposporous development starts in a nucellar cell in contact with the sexually fated megaspore (Juranić et al., 2018) which acquires a gametogenetic fate and develops into an unreduced female gametophyte (Asker and Jerling, 1992). The germline in these apomicts often degenerates before or during the meiotic division. In some cases, the developing meiotic female gametophyte can degenerate or develop till maturity and produce a sexual seed.

5.3.2.1.1 The Hieracium Type

This type of unreduced female gametophyte development is the only type of aposporous development observed among daisies and is present in 27 species from different subfamilies (Table 5.1). Within the subfamily Asteroideae, the Hieracium type occurs in species of *Calotis* spp. (1 species; Davis, 1968), in *Campovassouria* spp. (1 species; Coleman and Coleman, 1984), in *Campuloclinium* spp. (1 species; Farco et al., 2012), in *Minuria* spp. (1 species; Davis, 1964a; Watanabe et al., 1996), in *Parthenium* spp. (2 species; Esau, 1944, 1946; Gerstel et al., 1953; Cruz et al., 2017). In the subfamily Cichorioideae, it occurs in species of *Crepis* spp. (8 species; Babcock and Stebbins, 1938; Stebbins and Jenkins, 1939), in *Hieracium* spp. (2 species; Battaglia, 1947d; Christoff and Christoff, 1948), and in *Pilosella* spp. (11 species; Rosenberg, 1906, 1907, 1930; Haberlandt, 1921a, 1921b; Bergman, 1941; Christoff, 1942; Gustafsson, 1946, 1947a, 1947b; Turesson and Turesson, 1960; Skalińska, 1967, 1973, 1976; Skalińska and Kubien, 1972; Gadella, 1982; Bicknell, 1994, 1997; Fehrer et al., 2007; Krahulec et al., 2004; Krahulcová and Krahulec, 1999, 2000; Krahulcová et al., 2001, 2013; Rotreklová et al. 2002; Křišťálová et al., 2010; Okada et al., 2011, 2013; Šingliarová et al., 2011; Kotani et al., 2014).

5.3.2.2 The Diplosporous Development

Diplospory has been recorded in 40.3% of all species and subspecies with gametophytic apomixis (n=121; Table 5.1). The development of an unreduced female gametophyte through diplospory occurs with the interruption of meiosis steps (either entry into meiosis or its progression) and the subsequent growth of an unreduced megaspore that develops into a mature gametophyte (Asker and Jerling, 1992). In occasional cases, the germline may progress through meiosis to first generate reduced megaspores and, subsequently, a reduced female gametophyte and a sexual seed after fertilization.

5.3.2.2.1 The Antennaria Type

This type of development is recorded for 45 species (Table 5.1). Within the subfamily Asteroideae, the Antennaria type is found in 20 species and subspecies, including species from the genera *Ageratina* spp. (1 species; Holmgren, 1919), *Antennaria* spp. (6 species and subspecies; Juel, 1898, 1900; Stebbins, 1932; Bergman, 1935c, 1937; Bayer and Stebbins, 1983, 1987; Bierzychudek, 1985), *Ayapana* spp. (1 species; Mukherjee and Desai, 1980), *Brachyscome* spp. (2 species; Davis, 1963, 1964b), *Chromolaena* spp. (2 species; Pullaiah, 1976, 1982; Maheswari Devi and Pullaiah, 1977a; Coleman, 1989; Bertasso-Borges and Coleman, 2005), *Erigeron* spp. (2 species; Tahara, 1915; Holmgren, 1919; McDonald, 1927), *Eupatorium* spp. (3 species; Pullaiah, 1976; Sullivan, 1976; Rozenblum et al., 1988); *Leontopodium* spp. (1 species; Sokolowska-Kulzyca, 1959; Maugini, 1962); *Parthenium* spp. (2 species; Esau, 1944, 1946; Gerstel et al., 1953; Cruz et al., 2017). In the subfamily Cichorioideae, it is observed exclusively in the genus *Hieracium* spp. (17 species and 3 subspecies; Gentscheff, 1937; Gentscheff and Gustafsson, 1940; Skawinska, 1962; Mráz et al., 2009; Hand et al., 2015).

5.3.2.2.2 The Taraxacum Type

The Taraxacum type of female gametophyte development is the most frequent, being recorded in 41 species and subspecies of daisies (Table 5.1). Within the subfamily Asteroideae it has been found in *Blumea* spp. (1 species; Chennaveeraiah and Patil, 1971). The remaining occurrences are among species of three genera within the subfamily Cichorioideae, including *Chondrilla* spp. (5 species; Rosenberg, 1912; Poddubnaja-Arnoldi, 1933, 1964; Bergman, 1944, 1950; Battaglia, 1949; Musiał and Kościńska-Pająk, 2019), *Hieracium* spp. (10 species; Rosenberg, 1927; Gentscheff, 1937; Gentscheff and Gustafsson, 1940; Bergman, 1935a, 1941; Hand et al., 2015), and *Taraxacum* spp. (25 species; Raunkiaer, 1903; Juel, 1904; Ikeno, 1910; Schkorbatow, 1912; Osawa, 1913; Sears, 1917; Stork, 1920; Rosenberg, 1930; Poddubnaja-Arnoldi and Dianova, 1964; Battaglia, 1948; Cooper and Brink, 1949; Małecka, 1967; Richards, 1970; Kaul and Singh, 1980; Menken and Morita, 1989; Vašut et al., 2005; Záveský et al., 2007; Mártonfiová et al., 2010; Musiał et al., 2013; Vašut and Majeský, 2015; Majeský et al., 2015; Janas et al., 2016).

5.3.2.2.3 The Ixeris Type

This is the less frequent type of female gametophyte development observed among daisies. It is found in 12 species and varieties of the subfamily Asteroideae, in the genera *Erigeron* spp. (2 species; Tahara, 1915; Holmgren, 1919; Carano, 1921, 1924; McDonald, 1927; Fagerlind, 1947; Battaglia, 1950; Beals, 1970; Mehra and Gill, 1972), *Rudbeckia* spp. (5 species; Battaglia, 1945, 1946a, 1946b, 1946c, 1947a, 1952, 1955; Solntzeva, 1973a, 1973b; Movsesyan and Kasparova, 1977a, 1977b; Kościńska-Pająk et al., 2014) and *Townsendia* spp. (5 species; Beaman, 1957), and in one species of the subfamily Cichorioideae, in the genus *Ixeridium* spp. (1 species; Okabe, 1932).

5.3.2.3 Variations in Female Gametophyte Ontogeny and Anatomy

Variations in the types of female gametophyte development have been recorded in different species, although they are less frequent than in grasses (Chapter 4). In species of different genera, the occurrence of meiotic and apomeiotic female gametophytes in ovules of the same or different individuals has been recorded directly by embryological observation or indirectly by FCSS or progeny tests. Yet, the simultaneous occurrence of Polygonum type and other types of diplosporous development in the same ovule is not possible.

Unlike grass species, in which development by monopolar gametogenesis produces unreduced four-nucleate female gametophytes, in daisies bipolar development of either meiotic or apomeiotic pathways during gametogenesis produces eight-nucleate female gametophytes of similar anatomy. Therefore, distinguishing between alternative developmental pathways requires a thorough evaluation of earlier developmental stages during germline differentiation and sporogenesis (see Chapter 2). In such cases, recognizing the occurrence of apospory versus diplospory is relatively easy, as aposporous development begins in the nucellar tissue, independent of the germline. However, evaluating the steps of meiosis to classify alternative pathways of diplospory is technically challenging. Meiosis is often elusive and detailed information about how meiosis proceeds in apomicts (but also in sexuals) is difficult to obtain. In different species of the genus *Hieracium*, analysis of an adequate number of ovules and developmental stages shows that the Antennaria type is usually the most frequent, with the concurrent development of the Taraxacum type in different ovules of the same material (e.g., Ostenfeld, 1906; Stebbins, 1932; Gustafsson, 1934a, 1934b; Battaglia, 1991). Thus, defining between types of diplosporous developments is not straightforward, specially between the Antennaria type and the Ixeris type, and requires critical evaluation of large numbers of ovules and other features of reproductive development in a species (see e.g. case studies in *Erigeron* spp. where alternative developments are likely to occur in the same or related individuals; Holmgren, 1919; Carano, 1921; Fagerlind, 1947; Battaglia, 1950).

A tendency to apospory (rather than its regular occurrence) is a common feature among species in which apomixis is not yet confirmed (Table 5.2). The simultaneous occurrence of aposporous and diplosporous female gametophytes has rarely been observed in daisy ovules. For example, in species of *Taraxacum* section *Palustria* Dahlstedt, aposporous-like female gametophyte developments were observed in otherwise fully diplosporous species (Małecka, 1973; van Baarlen et al., 2002). In *Parthenium argentatum* A. Gray and *P. incanum* Kunth, Powers and Rollins (1945) and Esau (1946) pointed out the spontaneous formation of Hieracium type gametophytes in ovules of species showing mainly Antennaria type gametophytes. Since these species are pseudogamous and aposporous gametophytes are located toward the chalaza, post-pollination analyses show a consistent lack of endosperm development in aposporous gametophytes but not in diplosporous gametophytes (Powers and Rollins, 1945; Esau, 1946).

Furthermore, in daisy species, the presence of multiple, disorganized, multinucleate and non-cellularized female gametophytes at anthesis is unusual (e.g., Esau, 1946; Małecka, 1973). Interestingly, female gametophytes with an abnormal number of nuclei and anatomy were observed in *Antennaria carpatica* (Wahlenb.) Bluff & Fingerh. (Bergman, 1951), a sterile species showing altered ovule development and the occurrence of aposporous (Hieracium type) and diplosporous female gametophytes of Taraxacum-like type (Table 5.2) instead of the common Antennaria type observed in all other apomictic species of the genus.

6 Apomictic Roses

6.1 INTRODUCTION

Gametophytic apomixis in Angiosperms was identified for the first time in the family Rosaceae Juss. Murbeck (1895) observed that *Alchemilla* species lacking pollen displayed a good reproductive ability and a remarkable morphological constancy. Then, he was probably the first botanist who designed an experiment to test *a priori* his hypothesis about the vegetative development of the embryo among *Alchemilla* species (Murbeck, 1897). He initiated several years earlier (in 1894) a study at the Botanical Garden of Lund University by cultivating four morphologically different and pollen-sterile *Alchemilla* species, i.e., *A. alpina* L., *A. sericata* Rchb., *A. glabra* (DC) Poir. [corresponding to *A. glabra* Poir. =*A. fissa* Günther & Schummel (De Candolle, 1893, p. 491)], and *A. pubescens* Lam. [=*A. hybrida* (L.) L.], together with the pollen-producer *A. speciosa* Buser. Murbeck (1897, p. 276) found out all species did produce normal amounts of seeds, and the offspring of each species grown in the garden and in open land were of maternal origin. In the coming years, Murbeck (1901, 1902) carried out cytological evaluations and provided the main embryological details about the development of unreduced female gametophytes and apomictic seeds in several *Alchemilla* species, including *A. arvensis* (L.) Scop. [≡*Aphanes arvensis* L.].

Beside a few studies carried out on different species and genera of roses (Strasburger, 1905; Böös, 1917, 1920, 1924; Müntzing, 1928, 1931; Popoff, 1935; Shimatomai, 1935), it was not until the 1940s that the interest on embryology and apomixis research in the Rosaceae grew. In cases like genus *Rubus* spp., apomixis was first documented through extensive crossing experiments (Lidforss, 1905, 1907a, 1907b) and later further corroborated on the basis of cytological and embryological studies (e.g., Crane, 1940; Gustafsson, 1943; Christen, 1950; Pratt and Einset, 1955; Haskell, 1960; Czapik, 1981a, 1981b). Apomixis in this family is based on the occurrence of aposporous (Hieracium type) and diplosporous (Antennaria type) pathways described for the Asteraceae family (see Chapter 5 and Table 6.1). Most roses display a remarkable embryological feature, predominantly visible among polyploids and apomicts, i.e., the presence of a multicellular female archesporium and the consequent development of multiple archespores in the ovules. This particularity has different consequences; the most relevant regards the categorization of apomictic plants (see details below).

Since the first observations in relation to apomixis in roses, many different studies on the reproductive biology have increased the list of apomictic taxa up to more than 290 species, subspecies, varieties and cultivars (Table 6.1). The present chapter gathers current information on species-level apomixis in the Rosaceae family.

6.2 CONSIDERATIONS ABOUT THE OCCURRENCE AND TYPE OF APOMIXIS IN ROSES

All collected records are based on original reports and previous lists of apomictic species (Stebbins, 1941; Nygren, 1954a, 1954b, 1967; Campbell et al., 1991; Asker and Jerling, 1992; Naumova, 1993; Carman, 1995; Dickinson et al., 2007) and regional floras (e.g., Czapik and Kościńska-Pająk, 2000; Richards, 2003; Gregor, 2004a, 2004b, 2013; Brožová et al., 2019). Several particularly old studies were accessed through the Biodiversity Heritage Library (BHL; https://www.biodiversitylibrary.org/) and the Academic Archive Online (DiVA, http://www.diva-portal.org/), or through different university repositories like the Lund University Publications (LUP; https://lup.lub.lu.se/search) and in the case of the Murbeck (1901) study by taking advantage of the eBooks on Demand service from the Uppsala University library (https://ub.uu.se/anvand-biblioteket/). Whenever possible, studies on

DOI: 10.1201/9781003088561-6

the occurrence of apomixis in different species were critically re-evaluated. As with the other families, species' taxonomy and nomenclature had been revised following the International Plant Names Index (IPNI; https://www.ipni.org/), Tropicos (https://www.tropicos.org/), The Plant List (http://www.theplantlist.org/), the Plants of the World Online (http://www.plantsoftheworldonline.org/), and the World Flora Online (WFO; http://www.worldfloraonline.org/). For some difficult cases, we used online resources and tools like the Integrated Taxonomic Information System (ITIS; https://www.itis.gov/), the Global Name Resolver (http://resolver.globalnames.org/) or the Taxonomic Name Resolution Service (TNRS; http://tnrs.iplantcollaborative.org/TNRSapp.html) (see Wagner, 2016 for a review of some of these tools).

We have included basic information about species names, ploidy of studied materials and the type of apomictic reproduction in Tables 6.1 and 6.2. As with the other families, a number of rose species had been subjected to changes in their nomenclature, as indicated in Table 6.1. Most cases involve synonyms, either as new species, subspecies and varieties (e.g., *Cotoneaster obscurus* var. *cornifolius* Rehder & E.H. Wilson is now placed under the name *Cotoneaster foveolatus* Rehder & E.H. Wilson, or *Rubus taeniarum* Lindeb. is now *Rubus infestus* Weihe) or placed under new genera [e.g., *Alchemilla arvensis* (L.) Scop. is now *Aphanes arvensis* L., or *Parageum montanum* (L.) Nakai & H. Hara ex H. Hara is now *Geum montanum* L.].

Several cases required specific considerations. For example, when author names were not published or the original work could not be accessed, the accepted names were considered only if ambiguities could be discarded [e.g., *Potentilla inclinata* Vill. (=*Potentilla canescens* Besser); *Potentilla incana* G. Gaertn., B.Mey. & Scherb. (=*Potentilla arenaria* Borkh.)]. Species cited as apomict but holding invalid names like *Potentilla fulgens* Wall. ex Hook were not placed in Table 6.1 or 6.2. The name *Amelanchier alnifolia* var. *pumila* (Nutt.) A. Nelson seems to be an invalid name following the original description of *Amelanchier canadensis* var. *pumila* Torr. & A. Gray [Fl. N. Amer. 1(3): 474 (1840)]. Thus, *Amelanchier alnifolia* var. *pumila* (Torr. & A. Gray) C.K. Schneid is considered in Table 6.1. Likewise, species known to reproduce apomictically but referred as inedit (e.g., *Rubus jari-cimrmanii* ined., *R. vatavensis* ined.; Šarhanová et al., 2012) were only included in our table of apomictic species if published [*R. jarae-cimrmanii* M. Lepší, P.Lepší, Trávn. & Žíla, *Preslia* 90(4): 405 (2018); *R. vatavensis* Žíla & Trávn., *Preslia* 90(4): 399 (2018)].

Rejected or unresolved names like *Cotoneaster froebelii* Vilm.; *Potentilla dissecta* Pursh, *Potentilla degenii* Siegfr. ex Th. Wolf, *Rubus* × *proteus*, *Rubus hartmannii* Sudre or *Rubus pseudopallidus* Gust., for which evidence of apomixis has been gathered (Shimatomai, 1935; Sax, 1954; Spies and du Plessis, 1985; Kraft and Nybom, 1995; Werlemark and Nybom, 2003) are not included in Table 6.1 or 6.2. Accepted hybrid names (like *Sorbus* × *avonensis* T.C.G. Rich, a putative interspecific hybrid, Ludwig et al., 2013; or ×*Sorbocotoneaster pozdnjakovii* Pojark., a putative intergeneric hybrid, Asbaganov et al., 2015) were considered as natural apomicts.

In recent taxonomic papers (Sennikov and Kurtto, 2017; Sennikov, 2018), a comprehensive taxonomic revision of *Sorbus* s.l. has included not only phylogenetic affinities but also hybridization history and reproductive modes, given former subgenera new generic names. Yet here we followed the traditional delimitation and provisionally refrained from using the new taxonomic treatment because no consensus exists, and by adopting the new circumscription we might split species aggregates. For instance, if *Sorbus teodori* Liljef. and *S. meinichii* (Lindeb. ex Hartm.) Hedl. are transferred to *Hedlundia teodori* (Liljef.) Sennikov & Kurtto (Sennikov and Kurtto, 2017), two newly described nothospecies within the *Soraria* subgenus (i.e., *Sorbus faohraei* Hedrén & J. Levin and *Sorbus atrata* Hedrén & J.Levin, Levin et al., 2019) will be separated.

In the Rosaceae, the taxonomic treatment of apomictic taxa is not less complex than in the Asteraceae (see e.g. Dickinson, 2018). The use of alternative species concepts is often accepted. For instance, to describe the complexity of *Cotoneaster integerrimus* s.l. in Europe, taxonomists use several narrow species concepts (Hrabětová-Uhrová, 1961, 1962; Baranec, 1992; Fryer and Hylmö, 2001) and a few broad species concepts (e.g., Dickore and Kasparek, 2010; Sennikov, 2010). Consequently, comprehensive floras using a narrow species concept can treat the same taxa as separate microspecies from different parts of Europe (Kurtto et al., 2013). Likewise, in highly diverse genera like *Rubus* spp., alternative taxonomic approaches are being used in both sides of the Atlantic (Sochor et al., 2018). While European taxonomists follow a Weberian concept (Weber, 1996) that considers a rather narrow species delimitation and treats morphologically stable genotypes with an extensive distribution as a species, North American taxonomists have rejected similar detailed treatments (Bailey, 1941–1945; Davis et al., 1967) and are now using a broader species concept that circumscribe different stable genotypes sharing morphological similarities into a species (Alice et al., 2015; Sochor et al., 2018).

The observed taxonomic complexity within genera containing apomictic species in the Rosaceae (like in other families) is associated with the occurrence of (usually a few) sexual cytotypes that hybridize and create facultative apomictic polyploids which in turn can hybridize forming intermediate types (see e.g. Dickinson and Phipps, 1986; Campbell and Dickinson, 1990; Campbell and Wright, 1996; Campbell et al., 1999; Krahulcová et al., 2013; Burgess et al., 2014; Sochor et al., 2015; Lepši et al., 2009), and is the cause, in many cases, of the lack of reliable infrageneric phylogenies.

Majeský et al. (2017) reviewed the taxonomic treatments of apomictic taxa in the Asteraceae and Rosaceae and provided five criteria to recognize an apomictic species. Despite the substantial burden of work imposed on taxonomists if following these criteria (discussed in Dickinson, 2018) it would be of help standardizing taxonomic treatments (both among genera and families) to describe species in plant groups where apomixis occurs. We do understand using the evolutionary species concept of Simpson (1961) that considers a species as a lineage evolving separately from others and holding its own unitary evolutionary role and tendencies is recommended from a theoretical viewpoint as it solves problems, e.g., associated with hybridization (Majeský et al., 2017). Yet, from a practical perspective, even when agamic lineages or even single clones may fulfil the criteria of the evolutionary species concept (Hörandl, 1998), delimitating evolutionary units might be technically demanding considering that most apomicts are facultative and lineages can maintain intermittent sex as revealed on phylogenetic networks (e.g., Majeský et al., 2012; Krak et al., 2013; Tyler and Jönsson, 2013; Hodač et al., 2014; Hamston et al., 2018). Here, we follow a procedure alike with the other families and agreed with the taxonomic treatments used by each specialist and the accepted species concepts. Microspecies and infraspecific delimitations are taken as valid taxonomic subdivisions of apomictic complexes.

Whenever possible, ploidy levels were included following data from the studied materials (see Tables 6.1 and 6.2). In some cases, the ploidy of a species was taken from records available elsewhere only when a single ploidy level was documented for the species (these references are not listed in Tables 6.1 and 6.2).

We tried to mention the first study reporting on apomixis, but there is a chance that we may have overlooked the original records in some cases. As with grasses and daisies, seed set analyses using emasculation/castration experiments on taxa were listed in Table 6.1 but would require embryological confirmation. Castration tests showing abundant fruit set by destyled flowers was considered as evidence of apomixis rather than parthenocarpy only when accompanied by proof of apomictic seed formation in related species [e.g., in triploids *Cotoneaster racemiflorus* (Desf.) K. Koch, *C. zabelii* C.K. Schneid., or in *C. tomentosus* (Aiton) Lindl., Sax, 1954] or by evidence of apomixis in later studies on the same species [e.g., in *Aronia prunifolia* (Marshall) Rehder, Hardin, 1973; Mahoney et al., 2019].

Embryological analyses in roses are challenged by the presence of a multicellular female archesporium. In most genera, during the development of the ovule, it is possible to distinguish one primary megaspore mother cell and several secondary megaspore mother cells. While in species from genera like *Alchemilla* spp., *Cotoneaster* spp., *Rubus* spp., or *Sorbus* spp. the (anatomical and cytological) differences between both cell types are – to a certain extent – noticeable (Murbeck, 1901; Böös, 1924; Liljefors, 1934; Christen, 1950; Sax, 1954), the recognition of the primary and secondary megaspore mother cells in species from other genera like *Potentilla* spp. or *Sanguisorba* spp. is not obvious (Rutishauser, 1943a, 1943b, 1943c; Håkansson, 1946; Nordborg, 1967).

This poses two constraints to the reproductive characterization of rose species developing through multicellular archesporia. The first constraint is technical-operational and relates both to the difficulty of obtaining good quality ovule preparations and to the operator's capacity to distinguish whether a growing female gametophyte arises from a primary or a secondary megaspore cell or from a cell of the nucellus. The second one is conceptual and relates to how to consider these secondary megaspore mother cells. Since female gametophytes arising from secondary mother cells do not enter meiosis and go from an early meiotic prophase directly into the first mitosis of gametogenesis and produce structurally similar mature female gametophytes irrespective of its origin, different authors have suggested to consider this as apospory (e.g., Fagerlind, 1944a, 1944b; Liljefors, 1953; Hjelmqvist, 1962). Yet, since these secondary cells might go throughout meiosis and develop reduced female gametophytes like sometimes does the primary megaspore mother cell (Böös, 1924; Liljefors, 1934), such development shall be considered as diplosporous (Gustafsson, 1946, p. 26; Christen, 1950; Pratt and Einset, 1955).

Consequently, in roses, the categorization of the apomictic development into aposporous or diplosporous depends – in a way – upon the experience of the authors and the critical assessment of materials and interpretation of results in each study. Here, we maintain the main conclusions extracted on each study and considered developments from the archesporium as diplosporous (Antennaria type) and those from the surrounding nucellar tissue as aposporous (Hieracium type) (Table 6.1).

The cytological and embryological complexity in the family can be exemplified by species of the genus *Rubus* spp. While in species like *R. apetalus* Poir. the diploid cytotype shows a normal sexual development (Polygonum type) and the polyploids show aposporous apomixis, in other species like *Rubus cuneifolius* Pursh or *R. pascuus* L.H. Bailey diploids display high proportions of degenerating female gametophytes and polyploids show varied proportions of facultative apomixis, including a complete lack of activation of the apomictic pathway in some genotypes and reduced fertility in *Rubus longepedicellatus* (Gust.) C.H. Stirt. (Spies and du Plessis, 1986). Yet, the interspecific hybrid *Rubus transvaliensis* × *R. longepedicellatus* presented only sexual female gametophytes (Spies and du Plessis, 1986).

Further evidence of such embryological complexity is illustrated by the studies by Dowrick (1961, 1966) who finds no evidence of apomixis in tetraploid *Rubus caesius* L. and suggest the occurrence of automixis or fusion of cleavage division nuclei. The author also cites the work by Christen (1950) indicating that he was unable to find either aposporic or diplosporic female gametophytes in *R. caesius* (Dowrick, 1966, p. 252). Nonetheless, Christen presents a detailed description of both diplosporous Antennaria type and aposporous Hieracium type developments of female gametophytes supported by several figures (Christen, 1950, pp. 157–166). Since other circumstantial evidence points towards the occurrence of functional apomixis in the species (e.g., Lidforss, 1914), we included this and other taxa with similar data in our Table 6.1.

In some cases, species were described as with dominating diplospory (e.g., *Cotoneaster bullatus* Bois, *Rubus caesius* L., *R. nessensis* Hall) or apospory (e.g., *Cotoneaster acutifolius* var. *villosulus* Rehder & E.H. Wilson, *Rubus grabowskii* Weihe ex Günther, Grab. & Wimm., *R. vestitus* Weihe & Nees) depending upon the observed proportions of female gametophytes originating from cells of the archesporium or cells surrounding the female archesporium (Christen, 1950; Hjelmqvist, 1962). Since both reproductive pathways are expected to be functional, we categorized these cases as being diplosporous and aposporous. Hence, unlike grasses with monosporic development where multiple gametophytes can be found only in aposporic species, in roses the observation of multiple gametophytes cannot be used as an indication of apospory unless it is accompanied by evidence of the appearance of initial of apospory cells (Muniyamma and Phipps, 1979; Dickinson and Phipps, 1986; Dickinson et al., 1996).

The frequent observation of abnormal pollen development in different roses was not considered as evidence for apomixis alone but when accompanied by embryological evaluation of the female gametophyte development in at least one of the species, castration experiments, flow cytometry analysis or progeny tests (e.g., in *Alchemilla* spp., Murbeck, 1901; Strasburger, 1905; in *Sorbus* spp., Lepši et al., 2009). Likewise, species cited as apomictic but without any other evidence that ploidy or hybrid nature (e.g., *Potentilla uniflora* Ledeb. and *P. hookeriana* Lehm., Eriksen, 1996; *Sorbus dolomiticola*

TABLE 6.1
Table of Apomictic Species (A) within The Family Rosaceae Juss., Sorted Alphabetically According to Subfamilies

Species	Ploidy	Type of Apomixis[a]	Reference
AmygdaloideaeArnott			
Sorbus admonitor M.C.F. Proctor	*4x*	A	Hamston et al. (2018)
Sorbus albensis M. Lepší, Boublík, P. Lepší & Vít	*3x*	A; Ps	Lepší et al. (2009)
Sorbus alnifrons Kovanda	*3x*	A; Ps	Lepší et al. (2009)
Sorbus anglica Hedl.	*4x*	A	Robertson et al. (2004)
Sorbus aria (L.) Crantz	*3x, 4x*	A; Ap (Hi); Di (An); Ps	Liljefors (1953, 1955), Jankun (1993), Hajrudinović et al. (2015)
Sorbus arranensis Hedl.	*3x*	A; Ap (Hi); Di (An); Ps	Liljefors (1953), Robertson et al. (2004)
Sorbus atrata Hedrén & J. Levin	*3x*	A; Ps	Levin et al. (2019)
Sorbus atrimontis Bernátová & Májovský	*4x*	A	Uhrinová et al. (2017)
Sorbus × *avonensis* T.C.G.Rich	*3x*	A; Ps	Ludwig et al. (2013)
Sorbus austriaca Hedl.	*4x*	A; Ps	Hajrudinović et al. (2015)
Sorbus barrandienica P. Vít, M. Lepší et P. Lepší	*3x*	A; Ps	Vít et al. (2012), Lepší et al. (2009)
Sorbus bohemica Kovanda	*3x*	A; Ap (Hi); Di (An); Ps	Jankun and Kovanda (1987)
Sorbus borbasii Jáv.	*4x*	A; Ps	Hajrudinović et al. (2015)
Sorbus bosniaca Hajrud., Frajman, Schönsw. & Bogunić	*4x*	A; Ps	Hajrudinović et al. (2015)
Sorbus bristoliensis Wilmott	*3x*	A; Ps	Robertson et al. (2010), Ludwig et al. (2013)
Sorbus caeruleomontana Bernátová & Májovský	*4x*	A	Uhrinová et al. (2017)
Sorbus cambrensis M. Proctor	*5x*	A; Ps	Lepší et al. (2009)
Sorbus cashmiriana Hedl.	*4x*	A; Ps	McAllister and Gillham (1980, in Campbell et al., 1991)
Sorbus chamaemespilus (L.) Crantz	*4x*	A; N; Ap (Hi); Ps	Liljefors (1934, 1953), Jankun (1993), Uhrinová et al. (2017)
Sorbus collina M.Lepší, P.Lepší & N.Mey.	*4x*	A; Ps	Lepší et al. (2009)
Sorbus croceocarpa P.D. Sell	*4x*	A	Nelson-Jones et al. (2002), Robertson et al. (2010)
Sorbus cucullifera M. Lepší & P. Lepší	*3x*	A; Ps	Lepší et al. (2009)
Sorbus dacica Borbás	*4x*	A; Ps	Németh et al. (2020)
Sorbus danubialis (Javorka) Kárpáti	*4x*	A; Ps	Lepší et al. (2009)
Sorbus decipiens (Bechst.) G. Kirchn.	*4x*	A	Robertson et al. (2010)
Sorbus devoniensis E.F.Warb.	*4x*	A	Hamston et al. (2018)
Sorbus eminens E.F.Warb.	*4x*	A; Ps	Robertson et al. (2010), Ludwig et al. (2013)
Sorbus eximia Kovanda	*3x*	A; Ap; Di; Ps	Lepší et al. (2009)

(Continued)

TABLE 6.1 (Continued)
Table of Apomictic Species (A) within The Family Rosaceae Juss., Sorted Alphabetically According to Subfamilies

Species	Ploidy	Type of Apomixis[a]	Reference
Sorbus faohraei Hedrén & J. Levin	*3x*	A; Ps	Levin et al. (2019)
Sorbus forrestii McAll. & Gillham	*4x*	A; Ps	McAllister (1986, in Campbell et al., 1991)
Sorbus gemella Kovanda	*3x*	A; Ps	Lepší et al. (2009)
Sorbus graeca (Spach) Lodd. ex S. Schauer [=*Sorbus obtusifolia* (DC.) Hedl.]	*4x*	A; Ap (Hi); Di (An)	Liljefors (1953)
Sorbus haljamovae Bernátová & Májovský	*4x*	A	Uhrinová et al. (2017)
Sorbus hibernica E.F. Warb.	*3x*	A	Cowan et al. (2008)
Sorbus × houstoniae T.C.G. Rich	*4x*	A; Ps	Ludwig et al. (2013)
Sorbus hybrida L.	*4x*	A; Ap (Hi); Di (An); Ps	Liljefors (1953), Lepší et al. (2009)
Sorbus intermedia (Ehrh.) Pers.	*4x*	A; N; Ap (Hi); Di (An); Ps	Liljefors (1953), Jankun (1994), Ludwig et al. (2013), Lepší et al. (2009)
Sorbus latifolia (Lam.) Pers.	*4x*	A; Ps	Ludwig et al. (2013), Lepší et al. (2009)
Sorbus legrei Cornier	*3x*	A	Cornier (2009)
Sorbus leighensis T.C.G. Rich	*3x*	A; Ps	Robertson et al. (2010), Ludwig et al. (2013)
Sorbus leptophylla E.F. Warb.	*4x*	A	Proctor and Groenhof (1992)
Sorbus leyana Wilmott	*3x*	A	Proctor and Groenhof (1992)
Sorbus margaretae M.C.F. Proctor	*4x*	A	Hamston et al. (2018)
Sorbus margittaiana (Jáv.) Kárpáti	*4x*	A; Ap (Hi)	McAllister (1986), Uhrinová et al. (2017)
Sorbus meinichii (Lindeb. ex Hartm.) Hedl.	*3x, 4x*	A; Ap (Hi); Di (An)	Liljefors (1953)
Sorbus milensis M. Lepší, Boublík, P. Lepší & Vít	*3x*	A; Ps	Lepší et al. (2009)
Sorbus minima (Ley) Hedl.	*3x*	A	Proctor and Groenhof (1992)
Sorbus montisalpae Bernátová & Májovský	*4x*	A	Uhrinová et al. (2017)
Sorbus moravica M. Lepší & P. Lepší	*3x*	A; Ps	Lepší et al. (2009)
Sorbus mougeotii Soy.-Will. & Godr.	*4x*	A; Ap (Hi); Di (An); Ps	Liljefors (1953), Lepší et al. (2009)
Sorbus omissa Velebil	*3x*	A; Ps	Lepší et al. (2009)
Sorbus pauca M. Lepší & P. Lepší	*4x*	A; Ps	Lepší et al. (2013)
Sorbus paxiana Jáv.	*4x*	A; Ps	Németh et al. (2020)
Sorbus pekarovae Májovský & Bernátová	*4x*	A	Uhrinová et al. (2017)
Sorbus pontis-satanae M. Lepší & P. Lepší	*3x*	A; Ps	Lepší et al. (2009)
Sorbus porrigentiformis E.F. Warb.	*4x*	A; Ps	Robertson et al. (2010), Ludwig et al. (2013), Hamston et al. (2018)

TABLE 6.1 (Continued)

Species	Ploidy	Type of Apomixis[a]	Reference
Sorbus portae-bohemicae M. Lepší, P. Lepší, Vít & Boublík	*3x*	A; Ps	Lepší et al. (2009)
Sorbus pseudofennica E.F. Warb.	*4x*	A; Ap; Ps	Robertson et al. (2004)
Sorbus pseudomeinichii Ashley Robertson	*3x*	A	Robertson (2006)
Sorbus ratisbonensis N. Mey.	*3x*	A; Ps	Faltermeier et al. (2016)
Sorbus rhodanthera Kovanda	*3x*	A; Ps	Lepší et al. (2009)
Sorbus rupicola (Syme) Hedl. [=*Sorbus salicifolia* (Hartm.) Hedl.]	*4x*	A; Ap (Hi); Di (An); Ps	Liljefors (1953), Pellicer et al. (2012), Hamston et al. (2018)
Sorbus stenophylla M. Proctor	*5x*	A; Ps	Lepší et al. (2009)
Sorbus subcuneata Wilmott	*3x*	A; Ps	Hamston et al. (2017)
Sorbus subsimilis Hedl.	*4x*	A; Ap (Hi); Di (An); Ps	Liljefors (1953)
Sorbus sudetica (Tausch) Bluff, Ness & Schauer	*4x*	A; Ap (Hi); Ps	Jankun and Kovanda (1986), Uhrinová et al. (2017)
Sorbus takhtajanii Gabrieljan	*4x*	A; Ap	Campbell et al. (1991)
Sorbus tamamschjanae Gabrieljan	*3x*	A; Ap	Campbell et al. (1991)
Sorbus thayensis M. Lepší & P. Lepší	*4x*	A; Ps	Lepší et al. (2009)
Sorbus teodori Liljef.	*3x*	A; Ap (Hi); Di (An); Ps	Liljefors (1953)
Sorbus vexans E.F.Warb.	*4x*	A	Hamston et al. (2018)
Sorbus whiteana T.C.G. Rich & L. Houston	*3x*	A; Ps	Robertson et al. (2010), Ludwig et al. (2013)
Sorbus wilmottiana E.F. Warb.	*3x*	A; Ps	Robertson et al. (2010), Ludwig et al. (2013)
Sorbus zuzanae Májovský & Bernátová	*3x, 4x*	A	Uhrinová et al. (2017)
Maloideae C.Weber			
Amelanchier alnifolia (Nutt.) Nutt. ex M. Roem. [sub *Amelanchier alnifolia* (Nutt.) Nutt. var. *alnifolia*]	*4x*	A; Ps	Burgess et al. (2014)
Amelanchier alnifolia var. *pumila* (Torr. & A. Gray) C.K. Schneid.	*4x*	A; Ps	Burgess et al. (2014)
Amelanchier arborea (F. Michx.) Fernald	*3x, 4x*	A; Ps	Burgess et al. (2014)
Amelanchier bartramiana (Tausch) M.Roem.	*3x, 4x*	A; Ps	Burgess et al. (2014)
Amelanchier canadensis (L.) Medik.	*4x*	A; Ap (Hi); Ps	Campbell et al. (1987), Burgess et al. (2014)
Amelanchier canadensis (L.) Medik. var. *canadensis*	*4x*	A; Ps	Burgess et al. (2014)
Amelanchier erecta Blanch.	*4x*	A; Ap (Hi); Ps	Campbell et al. (1997)
Amelanchier gaspensis (Wieg.) Fern. & Weath.	*4x*	A; Ps	Burgess et al. (2014)
Amelanchier laevis Wiegand	*4x*	A; Ap (Hi); Ps	Campbell et al. (1985), Weber and Campbell (1989), Burgess et al. (2014)

(Continued)

TABLE 6.1 (Continued)
Table of Apomictic Species (A) within The Family Rosaceae Juss., Sorted Alphabetically According to Subfamilies

Species	Ploidy	Type of Apomixis[a]	Reference
Amelanchier nantucketensis E.P. Bicknell [=*Amelanchier stolonifera* f. *micropetala* (B.L. Rob.) Rehder]	*4x*	A; Ap (Hi); Ps	Campbell et al. (1987), Burgess et al. (2014)
Amelanchier × *neglecta* Eggl. ex G.N. Jones	*4x*	A; Ap (Hi); Ps	Weber and Campbell (1989)
Amelanchier sanguinea (Pursh) DC.	*3x, 4x*	A; Ps	Burgess et al. (2014)
Amelanchier spicata (Lam.) K. Koch	*3x, 4x*	A; Ps	Burgess et al. (2014)
Amelanchier stolonifera Wiegand	*4x*	A; Ap (Hi); Ps	Campbell et al. (1987)
Amelanchier utahensis Koehne	*3x, 4x*	A; Ps	Burgess et al. (2014)
Cotoneaster acutifolius Turcz.	*3x, 4x*	A; Ps	Sax (1954)
Cotoneaster acutifolius var. *villosulus* Rehder & E.H. Wilson	*4x*	A; Ap (Hi); Di (An); Ps	Sax (1954), Hjelmqvist (1962)
Cotoneaster adpressus Bois	*3x*	A; Ps	Sax (1954)
Cotoneaster adpressus var. *praecox* (M. Vilm. ex Bois & Berthault) Bois & Berthault	*3x*	A; Ps	Sax (1954)
Cotoneaster ambiguus Rehder & E.H. Wilson	*3x*	A; Ps	Sax (1954)
Cotoneaster apiculatus Rehder & E.H. Wilson	*3x*	A; Ps	Sax (1954)
Cotoneaster bullatus Bois	*3x*	A; Ap (Hi); Di (An); Ps	Sax (1954), Hjelmqvist (1962)
Cotoneaster bullatus var. *floribundus* (Stapf) L.T. Lu & Brach [=*Cotoneaster bullatus* f. *floribundus* (Stapf) Rehder & E.H. Wilson]	*3x*	A; Ps	Sax (1954)
Cotoneaster dielsianus E. Pritz.	*3x*	A; Ps	Sax (1954)
Cotoneaster divaricatus Rehder & E.H. Wilson	*3x*	A; Ps	Sax (1954)
Cotoneaster horizontalis Decne.	*3x*	A; Ps	Sax (1954)
Cotoneaster integerrimus Medik. (=*Cotoneaster canescens* Vestergr. ex B. Hylmö; =*Cotoneaster scandinavicus* B. Hylmö; =*Cotoneaster juranus* Gand.)	*4x*	A; Ps	Sax (1954), Bartish et al. (2001), Nybom and Bartish (2007), Mahmutović et al. (2015), Macková et al. (2020), Bogunić et al. (2021)
Cotoneaster ellipticus (Lindl.) Loudon (=*Cotoneaster lindleyi* Steud.)	*3x*	A; Ps	Sax (1954)
Cotoneaster moupinensis Franch.	*3x*	A; Ps	Sax (1954)
Cotoneaster multiflorus Bunge	*3x, 4x*	A; Ps	Sax (1954)
Cotoneaster granatensis Boiss. [=*Cotoneaster multiflorus* var. *granatensis* (Boiss.) Wenz.]	*3x*	A	Sax (1954)

TABLE 6.1 (Continued)

Species	Ploidy	Type of Apomixis[a]	Reference
Cotoneaster niger (Wahlb.) Fr.	*4x*	A	Ploompuu (1999, in Nybom and Bartish, 2007)
Cotoneaster nitens Rehder & E.H. Wilson	*3x*	A; Ap (Hi); Di (An); Ps	Sax (1954), Hjelmqvist (1962), Flinck and Hylmö (1962)
Cotoneaster obscurus Rehder & E.H. Wilson	*3x*	A; Ap (Hi); Di (An); Ps	Sax (1954), Hjelmqvist (1962)
Cotoneaster foveolatus Rehder & E.H. Wilson (=*Cotoneaster obscurus* var. *cornifolius* Rehder & E.H. Wilson)	*4x*	A	Sax (1954)
Cotoneaster racemiflorus (Desf.) K. Koch (=*Cotoneaster racemiflorus* var. *desfontainesii* Zabel)	*3x*	A; Au	Sax (1954)
Cotoneaster soongoricus (Regel & Herder) Popov [=*Cotoneaster racemiflorus* var. *soongoricus* (Regel) C.K. Schneid.]	*3x, 4x*	A; Ap (Hi); Di (An); Ps	Sax (1954), Hjelmqvist (1962)
Cotoneaster silvestrii Pamp. (=*Cotoneaster racemiflorus* var. *veitchii* Rehder & E.H. Wilson)	*3x*	A	Sax (1954)
Cotoneaster roseus Edgew.	*3x*	A; Ap (Hi); Di (An); Ps	Sax (1954), Hjelmqvist (1962)
Cotoneaster simonsii Baker	*3x*	A; Ps	Sax (1954)
Cotoneaster tenuipes Rehder & E.H. Wilson	*3x*	A; Ps	Sax (1954)
Cotoneaster tomentosus (Aiton) Lindl. [=*Cotoneaster nebrodensis* (Guss.) K.Koch]	*3x, 4x, 5x*	A; Ps; Au	Sax (1954), Macková et al. (2020)
Cotoneaster wardii W.W.Sm.	*3x*	A; Ps	Sax (1954)
Cotoneaster zabelii C.K.Schneid.	*3x*	A; Au	Sax (1954)
Crataegus calpodendron (Ehrh.) Medik.		A; Ap (Hi); Ps	Muniyamma and Phipps (1984b)
Crataegus chrysocarpa Ashe	*4x*	A; Ap (Hi); Ps	Muniyamma and Phipps (1984b); Smith and Phipps (1988)
Crataegus crus-galli L.	*3x, 4x*	A; Ap (Hi); Ps	Talent and Dickinson (2007a, 2007b, 2007c)
Crataegus crus-galli var. *pyracanthifolia* Aiton	*4x*	A; Ps	Talent and Dickinson (2007a, 2007b, 2007c)
Crataegus dissona Sarg.	*3x*	A; Di (Ta); Ps	Muniyamma and Phipps (1984a)
Crataegus dodgei Ashe [=*Crataegus flavida* Sarg.; *Crataegus dodgei* var. *flavida* (Sarg.) P.G.Sm. & J.B. Phipps]	*3x, 4x*	A; Ap (Hi); Ps	Muniyamma and Phipps (1984b), Smith and Phipps (1988)
Crataegus douglasii Lindl.	*3x, 4x*	A; Ap (Hi); Ps	Campbell et al. (1991), Dickinson et al. (1996), Lo et al. (2013)
Crataegus fontanesiana (Spach) Steud.	*3x, 4x*	A; Ap (Hi); Ps	Dickinson and Phipps (1986)
Crataegus grandis Ashe	*3x*	A; Ap (Hi); Ps	Dickinson and Phipps (1986)

(*Continued*)

TABLE 6.1 (Continued)
Table of Apomictic Species (A) within The Family Rosaceae Juss., Sorted Alphabetically According to Subfamilies

Species	Ploidy	Type of Apomixis[a]	Reference
Crataegus laevigata (Poir.) DC. (=*Crataegus palmstruchii* Lindm.)	*3x*	A; Ap	Ptak (1989, in Czapik and Kościńska-Pająk, 2000)
Crataegus lumaria Ashe	*3x, 4x*	A; Ap (Hi); Ps	Smith and Phipps (1988)
Crataegus macracantha Lodd. ex Loudon		A; Ap (Hi); Ps	Dickinson and Phipps (1986), Talent and Dickinson (2007a, 2007b, 2007c)
Crataegus macrosperma Ashe	*4x*	A; Ap (Hi); Ps	Muniyamma and Phipps (1984b)
Crataegus persimilis Sarg. [=*Crataegus prunifolia* (Poir.) Pers.]	*4x*	A; Ap (Hi); Ps	Dickinson and Phipps (1986)
Crataegus pruinosa (H.L. Wendl.) K. Koch	*3x*	A; Ap (Hi); Ps	Muniyamma and Phipps (1979), Muniyamma and Phipps (1984b)
Crataegus rhipidophylla var. *lindmanii* (Hrabětová) K.I.Chr. (=*Crataegus lindmanii* Hrabětová)	*3x*	A; Ap	Ptak (1989, in Czapik and Kościńska-Pająk, 2000)
Crataegus schuettei Ashe var. *basilica* (Beadle) J.B. Phipps	*3x*	A; Ap (Hi); Ps	Muniyamma and Phipps (1984b)
Crataegus schuettei Ashe var. *cuneata* Kruschke		A; Ap (Hi); Ps	Muniyamma and Phipps (1984b)
Crataegus submollis Sarg.	*4x*	A; Ps	Talent and Dickinson (2007a, 2007b, 2007c
Crataegus subsphaerica Gand.	*3x, 4x*	A; Ps	Vašková and Kolarčik (2019)
Crataegus succulenta Schrad. ex Link (sub *Crataegus succulenta* Link)	*3x*	A; Ap (Hi); Ps	Muniyamma and Phipps (1984b)
Crataegus suksdorfii (Sarg.) Kruschke (=*Crataegus douglasii* var. *suksdorfii* Sarg.)	*3x, 4x*	A; Ap (Hi); Ps	Campbell et al. (1991), Dickinson et al. (1996), Lo et al. (2010, 2013)
Malus hupehensis (Pamp.) Rehder	*3x*	A; Ap (Hi); Ps	Dermen (1936), Campbell (1960), Campbell and Wilson (1962), Sampson (1969), Olien (1987), Zhou et al. (1995), Ur-Rahman (1997), Ranney et al. (2004)
Malus coronaria (L.) Mill.	*4x*	A; Ps	Brown (1975), Kron and Husband (2009)
Malus lancifolia Rehder	*3x, 4x*	A	Brown (1975)
Malus platycarpa Rehder	*3x, 4x*	A	Brown (1975)
Malus rockii Rehder	*4x*	A	Sax (1959)
Malus sargentii Rehder	*3x, 4x*	A; Ap (Hi); Ps	Sax (1959), Schmidt (1964, 1977), Sampson (1969), Li (2001), Ranney et al. (2004)
Malus sieboldii (Regel) Rehder	*3x, 4x*	A; Ap (Hi); Ps	Olden (1953), Hjelmqvist (1957); Schmidt (1964), Sampson (1969), Li (2001), Bisognin et al. (2009)
Malus sikkimensis (Wenz.) Koehne ex C.K.Schneid.	*3x*	A	Sax (1949, 1959), Sampson (1969)

TABLE 6.1 (Continued)

Species	Ploidy	Type of Apomixis[a]	Reference
Malus toringoides (Rehder) Hughes	*3x*	A; Ap (Hi); Ps	Sax (1949, 1959), Campbell and Wilson (1962), Liao and Li (1997), Ur-Rahman (1997)
Malus xiaojinensis M.H. Cheng & N.G. Jiang	*4x*	A; Ap; Ps	Cheng et al. (1984), Yang and Zhou (2009), Wang et al. (2012)
× *Sorbocotoneaster pozdnjakovii* Pojark.	*3x, 4x*	A; Ps	Asbaganov et al. (2015)
Rosoideae Juss. ex Arn.[§]			
Alchemilla alpina L.		A; Ap (Hi); Di (An); Au	Murbeck (1897, 1901)
Alchemilla acutiloba Opiz (=*Alchemilla acutangula* Buser)		A; N; Ap (Hi); Di (An); Au	Murbeck (1901, 1902), Izmaiłow (1994a, 1994b)
Alchemilla babiogorensis Pawł	*131–141, 145–156*	A; N; Ap; Di	Izmaiłow (1986, in Czapik and Kościńska-Pająk, 2000)
Alchemilla eugenii Pawł	*162–170*	A; N; Ap; Di	Izmaiłow (1986, in Czapik and Kościńska-Pająk, 2000)
Alchemilla fallax Buser		A; Au	Strasburger (1905)
Alchemilla flabellata Buser	*101–109*	A; N; Ap; Di	Izmaiłow (1994a, in Czapik and Kościńska-Pająk, 2000)
Alchemilla filicaulis Buser	*106*	A; Ap (Hi)	Glazunova (1987)
Alchemilla fissa Günther & Schummel [sub *Alchemilla glabra* (DC) Poir. =*Alchemilla glabra* Poir.; = *Alchemilla pyrenaica* Dufour]	*114–120, 145–156*	A; N; Ap; Di	Murbeck (1897), Izmaiłow (1986, in Czapik and Kościńska-Pająk, 2000)
Alchemilla fissimima Buser		A	Strasburger (1905)
Alchemilla gemmia Buser		A	Strasburger (1905)
Alchemilla giewontica Pawł	*101–109*	A; N; Ap; Di	Izmaiłow (1986, in Czapik and Kościńska-Pająk, 2000)
Alchemilla glabra Neygenf. (=*Alchemilla alpestris* F.W. Schmidt)	*106*	A; Ap (Hi); Di; Ps; Au	Murbeck (1901), Turesson (1943, 1956), Lundh-Almestrand (1958), Baldjiev and Robeva-Davidova (2008)
Alchemilla glaucescens Wallr.	*106*	A; Ps	Turesson (1943, 1956), Izmaiłow (1994a)
Alchemilla glomerulans Buser	*106*	A	Turesson (1943, 1956), Lundh-Almestrand (1958)
Alchemilla gorcensis Pawł	*98–105*	A; N; Ap; Di	Izmaiłow (1986, in Czapik and Kościńska-Pająk, 2000)
Alchemilla hybrida (L.) L. (=*Alchemilla minor* Huds.; =*Alchemilla pubescens* Lam.)		A; Ap (Hi); Di (An); Au	Murbeck (1897, 1901)
Alchemilla incisa Buser	*102–108*	A; N; Ap; Di	Izmaiłow (1986, in Czapik and Kościńska-Pająk, 2000)
Alchemilla jasiewiczii Pawł	*160–165*	A; N; Ap; Di	Izmaiłow (1986, in Czapik and Kościńska-Pająk, 2000)
Alchemilla leptoclada Buser		A; Au	Strasburger (1905)

(Continued)

TABLE 6.1 (Continued)
Table of Apomictic Species (A) within The Family Rosaceae Juss., Sorted Alphabetically According to Subfamilies

Species	Ploidy	Type of Apomixis[a]	Reference
Alchemilla monticola Opiz (=*Alchemilla pastoralis* Buser; =*Alchemilla gracilis* Opiz)	*101–106*	A; N; Ap (Hi); Di (An); Ps; Au	Murbeck (1901), Lundh-Almestrand (1958), Glazunova and Myatlev (1990), Izmaiłow (1994a)
Alchemilla oculimarina Pawł	*98–106, 150–161*	A; N; Ap; Di	Izmaiłow (1986, in Czapik and Kościńska-Pająk, 2000)
Alchemilla orbiculata Ruiz & Pav.	*106*	A; Ap (Hi); Di (An); Au	Böös (1917, 1920)
Alchemilla sabauda Buser		A	Strasburger (1905)
Alchemilla sericata Rchb.	*92–95*	A; Ap (Hi)	Murbeck (1897, 1901)
Alchemilla sericoneuroides Pawł	*149–160*	A; N; Ap; Di	Izmaiłow (1986, in Czapik and Kościńska-Pająk, 2000)
Alchemilla speciosa Buser	*66–76*	A; Au	Murbeck (1901), Strasburger (1905)
Alchemilla splendens H. Christ		A	Strasburger (1905)
Alchemilla subcrenata Buser	*106*	A; N	Murbeck (1901, 1902), Turesson (1943, 1956), Lundh-Almestrand (1958)
Alchemilla trullata Buser		A; Au	Strasburger (1905)
Alchemilla vulcanica Cham. & Schltdl.		A; Ap (Hi); Di (An); Au	Böös (1917, 1920)
Alchemilla vulgaris L.		A; Ap; Di	Strasburger (1905), Turesson (1956), Glazunova (1984)
Aphanes arvensis L. [≡*Alchemilla arvensis* (L.) Scop.; =*Alchemilla aphanes* Leers]	*6x*	A; Ap (Hi); Di (An); Ps	Murbeck (1901), Böös (1924), Gudjonsson (1941), Hjelmqvist (1959), Izmaiłow (1999)
Aronia arbutifolia (L.) Pers.	*4x*	A; Ps	Hovmalm et al. (2004), Mahoney et al. (2019)
Aronia melanocarpa (Michx.) Elliott	*4x*	A; Ps	Hovmalm et al. (2004), Mahoney et al. (2019)
Aronia mitschurinii A.K. Skvortsov & Maitul.	*4x*	A; Ps	Mahoney et al. (2019)
Aronia prunifolia (Marshall) Reheder	*3x, 4x*	A; Ps	Hardin (1973), Mahoney et al. (2019)
Chaenomeles japonica (Thunb.) Lindl. ex Spach	*2x*	A; N	Daskalova (2005)
Drymocallis geoides (M. Bieb.) Soják (≡*Potentilla geoides* M. Bieb.)	*2x*	A; N	Popoff (1935)
Geum montanum L. [≡*Parageum montanum* (L.) Nakai & H. Hara ex H. Hara]	*6x*	A	Krch (2002)
Lachemilla hirta (L. M. Perry) Rothm		A; Ps	Samaniego et al. (2018)
Lachemilla hispidula (L. M. Perry) Rothm	*3x, 4x*	A; Ps	Samaniego et al. (2018)
Potentilla adscharica Sommier & Levier	*6x*	A; Ap (Hi); Ps	Christoff and Papasova (1943), Dobeš et al. (2015)
Potentilla alpicola La Soie	*5x*	A	Paule et al. (2012)

TABLE 6.1 (Continued)

Species	Ploidy	Type of Apomixis[a]	Reference
Potentilla argentea L.	*4x, 5x, 6x, 8x*	A; Ap (Hi); Ps; Au	Müntzing (1928), Popoff (1935), Christoff and Papasova (1943), Rutishauser (1943a, 1943b, 1943c, 1945, 1946, 1947, 1948), Håkansson (1946), Asker (1967, 1970b, 1970c, 1971), Kausche (2010), Paule et al. (2011)
Potentilla argyrophylla Wall. ex Lehm.	*9x*	A; Di (An); Ps	Gentscheff (1938), Christoff and Papasova (1943)
Potentilla atrosanguinea G. Lodd. ex. D.Don		A	Dobeš et al. (2015)
Potentilla carniolica A. Kern.		A; Ps	Kausche (2010)
Potentilla chamissonis Hultén		A; Ps	Nyléhn et al. (2003)
Potentilla collina Wibel	*6x*	A; Ps	Müntzing (1928, 1958), Paule et al. (2012)
Potentilla crantzii (Crantz) Beck ex Fritsch (=*Potentilla levieri* Siegfr. & R. Keller ex R. Keller)	*6x, 9x*	A; Ap (Hi); Di (An); Ps	Müntzing (1928, 1958), Håkansson (1946), Skalińska and Czapik (1958), Czapik (1961), Smith (1963b); Hörandl et al. (2011), Dobeš et al. (2015)
Potentilla crantzii subsp. *gelida* (C.A. Mey.) Soják (=*Potentilla gelida* C.A. Mey.)	*6x*	A; Ps	Shimatomai (1935)
Potentilla chrysantha (Zoll. & Moritzi) Trevir.	*6x*	A; Ps	Shimatomai (1935), Christoff and Papasova (1943)
Potentilla kurdica Boiss. & Hohen. (sub *Potentilla curdica*)	*8x*	A	Popoff (1935)
Potentilla detommasii Ten.		A; Ps	Dobeš et al. (2015)
Potentilla gracilis Douglas ex Hook.	*11x*	A; Ps	Christoff and Papasova (1943)
Potentilla gracilis var. *fastigiata* (Nutt.) S. Watson (=*Potentilla nuttallii* Lehm.)		A; Ps	Kausche (2010)
Potentilla grandiflora L.		A; Ps	Kausche (2010), Dobeš et al. (2015)
Potentilla hippiana Lehm.		A; Ps	Kausche (2010), Dobeš et al. (2015)
Potentilla hirta L.	*4x*	A; Di (An); Ps	Popoff (1935), Shimatomai (1935), Christoff and Papasova (1943)
Potentilla incana G. Gaertn., B. Mey. & Scherb. (=*Potentilla arenaria* Borkh.)	*6x*	A; Ap (Hi); Ps	Czapik (1962), Rutishauser (1948), Asker (1986), Kausche (2010), Dobeš et al. (2013a)
Potentilla inclinata Vill. (=*Potentilla canescens* Besser)	*6x*	A; Ap (Hi); Ps	Christoff and Papasova (1943), Rutishauser (1943a, 1943b, 1943c, 1945a, 1945b, 1946, 1947, 1948)
Potentilla insularis Soják		A; Ps	Nyléhn et al. (2003)
Potentilla intermedia L.	*8x*	A; Ap (Hi); Ps	Shimatomai (1935), Asker (1970a)
Potentilla lineata Trevir. (=*Potentilla fulgens* Wall. ex Hook.)		A; Ps	Dobeš et al. (2015)
Potentilla multifida L.	*4x*	A; N; Ps	Popoff (1935), Kausche (2010), Dobeš et al. (2015)

(Continued)

TABLE 6.1 (Continued)
Table of Apomictic Species (A) within The Family Rosaceae Juss., Sorted Alphabetically According to Subfamilies

Species	Ploidy	Type of Apomixis[a]	Reference
Potentilla nepalensis Hook.	*4x*	A; Di (An); Ps	Gentscheff (1938)
Potentilla neumanniana Rchb.	*6x, 7x, 8x*	A; Ps	Rutishauser (1943a, 1943b), Smith (1963a)
Potentilla nevadensis Boiss.	*4x*	A; Ps	Shimatomai (1935)
Potentilla nivea L.	*6x, 7x*	A; Di; Ps	Eriksen and Fredrikson (2000)
Potentilla nivea L. ssp. *subquinata* (Lange) Hultén		A	Nyléhn et al. (2003)
Potentilla norvegica L.	*10x*	A; Ap (Hi); Ps	Asker (1970a), Dobeš et al. (2013)
Potentilla ornithopoda Tausch		A; Ps	Kausche (2010), Dobeš et al. (2015)
Potentilla pamirica T. Wolf		A; Ps	Brožová et al. (2019)
Potentilla pensylvanica L.	*4x*	A; Ps	Kausche (2010)
Potentilla praecox F.W. Schultz	*6x*	A; Ap (Hi); Ps	Rutishauser (1943a, 1943b, 1943c, 1945a, 1945b, 1946, 1947, 1948), Håkansson (1946)
Potentilla puberula Krašan (=*Potentilla pusilla* Host)	*5x–9x, 12x*	A; Ps	Paule et al. (2012), Dobeš et al. (2013a, 2013b, 2018), Nardi et al. (2018)
Potentilla pulcherrima Lehm.	*12x*	A; Ps	Kausche (2010), Dobeš et al. (2015)
Potentilla recta L.	*4x, 6x*	A; Ap (Hi); Ps	Popoff (1935), Christoff and Papasova (1943), Kausche (2010), Dobeš et al. (2015)
Potentilla reptans L.		A; N	Souèges (1935)
Potentilla rigona Th. Wolf	*7x*	A; Ps	Rossa (2013)
Potentilla semilaciniosa Borbás		A; Ps	Dobeš et al. (2015)
Potentilla sericea L.		A; Ps	Brožová et al. (2019)
Potentilla sojakii Dikshit & Panigrahi		A; Ps	Brožová et al. (2019)
Potentilla supina L.	*6x*	A; Ap	Dikshit and Panigrahi (1998)
Potentilla tabernaemontani Asch.	*5x, 6x, 7x, 8x, 9x, 12x*	A; Ap; Di; Ps	Müntzing (1928, 1958), Smith (1963a), Asker (1985)
Potentilla thuringiaca Bernh. ex Link	*9x*	A; Ps	Dobeš et al. (2013)
Potentilla verna L.	*6x*	A; Ap (Hi); Di (An); Ps	Rutishauser (1948)
Potentilla venusta Soják		A; Ps	Dobeš et al. (2015)
Potentilla villosa Pall. ex Pursh		A; Ps	Kausche (2010)
Potentilla virgata Lehm. (=*Potentilla dealbata* Bunge)	*5x*	A; Ps	Christoff and Papasova (1943)
Rubus alceifolius Poir.	*4x*	A; Ap; Ps	Amsellem et al. (2001, 2002)
Rubus alumnus L.H. Bailey (=*Rubus bellobatus* L.H. Bailey)	*3x, 4x*	A; Ap (Hi); Ps	Einset (1951), Pratt and Einset (1955)
Rubus apetalus Poir.	*4x*	A; Ap (Hi); Ps	Spies and du Plessis (1986)
Rubus austromoravicus Holub	*3x*	A; Ps	Šarhanová et al. (2012)
Rubus austroslovacus Trávn.	*3x*	A; Ps	Šarhanová et al. (2012)
Rubus bifrons Vest	*4x*	A; Ap (Hi); Di (An); Ps	Christen (1950), Šarhanová et al. (2012, 2017)

TABLE 6.1 (Continued)

Species	Ploidy	Type of Apomixis[a]	Reference
Rubus bohemiicola Holub & Palek ex Holub	*4x*	A; Ps	Šarhanová et al. (2012)
Rubus bregutiensis A.Kern. ex Focke	*4x*	A; Ap (Hi); Di (An); Ps	Christen (1950)
Rubus caesius L.	*4x*	A; Ap (Hi); Di (An); Ps	Lidforss (1914), Berger (1953), Christen (1950, 1952), Gerlach (1965), Dowrick (1961, 1966), Czapik (1981a)
Rubus canadensis L.	*3x*	A; Ap (Hi); Ps	Einset (1951), Pratt and Einset (1955)
Rubus crispomarginatus Holub	*3x*	A; Ps	Šarhanová et al. (2012)
Rubus cuneifolius Pursh	*4x*	A; Ap (Hi); Ps	Spies and du Plessis (1986)
Rubus elatior Focke ex Gremli	*3x*	A; Ps	Šarhanová et al. (2012)
Rubus epipsilos Focke	*4x*	A; Ps	Šarhanová et al. (2012)
Rubus flagellaris Willd. (=*Rubus geophilus* Blanch.)	*4x, 7x, 9x*	A; Ap (Hi); Ps	Einset (1951), Pratt and Einset (1955), Spies and du Plessis (1986)
Rubus flos-amygdalae Trávn. et Holub	*3x*	A; Ps	Šarhanová et al. (2012)
Rubus frondosus Bigelow (=*Rubus cardianus* L.H. Bailey)	*5x*	A; Ps	Einset (1951)
Rubus grabowskii Weihe ex Günther, Grab. & Wimm. (=*Rubus thyrsoideus* Wimm.; =*Rubus thyrsanthus* Focke)	*3x*	A; Ap (Hi); Di (An); Ps	Christen (1950), Šarhanová et al. (2012)
Rubus gracilis J. Presl & C. Presl (sub *Rubus gracilis* J. Presl & C. Presl var. *gracilis*)	*4x*	A; Ps	Kraft et al. (1996)
Rubus guttiferus Trávn. & Holub	*3x*	A; Ps	Šarhanová et al. (2012)
Rubus hartmanii Gand. (=*Rubus fuscus* Weihe?)	*4x*	A; Ps	Kraft and Nybom (1995)
Rubus henrici-egonis Holub	*3x*	A; Ps	Šarhanová et al. (2012)
Rubus idaeus L.	*3x*	A; Ap (Hi); Ps	Petrov (1939), Christen (1952), Pratt and Einset (1955), Pratt et al. (1958)
Rubus indusiatus Focke	*4x*	A; Ps	Šarhanová et al. (2012)
Rubus infestus Weihe (=*Rubus taeniarum* Lindeb.)	*4x*	A; Ps	Kraft and Nybom (1995), Werlemark and Nybom (2003)
Rubus insularis F. Aresch. [=*Rubus gracilis* subsp. *insularis* (F. Aresch.) H.E. Weber]	*4x*	A; Ps	Nybom (1985), Kraft et al. (1996)
Rubus jarae-cimrmanii M. Lepší, P. Lepší, Trávn. & Žíla	*4x*	A; Ps	Šarhanová et al. (2012)
Rubus laciniatus Willd.	*4x*	A; Ps	Werlemark and Nybom (2003)
Rubus meracus L.H. Bailey	*7x*	A; Ap (Hi); Ps	Einset (1951), Pratt and Einset (1955)
Rubus mercieri Genev.	*3x*	A; Ap (Hi); Di (An); Ps	Christen (1950)
Rubus montanus Lib. ex Lej.	*3x*	A; Ps	Šarhanová et al. (2012)
Rubus muenteri T.Marsson (=*Rubus scheutzii* Lindeb.)	*4x*	A; Ps	Kraft and Nybom (1995)
Rubus muhelicus Danner	*4x*	A; Ps	Šarhanová et al. (2012)
Rubus nessensis Hall (=*Rubus suberectus* And.)	*4x*	A; Ap (Hi); Di (An); Ps	Christen (1950), Nybom (1995), Kraft et al. (1996)

(Continued)

TABLE 6.1 (Continued)
Table of Apomictic Species (A) within The Family Rosaceae Juss., Sorted Alphabetically According to Subfamilies

Species	Ploidy	Type of Apomixis[a]	Reference
Rubus nitidiformis Sudre (=*Rubus nitidoides* W.C.R. Watson)	*4x*	A; Ap; Ps	Crane and Thomas (1939), Thomas (1940)
Rubus noveboracus L.H. Bailey	*5x*	A; Ps	Einset (1951)
Rubus parthenocissus Trávn. & Holub	*3x*	A; Ps	Šarhanová et al. (2012)
Rubus pascuus L.H. Bailey	*3x, 4x*	A; Ap (Hi); Ps	Spies and du Plessis (1986)
Rubus pedemontanus Pinkw. (=*Rubus bellardi* Weihe)	*4x, 5x*	A; Ap; Ps	Gustafsson (1943), Czapik (1987), Nybom (1988)
Rubus pensilvanicus Poir. (=*Rubus abactus* L.H. Bailey)	*3x, 5x*	A; Ap (Hi); Ps	Einset (1951), Pratt and Einset (1955)
Rubus pericrispatus Holub & Trávn.	*3x*	A; Ps	Šarhanová et al. (2012)
Rubus perperus H.E. Weber	*3x*	A; Ps	Šarhanová et al. (2012)
Rubus phyllostachys P.J. Müll.	*3x*	A; Ps	Šarhanová et al. (2012)
Rubus plicatus Weihe & Nees	*4x*	A; Ps	Kraft et al. (1996)
Rubus polyanthemus Lindeb.	*4x*	A; Ps	Werlemark and Nybom (2003)
Rubus praecox Bertol. (=*Rubus procerus* P.J. Müll. ex Boulay)	*4x*	A; Ap; Di; Ps	Berger (1953), Šarhanová et al. (2012)
Rubus radula Weihe	*4x*	A; Ps	Šarhanová et al. (2012)
Rubus recurvans Blanch. (=*Rubus wiegandii* L.H. Bailey)	*5x*	A; Ps	Einset (1951)
Rubus rosa L.H. Bailey	*3x, 4x*	A; Ps	Einset (1951)
Rubus rudis Weihe	*4x*	A; Ap; Di; Ps	Berger (1953)
Rubus saxatilis L.	*4x*	A; Ap; Di	Czapik (1981b)
Rubus scissus W.C.R. Watson	*4x*	A; Ps	Nybom (1995), Kraft et al. (1996)
Rubus septentrionalis W.C.R. Watson	*4x*	A; Ps	Kraft et al. (1996)
Rubus sulcatus Vest	*4x*	A; Ps	Nybom (1995)
Rubus uvidus L.H. Bailey (=*Rubus localis* L.H. Bailey)	*3x*	A; Ap (Hi); Ps	Einset (1951), Pratt and Einset (1955)
Rubus vatavensis Žíla & Trávn.	*4x*	A; Ps	Šarhanová et al. (2012)
Rubus vestitus Weihe & Nees	*4x*	A; Ap (Hi); Di (An); Ps	Christen (1950), Werlemark and Nybom (2003)
Rubus vitifolius Cham. & Schltdl.	*8x*	A; Ps	Crane and Thomas (1939)
Sanguisorba minor Scop.	*4x, 8x*	A; Ap (Hi); Di (An); Ps	Nordborg (1967)

[a]For species with sporophytic apomixis, N: nucellar embryony. For species with gametophytic apomixis, Ap ( ): apospory (type of female gametophyte development); Di ( ): diplospory (type of female gametophyte development); Ps: pseudogamous endosperm development; Au: autonomous endosperm development.

Hi: Hieracium type; An: Antennaria type; Ta: Taraxacum type.

§ Chromosome number rather than ploidy is given for 19 species of *Alchemilla*.

Mikoláš, Mikoláš, 1996; *S. barabitsii* C. Németh and *S. polgariana* C. Németh, Németh, 2012) are not included in Table 6.1 or 6.2. Taxonomic reviews indicating the occurrence of many apomictic taxa, either as hybridogenous species or nothotaxa, deserve a special mention. For example, Sennikov and Kurtto (2017) report 191 apomictic taxa in its phylogenetic checklist of *Sorbus* s.l. Yet, studies about the reproductive biology of each taxon were not cited and, in several cases, evidence on apomixis is circumstantial and has not been experimentally checked. Thus, only species for which apomixis has been verified through any of the approaches mentioned on Chapter 2 are included in Table 6.1.

Reports of polyembryony based on apogamety were not included in our list of apomictic species [e.g., for *Alchemilla hoppeana* (Rchb.) Dalla Torre, Lebègue, 1952; *Fragaria vesca* L., Sharma and Thorpe, 1995; *Sibbaldia procumbens* L., Czapik, 1990; *Sorbus bohemica* Kovanda, Jankun and Kovanda, 1987]. Even though apogamy is certainly not a case of sporophytic apomixis, the mentioned cases were placed in Table 6.2 to encourage further studies. Possible cases of experimental induction of apomixis [e.g., in *Rosa canina* L., El Mokadem et al., 2001; or in *Pyrus pyrifolia* (Burm. f.) Nakai, Inoue et al., 2002] or rescue of apomictic plants from progenies of interspecific or intergeneric crosses [e.g., between *Fragaria* × *ananassa* Duchesne × *Potentilla* spp., Niemirowicz-Szczytt, 1984b; or between *Pyrus communis* L. × *Chaenomeles japonica* (Thunb.) Lindl., Jafarova et al., 2013] were not included in our lists of apomictic species.

Molecular evidence for clonality was based on tests using five or more individuals (Hovmalm et al., 2004; Robertson et al., 2004, 2010; Ludwig et al., 2013; Burgess et al., 2014, Uhrinová et al., 2017). However, molecular studies suggesting the occurrence of apomixis based on interspecific comparisons of single or few individuals from different localities were not considered as evidence (e.g., Bartish et al., 2001). *Rubus hartmanii* Gand. and *R. fuscus* Weihe are here treated as separate species in agreement with the consulted databases. Yet, a previous study showed significant biometrical but no genetic differences among individuals from these taxa (Kraft and Nybom, 1995) which might suggest they belong to a single species.

Despite indications that diploid *Potentilla argentea* L. is apomict (Müntzing, 1928; Rutishauser, 1943a; Håkansson, 1946; Hunziker, 1954; Asker, 1967, 1970c, 1971), recent studies using molecular markers (isozymes and RAPD, AFLP, cpDNA) and flow cytometry had unquestionably shown that diploid cytotypes reproduce sexually in nature, primarily by self-fertilization but also through outcrossing (Holm and Ghatnekar, 1996a, 1996b; Holm et al., 1997; Paule et al., 2011). Apomixis only occurs in the polyploid cytotypes of *P. argentea*.

Several new apomictic roses had been detected using FCSS (e.g., Kausche, 2010; Šarhanová et al., 2012; Burgess et al., 2014; Dobeš et al., 2015, 2018; Lepši et al., 2019). Evidence of apomixis in FCSS analyses was considered from as a few as one evaluated seed only when linked to supporting evidence in related species [e.g., in triploids *Amelanchier arborea* (F. Michx.) Fernald and *A. utahensis* Koehne, Burgess et al., 2014]. For these new apomicts embryological studies are needed before determining the type of female gametophyte development. For roses in general, the frequent occurrence of multiple archespores demands more exhaustive embryological evaluations than in the other families to accurately assess the existence of apospory and/or diplospory.

Naumova (1993), based on a study by Gentscheff (1938), cites the experimental hybrid *Potentilla nepalensis* × *argyrophylla* as reproducing through nucellar polyembryony. Yet, Gentscheff studied the reproductive biology of *Potentilla nepalensis* Hook., *P. argyrophylla* Wall. ex Lehm., and the hybrid *P. nepalensis* × *P. splendens*. While the two species were found to be apomict (see Table 6.1), the hybrid was sterile due to lack of female and male gametophyte development. Interestingly, the hybrid produced seeds when pollinated with *P. nepalensis* through the formation of nucellar embryos (Gentscheff, 1938, pp. 405–406). As this is an experimental hybrid, it was not included in our lists of apomictic species (Table 6.1 and 6.2).

We apologize for any missing data or errors resulting from lack of access to certain works or from our language limitations (e.g., Hrabětová-Uhrová, 1961; Glazunova, 1983; Kolesnik, 1994; Liao and Li, 1997).

6.2.1 Doubtful Cases of Apomixis in Roses

The occurrence of apomixis in several species of roses (and five *Fragaria* × *ananassa* cultivars) is to be confirmed (*n*=45; Table 6.2). In all these cases, evidence on apomixis is circumstantial and specific analysis or experiments are needed to corroborate the formation of asexual seeds. Around 78% of the doubtful cases listed in Table 6.2 had a ploidy record, most involve tetraploids (41% of species with a ploidy record) and diploids (25% of species with a ploidy record). As with taxa from other plant families, observations in diploids are likely to be based on the spontaneous occurrence of apomixis-like developments lacking functional apomixis, a common phenomenon in apomictic systems with ploidy series and sexual cytotypes (Hojsgaard and Hörandl, 2019). In these taxa, a critical assessment of the reproductive biology is required before categorizing these species as apomicts.

For example, diploid materials of *Potentilla arguta* Pursh and *P. aurea* L. were considered as apomictic by Popoff (1935) and Shimatomai (1935) based on experimental crossings (to tetraploid *P. recta* L. and hexaploid *P. alpestris* Haller f., respectively) and (morphological) progeny tests. However, no details about the progeny test were discussed, nor the ploidy of the offspring was evaluated. Yet, alike studies in diploid *P. argentea* L. indicating apomixis were later refuted through flow cytometry and molecular markers (see details in the previous Section), pointing out the likelihood of methodological pitfalls like unintended self-pollination and mentor effects promoting selfing, dominance of maternal markers, or limited geographical sampling as possible explanations (see Paule et al., 2011). Consequently, these species were temporarily placed in Table 6.2 for confirmation. In another *Potentilla* sp., Müntzing (1928) cites apospory for *P. silvestris* based on Forenbacher (1913). However, the author of the species name was not mentioned, and the original work could not be accessed. Since *Potentilla sylvestris* Neck. is a synonym of *Potentilla erecta* (L.) Raeusch., but *Potentilla sylvestris* P. Renault is a synonym of *Potentilla anglica* Laichard., we opted by placing this species in the quarantine list for confirmation (Table 6.2).

Spontaneous observation of apomictic female gametophytes in diploids from species like *Waldsteinia geoides* Willd. (Czapik, 1985), *Chaenomeles japonica* (Thunb.) Lindl. ex Spach (Daskalova, 2005) or *Sibbaldia procumbens* L. (Czapik, 1990) without information on their functionality, and in the case of the later, with evidence of sexuality (Dobeš et al., 2015, only 3 seeds assayed by flow cytometry), were listed in our quarantine table (Table 6.2). Similarly, species presumed apomictic based on circumstantial evidence from comparative morphology, ploidy and/or phylogenetic analyses, like tetraploid *Amelanchier cusickii* Fernald. (Cushman et al., 2017) or in *Photinia villosa* (Thunb.) DC. (Leonard, 2011), were also listed in Table 6.2.

In *Sorbus* spp., 12 out of 24 polyploid species categorized as apomicts by McAllister and Gillham (1980) and McAllister (1986) were placed in Table 6.2 because, from the comprehensive review on the reproductive biology of the Maloid subfamily by Campbell et al. (1991), it follows that no detailed evidence supporting apomixis was presented in those studies. Experimental confirmation of apomixis would be required for those taxa.

The occurrence of apomixis in diploids *Pyrus* and *Malus* hybrids might occur as a non-recurrent phenomenon (see Sherman et al., 1978; Ranney et al., 2004; Phillips et al., 2016). Evaluating triploid *Pyrus* hybrids from different interspecific crosses, Phillips et al. (2016) observed the formation of a few seeds through unreduced female gametes with and without fertilization of the central cell. Yet, most of seeds were of sexual origin and remain to be tested if apomixis can be fixed and, perhaps, used to increase relative fertility in these hybrids.

In the case of *Malus* spp., unfortunately, several relevant studies were not accessed (e.g., Brown, 1975; Mao et al., 1995; Qu et al., 2008) and therefore, in those cases we cannot know which species were examined and found to be apomicts.

Malus lancifolia Rehder and *M. platycarpa* Rehder, two polyploid species categorized as apomicts by Brown (1975, listed in Campbell et al., 1991), were here placed in Table 6.2 as it was done with a dozen of *Sorbus* species mentioned above, due to the apparent lack of experimental evidence supporting apomixis. Even when these species are likely apomicts, we want to motivate researchers to confirm field observations. Likewise, the tetraploid species *Malus coronaria* (L.) Mill. was

TABLE 6.2
Table of Doubtful Apomictic Species (A) within the Family Rosaceae Juss., Sorted Alphabetically According to Subfamilies

Species	Ploidy	Type of Apomixis[a]	Reference
Maloideae C.Weber			
Amelanchier cusickii Fernald	*4x*	A; Ps	Cushman et al. (2017)
Crataegus laevigata × *C. subsphaerica*	*3x*	A; Ps	Vašková and Kolarčik (2019)
Cotoneaster laxiflorus Jacq. ex Lindl. (=*Cotoneaster melanocarpus* G. Lodd.)	*4x*	A	Sax (1954)
Cotoneaster melanocarpus var. *laxiflorus* (Jacq. ex Lindl.) C.K. Schneid.	*3x*	A	Sax (1954)
Cotoneaster melanocarpus var. *commixtus* C.K. Schneid.	*4x*	A	Sax (1954)
Malus × *domestica* Borkh.	*2x*	A	Sherman et al. (1978), Castro et al. (2016)
Pyrus pyrifolia (Burm. f.) Nakai	*2x*	A	Inoue et al. (2002)
Pyrus hybrids	*3x*	A; Ps; Au	Phillips et al. (2016)
Photinia glabra (Thunb.) Maxim. (=*Photinia serrulata* Lindl.)	*4x*	A; Ps	Leonard (2011)
Photinia villosa (Thunb.) DC.	*4x*	A; Ps	Leonard (2011)
Sorbus californica Greene (=*Sorbus cascadensis* G. Jones)	*4x*	A	McAllister (1986, in Campbell et al., 1991)
Sorbus decora (Sarg.) C.K. Schneid.	*4x*	A	McAllister (1986, in Campbell et al., 1991)
Sorbus discolor (Maxim.) Maxim. (=*Sorbus hupehensis* C.K. Schneid.	*4x*	A; Ps	McAllister (1986, in Campbell et al., 1991)
Sorbus fruticosa Steud.		A	McAllister (1986, in Campbell et al., 1991)
Sorbus hajastana Gabrieljan	*4x*	A	McAllister (1986, in Campbell et al., 1991)
Sorbus lancastriensis E.F. Warb.		A	McAllister (1986, in Campbell et al., 1991)
Sorbus oligodonta (Cardot) Hand.-Mazz. [=*Sorbus glabrescens* (Cardot) Hand.-Mazz.]	4x	A	McAllister and Gillham (1980, in Campbell et al., 1991)
Sorbus poteriifolia Hand.-Mazz.		A	McAllister (1986, in Campbell et al., 1991)
Sorbus prattii Koehne		A	McAllister (1986, in Campbell et al., 1991)
Sorbus reducta Diels		A	McAllister (1986, in Campbell et al., 1991)
Sorbus sitchensis M. Roem.		A	McAllister (1986, in Campbell et al., 1991)
Sorbus subfusca (Ledeb. ex Nordm.) Boiss.	*2x*	A	McAllister (1986, in Campbell et al., 1991)
Sorbus udvardyana Somlyay & Sennikov		A	Somlyay and Sennikov (2014)
Rosoideae Juss. ex Arn.			
Alchemilla alpina L.		A; N; Au	Murbeck (1902)
Alchemilla glabra Neygenf. (=*Alchemilla alpestris* F.W. Schmidt)	*106*	A; N	Murbeck (1902)
Alchemilla hoppeana (Rchb.) Dalla Torre		A; Ps	Lebègue (1952a, 1952b)

(Continued)

TABLE 6.2 (Continued)

Species	Ploidy	Type of Apomixis[a]	Reference
Chaenomeles japonica (Thunb.) Lindl. ex Spach	*2x*	A; Di	Daskalova (2005)
Fragaria × *ananassa* (Duchesne ex Weston) Duchesne ex Rozier [sub *Fragaria* × *ananassa* Duchesne (cv. 'Selva')]	*8x*	A; Di (An); Ps	Dziadczyk et al. (2011), Leszczuk et al. (2018)
Fragaria × *ananassa* (Duchesne ex Weston) Duchesne ex Rozier [sub *Fragaria* × *ananassa* Duchesne (cv. 'Mieze Schindler')]	*8x*	A; Di (An, Ta); Ps	Kunz and Gröber (1988)
Fragaria × *ananassa* (Duchesne ex Weston) Duchesne ex Rozier [sub *Fragaria* × *ananassa* Duchesne (var. 'Freja')]	*8x*	A; Ap; Di; Ps	Niemirowicz-Szczytt (1984a, 1984b, 1984c)
Fragaria × *ananassa* (Duchesne ex Weston) Duchesne ex Rozier [sub *Fragaria* × *ananassa* Duchesne (var. 'Senga Sengana')]	*8x*	A; Ap (Hi); Di (An); Ps	Leszczuk et al. (2018)
Fragaria × *ananassa* (Duchesne ex Weston) Duchesne ex Rozier [sub *Fragaria* × *ananassa* Duchesne (var. 'Mount Everest')]	*8x*	A; Ap (Hi); Di (An); Ps	Leszczuk et al. (2018)
Fragaria moschata Weston	*6x*	A; Di (An, Ta); Ps	Kunz and Gröber (1988)
Fragaria chiloensis (L.) Mill.	*8x*	A; Ps	Waldo (1930)
Fragaria vesca L.	*2x*	A; N	Lebègue (1952a, 1952b), Sharma and Thorpe (1995)
Potentilla arguta Pursh	2x	A	Popoff (1935)
Potentilla aurea L.		A; N; Ps	Shimatomai (1935), Lebègue (1950), Kausche (2010)
Potentilla erecta (L.) Raeusch.	*4x*	A; Ap	Forenbacher (1913)
Rosa dumalis	*5x*	A; Ps	Werlemark (1999, 2000), Nybom et al. (2004)
Rosa rubiginosa	*5x*	A; Ps	Werlemark (2000), Nybom et al. (2004)
Rosa villosa L.	*4x*	A; Ps	Nybom et al. (2004)
Sanguisorba officinalis L.	*4x, 8x*	A; Ap; Di; Ps	Nordborg (1963)
Sibbaldia procumbens L.	*2x*	A; Au	Czapik (1990)
Waldsteinia geoides Willd.	*2x*	A; Ap; Di; Ps	Czapik (1985)

[a]For species with sporophytic apomixis; N: nucellar embryony. For species with gametophytic apomixis; Ap ( ): apospory (type of female gametophyte development); Di ( ): diplospory (type of female gametophyte development); Ps: pseudogamous endosperm development; Au: autonomous endosperm development.

Hi: Hieracium type; An: Antennaria type; Ta: Taraxacum type.

also categorized as apomict by Brown (1975) and recently confirmed by flow cytometry analysis by Kron and Husband (2009).

Malus xiaojinensis M.H. Cheng & N.G. Jiang was momentarily placed in Table 6.1. Despite embryological and genetic evidence suggesting the occurrence of apomixis (Cheng et al., 1984; Yang and Zhou, 2009; Wang et al., 2012), the studies by Cheng et al. (1984) and Yang and Zhou (2009) could not be accessed. While in the abstract of the latter it is not clear if the unreduced female gametophytes are functional, in the study by Wang et al. (2012) the progeny from a cross between tetraploid *Malus xiaojinensis* × diploid *M. baccata* is classified as apomictic based on the absence of a homozygous SSR marker carried by the diploid male parent, and yet 2x, 3x, and 4x plants are described among the asexual offspring. A confirmation of the occurrence of functional apomixis in this species will help resolving these uncertainties.

The genus *Fragaria* L. requires a detailed examination. Waldo (1930, cited in Kunz and Gröber, 1988, p. 326) presented the first indication for apomixis after observing the occurrence of maternal offspring in a cross between *Fragaria chiloensis* (L.) Mill. × *F. cuneifolia* Nutt. ex Howell. Evidence of the parallel occurrence of sexual and apomictic pathways has been proven through detailed embryological analyses showing the occurrence of multicellular archespores and multiple female gametophytes in ovules of *Fragaria moschata* Weston and several *Fragaria* × *ananassa* cultivars (Kunz and Gröber, 1988; Dziadczyk et al., 2011; Leszczuk et al., 2018). Yet, in all roses in which apomixis has been proven the primary megaspore mother cell often aborts in the developing ovule and female gametophytes arise from other cells in the multicellular archesporium (either secondary megaspore or nucellar cells). In contrast, in all the cases of *Fragaria* species mentioned above the primary megaspore is functional and develops a Polygonum type functional female gametophyte in most ovules, while parallel apomictic developments occur in between 14% and 50% of ovules (e.g., Dziadczyk et al., 2011; Leszczuk et al., 2018). Furthermore, different studies suggest that sexuality, rather than apomixis, is functional in the genus. Staudt (1970) found only hybrid offspring in crosses between octoploid *Fragaria* × *ananassa* cultivars (including cultivars with embryological evidence of apomixis) and diploids *F. nubicola* (Hook. f.) Lindl. ex Lacaita and *F. vesca* L. Dobeš et al. (2015) presented evidence of sexual seed formation in five species of *Fragaria* (without providing species names), and Nosrati et al. (2010) using molecular markers found no clonal individuals among matroclinal offspring from crosses between different *Fragaria* species and *Fragaria* × *ananassa* cultivars (the study included different cultivars than those evaluated embryologically).

Thus, the occurrence of facultative apomixis in *Fragaria* spp. cannot be determined with certainty (see Table 6.2) and thorough analyses involving simultaneous use of embryological, cytological (FCSS) and molecular approaches are needed.

A comprehensive examination of the rich literature on the induction of apomixis in species like *Fragaria vesca* L., *Fragaria* × *vescana* Rud.Bauer & A. Bauer, *Fragaria* × *ananassa* cultivars after intergeneric crosses to species of related genera like *Potentilla* spp. or *Duschesnea* spp. (e.g., Niemirowicz-Szczytt, 1984b; Sukhareva et al., 2002; Baturin, 2007, 2009) would help us understand the relevance of apogamy and autogamy within the genus *Fragaria* L. Unfortunately, several relevant papers by Russian authors were unavailable in online repositories (e.g., Beridze, 1964; Konstantinov, 1966; Koloteva et al., 1978; Sukhareva, 1978; Berzenko, 1979; Baturin, 1997).

Another reproductively exciting genus is *Rosa* L. The occurrence of apomixis within the genus *Rosa* L. sect. *Caninae* has been much debated (see MacPhail and Kevan, 2009). Autonomous apomixis has been suggested for several *Rosa* spp. based on the formation of variable proportions of seeds (5–78%) after castration experiments (Matsson, 1912; Schwertschlager, 1915, cited in Täckholm, 1922, p. 265; Wissemann and Hellwig, 1997), and by the detection of low levels of genetically uniform offspring using different molecular markers (Werlemark, 2000; Nybom et al., 2004, 2005).

Yet, in cytological studies, researchers had been unable to find evidence of apomictic development and formation of unreduced female gametophytes (Fagerlind, 1940a, 1940b, 1945a, 1945b; Gustafsson and Håkansson, 1942). Indirect support obtained through flow cytometry and molecular markers (Crespel and Gudin, 2003; Nybom, 2007) points to the occurrence of non-apomictic unreduced female gametophytes. All species in the *Rosa canina* aggregate reproduce sexually due to a unique asymmetrical meiosis plus balanced heterogamy (Richards, 1997; Nybom, 2007). In this distinctive breeding system, the 'canina meiosis' in polyploids involve only two sets of chromosomes forming bivalents and being regularly distributed to male and female gametes, whereas the remaining chromosomes form univalents and are exclusively transmitted by the egg cells (Täckholm, 1920, 1922; Lunerová et al., 2020). Consequently, the contribution of male and female gametes to the chromosome complement of the zygote is not only uneven, but a high proportion of the maternal chromosomes (66.7% in tetraploids, 75% in pentaploids and 80% in hexaploids) are non-recombinant and thus, it results in a matroclinal inheritance of genetic material. This might explain the occurrence of maternal offspring in progeny tests using morphological traits (e.g., Gustafsson, 1937), while genetically uniform offspring (e.g., Werlemark, 2000; Nybom et al., 2004) might be a consequence of the high level of genetic homogeneity observed in the investigated species of the section *Caninae* (Olsson, 1999; Nybom, 2007) and the capability to reveal polymorphisms in genetic markers necessarily linked to the one set of maternal and paternal chromosomes capable of recombination and segregation in the progeny.

Although the occurrence of genetically uniform progeny should represent convincing evidence for apomixis, the very particular breeding system of dogroses and the fact that the offspring are expected to be very homogeneous, regardless of whether derived from apomixis, selfing or from intraspecific cross-pollination (Nybom et al., 2004), do not allow to conclude that dogroses follow any of the developmental types of apomixis observed among angiosperms. Apomixis in hemisexual species is difficult to test (see further comments in Chapter 5) and dogroses suspected to be apomictic due to the formation of genetically uniform offspring were here listed in our quarantine Table 6.2.

Potentilla aurea L. was included in the list of species with nucellar polyembryony (Naumova, 1993) based on a short report by Lebègue (1950). After revising the original paper it seems likely that the report refers to a rare case, and it can be deduced from the text and the figure in it that the two embryos are associated with separate endosperms and were originated from two female gametophytes. Thus, we placed the species in our quarantine list (Table 6.2).

6.3 PATHWAYS TO ASEXUAL SEEDS IN ROSES

A total of 298 taxa, including 289 rose species, 2 subspecies and 8 varieties have at least one record of apomixis, either sporophytic or gametophytic apomixis (see Chapter 2).

The pathways of apomictic development are known only for 51% of these species, the remaining 49% of cases have been determined without details on the type of apomixis (mostly through progeny tests). Information about the development of the endosperm is collected for about 80.1% of all apomictic roses. Most recorded apomicts show pseudogamous endosperm development, and only 5% of species show autonomous endosperm development. All species but two are polyploid, mostly tetraploids (n=133) and triploids (n=108) (Table 6.1). Other ploidies among apomictic roses are pentaploid (n=13), hexaploid (n=18), heptaploid (n=7), octoploid (n=8) and nonaploid (n=6). More rare ploidies are decaploid (n=1) and dodecaploid (n=3) (Table 6.1). No diploid apomicts have been reported with compelling evidence. Reports of apomixis in diploids connected to embryological studies showing apomictic development but without confirmation of the formation of apomictic seeds are placed in our quarantine table (Table 6.2). Cases of apomictic seed formation in diploids represent sporadic events of apomictic developments rather than established asexual diploid lineages (e.g., in *Amelanchier* spp., Burgess et al., 2014). Earlier reports of apomixis in diploids (e.g., in *Sorbus eximia* Kovanda, Jankun and Kovanda, 1988) later rejected through morphological, molecular, and cytological analyses (Vít et al., 2012) are not listed.

The diversity of unreduced female gametophyte developments observed among aposporous and diplosporous roses is narrower than in daisies, involving almost exclusively bipolar Antennaria type or Hieracium type developments. No monopolar developments (e.g., Panicum type, or Eragrostis type) were listed for the Rosaceae family.

6.3.1 Sporophytic Apomixis in Roses

The formation of adventive embryos is exceedingly rare in the family Rosaceae. It has been observed in 17 species within the subfamily Rosoideae Juss. ex Arn., in all cases arising from the nucellar tissue often in species with gametophytic apomixis (Table 6.1). However, in all cases, the reports refer to polyembryos being develop in one or few ovules, making clear that these are spontaneous cases and not a regular mechanism of apomixis as observed in species from other genera (e.g., in *Citrus* spp.; *Euonymus* spp., see Chapter 7). The taxa belong to the genera *Alchemilla* spp. (11 species; Murbeck, 1902; Izmaiłow, 1986), *Chaenomeles* spp. (1 species; Daskalova, 2005), *Drymocallis* spp. (1 species; Popoff, 1935), *Potentilla* spp. (2 species; Popoff, 1935; Souèges, 1935) and *Sorbus* spp. (2 species; Jankun, 1993, 1994).

6.3.2 Gametophytic Apomixis in Roses

In total, 298 rose species, subspecies and varieties are recorded as apomict by direct or indirect evidence of the formation of unreduced gametophytes and clonal offspring, including two species also showing evidence of sporophytic apomixis. Knowledge about apomictic developmental pathways coincides with the values mentioned above, being known in 51% of taxa, while in the other 49% of taxa asexual seed formation is established without details on the type of apomixis (mostly through progeny tests).

6.3.2.1 The Aposporous Development

Apospory has been recorded in around 36.6% of all species with gametophytic apomixis (*n*=109; Table 6.1). The aposporous development in roses starts in a nucellar cell close to the primary or the secondary megaspores of the multicellular archesporium which acquires a gametogenetic fate and develops into an unreduced female gametophyte (Asker and Jerling, 1992). The primary megaspore in these apomicts often degenerates before or during the meiotic division, and the female gametophyte develops from an aposporous initial cell. In a high number of cases (*n*=48; Table 6.1), the female gametophyte can be initiated from an aposporous cell or from one of the secondary megaspore cells. In few cases, the primary megaspore cell can develop into a mature female gametophyte and produce a sexual seed.

6.3.2.1.1 The Hieracium Type

This type of unreduced female gametophyte development is the only type of aposporous development observed among roses and is present in 84 species from three different subfamilies (Table 6.1). Within the Rosoideae, the Hieracium type occur in species of *Alchemilla* spp. (9 species; Murbeck, 1897, 1901, 1902; Strasburger, 1905; Böös, 1917, 1920; Turesson, 1943, 1956; Lundh-Almestrand, 1958; Glazunova, 1987; Baldjiev and Robeva-Davidova, 2008), in *Aphanes* spp. (1 species; Murbeck, 1901; Böös, 1924; Gudjonsson, 1941; Hjelmqvist, 1959; Izmaiłow, 1999), in *Geum* spp. (1 species; Krch, 2002), in *Potentilla* spp. (10 species; Müntzing, 1928, 1958; Popoff, 1935; Shimatomai, 1935; Gentscheff, 1938; Christoff and Papasova, 1943; Rutishauser, 1943a, 1943b, 1943c, 1945a, 1945b, 1946, 1947, 1948; Håkansson, 1946; Skalińska and Czapik, 1958; Czapik, 1961, 1962; Smith, 1963a; Asker, 1967, 1970a, 1970b, 1970c; Eriksen and Fredrikson, 2000), in *Rubus* spp. (17 species; Lidforss, 1914; Petrov, 1939; Thomas, 1940; Gustafsson, 1943; Christen, 1950, 1952; Einset, 1951; Berger, 1953; Pratt and Einset, 1955; Pratt et al., 1958; Gerlach, 1965; Dowrick, 1961, 1966; Czapik, 1981b, 1987; Spies and du Plessis, 1986), and in *Sanguisorba* spp. (1 species; Nordborg, 1967).

In the subfamily Maloideae C.Weber, it occurs in species of *Amelanchier* spp. (13 species and 2 varieties; Campbell et al., 1985, 1987, 1997; Weber and Campbell, 1989; Burgess et al., 2014), in *Cotoneaster* spp. (6 species; Sax, 1954; Hjelmqvist, 1962; Flinck and Hylmö, 1962), in *Crataegus* spp. (15 species and 1 variety; Muniyamma and Phipps, 1979, 1984a, 1984b; Dickinson and Phipps, 1986; Smith and Phipps, 1988; Campbell et al., 1991; Dickinson et al., 1996), in *Malus* spp. (4 species; Dermen, 1936; Sax, 1949, 1959; Olden, 1953; Hjelmqvist, 1957; Schmidt, 1964; Liao and Li, 1997; Li, 2001; Liu et al., 2006; Yang and Zhou, 2009).

Last, in the subfamily Amygdaloideae Arnott, the Hieracium type development occurs in *Sorbus* spp. (14 species; Liljefors, 1934, 1953, 1955; Jankun and Kovanda, 1986, 1987, 1988; McAllister, 1986; Campbell et al., 1991; Jankun, 1993; Ludwig et al., 2013).

6.3.2.2 The Diplosporous Development

Diplospory has been recorded in 18.8% of all species and subspecies with gametophytic apomixis (*n*=56; Table 6.1); all but 15 cases are concurrent with apospory. Thus, the development of an unreduced female gametophyte can go through diplospory or apospory. In roses, unlike in other families, diplospory almost always starts from a secondary megaspore cell from the multicellular archesporium that skip meiosis and develops an unreduced female gametophyte (Asker and Jerling, 1992). In occasional cases, the primary megaspore mother cell may progress throughout meiosis and produce reduced megaspores, a reduced female gametophyte and a sexual seed after fertilization.

6.3.2.2.1 The Antennaria Type

This type of development is recorded in 37 species (Table 6.1). Within the subfamily Rosoideae, the Antennaria type is found in 20 species, including species from the genus *Alchemilla* spp. (6 species; Murbeck, 1897, 1901; Böös, 1917, 1920; Lundh-Almestrand, 1958; Glazunova and Myatlev, 1990; Baldjiev and Robeva-Davidova, 2008), *Aphanes* spp. (1 species; Murbeck, 1901; Böös, 1924; Gudjonsson, 1941; Hjelmqvist, 1959; Izmaiłow, 1999), *Potentilla* spp. (5 species; Müntzing, 1928, 1958; Popoff, 1935; Shimatomai, 1935; Gentscheff, 1938; Christoff and Papasova, 1943; Håkansson, 1946; Rutishauser, 1948; Skalińska and Czapik, 1958; Czapik, 1961; Smith, 1963b), *Rubus* spp. (7 species; Christen, 1950, 1952; Berger, 1953; Gerlach, 1965; Dowrick, 1961, 1966; Czapik, 1981a, 1981b), and in *Sanguisorba* spp. (1 species; Nordborg, 1967).

In the subfamily Maloideae, the Antennaria type development is found in *Cotoneaster* spp. (6 species; Sax, 1954; Hjelmqvist, 1962; Flinck and Hylmö, 1962), and in the subfamily Amygdaloideae in *Sorbus* spp. (11 species; Liljefors, 1953, 1955; Jankun and Kovanda, 1987; McAllister, 1986; Campbell et al., 1991; Jankun, 1993).

6.3.2.2.2 The Taraxacum Type

The Taraxacum type of female gametophyte development is exceedingly rare in roses, being recorded in 1 species of *Crataegus* spp. of the subfamily Maloideae (Muniyamma and Phipps, 1984a) (Table 6.1).

6.3.2.3 Variations in Female Gametophyte Anatomy and Maturity

Variations on the types of female gametophyte development had been recorded in different species, yet they are less frequent than among daisies (Chapter 5). In species from different genera, the occurrence of meiotic and apomeiotic female gametophytes in ovules of the same or different individuals has been recorded directly by embryological observation or indirectly by FCSS or progeny tests. Unlike species with monosporic development, in species with a multicellular female archesporium the simultaneous occurrence of the Polygonum type and diplosporous types of development in a single ovule is possible.

In roses, the gametogenetic programme goes through a bipolar pattern of development in both the meiotic and the apomeiotic pathways, producing eight-nucleate female gametophytes of similar anatomy. Thus, recognizing alternative developmental pathways requires the evaluation of earlier developmental stages, during sporogenesis (see Chapter 2). The occurrence of a multicellular archesporium makes harder identifying apospory or diplospory (see e.g. Nordborg, 1967), mostly because delimitating the archesporial and the nucellar tissues can be tricky. In addition, classifying the alternative diplosporous pathways is technically challenging (see Chapter 5). Like in Asteraceae, a proper examination of meiosis and the apomictic pathways in roses requires the analysis of an appropriate (possibly large) number of ovules and developmental stages.

Despite the concurrence of apospory and diplospory in single ovules of many rose species (Table 6.1), only one developmental pathway often goes throughout the gametogenesis and produces either an unreduced Hieracium type or Antennaria type female gametophyte, or (less frequently) a reduced Polygonum type female gametophyte. Ovules carrying multiple, unorganized female gametophytes as well as other embryologically weird cases have been reported. As examples, different *Rubus* species show deviations to the above patterns of gametophyte developments, including displacement of nuclei, unusual number of nuclei, and abnormal variation in the size of nuclei (e.g., Pratt and Einset, 1955). In *Malus* spp., Dermen (1936) describes the regular occurrence of structurally complex female gametophytes, resulting from the fusion of two or more aposporic gametophytes. The occurrence of variations to the regular mechanism of apomixis like apogametic embryos and autonomous development of the endosperm in triploid plants of *Malus hupehensis* (Pamp.) Rehder are also pointed out (Dermen 1936, p. 95).

7 Apomixis in Other Families of Angiosperms

7.1 INTRODUCTION

Studies dealing indirectly with the occurrence of apomixis in different plant families date back to the eighteenth century. Every seed carries a single embryo. In numerous species, the phenomenon of polyembryony occurs, whereby individual seeds can carry more than one embryo, in many instances originated through sporophytic apomixis. Since polyembryony is a simple phenotypic character for sporophytic apomixis, it was the first apomictic mechanism indirectly detected by the occurrence of twin embryos in orange (Leeuwenhoek, 1719). Many more cases were followed by several works (e.g., Müller, 1847; Schacht, 1850; Hofmeister, 1858; Dickie, 1875; Strasburger, 1877), including the classic study of Braun (1860) evaluating about 60 species with polyembryony. Even though some of those studies included cytological observations, the developmental characteristics of polyembryonic seeds associated with apomixis (adventitious embryony) were described in detail by Strasburger (1878) through embryological studies in *Citrus* spp. (Rutaceae Juss.), in *Coelebogyne* spp. (=*Alchornea* Sw., Euphorbiaceae Juss.), in *Funkia* spp. (=*Hosta* Tratt., Asparagaceae Juss.), and in *Nothoscordum* spp. (Amaryllidaceae J.St.-Hil.), which showed that extra embryos can arise from nucellar tissue in the ovule. The case of *Alchornea aquifolia* (Js.Sm.) Domin (=*Coelebogyne ilicifolia* Js.Sm.) is of particular interest since Smith (1841) found the first direct evidence of agamospermy in a single female plant from Australia cultivated in England (at Kew Gardens) which produced viable seeds without apparent action of male gametophytes.

Later studies and revisions (e.g., Biermann, 1896; Ganong, 1898a, 1898b; Leavitt, 1901; Osawa, 1913; Duthie, 1915; Ensign, 1919; Carano, 1925; Archibald, 1939; Maheshwari, 1950; Lebègue, 1952a, 1952b; Wirth and Withner, 1959; Johri, 1965; Yakovlev, 1967; just to mention a few) provided detailed information on the many ways a plant can produce polyembryonic seeds, including by the fusion of ovules, the development of multiple female gametophytes in the same ovule, embryogenesis in cells of somatic tissues of the ovule or in cells of female gametophytes other than the egg-cell, or divisions of the proembryo (see below and Chapter 2). As we deal with cases of apomixis *sensu stricto* (i.e., where seed embryos are genetic copies of the mother), several taxa that show a polyembryonic development that does not meet this requirement are not considered here and are mentioned only if necessary. Adventitious embryos arising from cells in the nucellus or integuments in the ovule represent cases of apomixis and have been found in species from numerous plant families. For example, nucellar embryony is common among species of *Sarcococca* Lindl. (Buxaceae Dumort.), *Opuntia* Mill. (Cactaceae Juss.), *Gymnadenia* R.Br. (Orchidaceae Juss.), or *Citrus* L. (Rutaceae Juss.), while integumentary embryony is regularly found among species of *Euonymus* L. (Celastraceae R.Br.) or *Garcinia* L. (Clusiaceae Lindl.), and both types of adventitious embryony occur simultaneously in species of *Anemopaegma* Mart. ex Meisn. (Bignoniaceae Juss.) and *Vincetoxicum* Wolf (Apocynaceae Juss.).

In addition to the numerous cases of sporophytic apomixis, more cases of gametophytic apomixis in different plant families are reported and added to those in previous chapters. These cases follow the main apomictic pathways described for species of genera in the family Poaceae (Chapter 4) and the family Asteraceae (Chapter 5). Examples of aposporous development are found in species of *Hypericum* L. (Hypericaceae Juss.) and *Ranunculus* L. (Ranunculaceae Juss.), diplospory is found frequently among species of *Boehmeria* Jacq. (Urticaceae Juss.) and *Limonium* Mill.

DOI: 10.1201/9781003088561-7

(Plumbaginaceae Juss.), and both types occur among species of *Boechera* Á. Löve & D. Löve (Brassicaceae Burnett) and *Elatostema* J.R. Forst. & G. Forst. (Urticaceae Juss.).

For this chapter we have grouped plant species with evidence of apomixis from all angiosperm families in which the number of apomictic species recorded is low or restricted to a single or few genera. Studies on the reproductive biology of plant families not listed in Chapters 4, 5 and 6 include a total 577 species and subspecies (Table 7.1). Thus, the present chapter brings together current information on apomixis at the species level in many angiosperm families.

7.2 CONSIDERATIONS ON THE OCCURRENCE AND TYPE OF APOMIXIS IN DISTINCT FAMILIES

Collected records are based on original reports and previous lists of apomictic species (Stebbins, 1941; Nygren, 1954a, 1954b, 1967; Davis, 1967; Tisserat et al., 1979; Lakshmanan and Ambegaokar, 1984; Pullaiah, 1984; Asker and Jerling, 1992; Naumova, 1993; Carman, 1995; Noyes, 2007; dos Santos et al., 2012) as well as apomictic cases in regional floras (e.g., Richards, 2003; Gregor, 2004a, 2004b, 2013). Several particularly old studies were accessed through the Biodiversity Heritage Library (BHL; https://www.biodiversitylibrary.org/) and the Academic Archive Online (DiVA, http://www.diva-portal.org/). Similarly, different open access research pieces were accessed directly on journal's webpages or through online repositories like CORE (https://core.ac.uk/), The Internet Archive (https://archive.org/), the Utrecht University Repository (https://dspace.library.uu.nl/). Whenever possible, studies on the occurrence of apomixis in each species have been checked and their apomictic status reconsidered. Species' taxonomy and nomenclature had been revised following the International Plant Names Index (IPNI; https://www.ipni.org/), Tropicos (https://www.tropicos.org/), the Plants of the World Online (http://www.plantsoftheworldonline.org/), the World Flora Online (WFO; http://www.worldfloraonline.org/), sites for specific families (e.g., Solanaceae Source – A global taxonomic resource for the nightshade family, http://solanaceaesource.org/; the BrassiBase, https://brassibase.cos.uni-heidelberg.de/), and sites for regional floras (e.g., the Vascular Plants of the Americas; Ulloa et al., 2017, 2018; the Flora del Cono Sur, http://conosur.floraargentina.edu.ar/; the Flora Ibérica, http://www.floraiberica.es/eng/index.php; the online atlas of the British and Irish Flora, https://www.brc.ac.uk/plantatlas/; the Flora of Italy, http://dryades.units.it/floritaly/index.php; the Flora of Greece, http://portal.cybertaxonomy.org/flora-greece/).

This chapter follows the procedures in Chapters 4, 5 and 6. We have included basic information about species names, ploidy of studied materials, and the type of apomictic reproduction in Tables 7.1 and 7.2. Since sporophytic and gametophytic apomixis coexists in species of genera from several plant families, we refrained from dividing them into two tables. A number of species in different families had been subjected to changes in their nomenclature. Such changes are indicated in Tables 7.1 and 7.2. Most cases involve synonyms, either as new species or subspecies [e.g., *Ardisia polytoca* A. Braun & Bouché is now placed under the name *Ardisia compressa* Kunth, Primulaceae Batsch ex Borkh.; *Balanophora globosa* Jungh. is now placed under *Balanophora fungosa* var. *globosa* (Jungh.) B. Hansen, Balanophoraceae Rich.; *Nothoscordum fragrans* (Vent.) Kunth is now under *Nothoscordum gracile* (Aiton) Stearn, Amaryllidaceae J.St.-Hil.; *Sarcococca pruniformis* Lindl. is now *Sarcococca saligna* (D. Don) Müll. Arg., Buxaceae Dumort.], or placed under new genera [e.g., *Cooperia pedunculata* Herb. is now *Zephyranthes drummondii* D. Don, Amaryllidaceae J.St.-Hil.; *Funkia ovata* Spreng. is now *Hosta ventricosa* Stearn, Asparagaceae Juss.; *Corunastylis apostasioides* Fitzg. is now *Genoplesium apostasioides* (Fitzg.) D.L. Jones & M.A. Clem., Orchidaceae Juss.; *Clidemia hirta* (L.) D. Don is now under the name *Miconia crenata* (Vahl) Michelang., Melastomataceae Juss.; or *Bombax kimuenzae* De Wild. & T. Durand and *Bombacopsis glabra* (Pasq.) Robyns are under the name *Pachira glabra* Pasq., Malvaceae Juss.]. A few species were moved into new genera [e.g., *Boechera laevigata* (Muhl. ex Willd.) Al-Shehbaz is now under the name *Borodinia laevigata* (Muhl. ex Willd.) P.J. Alexander & Windham, Brassicaceae Burnett].

In some cases, we have followed the online databases (e.g., *Nigritella* Rich. merges with *Gymnadenia* R.Br., Orchidaceae Juss.), even though such taxonomic changes are not accepted by all scientists and there is preliminary evidence against them (e.g., Hedrén et al., 2018 provide molecular evidence that could support keeping the two above genera separate).

A few cases required specific considerations. For example, *Skimmia lavalei*, for which Desai (1961) reported apospory, or *Epilobium dulcis*, having nucellar embryony (Naumova, 1993), were not found in the online taxonomic databases. Likewise, *Esenbeckia jussueni* does not appear in the surveyed taxonomic databases but it is cited by Naumova (1993) with nucellar embryony following Mauritzon (1935) who refers to the occurrence of polyembryony in this species based on Braun (1860). The same for two cultivated materials, i.e., *Aglaonema pseudobracteatum*, a name likely corresponding to a cultivated variety of *A. commutatum* Schott, Araceae Juss., cited as apomict by Veyret (1965), and the cultivated daylily *Hemerocallis mexicana*, Xanthorrhoeaceae Dumort. (Naumova, 1993, following Braun, 1860) which are not included in our lists. When the author of a species name was not available in the original study or the study was not accessed and several authors were possible, we followed the accepted species name and priority given by IPNI, the International Plant Names Index [e.g., in *Murraya paniculata* (L.) Jack, Rutaceae Juss.]

Illegitimate species names (e.g., *Eugenia michelii* Miq. and *Eugenia bancana*, van der Pijl, 1934) were replaced whenever possible. Six unresolved nomenclatural names and three unpublished names cited elsewhere as apomicts were not included in the table of apomictic species [i.e., *Limonium arenosum* Erben; *Limonium confusum* (Gren. & Godr.) O. Kuntze; *Limonium cyprium* (Meikle) Hand & Buttler; *Limonium longispicatum* Erben; *Beta rubra* Delile; *Citrus spaerocarpa* Tanaka; ined. *Ranunculus polonicus* W. Koch, *Ranunculus distentus* W. Koch in Rutishauser and Hunziker (1950), *Ranunculus laeteviridis* in Nygren (1954a, 1954b)].

Among the different plant families, alternative taxonomic treatments of apomictic taxa are often considered. In the diverse genera *Limonium* spp. (Plumbaginaceae Juss.) distinct authors use slightly different species' circumscription systems (Cowan et al., 1998). While most authors recognize all agamospermous taxa at specific rank and segregate variable agamospermous taxa into narrowly distinguished species (see e.g. Erben, 1978, 1988, 1989, 1993; Pignatti, 1972; Artelari, 1989a, 1989b, 1989c; Brullo, 1978; Brullo and Erben, 1989), others allow agamospermous species to be variable and recognize different ranks like varieties, subspecies, species and informal supraspecific groups based on morphology, palynology and cytology (e.g., Aimé and Roiron, 1985; Ingrouille and Stace, 1986; Bolôs and Vigo, 1996). According to Brullo and Erben (1989), since sexual and agamospermous lineages in *Limonium* spp. have different evolutionary origin and undergo distinct processes of speciation and arrangement of two genome sets (x=8 and x=9), the microspecies concept is the most practical one to solve the taxonomic problems.

Similar approaches considering morphological and genetic variability and differential rates of sexuality are being taken on other genera with apomixis and hybridization (e.g., Windham and Al-Shehbaz, 2006, 2007a, 2007b; Hörandl, 2018; Karbstein et al., 2020). From a taxonomic point of view, there is a propensity to accept the concept of microspecies to resolve apomictic groups considering morphologically, cytologically and/or phylogenetically related biotypes that reproduce asexually as valid taxonomic subdivisions. However, the debate on the taxonomic treatment of apomictic taxa is unresolved (see Majeský et al., 2017 and the comments in Chapter 4, 5 and 6) and therefore, for each genus, we have followed the taxonomic concepts accepted by specialists (i.e., the authors of the article on each case).

Whenever possible, ploidy levels from the studied materials were included (see Tables 7.1 and 7.2). However, in cases where ploidy was not evaluated or provided in the original study, the ploidy level was added only when a single ploidy record was available in the literature (without providing references). In a few cases where ploidy could not be determined despite chromosomal data, ploidy was inferred using relevant information. For example, in *Eriotheca pubescens* (Mart. & Zucc.) Schott & Endl. (Malvaceae Juss.), a tree with 2n=276 chromosomes, hexaploidy was inferred (Table 7.1) based on the existence of the sexual counterpart *E. gracilipes* (K. Schum.) A. Robyns holding $2n$=92 chromosomes (Oliveira et al., 1992).

In each case, we tried to mention the first study reporting apomixis. Seed set analyses using emasculation/castration experiments or pollen exclusion tests on taxa were listed in Table 7.1 but would require embryological confirmation (see Appendix 1.1). Likewise, embryology-based studies are placed in Table 7.1, and those in doubt are discussed when appropriate (see also Appendix 1.2, 1.3, 1.4, 1.5, 1.6, 1.7, 1.8). Cases of odd-ploidy taxa displaying high seed set but lacking experimental pollen exclusion or castration tests are not listed in Tables 7.1 or 7.2 [e.g., species of *Parrya* R. Brown and *Erysimum* L., Brassicaceae Burnett, Mulligan (1966); *Listera borealis* Morong ≡*Neottia borealis* (Morong) Szlach., Orchidaceae Juss., Simon (1968)]. Similar cases for which there is evidence of apomixis in related species are listed in our quarantine Table 7.2 [e.g., species of *Elatostema* J.R. Forst. & G. Forst., Urticaceae Juss., Fu et al. (2017); *Miconia ruddae* (Wurdack) Michelang. ≡*Clidemia ruddae* Wurdack, Melastomataceae Juss., Almeda and Chuang (1992)].

Evidence of experimental apomixis obtained either by subjecting plants to different laboratory conditions or from artificial hybrids was not considered here. For example, Halperin and Wetherell (1964) found the formation of large numbers of adventive embryos in carrot (*Daucus carota* L., Apiaceae Lindl.) after using different tissue culture media. Bogomolov (2005) obtained sugar beet apomictic gamma-lines by pollinating male-sterile sugar beet plants with gamma-irradiated pollen of the wild beets *Beta corolliflora* and *B. trigyna* (Amaranthaceae Juss.). Ellerstrom and Zagorcheva (1977) and Ellerstrom (1983) reported high frequency of apospory (70% of ovules) in *Raphanus* L. × *Brassica* L. hybrids (Brassicaceae Burnett). Other examples include the study by Mullins and Srinivasan (1976) in grapevine (*Vitis vinifera* L., Vitaceae Juss.), Pliego-Alfaro and Murashige (1988) in avocado (*Persea americana* Mill., Lauraceae Juss.), Shilpa and Nadgauda (2000) in cashew-nut (*Anacardium occidentale* L., Anacardiaceae R.Br.), Taşkin et al. (2009) in *Boechera holboellii* (Hornem.) Á. Löve & D. Löve (Brassicaceae Burnett), Vila et al. (2009) in *Cedrela fissilis* Vell. (Meliaceae Juss.), Chen et al. (2018) in cassava (*Manihot esculenta* Crantz, Euphorbiaceae Juss.) or Heslop-Harrison (1959) in *Dactylorhiza* Neck. ex Nevski (=*Dactylorchis* Verm.) and Veyret (1972) in *Cynorkis* Thouars interspecific hybrids (Orchidaceae Juss.).

This also excludes previous reports of apomixis in cassava plants derived from interspecific crosses between *Manihot* species (Nassar, 1994, 2006; Nassar et al., 1998; Freitas and Nassar, 2013). The nature of apomixis in cassava is suggested to be aposporous but embryological evidence is unclear (e.g., Nassar, 2002; Nassar et al., 2000, 2011), and the low rates of clonality (<2%) detected in diploid plants using molecular markers and the spontaneous occurrence of multiple seedlings suggest a case of adventitious embryony (Grattapaglia et al., 1996). Verification of apomixis in natural populations of *Manihot esculenta* Crantz (Table 7.2) and other species of the genus will help to clarify this case.

Categorization of apomictic taxa based on pollen morphology and size were considered only when support from other analyses was available. For example, in *Boechera* spp. there exists a correlation between ploidy, pollen morphology and size, and reproductive mode (Dobeš et al., 2004a, 2004b). Yet, species categorized as apomictic based on morphological or anatomical characters (Mulligan and Findlay, 1970; Decker, 2006; Windham and Al-Shehbaz, 2006, 2007a, 2007b) were listed only if additional evidence supporting or probing apomixis was available (e.g., Jordon-Thaden and Koch, 2012; Mau et al., 2015; Carman et al., 2019).

Despite the frequent observation of abnormal pollen development and asynapsis in apomictic species from different families, indications about occurrence of apomixis based on observations of meiotic abnormalities during the development of the male gametophyte and poor pollen quality (e.g., in species of *Plectranthus* L'Hér., Lamiaceae Martinov, Thoppil and Jose, 1996), based on biosystematics or biogeographic studies involving ploidy distributions and/or dimorphic pollen and stigma combinations (e.g., in *Boehmeria* Jacq., Yahara, 1983, 1990; in *Limonium* Mill., Baker, 1953; Erben, 1978; Ingrouille and Stace, 1985; Georgakopoulou et al., 2006; Kouzali et al., 2012; Caperta et al., 2017), were considered only when complemented by embryological evaluation of the female gametophyte development in at least one of the species (e.g., in *Boehmeria* spp., Zhang et al., 1997; Tang et al., 2016; in *Limonium* spp., D'Amato, 1949; Hjelmqvist and Grazi, 1964; Róis et al., 2016).

TABLE 7.1
Table of Species with Apomixis (A) from all Angiosperm Families Except Poaceae Barnhart, Asteraceae Giseke, and Rosaceae Juss., Alphabetically Sorted (M: Monocot; D: Dicot)

Species	Ploidy	Type of Apomixis[a]	References
Amaranthaceae Juss. (D)			
Aerva javanica (Burm. f.) Juss. (=*Aerva tomentosa* Forssk.)	*4x*	A; Di (An); Au	Sachar and Murgai (1958), Khan et al. (1970)
Beta corolliflora Zosimovic ex Buttler	*4x*	A; N; I; Ps	Jassem and Jassem (1970)
Beta lomatogona Fisch. & C.A. Mey.	*4x*	A; Ap (Hi); N; I; Au	Barocka (1966), Filutowicz and Dalke (1972), Jassem and Jassem (1970), Jassem (1976, 1990)
Beta trigyna Waldst. & Kit.	*6x*	A; Ap (Hi); Di (An); N; I; Au	Barocka (1966), Filutowicz and Dalke (1972), Jassem and Jassem (1970), Jassem (1976, 1990)
Amaryllidaceae J.St.-Hil. (M)			
Allium nutans L.	*5x*	A; Di	Håkansson (1951), Håkansson and Levan (1957)
Allium ramosum L. (=*Allium odorum* L.)	*4x*	A; Di; I; Ps	Tretjakow (1895), Modilewsky (1925, 1931), Håkansson (1951), Håkansson and Levan (1957), Kojima and Nagato (1992a, 1992b, 1997), Kojima et al. (1991), Nakazawa et al. (2006), Yamashita et al. (2012)
Allium rotundum L.		A; Au	Gvaladze (1976)
Allium schoenoprasum L.		A; Di	Chochlov et al. (1978)
Allium senescens L.	*6x*	A; Di; Ps	Kim et al. (1999)
Allium tuberosum Rottler ex Spreng.	*4x*	A; Di; AE; Ps	Gohil and Kaul (1981), Kojima and Kawaguchi (1989), Kojima et al. (1991), Kojima and Nagato (1992a, 1992b)
Calostemma purpureum R. Br.		A; N; Au	Schlimbach (1924)
Zephyranthes chlorosolen (Herb.) D. Dietr. (=*Zephyranthes brazosensis* Traub; =*Cooperia drummondii* Herb.; =*Zephyranthes herbertiana* D. Dietr.)	*8x, 10x, 12x*	A; Di (An); Ps	Coe (1953), Raina and Khoshoo (1972), Church (1916), Crane (2019)
Zephyranthes drummondii D. Don (=*Cooperia pedunculata* Herb.)	*8x*	A; Di (An); Ps	Coe (1953), Crane (2019)
Zephyranthes macrosiphon Baker	*8x*	A; Di (An); Ps	Solntzeva (1978)
Zephyranthes smallii (Alexander) Traub	*9x, 16x*	A; Ps	Flagg and Flory (1976)
Zephyranthes sulphurea Hort.	*8x*	A; Ps	Raina and Khoshoo (1972)
Zephyranthes tubispatha (L'Hér.) Herb. [≡*Habranthus tubispathus* (L'Hér.) Traub; =*Zephyranthes texana* Herb.]	*4x*	A; Di (An); Ps; Au	Pace (1913), Brown (1951), Gianini Aquino et al. (2019), Gianini Aquino (2021)
Nothoscordum gracile (Aiton) Stearn [=*Nothoscordum fragrans* (Vent.) Kunth]	*3x*	A; N; Ps	Strasburger (1878), Messeri (1931), D'Amato (1940), Håkansson (1953)
Anacardiaceae R.Br. (D)			
Astronium fraxinifolium Schott		A; AE; Au	Allem (1991), Salomão and Allem (2001)
Mangifera indica L.	*2x*	A; N	Strasburger (1878), Juliano (1934, 1937), Sachar and Chopra (1957), Martínez Ochoa et al. (2012)

(Continued)

TABLE 7.1 (Continued)
Table of Species with Apomixis (A) from all Angiosperm Families Except Poaceae Barnhart, Asteraceae Giseke, and Rosaceae Juss., Alphabetically Sorted (M: Monocot; D: Dicot)

Species	Ploidy	Type of Apomixis[a]	References
Mangifera odorata Griff.	2*x*	A; N	Juliano (1934, 1937), Sachar and Chopra (1957)
Tapirira guianensis Aubl.		A; AE; Au	Salomão and Allem (2001), Lenza and Oliveira (2005)
Apiaceae Lindl. (D)			
Ammi majus L.	2*x*	A; I	Naumova (1993)
Apocynaceae Juss. (D)			
Vincetoxicum hirundinaria Medik. [=*Cynanchum vincetoxicum* (L.) Pers.; =*Vincetoxicum officinale* Moench; =*Cynanchum medium* (Decne.) K. Schum.][7]	2*x*	A; N; I	Seefeldner (1912), Kordyum (1961)
Vincetoxicum hirundinaria subsp. *cretaceum* (Probed.) Markgr. [=*Vincetoxicum cretaceum* (Pobed.) O. D. Wissjul.]		A; AE	Naumova and Yakovlev (1972)
Vincetoxicum hirundinaria subsp. *stepposum* (Pobed.) Markgr. [=*Vincetoxicum stepposum* (Pobedimova in Komarov) Á. Löve & D. Löve]		A; AE	Naumova (1993)
Vincetoxicum nigrum (L.) Moench [≡*Cynanchum nigrum* (L.) Pers.; =*Cynanchum louiseae* Kartesz & Gandhi][7]	2*x*	A; N; I	Guignard (1922), von Hausner (1976)
Vincetoxicum rossicum (Kleopow) Barbar.		A; AE	von Hausner (1976), Hotchkiss et al. (2008)
Araceae Juss. (M)			
Aglaonema commutatum Schott[5]	6*x*	A; Ap (Hi)	Pfitzer (1957, 1962)
Aglaonema crispum (Pitcher & Manda) Nicolson (=*Aglaonema roebelinii* Pitcher & Manda)[5]	3*x*	A; Ap (Hi)	Pfitzer (1962)
Spathiphyllum patinii (R. Hogg) N.E.Br.		A; N	Schürhoff and Jüssen (1925)
Asparagaceae Juss. (M)			
Hosta ventricosa Stearn (=*Hosta coerulea* Tratt.; =*Funkia ovata* Spreng.)	6*x*	A; N; Au	Braun (1860), Strasburger (1878), Fagerlind (1944a, 1944b), Hu (1963)
Hosta ovata Spring	4*x*	A; N; Au	Strasburger (1878), Fagerlind (1944a, 1944b)
Ornithogalum gussonei Ten.		A; Ap	Zabińska (1972, in Gregor, 2013)
Ornithogalum umbellatum L.	3*x*, 5*x*	A; N	Naumova (1993), Gregor (2013)
Balanophoraceae Rich. (D)			
Balanophora fungosa var. *globosa* (Jungh.) B. Hansen (=*Balanophora globosa* Jungh.)		A; Di (Ta); Au	Zweifel (1939), Fagerlind (1945a, 1945b)
Balanophora japonica Makino	6*x*	A; Di (Ta)	Kuwada (1928)
Ombrophytum subterraneum (Aspl.) B. Hansen (≡*Juelia subterranea* Aspl.)		A; Au	Asplund (1928)
Betulaceae Gray (D)			
Alnus incana (L.) Moench		A; N	Naumova (1993)
Alnus incana subsp. *rugosa* (Du Roi) R.T. Clausen [=*Alnus rugosa* (Du Roi) Spreng.]	4*x*	A; Di (Ta); N; Au	Woodworth (1929, 1930)

TABLE 7.1 (Continued)

Species	Ploidy	Type of Apomixis[a]	References
Alnus alnobetula subsp. *sinuata* (Regel) Raus [=*Duschekia kamtschatica* (Callier) Pouzar]		A; Di; N; Au	Naumova (1993)
Betula pendula Roth		A; N; Au	Naumova (1993)
Betula pendula var. *oycowiensis* (Besser) Dippel (=*Betula oycowiensis* Besser)		A; N; Au	Korchagina (1981), Naumova (1993)
Biebersteiniaceae Schnizl. (D)			
Biebersteinia odora Stephan ex Fisch.		A; Ps	Brožová et al. (2019)
Bignoniaceae Juss. (D)			
Anemopaegma acutifolium DC.	*4x*	A; N; I; Ps	Firetti-Leggieri et al. (2013), Sampaio et al. (2013a, 2013b)
Anemopaegma arvense (Vell.) Stellfeld ex De Souza	*4x*	A; N; I; Ps	Firetti-Leggieri et al. (2013)
Anemopaegma glaucum Mart. ex DC.	*4x*	A; N; I; Ps	Firetti-Leggieri et al. (2013)
Anemopaegma scabriusculum Mart. ex DC.	*4x*	A; AE; Ps	Firetti-Leggieri et al. (2013)
Handroanthus chrysotrichus (Mart. ex DC.) Mattos [≡*Tabebuia chrysotricha* (Mart. ex DC.) Standl.]	*4x*	A; N; Ps	Piazzano (1998), Souza et al. (2005), Bittencourt and Moraes (2010)
Handroanthus ochraceus (Cham.) Mattos [≡*Tabebuia ochracea* (Cham.) Standl.]	*4x*	A; N; I; Ps	Salomão and Allem (2001), Costa et al. (2004), Bittencourt and Moraes (2010), Mendes et al. (2018)
Handroanthus serratifolius (Vahl) S.O. Grose	*6x*	A; N; I; Ps	Alves et al. (2016), Braga et al. (2019)
Brassicaceae Burnett (D)			
Boechera breweri (S. Watson) Al-Shehbaz	*2x*	A	Mau et al. (2015)
Boechera carrizozoensis P.J. Alexander	*2x*	A	Alexander et al. (2015)
Boechera centrifendleri P.J. Alexander	*2x*	A	Alexander et al. (2015)
Boechera cobrensis (M.E. Jones) Dorn	*2x*	A	Mau et al. (2015)
Boechera collinsii (Fernald) Á. Löve & D. Löve	*2x, 3x*	A; Di (Ta); Au	Mau et al. (2015)
Boechera crandallii (B.L. Rob.) W.A. Weber	*2x*	A; Ap; Au	Aliyu et al. (2010), Mateo de Arias (2015), Mau et al. (2015)
Boechera divaricarpa (A. Nelson) Á. Löve & D. Löve	*2x, 3x*	A; Ps; Au	Vorobik (1985), Schranz et al. (2005), Aliyu et al. (2010), Mau et al. (2015)
Boechera fendleri (S. Watson) W.A. Weber	*3x*	A; Di (Ta); Au	Mau et al. (2015)
Boechera fernaldiana (Rollins) W.A. Weber (sub *Boechera fernaldiana* subsp. *fernaldiana*)	*2x, 3x*	A	Mau et al. (2015)
Boechera formosa (Greene) Windham & Al-Shehbaz	*2x*	A; Ps; Au	Aliyu et al. (2010)
Boechera glareosa Dorn	*2x*	A; Ps; Au	Aliyu et al. (2010)
Boechera gracilipes (Greene) Dorn	*2x*	A; Ap (Hi)	Mateo de Arias (2015), Carman et al. (2019)
Boechera gunnisoniana (Rollins) W.A. Weber	*3x*	A; Di (Ta); Ps	Roy (1995), Taşkin et al. (2004), Carman et al. (2019)
Boechera holboellii (Hornem.) Á. Löve & D. Löve (≡*Arabis holboellii* Hornem.)	*2x, 3x, 4x*	A; Di (Ta); Ps; Au	Böcher (1951), Roy (1995), Naumova et al. (2001), Schranz et al. (2005), Aliyu et al. (2010)

(Continued)

TABLE 7.1 (Continued)

Table of Species with Apomixis (A) from all Angiosperm Families Except Poaceae Barnhart, Asteraceae Giseke, and Rosaceae Juss., Alphabetically Sorted (M: Monocot; D: Dicot)

Species	Ploidy	Type of Apomixis[a]	References
Boechera inyoensis (Rollins) Al-Shehbaz	*3x*	A	Rollins and Rüdenberg (1971), Mau et al. (2015)
Boechera lemmonii (S. Watson) W.A. Weber	*2x*	A; Au	Aliyu et al. (2010), Mau et al. (2015)
Boechera lignifera (A. Nelson) W.A. Weber (≡*Arabis lignifera* A. Nelson)	*2x*	A; Ap (Hi); Ps; Au	Roy (1995), Aliyu et al. (2010), Mateo de Arias (2015)
Boechera lyallii (S. Watson) Dorn	*3x*	A; Di (Ta); Au	Vorobik (1985), Mulligan (1996), Mau et al. (2015)
Boechera macounii (S. Watson) Windham & Al-Shehbaz		A	Mau et al. (2015)
Boechera microphylla (Nutt.) Dorn	*2x*	A; Ap (Hi); Di (Ta); Au	Böcher (1969), Carman (2007), Mateo de Arias (2015), Mau et al. (2015)
Boechera pallidifolia (Rollins) W.A. Weber [=*Boechera thompsonii* (S.L. Welsh) N.H. Holmgren]	*2x, 3x*	A; Ap (Hi); Ps; Au	Aliyu et al. (2010), Mau et al. (2015), Carman et al. (2019)
Boechera pauciflora (Nutt.) Windham & Al-Shehbaz		A	Mau et al. (2015)
Boechera pendulina (Greene) W.A. Weber	*2x*	A	Mau et al. (2015)
Boechera pendulocarpa (A. Nelson) Windham & Al-Shehbaz	*2x, 3x*	A	Lovell et al. (2013), Mau et al. (2015)
Boechera perennans (S. Watson) W.A. Weber	*2x*	A	Mau et al. (2015)
Boechera pinetorum (Tidestr.) Windham & Al-Shehbaz	*3x*	A	Mau et al. (2015)
Boechera polyantha (Greene) Windham & Al-Shehbaz	*2x*	A; Ps; Au	Aliyu et al. (2010), Lovell et al. (2013)
Boechera porphyrea (Wooton & Standl.) Windham, Al-Shehbaz & P.J. Alexander	*3x*	A	Alexander et al. (2015)
Boechera puberula (Nutt.) Dorn	*3x*	A; Au	Mulligan (1996), Mau et al. (2015)
Boechera pulchra (M.E. Jones ex S. Watson) W.A. Weber	*3x*	A; Ap (Hi)	Rollins and Rüdenberg (1971), Mateo de Arias (2015)
Boechera retrofracta (Graham) Á. Löve & D. Löve	*2x, 3x*	A; Ap (Hi); Ps; Au	Aliyu et al. (2010), Lovell et al. (2013), Mateo de Arias (2015), Mau et al. (2015), Carman et al. (2019)
Boechera sanluisensis P.J. Alexander	*2x*	A	Alexander et al. (2015)
Boechera sparsiflora (Nutt.) Dorn (=*Arabis columbiana* Macoun)	*3x*	A; Di (Ta); Au	Raven et al. (1965), Böcher (1969), Vorobik (1985), Mulligan (1996), Mau et al. (2015)
Boechera spatifolia (Rydb.) Windham & Al-Shehbaz	*2x, 3x*	A	Lovell et al. (2014, 2017), Mau et al. (2015)
Boechera stricta (Graham) Al-Shehbaz (=*Arabis drummondii* A. Gray)	*2x, 3x*	A; Di (Ta); Ps; Au	Mulligan (1996), Aliyu et al. (2010), Mau et al. (2015), Rojek et al. (2018)
Boechera suffrutescens (S. Watson) Dorn	*2x*	A; Ps; Au	Aliyu et al. (2010)
Boechera williamsii (Rollins) Dorn	*2x*	A; Ps; Au	Aliyu et al. (2010), Mau et al. (2015)
Boechera zephyra P.J. Alexander	*2x*	A	Alexander et al. (2015)
Borodinia laevigata (Muhl. ex Willd.) P.J. Alexander & Windham [≡*Boechera laevigata* (Muhl. ex Willd.) Al-Shehbaz]	*2x*	A; Ap (Hi)	Mau et al. (2015), Carman et al. (2019)

TABLE 7.1 (Continued)

Species	Ploidy	Type of Apomixis[a]	References
Borodinia canadensis (L.) P.J. Alexander & Windham [≡*Boechera canadensis* (L.) Al-Shehbaz]	*2x*	A	Mau et al. (2015)
Draba oligosperma Hook.	*8x*	A; Ap; Au	Jordon-Thaden and Koch (2012)
Iberis saxatilis L. (sub *Iberis saxatilis* subsp. *saxatilis*)	*3x*	A; Ap; Di; Ps; Au	Yurukova Grancharova et al. (2004)
Phoenicaulis cheiranthoides Nutt.	*2x, 3x, 4x*	A; Ap	Mandáková et al. (2021)
Polyctenium fremontii (S. Watson) Greene	*4x*	A; Ap (Hi)	Mandáková et al. (2020)
Sandbergia whitedii (Piper) Greene	*3x*	A; Ap (Hi)	Mandáková et al. (2020)
Burmanniaceae Blume (M)			
Burmannia coelestis D. Don		A; Di (An)	Ernst and Bernard (1912)
Burseraceae Kunth (D)			
Commiphora leptophloeos (Mart.) J.B. Gillett		A; AE	Salomão and Allem (2001)
Commiphora wightii (Arn.) Bhandari		A; N; Au	Gupta et al. (1996), Kavane et al. (2021)
Garuga pinnata Roxb.	*2x*	A; N	Ghosh (1970)
Buxaceae Dumort. (D)			
Sarcococca confusa Sealy	*4x*	A	Hoskins and Contreras (2019)
Sarcococca hookeriana Baill.	*2x*	A; N; Au	Naumova (1980, 1993)
Sarcococca hookeriana var. *digyna* Franch. (=*Sarcococca humilis* Stapf, sub *Sarcococca humilis* Hort.)	*2x*	A; N; Au	Naumova (1980), Naumova and Willemse (1983a, 1983b)
Sarcococca ruscifolia Stapf	*4x*	A; N	Orr (1923), Naumova (1993)
Sarcococca saligna (D. Don) Müll. Arg. (=*Sarcococca pruniformis* Lindl.)		A; N; Au	Wiger (1930), Naumova (1993)
Sarcococca zeylanica Baill.		A; N	Naumova (1980, 1993)
Cactaceae Juss. (D)			
Cactus tenuis (DC.) Kuntze (=*Mammillaria tenuis* DC.)		A; N; Ps	Tiagi (1956)
Haageocereus tenuis F. Ritter	*3x*	A; AE	Arakaki et al. (2013)
Opuntia aurantiaca Lindl.		A; N; Au	Archibald (1939)
Opuntia dillenii (Ker Gawl.) Haw.		A; N; Au	Maheshwari and Chopra (1955)
Opuntia elata Link & Otto ex Salm-Dyck		A; N	Naumova (1978, 1993)
Opuntia ficus-indica (L.) Mill.		A; N	Montemartini (1899)
Opuntia glaucophylla H.L. Wendl.		A; N	Braun (1860, in Tisserat et al., 1979, p. 6)
Opuntia leucantha Link & Otto (sub *Opuntia leucantha* Link)		A; N	Montemartini (1899, in Tisserat et al., 1979, p. 6)
Opuntia humifusa (Raf.) Raf. (=*Opuntia rafinesquei* Engelm., sub *Opuntia rafinesquii* Engelm.)		A; N	Hull (1915)
Opuntia macrorhiza Engelm. (=*Opuntia tortispina* Engelm. & J.M. Bigelow, sub *Opuntia tortispina* Engelm.)	*6x*	A; N	Braun (1860, in Tisserat et al., 1979, p. 6)
Opuntia ficus-indica (L.) Mill. (=*Opuntia vulgaris* Mill.)		A; N; Ps	Ganong (1898a, 1898b)

(Continued)

TABLE 7.1 (Continued)
Table of Species with Apomixis (A) from all Angiosperm Families Except Poaceae Barnhart, Asteraceae Giseke, and Rosaceae Juss., Alphabetically Sorted (M: Monocot; D: Dicot)

Species	Ploidy	Type of Apomixis[a]	References
Calycanthaceae Lindl. (D)			
Calycanthus floridus var. *glaucus* (Willd.) Torr. & A. Gray (= *Calycanthus fertilis* Walter)[4]	*2x*	A; N; I; Au	Peter (1920), Schürhoff (1923)
Calycanthus floridus L.[4]	*2x*	A; N; I; Au	Peter (1920), Schürhoff (1923)
Calycanthus occidentalis Hook. & Arn.[4]	*2x*	A; N; I; Au	Peter (1920), Schürhoff (1923)
Chimonanthus praecox (L.) Link	*2x*	A; I	Naumova (1993)
Cannabaceae Martinov (D)			
Humulus lupulus L.		A; Au	Kerner (1898, in Kirchner, 1905), Wettstein (1925)
Capparaceae Juss. (D)			
Capparidastrum frondosum (Jacq.) Cornejo & Iltis (≡*Capparis frondosa* Jacq.)		A; I	Mauritzon (1934)
Casuarinaceae R.Br. (D)			
Allocasuarina distyla (Vent.) L.A.S. Johnson (≡*Casuarina distyla* Vent.)		A; Di; Au	Barlow (1959)
Allocasuarina nana (Sieber ex Spreng.) L.A.S. Johnson (≡*Casuarina nana* Sieber ex Spreng.)	*3x*	A; Di; Ps	Barlow (1958a, 1958b, 1959)
Allocasuarina pusilla (Macklin) L.A.S. Johnson (≡*Casuarina pusilla* Macklin)		A; Di; Ps	Barlow (1959)
Allocasuarina striata (Macklin) L.A.S. Johnson (≡*Casuarina striata* Macklin)		A; Di; Au	Barlow (1959)
Celastraceae R.Br. (D)			
Celastrus paniculatus Willd.	*2x*	A; I	Naumova (1993)
Celastrus scandens L.	*2x*	A; I	Naumova (1993)
Euonymus alatus (Thunb.) Siebold [=*Euonymus alatus* (Thunb.) Regel var. *apterus* Regel; =*Euonymus alatus* var. *apertus* Loes.]	*5x*	A; I	Brizicky (1964), Naumova (1971, 1979)
Euonymus yakushimensis Makino [=*Euonymus alatus* f. *angustatus* (Makino) H. Hara, sub *Euonymus alatus* f. *angustifolia*]		A; I	Naumova (1993)
Euonymus americanus L.		A; I; Ps	Braun (1860), Brizicky (1964), Naumova (1971, 1979)
Euonymus dielsianus Loes. ex Diels		A; I	Brizicky (1964), Naumova (1971, 1979)
Euonymus europaeus L.	*8x*	A; I	Naumova (1970a, 1971, 1979)
Euonymus fimbriatus Wall.		A; I	Naumova (1971, 1979)
Euonymus hamiltonianus Wall. (=*Euonymus yedoensis* Koehne)	*4x*	A; I	Naumova (1971, 1979)
Euonymus japonicus Thunb.		A; I	Copeland (1966)
Euonymus latifolius (L.) Mill.	*8x*	A; N; I; Ps	Braun (1860), Strasburger (1878), Andersson (1931), Brizicky (1964), Naumova (1971, 1979)

TABLE 7.1 (Continued)

Species	Ploidy	Type of Apomixis[a]	References
Euonymus macropterus Rupr.		A; I	Brizicky (1964), Naumova (1970b, 1978)
Euonymus maximowiczianus (Prokh.) Vorosch.	*8x*	A; I	Naumova (1971, 1979)
Euonymus miniatus Tolm.		A; I	Naumova (1971, 1979)
Euonymus nanus M.Bieb. (=*Euonymus koopmannii* Lauche)		A; I	Naumova (1971, 1979)
Euonymus oxyphyllus Miq.		A; I	Brizicky (1964), Naumova (1971, 1979)
Euonymus sachalinensis (F. Schmidt) Maxim. (=*Euonymus planipes* Koehne)	*8x*	A; I	Naumova (1970b)
Euonymus vagans Wall.		A; I	Brizicky (1964), Naumova (1971, 1979)
Euonymus verrucosoides Loes.		A; I	Brizicky (1964), Naumova (1971, 1979)
Clusiaceae Lindl. (D)			
Clusia alba Jacq.		A; N; Ps	Goebel (1900)
Clusia minor L.		A; AE	Maguire (1976), Hammel (1986)
Clusia rosea Jacq.		A; AE	Maguire (1976), Cheek and Lalla (2017)
Clusia criuva Cambess.		A; AE	Rodrigues-Correia et al. (2011)
Clusia fluminensis Planch. & Triana		A; AE	Rodrigues-Correia et al. (2011)
Garcinia atroviridis Griff. ex T. Anderson		A; Au	Pangsuban et al. (2009)
Garcinia cowa Roxb.	*2x*	A; I	Treub (1911a, 1911b)
Garcinia hombroniana Pierre	*2x*	A; I; Au	Richards (1990)
Garcinia livingstonei T. Anderson		A; I; Au	Puri (1939)
Garcinia mangostana L.	*4x*	A; I; Au	Sprecher (1919), Horn (1940), Lim (1974, 1984), Kaur et al. (1978)
Garcinia parvifolia (Miq.) Miq.		A; Di (Ta); I; Au	Ha et al. (1988)
Garcinia scortechinii King		A; I; Au	Ha et al. (1988), Thomas (1997)
Garcinia treubii Pierre		A; I	Treub (1911a, 1911b)
Colchicaceae DC. (M)			
Colchicum autumnale L.		A; N	Braun (1860, in Naumova (1993))
Combretaceae R.Br. (D)			
Combretum coccineum (Sonn.) Lam. [≡*Poivrea coccinea* (Sonn.) DC.]		A; N	Venkateswarlu (1947)
Combretum paniculatum Vent. (=*Combretum pincianum* Hook.)		A; N	Naumova (1993)
Convolvulaceae Juss. (D)			
Cuscuta campestris Yunck.		A; Au	Plitmann (2002)
Cucurbitaceae Juss. (D)			
Bryonia alba L.		A; Au	Engle (1988), Novak and Mack (1995, 2000)
Cucumis anguria L.	*4x*	A; Au	Zagorcheva (1985)
Cucumis ficifolius A. Rich.	*4x*	A; Au	Zagorcheva (1985)
Cucumis melo L. (=*Cucumis melo* L. var. *pubescens* Willd.)	*4x*	A; N	Singh (1955)
Cucumis metuliferus E. Mey. ex Naudin	*4x*	A; Ap (Hi); Di	Singh and Dathann (1972)
Luffa cylindrica (L.) M. Roem.	*2x*	A; Di (An)	Deshpande et al. (1984)

(Continued)

TABLE 7.1 (Continued)
Table of Species with Apomixis (A) from all Angiosperm Families Except Poaceae Barnhart, Asteraceae Giseke, and Rosaceae Juss., Alphabetically Sorted (M: Monocot; D: Dicot)

Species	Ploidy	Type of Apomixis[a]	References
Momordica charantia L. [=*Momordica charantia* var. *muricata* (Willd.) Chakrav.]		A; Di; N	Agrawal and Singh (1957), Bhuskute (1987)
Sicyos angulatus L.		A; N	Naumova (1993)
Telfairia occidentalis Hook.f.	2*x*	A; AE; Au	Odiyi and Uzo (2001), Fayeun et al. (2016)
Cunoniaceae R.Br. (D)			
Davidsonia pruriens F. Muell.		A; AE	Eliott et al. (2014)
Dipterocarpaceae Blume (D)			
Hopea latifolia Symington	3*x*	A; AE	Kaur et al. (1978)
Hopea mengarawan Miq.		A; AE	Maury (1968, in Kaur et al., 1978, p. 441)
Hopea odorata Roxb.	3*x*	A; AE	Maury (1968, in Kaur et al., 1978, p. 441), Kaur et al. (1986), Wickneswari et al. (1995)
Hopea subalata Symington	3*x*	A; N; Ps	Chan (1981), Kaur et al. (1986)
Shorea argentifolia Symington	2*x*	A; AE	Kaur et al. (1978, 1986)
Shorea agami P.S. Ashton	2*x*	A; N	Kaur et al. (1978, 1986)
Shorea macroptera Dyer	2*x*	A; AE	Kaur et al. (1986)
Shorea ovalis Blume subsp. *sericea* P.S. Ashton	4*x*	A; N; Ps	Chan (1981), Kaur et al. (1978, 1986), Ng et al. (2004)
Shorea ovalis Blume subsp. *ovalis*	4*x*	A; N; Ps	Kaur et al. (1986)
Shorea pauciflora King	2*x*	A; AE	Kaur et al. (1986)
Shorea resinosa Foxw.	3*x*	A; AE	Kaur et al. (1986)
Shorea robusta Gaertn.	2*x*	A; AE	Foxworthy (1932, in Kaur et al., 1978, p. 441)
Shorea smithiana Symington	2*x*	A; AE	Maury (1968, in Kaur et al. (1978), p. 441)
Stemonoporus oblongifolius Thwaites		A; AE; Ps	Murawski and Bawa (1994)
Elaeagnaceae Juss. (D)			
Elaeagnus rhamnoides (L.) A. Nelson (≡*Hippophae rhamnoides* L.)[8]	2*x*	A; Ap (Pa); N; Ps; Au	Mangla et al. (2015)
Erythroxylaceae Kunth (D)			
Erythroxylum undulatum Plowman	3*x*	A; Ap (Hi); Au	Berry et al. (1991)
Euphorbiaceae Juss. (D)			
Alchornea aquifolia (Js.Sm.) Domin [=*Alchornea ilicifolia* (Js.Sm.) Müll.Arg.; =*Coelebogyne ilicifolia* Js.Sm.]		A; AE; Au	Smith (1841)
Euphorbia dulcis L.	4*x*	A; N; Au	Hegelmaier (1911, 1913), Carano (1926), Cesca (1961), Kapil (1961)
Euphorbia platyphyllos L.		A; N	Naumova (1993)
Euphorbia rosea Retz.		A; N	Naumova (1993)
Mallotus japonicus (Spreng.) Müll. Arg.		A; N	Ventura (1934)
Fabaceae Juss. (D)			
Acacia aneura F. Muell. ex Benth.	3*x*, 4*x*, 5*x*	A; AE	Miller et al. (2002), Andrew et al. (2003)
Acacia ayersiana Maconochie	4*x*	A; AE	Andrew et al. (2003)

TABLE 7.1 (Continued)

Species	Ploidy	Type of Apomixis[a]	References
Acacia cowleana Tate		A	Moran et al. (1992)
Acacia holosericea A. Cunn. ex G. Don	*4x*	A	Moran et al. (1992)
Acacia minyura Randell	*3x, 4x*	A; AE	Andrew et al. (2003)
Cassia artemisioides Gaud. ex DC.	*3x, 4x*	A; N; Ps	Symon (1956), Randell (1970), Holman and Playford (2000)
Cassia circinnata Benth.	*4x*	A; N; Ps	Randell (1970)
Cassia eremophila A. Cunn. ex Vogel		A; AE	Symon (1956)
Cassia helmsii Symon	*3x, 4x*	A; AE	Randell (1970)
Cassia nemophila var. *coriacea* (Benth.) Symon	*3x, 4x*	A; N; Ps	Randell (1970)
Cassia nemophila var. *nemophila* Vogel	*3x, 4x*	A; AE	Randell (1970)
Cassia nemophila var. *zygophylla* (Benth.) Benth.	*3x, 4x*	A; AE	Randell (1970)
Cassia oligophylla F. Muell.	*3x, 4x*	A; AE	Randell (1970)
Cassia sturtii R.Br.	*3x, 4x*	A; N; Ps	Symon (1956), Randell (1970)
Genista salzmannii DC.	*2x, 4x*	A; Di (An); Au	Diana and Villa (1991)
Millettia peguensis Ali (=*Millettia ovalifolia* Kurz)	*2x*	A; N	Pal (1960), Naumova (1993)
Gentianaceae Juss. (D)			
Schultesia brachyptera Cham.		A; Au	Ramirez and Brito (1990)
Grossulariaceae DC. (D)			
Ribes biebersteinii Berland. ex DC.		A; N	Kobasnidze (1976, 1983, in Naumova, 1993)
Ribes nigrum L.	*2x*	A; N	Kobasnidze (1976, 1983, in Naumova, 1993)
Ribes rubrum L.	*2x*	A; N	Kobasnidze (1976, 1983, in Naumova, 1993)
Hypericaceae Juss. (D)			
Hypericum acmosepalum N. Robson	*4x*	A; Ps	Matzk et al. (2003)
Hypericum beanii N. Robson	*4x*	A; Ps	Matzk et al. (2003)
Hypericum bellum H.L. Li		A; Ps	Matzk et al. (2003)
Hypericum × *carpaticum* Mártonfi	*5x*	A; Ps	Mártonfi (2001)
Hypericum × *desetangsii* Lamotte [=*Hypericum* × *desetangsii* nothosubsp. *carinthiacum* (A. Frohl.) N. Robson]	*4x, 5x*	A; Ps	Lihová et al. (2000), Matzk et al. (2003)
Hypericum dyeri Rehder	*4x*	A; Ps	Matzk et al. (2003)
Hypericum forrestii (Chitt.) N. Robson	*4x*	A; Ps	Matzk et al. (2003)
Hypericum hookerianum Wight & Arn.	*4x*	A; Ps	Matzk et al. (2003)
Hypericum kamtschaticum Ledeb.	*4x*	A; Ps	Matzk et al. (2003)
Hypericum kouytchense H. Lév.	*4x*	A; Ps	Matzk et al. (2003)
Hypericum maculatum subsp. *obtusiusculum* (Tourlet) Hayek (=*Hypericum dubium* Leers)	*4x*	A; Ps	Matzk et al. (2003), Mártonfi (2008)
Hypericum oblongifolium Choisy	*4x*	A; Ps	Matzk et al. (2003)
Hypericum patulum Thunb.	*4x*	A; Ps	Matzk et al. (2003)

(Continued)

TABLE 7.1 (Continued)

Table of Species with Apomixis (A) from all Angiosperm Families Except Poaceae Barnhart, Asteraceae Giseke, and Rosaceae Juss., Alphabetically Sorted (M: Monocot; D: Dicot)

Species	Ploidy	Type of Apomixis[a]	References
Hypericum perforatum L.	*4x, 5x, 6x*	A; Ap (Hi); Ps	Noack (1939), Mártonfi et al. (1996), Matzk et al. (2001, 2003), Barcaccia et al. (2007), Galla et al. (2011, 2015)
Hypericum pseudohenryi N. Robson	*4x*	A; Ps	Matzk et al. (2003)
Hypericum scabrum L.		A; Ps	Matzk et al. (2003)
Hypericum yezoense Maxim.	*4x, 6x*	A; Ps	Matzk et al. (2003)
Juglandaceae DC. ex Perleb (D)			
Carya cathayensis Sarg.		A; N; Ps	Wang et al. (2010), Zhang et al. (2012)
Juglans regia L.[2]		A; Au	Zarubin (1948), Schanderl (1964), Terziĭski and Stefanova (1990), Valdeviesso (1990), Loiko (1990), Sartorius and Stösser (1997), Zhang et al. (2000), Asadian and Pieber (2005), Wu et al. (2007), Cosmulescu et al. (2012)
Lamiaceae Martinov (D)			
Thymus loscosii Willk.		A; Au	Orellana et al. (2005)
Lauraceae Juss. (D)			
Lindera glauca (Siebold & Zucc.) Blume		A; Au	Dupont (2002), Nakamura et al. (2021)
Magnoliaceae Juss. (D)			
Magnolia kwangsiensis Figlar & Noot. [=*Woonyoungia septentrionalis* (Dandy) Y.W. Law][3]		A; Ap	Zeng et al. (2003), Wang et al. (2004), Fu et al. (2012)
Malpighiaceae Juss. (D)			
Aspicarpa longipes A. Gray		A; N	Ritzerow (1907)
Aspicarpa hirtella Rich.		A; N	Ritzerow (1907)
Aspicarpa schininii W.R. Anderson (=*Aspicarpa lanata* Rose)		A; N	Ritzerow (1907)
Galphimia glauca Cav. [=*Thryallis glauca* (Cav.) Kuntze]	*4x*	A; N	Singh (1959), Naumova (1993)
Heteropterys laurifolia (L.) A. Juss. (≡*Banisteria laurifolia* L.)		A; N	Braun (1860), Subba Rao (1937)
Heteropterys eglandulosa A. Juss. [=*Heteropterys caerulea* (Lam.) DC.; =*Heteropterys anoptera* A. Juss., replaced synonym *Heteropteris caerulea* Juss.]	*2x*	A; N	Braun (1860), Naumova (1993)
Heteropterys syringifolia Griseb.		A; N	Braun (1860), Naumova (1993)
Hiptage benghalensis (L.) Kurz (=*Hiptage madablota* Gaertn.)		A; N; Au	Subba Rao (1937, 1940)
Stigmaphyllon emarginatum (Cav.) A. Juss.		A; N	Braun (1860), Naumova (1993)
Malvaceae Juss. (D)			
Eriotheca pubescens (Mart. & Z-ucc.) Schott & Endl.	*6x*	A; N; Ps	Oliveira et al. (1992), Salomão and Allem (2001), Martins and Oliveira (2003), Mendes-Rodrigues et al. (2005, 2018)

TABLE 7.1 (Continued)

Species	Ploidy	Type of Apomixis[a]	References
Eriotheca gracilipes (K. Schum.) A. Robyns	*6x*	A; N; Ps	Mendes-Rodrigues et al. (2018)
Pachira glabra Pasq. [≡*Bombacopsis glabra* (Pasq.) Robyns; =*Bombax kimuenzae* De Wild. & T. Durand; *Bombax oleagineum* Robyns]		A; N; Ps	Robins (1959), Duncan (1970), Naumova (1993)
Pachira oleagina Decne.	*3x*	A; N; Ps	Baker (1960)
Melanthiaceae Batsch ex Borkh. (M)			
Trillium undulatum Willd.	*2x*	A; N; Au	Jeffrey and Haertl (1939a, 1939b), Swamy (1948a), Naumova (1993)
Trillium reliquum J.D. Freeman[1]		A; N; Au	Gyer (1996), Gonzales (2005), Heckel and Leege (2007)
Melastomataceae Juss. (D)			
Aciotis acuminifolia (Mart. ex DC.) Triana		A; Au	Renner (1984, in Renner, 1989), Freire-Fierro (2002)
Comolia microphylla Benth.		A; Au	Hokche and Ramírez (2008)
Henriettea succosa (Aubl.) DC.		A; Au	Melo and Machado (1996)
Lavoisiera pulchella Cham.		A; Au	Maia et al. (2016)
Leandra aurea (Cham.) Cogn.		A; Di (An); Au	Caetano (2014), Maia et al. (2016)
Leandra australis (Cham.) Cogn.		A; Di; N; Au	Borges (1991), Goldenberg and Varassin (2001), Caetano and Cortez (2014), Maia et al. (2016)
Leandra melastomoides Raddi		A; Di (An); N; Au	Goldenberg and Varassin (2001)
Macairea parvifolia Benth.		A; Au	Hokche and Ramírez (2008)
Macairea theresiae Cogn.		A; Au	Renner (1984, in Renner, 1989)
Melastoma malabathricum L.	*2x*	A; AE	Subramanyam (1948)
Meriania urceolata Triana	*2x*	A; Au	Hokche and Ramírez (2008)
Miconia alata (Aubl.) DC.		A; Au	Renner (1984, in Renner, 1989)
Miconia albicans (Sw.) Steud.	*2x, 3x*	A; Di; AE; Au	Renner (1984, in Renner, 1989), Goldenberg and Shepherd (1998), Hokche and Ramírez (2008), Mendes-Rodrigues and Oliveira (2012), Caetano et al. (2013), Caetano and Cortez (2014), Maia et al. (2016), Dias et al. (2021)
Miconia alternidomatia Michelang. (=*Maieta poeppigii* Mart. ex Cogn.)	*3x*	A; Au	Renner (1989)
Miconia aplostachya (Bonpl.) DC.		A; Au	Hokche and Ramírez (2008)
Miconia argyrophylla DC.		A; Au	Renner (1984, in Renner, 1989)
Miconia bullatifolia Michelang. [=*Clidemia novemnervia* (DC.) Triana]		A; Au	Renner (1984, in Renner, 1989)
Miconia bullosa (DC.) Michelang. (=*Clidemia bullosa* DC.)	*2x*	A; Au	Melo et al. (1999)
Miconia campestris (Benth.) Triana		A; Au	Renner (1984, in Renner, 1989)
Miconia chartacea Triana		A; Au	Saraiva et al. (1996)

(Continued)

TABLE 7.1 (Continued)
Table of Species with Apomixis (A) from all Angiosperm Families Except Poaceae Barnhart, Asteraceae Giseke, and Rosaceae Juss., Alphabetically Sorted (M: Monocot; D: Dicot)

Species	Ploidy	Type of Apomixis[a]	References
Miconia ciliata (Rich.) DC.	*2x*	A; AE	Dent-Acosta and Breckon (1991, in Goldenberg and Shepherd, 1998), Hokche and Ramírez (2008)
Miconia crenata (Vahl) Michelang. [=*Clidemia hirta* (L.) D. Don]	*2x*	A; Ap (Hi); Di (Ta; An); AE; Au	Renner (1984, in Renner, 1989), Mendes-Rodrigues and Oliveira (2012), Caetano (2014), Caetano and Cortez (2014)
Miconia dependens (Pav. ex D. Don) Judd & Majure [=*Clidemia capitellata* (Bonpl.) D. Don.]	*2x*	A; AE; Au	Renner (1984, in Renner, 1989), Ramirez and Brito (1990), Melo et al. (1999), Mendes-Rodrigues and Oliveira (2012)
Miconia discolor DC.		A; N; Au	Borges (1991)
Miconia dodecandra Cogn.	*4x, 8x*	A; AE	Sobrevila and Arroyo (1982)
Miconia egensis Cogn.		A; Au	Renner (1984, in Renner, 1989)
Miconia epibaterium (DC.) Michelang. (=*Clidemia epibaterium* DC.)		A; Au	Renner (1984, in Renner, 1989)
Miconia fallax DC.	*2x, 4x*	A; Ap (Hi); N; Au	Saraiva et al. (1996), Goldenberg and Shepherd (1998), Mendes-Rodrigues and Oliveira (2012), Caetano and Cortez (2014), Caetano et al. (2018)
Miconia fenestrata (Benth.) Michelang. (=*Clidemia pustulata* DC.)	*2x*	A; Au	Hokche and Ramírez (2008)
Miconia ferruginata DC.		A; AE; Au	Mendes-Rodrigues and Oliveira (2012), dos Santos et al. (2012)
Miconia hyemalis A. St.-Hil. & Naudin		A; Au	Maia et al. (2016)
Miconia ibaguensis (Bonpl.) Triana	*4x*	A; AE; Au	Goldenberg (1994), Goldenberg and Shepherd (1998), Hokche and Ramírez (2008), Mendes-Rodrigues and Oliveira (2012)
Miconia lacunosa (Cogn.) R. Goldenb. (≡*Leandra lacunosa* Cogn.)		A; Au	Saraiva et al. (1996), Goldenberg and Shepherd (1998)
Miconia laevigata (L.) D. Don	*3x*	A; Au	Dent-Acosta and Breckon (1991), Mendes-Rodrigues and Oliveira (2012)
Miconia langsdorffii Cogn.		A; Au	Saraiva et al. (1996), Goldenberg and Shepherd (1998)
Miconia latecrenata (DC.) Naudin		A; Au	Goldenberg and Varassin (2001)
Miconia leaumbellata R. Goldenb. (=*Leandra involucrata* DC.)		A; Au	Saraiva et al. (1996)
Miconia leucocarpa DC.		A; Di (Ta, An); Au	Caetano (2014)
Miconia ligustroides (DC.) Naudin		A; AE; Au	Saraiva et al. (1996), Goldenberg and Shepherd (1998), Mendes-Rodrigues and Oliveira (2012), Maia et al. (2016)
Miconia mayeta (D. Don) Michelang. (=*Maieta guianensis* Aubl.)	*2x*	A; Au	Renner (1989)

TABLE 7.1 (Continued)

Species	Ploidy	Type of Apomixis[a]	References
Miconia microstachya (Naudin) R. Goldenb. [=*Leandra purpurascens* (DC.) Cogn.]		A; Au	Goldenberg and Varassin (2001), Maia et al. (2016)
Miconia minutiflora (Bonpl.) DC. (* this species correspond to *Miconia regelii* Cogn. according to Goldenberg and Shepherd, 1998)	*2x*	A; Au	Renner (1984, in Renner, 1989), Mendes-Rodrigues and Oliveira (2012)
Miconia petropolitana Cogn.		A; Au	Goldenberg and Varassin (2001)
Miconia pohliana Cogn.		A; Au	Goldenberg and Shepherd (1998)
Miconia polystachya (Naudin) R. Goldenb. [≡*Leandra polystachya* (Naudin) Cogn.]		A; Au	Maia et al. (2016)
Miconia prasina (Sw.) DC.	*3x*	A; Au	Renner (1989)
Miconia rariflora (Bonpl.) Michelang. (=*Clidemia fendleri* Cogn.)		A; Au	Sobrevila and Arroyo (1982)
Miconia regelii Cogn.		A; Au	Renner (1984, in Renner, 1989)
Miconia rubiginosa (Bonpl.) DC.	*3x*	A; AE; Au	Goldenberg and Shepherd (1998), Mendes-Rodrigues and Oliveira (2012)
Miconia rubra (Aubl.) Mabb. [≡*Clidemia rubra* (Aubl.) Mart.]		A; Au	Renner (1984, in Renner, 1989)
Miconia sellowiana Naudin		A; AE; Au	Saraiva et al. (1996), Mendes-Rodrigues and Oliveira (2012)
Miconia sericea (D. Don) Michelang. (≡*Clidemia sericea* D. Don)	*2x*	A	Hokche and Ramírez (2008)
Miconia spinulosa Naudin	*3x*	A; Au	Sobrevila and Arroyo (1982)
Miconia stenostachya DC.	*3x*	A; AE; Au	Renner (1984, in Renner, 1989), Saraiva et al. (1996), Goldenberg and Shepherd (1998), Mendes-Rodrigues and Oliveira (2012)
Miconia stephananthera Üle		A; Au	Ramirez and Brito (1990)
Miconia tomentosa (Rich.) D. Don ex DC.		A; Au	Renner (1984, in Renner, 1989)
Miconia trianae Cogn.		A; Au	Renner (1984, in Renner, 1989)
Microlicia benthamiana Triana		A; Au	Hokche and Ramírez (2008)
Microlicia fasciculata Mart. ex Naud.	*4x*	A; Ap; AE; Au	dos Santos et al. (2012), Viana et al. (2021)
Microlicia polystemma Naudin		A; AE; Au	Viana (2018), Viana et al. (2021)
Microlicia serpyllifolia D. Don		A; Au	Viana (2018)
Osbeckia hispidissima Wight		A; N	Subramanyam (1942)
Ossaea amygdaloides (DC.) Triana		A; Di (An); Au	Goldenberg and Varassin (2001), Caetano (2014)
Ossaea confertiflora (DC.) Triana		A; Di (An); Au	Goldenberg and Varassin (2001), Caetano (2014)
Pterolepis glomerata (Rottb.) Miq.		A; Au	Ramirez and Brito (1990)
Rhexia mariana L.		A; Au	Ehtheridge and Herr (1968)
Sandemania hoehnei (Cogn.) Wurdack		A; Au	Renner (1987)
Pleroma sellowianum (Cham.) P.J.F. Guim. & Michelang. [≡*Tibouchina sellowiana* (Cham.) Cogn.]	*4x*	A; Au	Maia et al. (2016)

(Continued)

TABLE 7.1 (Continued)
Table of Species with Apomixis (A) from all Angiosperm Families Except Poaceae Barnhart, Asteraceae Giseke, and Rosaceae Juss., Alphabetically Sorted (M: Monocot; D: Dicot)

Species	Ploidy	Type of Apomixis[a]	References
Tibouchina fraterna N.E. Br.		A; Au	Hokche and Ramírez (2008)
Trembleya parviflora (D. Don) Cogn.		A; Au	Souza-Silva (2000), dos Santos et al. (2012)
Meliaceae Juss. (D)			
Aphanamixis polystachya (Wall.) R. Parker [=*Amoora rohituka* (Roxb.) Wight & Arn.]		A; N; Au	Ghosh (1962), Ghosh et al. (1976), Rai (2014)
Lansium domesticum Corrêa		A; AE	Bernardo et al. (1961), Prakash et al. (1977), Salma and Razali (1987)
Moraceae Gaudich. (D)			
Dorstenia drakena L.		A; Au	Modilewsky (1908)
Dorstenia contrajerva L. (sub *Dorstenia contrayerva*)	2x	A; Au	Modilewsky (1908)
Dorstenia bonijesu Carauta & C.Valente		A; Au	Araujo et al. (2005)
Ficus auriculata Lour. (=*Ficus roxburghii* Steud.)		A; N	Naumova (1993)
Ficus johannis subsp. *afghanistanica* (Warb.) Browicz (=*Ficus afghanistanica* Warb.)		A; N	Naumova (1993)
Streblus taxoides (Roth) Kurz		A; N	Naumova (1993)
Myrtaceae Juss. (D)			
Callistemon lanceolatus[§] (Sm.) Sweet		A; AE	Polunina (1958, in Naumova, 1993)
Eugenia uniflora L.	2x, 3x	A; N; Ps	Silva et al. (2013)
Myrceugenia ovata (Hook. & Arn.) O.Berg (=*Eugenia cumingii* Hook. & Arn.)		A; I	Naumova (1993)
Myrcia bracteata (Rich.) DC. (≡*Eugenia bracteata*[§] Rich.)		A; AE	Naumova (1993)
Syzygium buxifolium Hook. & Arn.		A; AE	Merril and Perry (1938)
Syzygium cumini (L.) Skeels [≡*Eugenia cumini* (L.) Druce; =*Eugenia jambolana* Lam.; =*Syzygium caryophyllifolium* (Lam.) DC.]	2x, 4x, 6x	A; N	Tiwary (1926), Gurgel and Soubihe Sobrinho (1951), Narayanaswami and Roy (1960), Roy and Sahai (1962), Naumova (1993)
Syzygium pycnanthum Merr. & L.M. Perry [=*Eugenia densiflora* (Blume) DC.]		A; N; Ps	van der Pijl (1934)
Syzygium formosum (Wall.) Masam. (≡*Eugenia formosa*[§] Wall.)		A; AE	Naumova (1993)
Syzygium forrestii Merr. & L.M. Perry		A; AE	Merril and Perry (1938);
Syzygium fruticosum DC. [≡*Eugenia fruticosa* (DC.) Roxb.]		A; I	Roy (1961)
Syzygium hancei Merr. & L.M.Perry		A; AE	Merril and Perry (1938);
Syzygium jambos (L.) Alston (≡*Eugenia jambos* L.)	4x	A; N; I; Ps	Cook (1907), van der Pijl (1934), Merril and Perry (1938), Gurgel and Soubihe Sobrinho (1951), Roy (1953, in Batygina and Vinogradova, 2007, p. 133)
Syzygium malaccense (L.) Merr. & L.M. Perry (≡*Eugenia malaccensis* L.)	2x	A; I; Ps	van der Pijl (1934), Gurgel and Soubihe Sobrinho (1951)

TABLE 7.1 (Continued)

Species	Ploidy	Type of Apomixis[a]	References
Syzygium megacarpum (Craib) Rathakr. & N.C. Nair [=*Syzygium latilimbum* (Merr.) Merr. & L.M. Perry]		A; AE	Merril and Perry (1938)
Syzygium nervosum A. Cunn. ex DC. (=*Eugenia operculata* Roxb.)	*2x*	A; N; Ps	van der Pijl (1934)
Syzygium paniculatum Gaertn.	*4x*	A; N; Ps	Thurlby et al. (2011)
Syzygium polycephalum (Miq.) Merr. & L.M. Perry (≡*Eugenia polycephala* Miq.)		A; N; Ps	van der Pijl (1934)
Syzygium salicifolium (Wight) J. Graham (=*Eugenia heyneana* Duthie)		A; I	Naumova (1993)
Syzygium samarangense (Blume) Merr. & L.M. Perry (=*Eugenia javanica* Lam.)		A; N; Ps	van der Pijl (1934)
Syzygium siamense (Craib) Chantaran. & J. Parn.		A; AE; Ps	Chantaranothai and Parnell (1994)
Luma apiculata (DC.) Burret (=*Eugenia hookeri* Steud.)		A; N	Johnson (1936), Titow de Tschischow (1956)
Plinia cauliflora (Mart.) Kausel [≡*Eugenia cauliflora*[§] (Mart.) DC.; ≡*Myrciaria cauliflora* (Mart.) O. Berg; sub *Myrciaria cauliflora* L.]		A; AE	Traub (1939), Gurgel and Soubihe Sobrinho (1951), Naumova (1993)
Plinia edulis (Vell.) Sobral (≡*Eugenia edulis* Vell.)		A; AE	Gurgel and Soubihe Sobrinho (1951), Rotman (1982), Naumova (1993)
Plinia peruviana (Poir.) Govaerts (≡*Guapurium peruvianum* Poir.; =*Myrciaria trunciflora* O. Berg)		A; AE	Gurgel and Soubihe Sobrinho (1951), Naumova (1993)
Psidium cattleyanum Sabine[6]	*7x*	A; Di (An); Ps	Souza-Pérez and Speroni (2017), Souza-Pérez et al. (2021)
Ugni molinae Turcz. [=*Eugenia ugni* (Molina) Hook. & Arn.]	*2x*	A; I	Naumova (1993)
Ochnaceae DC. (D)			
Ochna atropurpurea DC.	*4x*	A; Ap (Hi); Di (An); I; Au	Ghosh (1964), Pauzé (1973), Naumova (1993)
Ochna multiflora DC.		A; Ap (Hi); Di (An); I; Au	Francini (1928), Naumova (1993)
Ochna serrulata Walp.	*3x*	A; Di (An); I; Au	Chiarugi and Francini (1930), Naumova (1993)
Orchidaceae Juss. (M)			
Cynorkis lilacina Ridl. (≡*Cynosorchis lilacina* Ridl.)		A; N; Au	Veyret (1967)
Cynorkis ridleyi T. Durand & Schinz		A; N; Au	Veyret (1972)
Cynorkis ampulacea H. Perrier ex Hermans	*4x*	A; N; Au	Veyret (1972)
Dactylorhiza insularis (Sommier ex Martelli) E. Nelson [sub *Dactylorhiza insularis* (Sommier ex Martelli) Landwehr]	*3x*	A; I	Diana (1997)
Epidendrum nocturnum Jacq.	*4x*	A; Di (An); I; Ps	Veyret (1982), Stort and Dos Santos Pavanelli (1985)

(Continued)

TABLE 7.1 (Continued)
Table of Species with Apomixis (A) from all Angiosperm Families Except Poaceae Barnhart, Asteraceae Giseke, and Rosaceae Juss., Alphabetically Sorted (M: Monocot; D: Dicot)

Species	Ploidy	Type of Apomixis[a]	References
Genoplesium apostasioides (Fitzg.) D.L. Jones & M.A. Clem. (≡*Corunastylis apostasioides* Fitzg.)	*8x*	A; Di (An); N; I; Au	Sorensen et al. (2009)
Gymnadenia archiducis-joannis (Teppner & E. Klein) Teppner & E. Klein (≡*Nigritella archiducis-joannis* Teppner & E. Klein)	*4x*	A; N; Au	Teppner and Klein (1985), Teppner (1996), Hedrén et al. (2000)
Gymnadenia buschmanniae (Teppner & Ster) Teppner & E. Klein (≡*Nigritella buschmanniae* Teppner & Ster)	*5x*	A	Hedrén et al. (2000)
Gymnadenia nigra (L.) Rchb. [≡*Nigritella nigra* (L.) Rchb.f. subsp. *nigra*]	*3x*	A; N; Au	Afzelius (1928, 1932), Chiarugi (1929), Teppner (1996)
Gymnadenia austriaca (Teppner & E. Klein) P. Delforge (≡*Nigritella nigra* subsp. *austriaca* Teppner & E. Klein)	*4x*	A; N; Au	Teppner (1996), Hedrén et al. (2000)
Gymnadenia austriaca var. *gallica* (E. Breiner & R. Breiner) P. Delforge (≡*Nigritella nigra* subsp. *iberica* Teppner & E. Klein)	*4x*	A; N; Au	Teppner (1996), Hedrén et al. (2000)
Gymnadenia conopsea (L.) R.Br.		A; N	Strasburger (1878, in Swamy, 1949, p. 220)
Gymnadenia runei (Teppner & E. Klein) Ericsson (≡ × *Gymnigritella runei* Teppner & E. Klein)	*4x*	A; Au	Hedrén et al. (2000)
Gymnadenia rubra Wettst. [=*Nigritella miniata* (Crantz) Janch.; =*Nigritella dolomitensis* (Teppner & E. Klein) Hedrén, E. Klein & Teppner; =*Nigritella hygrophila* W. Foelsche & Heidtke]	*4x*	A; N; Au	Teppner and Klein (1985), Teppner (1996), Hedrén et al. (2000, 2018), Hedrén and Lorenz (2019)
Gymnadenia stiriaca (Rech.) Teppner & E. Klein [≡*Nigritella stiriaca* (Rech.) Teppner & E. Klein]	*4x*	A; N; Au	Teppner and Klein (1985), Teppner (1996), Hedrén et al. (2000)
Gymnadenia widderi (Teppner & E. Klein) Teppner & E. Klein (≡*Nigritella widderi* Teppner & E. Klein)	*4x*	A; N; Au	Teppner (1996), Hedrén et al. (2000)
Habenaria malintana (Blanco) Merr.		A; Au	Zhang and Gao (2018)
Mormolyca cleistogama (Brieger & Illg) M.A. Blanco (≡*Maxillaria cleistogama* Brieger & Illg)		A; N	Illg (1977)
Crepidium acuminatum (D.Don) Szlach. [=*Malaxis wallichii* (Lindl.) Deb; =*Microstylis wallichii* Lindl.]		A; AE	Sood and Ram (1995)
Microtis orbicularis R.S. Rogers		A; Au	Bates (1983)
Microtis unifolia (G. Forst.) Rchb. f.	*2x, 4x*	A; Au	Bates (1983)
Pterygodium newdigateae Bolus		A; AE; Au	Duthie (1915)
Sacoila lanceolata (Aubl.) Garay var. *lanceolata*		A; I; Au	Catling (1987)

TABLE 7.1 (Continued)

Species	Ploidy	Type of Apomixis[a]	References
Sacoila lanceolata var. *paludicola* (Luer) Sauleda, Wunderlin & B.F. Hansen		A; I; Ps	Catling (1987)
Satyrium nepalense var. *ciliatum* (Lindl.) Hook.f. (=*Satyrium ciliatum* Lindl.)		A; Au	Huang et al. (2009), Lu et al. (2010)
Spiranthes casei Catling & Cruise var. *casei*		A; I; Au	Catling (1982)
Spiranthes casei var. *novaescotiae* Catling		A; I; Au	Catling (1982)
Spiranthes cernua (L.) Rich. [=*Spiranthes cernua* (L.) Rich. var. *cernua*]	2*x*, 3*x*	A; N; I; Au	Leavitt (1900), Swamy (1948b), Catling (1982), Schmidt and Antlfinger (1992)
Spiranthes floridana (Wherry) Cory		A; Au	Stewart (2007)
Spiranthes hongkongensis S.Y. Hu & Barretto	4*x*	A	Sun (1997)
Spiranthes magnicamporum Sheviak	2*x*	A; I; Au	Catling (1982)
Spiranthes ochroleuca (Rydb.) Rydb.	2*x*	A; I; Au	Catling (1982)
Spiranthes odorata (Nutt.) Lindl. [=*Spiranthes cernua* var. *odorata* (Nutt.) Correll]		A; Ps	Catling (1982)
Spiranthes parksii Correll		A; AE	Catling and McIntosh (1979)
Spiranthes romanzoffiana Cham.	2*x*, 3*x*	A	Leavitt (1900), Swamy (1948b), Catling and Catling (1991)
Spiranthes sinensis (Pers.) Ames [=*Spiranthes australis* (R. Br.) Lindl.]	2*x*	A; I; Au	Swamy (1948b)
Zeuxine strateumatica (L.) Schltr [=*Zeuxine sulcata* (Roxb.) Lindl. ex Wight]	4*x*	A; N; Au	Seshagiriah (1941), Swamy (1946), Sun (1997), Sun and Wong (2001), Kant and Verma (2012)
Zygopetalum maculatum (Kunth) Garay (=*Zygopetalum mackayi* Hook.)	3*x*, 4*x*	A; N; I; Ps	Hurst (1898), Suessenguth (1923), Afzelius (1959), Campacci et al. (2017), Moura et al. (2020)
Orobanchaceae Vent. (D)			
Cistanche mongolica Beck [=*Cistanche tubulosa* (Schenk) Hook. f.][6]	2*x*	A; Di; Ps; Au	Pazy (1998), Plitmann (2002)
Cistanche salsa (C.A. Mey.) Beck[6]		A; Di	Pazy and Plitman (1996), Pazy (1998)
Orobanche uniflora L.[6]	4*x*	A; Di; Ps	Jensen (1951)
Oxalidaceae R.Br. (D)			
Oxalis florida Salisb. [=*Oxalis dillenii* subsp. *filipes* (Small) G. Eiten]		A; Au	Lovett Doust et al. (1981)
Pandanaceae R.Br. (D)			
Pandanus tectorius Parkinson ex Du Roi		A; Au	Cox (1985, 1990)
Phyllanthaceae Martinov (D)			
Uapaca kirkiana Müll.-Arg.	2*x*	A; AE	Ngulube et al. (1998), Maliro and Kwapata (2000)
Plumbaginaceae Juss. (D)			
Limonium algarvense Erben	3*x*	A	Erben (1978), Ingrouille (1985)
Limonium angustibracteatum Erben	3*x*	A	Erben (1978)
Limonium aphroditae R. Artelari & Georgiou	3*x*	A	Artelari and Georgiou (1999)
Limonium aucheri (Girard) Greuter & Burdet	5*x*	A	Kouzali et al. (2012)
Limonium auriculae-ursifolium (Pourr.) Druce	3*x*	A; Au	Erben (1978), Ingrouille (1985)

(Continued)

TABLE 7.1 (Continued)

Table of Species with Apomixis (A) from all Angiosperm Families Except Poaceae Barnhart, Asteraceae Giseke, and Rosaceae Juss., Alphabetically Sorted (M: Monocot; D: Dicot)

Species	Ploidy	Type of Apomixis[a]	References
Limonium avei (De Not.) Brullo & Erben (=*Limonium longispicatum* Erben)	*3x*	A	Erben (1978), Kouzali et al. (2012)
Limonium binervosum (G.E.Sm.) C.E. Salmon	*3x, 4x*	A; Di (Ix); Au	Baker (1966), Erben (1978), Ingrouille (1982, in Cowan et al., 1998), Ingrouille and Stace (1985), Caperta et al. (2017)
Limonium catalaunicum (Willk. & Costa) Pignatti	*3x*	A	Erben (1978)
Limonium cavanillesii Erben		A	Palacios and González-Candelas (1997b)
Limonium × *coincyi* Sennen	*3x*	A	Erben (1978)
Limonium companyonis Kuntze	*3x*	A	Baker (1966), Erben (1981)
Limonium contortirameum (Mabille) Erben	*3x*	A; Di (Ix); Au	Masia (1985, in Diana, 1992, p. 188)
Limonium costae (Willk.) Pignatti	*3x*	A	Erben (1978)
Limonium cosyrense (Guss.) Kuntze	*3x*	A	Baker (1953, 1966)
Limonium cuspidatum (Delort.) Erben	*3x*	A	Erben (1978)
Limonium delicatulum (Girard) Kuntze	*3x*	A	Baker (1966), Erben (1978)
Limonium densiflorum (Guss.) Kuntze	*3x*	A	Baker (1966)
Limonium densissimum (Pignatti) Pignatti	*3x*	A	Erben (1978)
Limonium dodartii Kuntze [sub *Limonium dodartii* (Girard) O. Kuntze]	*4x*	A	Baker (1966), Erben (1978)
Limonium dodartiforme Ingr.	*4x*	A	Ingrouille and Stace (1985)
Limonium dufourii (Girard) Kuntze [sub *Limonium dufourei* (Girard) O. Kuntze]	*3x*	A	Baker (1966), Erben (1978), Palacios and González-Candelas (1997a), Palop-Esteban et al. (2007)
Limonium duriusculum (Girard) Fourr.	*3x*	A	Erben (1978)
Limonium eugeniae Sennen	*3x*	A	Erben (1978)
Limonium geronense Erben	*4x*	A	Erben (1978)
Limonium gibertii (Sennen) Sennen	*3x*	A	Erben (1978)
Limonium girardianum (Guss.) Kuntze [sub *Limonium girardianum* (Guss.) Fourr.]	*3x*	A	Baker (1966), Erben (1978), Ingrouille (1984)
Limonium gougetianum (Girard) Kuntze		A	Baker (1966)
Limonium latibracteatum Erben	*3x*	A	Erben (1978)
Limonium legrandii (Gaut. & Timb.-Lagr.) Erben	*3x*	A	Erben (1978)
Limonium loganicum Ingrouille	*4x*	A	Ingrouille and Stace (1985, 1986)
Limonium majus (Boiss.) Erben	*3x*	A	Erben (1978)
Limonium multiflorum Erben	*4x*	A; Di (Ix); Au	Erben (1978), Róis et al. (2016)
Limonium normannicum Ingrouille	*3x*	A; Au	Ingrouille (1985)
Limonium ocymifolium (Poir.) Kuntze	*5x*	A	Baker (1966)
Limonium oleifolium Mill. [≡*Statice oleifolia* (Mill.) Scop.; sub *Statice oleifolia* Scop. var. *confusa*; =*Limonium virgatum* (Willd.) Fourr.]	*3x*	A; Di (Ix); Au	D'Amato (1940, 1949), Erben (1978), Georgakopoulou et al. (2006), Kouzali et al. (2012)
Limonium paradoxum Pugsley	*4x*	A	Baker (1966), Ingrouille and Stace (1985)
Limonium parvum Ingr.	*4x*	A	Ingrouille and Stace (1985, 1986)
Limonium plurisquamatum Erben	*3x*	A	Erben (1978)

TABLE 7.1 (Continued)

Species	Ploidy	Type of Apomixis[a]	References
Limonium procerum (C.E. Salmon) Ingr.	*4x*	A	Ingrouille and Stace (1985)
Limonium recurvum C.E. Salmon	*3x*	A	Baker (1966), Ingrouille and Stace (1985)
Limonium ruizii (Font Quer) Fern.Casas	*3x*	A	Erben (1978)
Limonium silvestrei Aparicio	*3x*	A	Aparicio (2005)
Limonium subglabrum Erben	*3x*	A	Erben (1978)
Limonium supinum (Girard) Pignatti [=*Limonium salsuginosum* (Boiss.) Kuntze]	*3x*	A	Baker (1966), Erben (1978)
Limonium tibulatium Pignatti	*3x*	A; Di (Ix); Au	Marras (1985, in Diana, 1992, p. 188)
Limonium tournefortii (Girard) Erben	*3x*	A	Erben (1978)
Limonium transwallianum (Pugsley) Pugsley	*3x, 4x*	A; Di (Ix); Au	Hjelmqvist and Grazi (1964), Ingrouille and Stace (1985)
Limonium tremolsii (Rouy) Erben [sub *Limonium tremolsii* (Rouy) P. Fourn.]	*3x*	A	Erben (1978)
Limonium viciosoi (Pau) Erben	*3x*	A	Erben (1978)
Polygonaceae Juss. (D)			
Atraphaxis frutescens (L.) K. Koch		A; Ap (Hi); N	Edman (1931)
Oxyria digyna (L.) Hill		A; Ap	Edman (1929)
Rumex acetosa L.	*2x*	A; Au	Roth (1906, in Gregor, 2013)
Rumex alpestris Jacq. (=*Rumex arifolius* All.)	*2x*	A; Au	Roth (1906, in Gregor, 2013)
Rumex nivalis Hegetschw.		A; Au	Roth (1906, in Gregor, 2013)
Potamogetonaceae Bercht. & J. Presl (M)			
Potamogeton obtusifolius Mert. & W.D.J. Koch		A; Di (Ta)	Teryokhin (1994).
Primulaceae Batsch ex Borkh. (D)			
Ardisia compressa Kunth (=*Ardisia polytoca* A. Braun & Bouché)		A; N; Ps	Braun (1860), Kume (1959)
Ardisia coriacea Sw.		A; AE	Braun (1860)
Ardisia japonica (Thunb.) Blume	*4x*	A; AE	Braun (1860), Kume (1959)
Ardisia martiana Miq. (sub *Ardisia marteana* Miq.)		A; AE	Kume (1959)
Ardisia serrulata Sw.		A; AE	Braun (1860)
Ranunculaceae Juss. (D)			
Halerpestes lancifolia (Bertol.) Hand.-Mazz.		A	Brožová et al. (2019)
Ranunculus allemannii Braun-Blanq. ex W. Koch	*4x, 6x*	A; Ps	Koch (1933, 1939)
Ranunculus argoviensis W. Koch	*4x*	A; Ap (Hi); Ps	Koch (1939), Häflinger (1943), Nogler (1984, 1990)
Ranunculus auricomus L.	*3x, 4x, 5x, 6x*	A; Ap (Hi); Ps	Häflinger (1943), Rutishauser (1953, 1954), Vuillemin (1992)
Ranunculus biformis W. Koch	*4x*	A; Ap (Hi); Ps	Häflinger (1943)
Ranunculus braun-blanquetii Pignatti	*4x*	A	Masci et al. (1994)
Ranunculus calvescens (Harry Sm. ex Fagerstr. & G. Kvist) Ericsson	*5x*	A; Ps	Rousi (1956)
Ranunculus carpaticola Soó (=*Ranunculus carpaticola* × *R. cassubicifolius*)	*6x*	A; Ap (Hi); Ps	Hörandl and Greilhuber (2002), Paun et al. (2006a, 2006b), Hojsgaard et al. (2014b)

(Continued)

TABLE 7.1 (Continued)
Table of Species with Apomixis (A) from all Angiosperm Families Except Poaceae Barnhart, Asteraceae Giseke, and Rosaceae Juss., Alphabetically Sorted (M: Monocot; D: Dicot)

Species	Ploidy	Type of Apomixis[a]	References
Ranunculus cassubicus L.	*4x, 6x*	A; Ap (Hi); Ps	Häflinger (1943), Izmaiłow (1965), Paun et al. (2006a, 2006b)
Ranunculus hungaricus Soó	*4x, 5x, 6x*	A; Ps	Hörandl et al. (2009)
Ranunculus kuepferi Greuter & Burdet	*4x*	A; Ap (Hi); Ps	Huber (1988), Burnier et al. (2009), Cosendai and Hörandl (2010)
Ranunculus luminarius Pignatti ex Greuter	*4x*	A	Masci et al. (1994)
Ranunculus fallax (Wimm. & Grab.) Sloboda	*4x*	A; Ap; Ps	Rousi (1956)
Ranunculus fragifer W.Koch ex Dunkel	*4x*	A; Ap (Hi); Ps	Rutishauser and Hunziker (1950)
Ranunculus gracillimus (Julin) Ericsson	*4x*	A; Ap (Hi); Ps	Rutishauser and Hunziker (1950)
Ranunculus grossidens W. Koch ex Brodtb.	*4x*	A; Ap (Hi); Ps	Rutishauser and Hunziker (1950)
Ranunculus marsicus Guss. & Ten.	*5x, 6x, 7x*	A; Ap	Masci et al. (1994)
Ranunculus megacarpus W. Koch	*4x*	A; Ap (Hi); Ps	Häflinger (1943), Nogler (1971, 1973, 1984)
Ranunculus membranaceus Royle		A	Brožová et al. (2019)
Ranunculus monophyllus Ovcz.	*4x, 5x, 6x*	A; Ps	Rousi (1956)
Ranunculus oligandrus (Markl.) Ericsson	*4x, 6x*	A; Ps	Rousi (1956)
Ranunculus parnassifolius L.	*4x*	A; Ap (Hi); Ps	Vuille and Küpfer (1985), Huber (1988)
Ranunculus pseudocassubicus (Christ) Christ ex W. Koch	*4x*	A	Koch (1933), Häflinger (1943), Masci et al. (1994), Hörandl, (1998)
Ranunculus puberulus W. Koch	*4x*	A; Ap (Hi); Ps	Häflinger (1943)
Ranunculus stricticaulis W. Koch	*5x*	A; Ps	Koch (1939), Häflinger (1943)
Ranunculus variabilis Hörandl & Gutermann	*4x*	A; Ps	Hörandl et al. (2000, 2001)
Thalictrum aquilegifolium L. (sub *Thalictrum aquilegifolium* L. subsp. *aqueligifolium*)	*2x*	A; N	Daskalova (2004)
Thalictrum fendleri Engelm. ex A. Gray		A; Au	Day (1896, in Overton, 1902)
Thalictrum purpurascens L.		A; Au	Overton (1902, 1904)
Rhamnaceae Juss. (D)			
Pomaderris hamiltonii L.B. Moore	*3x*	A; Di (An); Ps	Harvey and Braggins (1985)
Pomaderris phylicifolia var. *ericifolia* L. Moore	*3x*	A; Di (An); Ps	Harvey and Braggins (1985)
Pomaderris phylicifolia var. *polifolia* L. Moore	*4x*	A; Di (An); Ps	Harvey and Braggins (1985)
Rubiaceae Juss. (D)			
Coccocypselum geophiloides Wawra		A; Au	Azevedo (2014)
Coccocypselum hasslerianum Chodat		A; Au	Azevedo (2014)
Coprosma robusta Raoul	*4x*	A; Au	Heenan et al. (2003)
Coprosma serrulata Hook.f. ex Buchanan	*4x*	A; Au	Heenan et al. (2002)
Coprosma waima A.P. Druce	*4x*	A; Au	Heenan et al. (2002)
Genipa americana L.		A; AE; Au	Crestana (1995), Salomão and Allem (2001)

TABLE 7.1 (Continued)

Species	Ploidy	Type of Apomixis[a]	References
Rutaceae Juss. (D)			
Aegle marmelos (L.) Corrêa	*2x*	A; N; Ps	Chakravarthy (1936), Johri and Ahuja (1956)
Citrus aurantiifolia (Christm.) Swingle	*2x*	A; N	Naumova (1993), Mondal and Saha (2014)
Citrus × *aurantium* L. (=*Citrus* × *taiwanica* Tanaka & Shimada; =*Citrus bigaradia* Risso & Poit.; =*Citrus vulgaris* Risso)	*2x*	A; N; Ps	Braun (1860), Biermann (1896), Osawa (1913), Strasburger (1878), Bacchi (1943), Naumova (1993)
Citrus hystrix DC.		A; N; Ps	Raja (2017)
Citrus jambhiri Lush.	*2x*	A; N	Frost (1926), Ashari et al. (1988)
Citrus japonica Thunb. [≡*Fortunella japonica* (Thunb.) Swingle; =*Fortunella crassifolia* Swingle; =*Fortunella margarita* (Lour.) Swingle; =*Citrus microcarpa* Bunge]	*2x*	A; N; Ps	Rangaswamy (1959, 1961), Wakana and Uemoto (1987), Cai et al. (1997)
Citrus juko Kokaya		A; AE	Naumova (1993)
Citrus junos Siebold ex Tanaka	*2x*	A; N	Woo et al. (2019)
Citrus limon (L.) Osbeck (=*Citrus limonum* Risso)	*2x*	A; N; Ps	Biermann (1896), Iglesias et al. (1974), Wakana and Uemoto (1987), Naumova (1993)
Citrus maxima (Burm.) Merr. [=*Citrus decumana* L.; =*Citrus grandis* (L.) Osbeck; =*Citrus obovoidea* Yu.Tanaka]	*2x*	A; N	Braun (1860), Ensign (1919), Banerji (1954), Woo et al. (2019), Naumova (1993)
Citrus medica L. (=*Citrus sudachi* Hort. ex Tanaka)	*2x, 3x, 4x*	A; N	Woo et al. (2019), Naumova (1993)
Citrus natsudaidai Hayata	*2x*	A; N	Furusato (1952), Yang (1968), Kobayashi (1979), Naumova (1993)
Citrus nobilis Lour.	*2x*	A; N; Ps	Osawa (1913), Naumova (1993)
Citrus paradisi Macfad.	*2x*	A; N	Bacchi (1943), Wakana and Uemoto (1987), Naumova (1993)
Citrus reshni Hort ex Tan.	*2x*	A; N; Ps	Andrade-Rodríguez et al. (2005)
Citrus reticulata Blanco [=*Citrus unshiu* Marcov.; =*Citrus sunki* (Hayata) Yu.Tanaka]	*2x*	A; N; Ps	Furusato et al. (1957), Yang (1968), Esen and Soost (1977), Kobayashi et al. (1978), Wakana and Uemoto (1987), Ashari et al. (1988), Cai et al. (1997), Wang et al. (2017)
Citrus sinensis (L.) Osbeck	*2x*	A; N; Ps	Iglesias et al. (1974), Kobayashi et al. (1978), Wakana and Uemoto (1987), Ashari et al. (1988), Naumova (1993)
Citrus taiwanica Tanaka & Shimada	*2x*	A; N	Xiang and Roose (1988)
Citrus tamurana Tanaka	*2x*	A; N; Ps	Miwa (1952, in Yang, 1968)
Citrus trifoliata L. [≡*Poncirus trifoliata* (L.) Raf.]	*2x, 3x, 4x*	A; N; Ps	Biermann (1896), Osawa (1913, in Batygina and Vinogradova, 2007, p. 133), Kobayashi et al. (1978), Wakana and Uemoto (1987), Ashari et al. (1988), Khan and Roose (1988), Hussain et al. (2011)
Citrus volkameriana Pasq.	*2x*	A; N; Ps	Xiang and Roose (1988), Andrade-Rodríguez et al. (2004)

(Continued)

TABLE 7.1 (Continued)
Table of Species with Apomixis (A) from all Angiosperm Families Except Poaceae Barnhart, Asteraceae Giseke, and Rosaceae Juss., Alphabetically Sorted (M: Monocot; D: Dicot)

Species	Ploidy	Type of Apomixis[a]	References
Esenbeckia pilocarpoides Kunth (=*Esenbeckia castanocarpa* Griseb., sub *Esenbeckia castanocarpum*)		A; N	Braun (1860, in Mauritzon, 1935, p. 346), Naumova (1993)
Galipea jasminiflora (A.St.-Hil.) Engl.		A; AE	Piedade and Ranga (1993)
Limonia acidissima L. [=*Feronia limonia* (L.) Swingle]	*2x*	A; AE	Banerjee and Pal (1958), Naumova (1993)
Murraya paniculata (L.) Jack (=*Murraya exotica* L.)	*2x*	A; N	Chakravarthy (1936)
Murraya koenigii (L.) Spreng.	*2x*	A; N	Chakravarthy (1935)
Ptelea trifoliata L.	*2x*	A; N	Schürhoff (1924), Mauritzon (1935), Naumova (1993)
Ruta patavina L.		A; N	Cappelletti (1929), Naumova (1993)
Skimmia japonica Thunb.		A; Ap	Desai (1961)
Toddalia asiatica (L.) Lam.	*2x*	A; N; Ps	Narmatha Bai and Lakshmanan (1982)
Triphasia trifolia (Burm.f.) P. Wilson (=*Triphasia aurantiola* Lour.)	*2x*	A; Ap; N	Braun (1860), Mauritzon (1935), Naumova (1993)
Zanthoxylum armatum DC. (=*Zanthoxylum alatum* Roxb.)	*4x, 8x*	A; N; Au	Mauritzon (1935), Naumova (1993)
Zanthoxylum americanum Mill.	*8x, 16x*	A; N; Au	Desai (1962), Naumova (1993)
Zanthoxylum bungeanum Maxim.	*4x*	A; N; Au	Desai (1962), Liu et al. (1987), Naumova (1993), Cai et al. (1997, 2000)
Zanthoxylum bungeanum Maxim. var. *bungeanum* (=*Zanthoxylum bungei* Planch. & Linden ex Hance)		A; N; Au	Longo (1908), Mauritzon (1935), Desai (1962), Naumova (1993)
Zanthoxylum oxyphyllum Edgew.	*4x*	A; N	Ghosh (1974)
Zanthoxylum piperitum DC.	*4x*	A; N; Au	Naumova (1993)
Zanthoxylum simulans Hance	*8x*	A; N; Au	Desai (1962), Liu et al. (1987)
Salicaceae Mirb. (D)			
Salix aurita L.	*2x, 4x*	A; Ap	Chochlov et al. (1978, in Gregor, 2013)
Salix purpurea L.	*2x*	A; Ap	Chochlov et al. (1978, in Gregor, 2013)
Salix viminalis L.	*2x*	A; Ap	Chochlov et al. (1978, in Gregor, 2013)
Saururaceae Rich. ex T. Lestib. (D)			
Houttuynia cordata Thunb.[3]	*6x?*	A; Di (An); Au	Shibata and Miyake (1908), Okabe (1930)
Smilacaceae Vent. (M)			
Maianthemum racemosum (L.) Link [≡*Smilacina racemosa* (L.) Desf.]		A; N; Ps; Au	McAllister (1913), Gorham (1952)
Solanaceae Adans. (D)			
Anisodus luridus Link ex Spreng. [≡*Scopolia lurida* (Link ex Spreng.) Dunal]	*4x*	A; AE	Naumova (1993)
Petunia axillaris (Lam.) Britton, Sterns & Poggenb. (=*Petunia nyctaginiflora* Juss.)		A; N	Banerji and Bhaduri (1933), Naumova (1993)
Petunia inflata R.E.Fr. (=*Petunia violacea* Chodat & Hassl.)	*2x*	A; AE	Naumova (1993)

TABLE 7.1 (Continued)

Species	Ploidy	Type of Apomixis[a]	References
Scopolia carniolica Jacq.		A; I	Glutchenko (1968)
Solanum nigrum L.		A; Di	Chochlov et al. (1978, in Gregor, 2013)
Withania somnifera (L.) Dunal	*4x*	A; AE	Banerji and Bhaduri (1933), Naumova (1993)
Symplocaceae Desf. (D)			
Symplocos uniflora (Pohl) Benth. (=*Symplocos klotzschii* Brand)		A; N	Ravishankar and Gulati (1985), Naumova (1993)
Tecophilaeaceae Leyb. (M)			
Cyanella hyacinthoides Royen ex L. (=*Cyanella capensis* L.)		A; N	de Vos (1950), Maheshwari and Sachar (1963)
Thymelaeaceae Juss. (D)			
Daphne papyracea Wall. ex G. Don (=*Daphne cannabina* Wall.)	*2x*	A; N	Venkateswarlu (1946)
Wikstroemia indica (L.) C.A. Mey. (=*Wikstroemia viridiflora* Wall. ex Meisn.)	*3x*	A; Di (Ta)	Winkler (1904, 1906), Fagerlind (1940a, 1940b)
Tropaeolaceae Juss. ex DC. (D)			
Tropaeolum majus L.	*2x*	A; N	Braun (1860), Naumova (1993)
Urticaceae Juss. (D)			
Boehmeria formosana Hayata	*3x*	A; Di; Au	Tang et al. (2016)
Boehmeria gracilis C.H. Wright	*3x*	A; Di; Au	Yahara (1986), Tang et al. (2016)
Boehmeria japonica (L.f.) Miq. (=*Boehmeria longispica* Steud.)	*3x, 4x, 5x*	A; Di; Au	Tang et al. (2016)
Boehmeria nivea (L.) Gaudich.	*3x*	A; Au	Okabe (1963)
Boehmeria silvestrii (Pamp.) W.T. Wang	*3x*	A; Di (Ta); Au	Yahara (1986, 1990), Zhang et al. (1997)
Boehmeria sieboldiana Blume	*3x*	A; Au	Yahara (1986)
Boehmeria spicata (Thunb.) Thunb.	*3x*	A; Au	Yahara (1986, 1990)
Boehmeria tricuspis (Hance) Makino	*3x*	A; Di (An); Au	Okabe (1963), Tang et al. (2016, 2017)
Elatostema acuminatum (Poir.) Brongn.	*4x*	A; Ap (Hi); N	Fagerlind (1944a, 1944b), Naumova (1993)
Elatostema eurhynchum Miq.	*4x*	A; Di; N	Fagerlind (1944a, 1944b), Naumova (1993)
Elatostema machaerophyllum Hallier f.	*5x*	A; Di (Ta); An	Fagerlind (1944a, 1944b)
Elatostema pedunculosum Miq.		A; Ap; AE	Fagerlind (1944a, 1944b)
Elatostema peltifolium (Ridl.) H.J.P. Winkl.	*5x*	A; Di (Ta)	Fagerlind (1944a, 1944b)
Elatostema penninerve H. Schroet.		A; Ap (Hi)	Fagerlind (1944a, 1944b)
Elatostema rostratum (Blume) Hassk.		A	Fagerlind (1944a, 1944b)
Elatostema sessile J.R. Forst. & G. Forst.	*3x*	A; Di (An); Au	Modilewsky (1908)
Elatostema sinuatum Hassk.	*6x*	A; Di	Fagerlind (1944a, 1944b)
Pellionia latifolia (Blume) Boerl. (≡*Elatostema latifolium* Blume ex H. Schroet.)	*4x*	A; Ap	Fagerlind (1944a, 1944b)
Pellionia repens (Lour.) Merr. [≡*Elatostema repens* (Lour.) Hallier f.]		A	Fagerlind (1944a, 1944b)
Procris frutescens Blume		A; Di (An)	Fagerlind (1944a, 1944b)

(Continued)

TABLE 7.1 (Continued)
Table of Species with Apomixis (A) from all Angiosperm Families Except Poaceae Barnhart, Asteraceae Giseke, and Rosaceae Juss., Alphabetically Sorted (M: Monocot; D: Dicot)

Species	Ploidy	Type of Apomixis[a]	References
Zygophyllaceae R.Br. (D)			
Zygophyllum coccineum L.	*2x*	A; N	Phatak (1971)
Zygophyllum fabago L.	*2x*	A; N	Phatak (1971)
Zygophyllum xanthoxylum (Bunge) Maxim.		A; Au	Chen and Zhao (2017)

[a]For species with sporophytic apomixis, AE: adventitious embryony; N: nucellar embryony; I: integumentary embryony. For species with gametophytic apomixis, Ap ( ): apospory (type of female gametophyte development); Di ( ): diplospory (type of female gametophyte development); Ps: pseudogamous endosperm development; Au: autonomous endosperm development.

Hi: Hieracium type; Pa: Panicum type; Ta: Taraxacum type; An: Antennaria type; Ix: Ixeris type.

[§] species with evidence of adventitious embryony for which the original reports were not accessed and author citation was not included in Naumova (1993), and two or more author name options are available [e.g., for *Eugenia bracteata* three options are possible: *Eugenia bracteata* (Willd.) Raeusch. ex DC., *Eugenia bracteata* Vell., *Eugenia bracteata* Rich., each name is synonymous with a different species]. In these cases, we followed IPNI and considered the legitimate names.

[1] see Appendix 1.1

[2] see Appendix 1.2

[3] see Appendix 1.3

[4] see Appendix 1.4

[5] see Appendix 1.5

[6] see Appendix 1.6

[7] see Appendix 1.7

[8] see Appendix 1.8

Preliminary occurrence of apomixis in taxa based on above analyses (lacking embryological data) from herbarium speciments (e.g., Ingrouille, 1984) was accepted only if corroborated by observations of fresh materials (either during field expeditions or during garden cultivation) and other experimental tests, like castration experiments, embryology or progeny tests – by morphological, cytological or molecular evaluation (e.g., Baker, 1966; Ingrouille, 1982; Ashari et al., 1988; Palacios and González-Candelas, 1997a; Kouzali et al., 2012; Róis et al., 2012; Conceição et al., 2021).

Crossing experiments indicating apomixis (i.e., maternal offspring) in *Beta lomatogona* Fisch. & C.A. Mey., Amaranthaceae Juss. (Barocka, 1966; Filutowicz and Dalke, 1972) are supported by embryology (Jassem and Jassem, 1970). In addition, *Beta trigyna* Waldst. & Kit. shows apospory, diplospory and adventitious embryony, while in *Beta corolliflora* Zosimovic ex Buttler the female gametophyte is of the Polygonum type, and apospory is observed in low frequency together with around 5% of nucellar and integumentary embryony (Jassem and Jassem, 1970; Jassem, 1976, 1990). The occurrence of apomixis in *Beta vulgaris* L. is based on different studies showing the occurrence of unreduced (recombinant) gametes (Allium type) (Maletskii and Maletskaya, 1996, in Szkutnik, 2010), seed formation under isolation of male sterile flowers and clonal offspring (based on segregation analysis of five isozyme systems) (Szkutnik, 2010). Yet, the average proportion of plants producing offspring expectedly apomictic in consecutive generations was extremely low compared to other *Beta* species, and therefore we placed this species in the quarantine group (Table 7.2).

In *Genista salzmannii* DC. (Fabaceae Juss.), an interesting case of apomixis has been recorded. According to Diana and Villa (1991), autonomous Antennaria type apomixis is frequent in a tetraploid and a diploid population. Yet, the rarity of functional apomixis occurring at diploid level (see Section 7.3) and the observation of extremely low pollen quality (compared to high pollen quality in tetraploids) which contrasts the pattern in most apomicts suggest that (1) either meiosis could occur rapidly, making it difficult to observe, or (2) diploids could have originated from apomictic polyploids as dihaploids. The last alternative was discussed by Diana and Villa (1991, p. 36). If this is the case, it might represent the first record of inverse polyploidization occurring in nature (so far, known dihaploids have been recovered only in experimental conditions, e.g., Kojima and Nagato, 1997; Noyes and Wagner, 2014). Experimental confirmation of apomixis in this cytotype would clarify the case.

Limonium spp. is a remarkably interesting genus. In addition to the occurrence of apomixis and tetraspory, polyploidy and hybridization combined to different genome sets and reproductive morphs shape the evolution of these species (Cowan et al., 1998). Several *Limonium* species are included in Table 7.1 on the basis of considerable circumstantial evidence in support of apomixis, including single self-incompatible pollen-stigma morphs observed in populations, low pollen fertility, odd ploidy and high seed set [e.g., *Limonium aphroditae* R. Artelari & Georgiou, Artelari and Georgiou (1999); *Limonium silvestrei* Aparicio, Aparicio (2005)]. However, corroboration through analyses of the development of the female gametophyte and progeny tests is recommended. The even ploidy cytotypes such as tetraploids and hexaploids are included in Table 7.1 only when direct evidence of apomixis is provided (e.g., Róis et al., 2016) because sexual dimorphisms and recombinant progenies had been proved in different species (e.g., Erben, 1979; Georgakopoulou et al., 2006; Palop-Esteban et al., 2011).

Species of *Allium* L., Amaryllidaceae J.St.-Hil., are categorized as diplosporous apomicts, even though the mechanism of formation of unreduced megaspores does not involve an altered meiosis. In these species, meiosis undergoes the usual steps, but due to a premeiotic doubling of chromosomes, recombination does not change the genetic information, posmeiotic nuclei have a somatic chromosomal constitution and a functional binucleated megaspore (bispory) develops an unreduced-like female gametophyte (Håkansson and Levan, 1957; Gohil and Kaul, 1981). This type of development is referred to as Allium type.

Indications of apomictic seed development based on personal observations are not included [e.g., in *Amorphophallus kiusianus* (Makino) Makino, Araceae Juss., Hetterscheid and Peng (1995)]. Likewise, suggestions of apomixis in isolated female plants producing fruits [e.g., in three *Pandanus* Parkinson species, Pandanaceae R.Br., Kurz (1867); Cox (1990); in *Corynocarpus laevigatus* J.R. Forst. & G. Forst., Corynocarpaceae Engl., or *Griselinia littoralis* (Raoul) Raoul, Cornaceae Bercht. ex J. Presl, among other species in Molloy (2019); see Appendix 1.9] or species categorized as apomictic based on unpublished results [e.g., in the *Callistemon rigidus-linearis-pinifolius* complex, Myrtaceae Juss., Rye (1979); in *Limonium inexpectans* L. Sáez & Rosselló, Plumbaginaceae Juss., Sáez and Roselló (1998)] were not included in our tables unless adequately verified. Suggestions of apomixis based on the lack of pollen tubes near female gametophytes, nucellar tissue and ovules in general [e.g., in *Epipactis phyllanthes* G.E.Sm., Orchidaceae Juss., Young (1952); in *Euphorbia esula* L., Euphorbiaceae Juss., Carmichael and Selbo (1999)], on the occurrence of non-functional, unreduced female gametophytes [e.g., in *Cliftonia monophylla* (Lam.) Britton ex Sarg., Cyrillaceae Lindl., Vijayaraghavan and Dhar (1978); or in *Adoxa moschatellina* L. Czapik (1976) and in *Sambucus racemosa* L., Tock (1981) in the Adoxaceae E.Mey. family], or the presence of structures resembling zygotes within unfertilized ovules in isolated and emasculated flowers but without apparent seed formation [e.g., in *Epipogium aphyllum* Sw., Orchidaceae Juss., Krawczyk et al. (2016)] are not considered as evidence of apomixis or parthenogenetic formation of embryos (Table 7.2).

The occurrence of polyembryos can be a character disclosing sporophytic apomixis if the extra embryos are developed from nucellar or integumentary tissues, or in rare cases from apomictic female gametophytes. In many studies, the occurrence of multiple seedlings indicating spontaneous polyembryony has been found at levels <1% [e.g., in *Borassus flabellifer* L., Arecaceae Bercht. & J. Presl, Veerasamy (1982); in *Nothapodytes nimmoniana* (J. Graham) Mabb., Icacinaceae Miers, Hombe Gowda et al. (2004); in *Putranjiva roxburghii* Wall., Putranjivaceae Endl., Thapliyal (2004); in *Vitis vinifera* L., Bouquet (1980); in *Dipterocarpus retusus* Blume, Singh and Thakur (2004); in

Calophyllum inophyllum L., Clusiaceae Lindt., Gunaga et al. (2011); in *Solanum lycopersicum* L. ≡*Lycopersicon esculentum* L., Solanaceae Adans., Marshall (1958)]. Yet, these studies lack an evaluation of early ovule ontogeny and thus, crucial evidence on the nature of the additional embryos is missing. For simplicity, these cases are not included here, but can be consulted elsewhere (Davis, 1966; Veerasamy, 1982; Singh et al., 2005).

Studies where the frequency of seeds carrying multiple embryos is higher – suggesting regular mechanisms of development – but still without evidence on the origin of accessory embryos [e.g., in *Inga vera* Willd. sbsp. *afinis*, Faria et al. (2004), in *Saraca asoca* (Roxb.) Willd., Fabaceae Juss., Singh et al. (2005); in *Cyrtopodium flavum* (Nees) Link & Otto ex Rchb. =*Cyrtopodium polyphyllum* (Vell.) Pabst ex F. Barros, Orchidaceae Juss., Pansarin et al. (2008); in *Madhuca longifolia* var. *latifolia* (Roxb.) A. Chev. =*Madhuca latifolia* (Roxb.) J.F. Macbr., Sapotaceae Juss., Chavan et al. (2014)] are placed in Table 7.2 for verification. A few cases indicating sporophytic apomixis (adventitious embryony) without direct confirmation but with substantial circumstantial evidence are placed in Table 7.1. For example, Maguire (1976) presented a preliminary report in two *Clusia* species suggesting the occurrence of adventitious embryony based on fruit production and formation of viable seeds in apparent absence of male flowers/trees. The author also mentions having sectioned material confirming the occurrence of adventitious embryos (Maguire, 1976, p. 242) but without providing further details. Despite pollinator exclusion tests discarded autonomous apomixis in *Clusia peninsulae* Hammel (Hochwallner et al., 2012), this agrees with pseudogamy as observed in *C. alba* Jacq., a species showing nucellar embryony (Goebel, 1900). Further circumstantial evidence suggests the presence of apomixis in *Clusia* spp. (van Tieghem, 1885; Vesque, 1893; Wood and Adams, 1976; Hammel, 1986; Rodrigues-Correia et al., 2011; Cheek and Lalla, 2017), where polyembryonate seedlings (e.g., up to three embryos in seeds of *C. rosea* Jacq.) have been recorded in germination tests and trees carrying only pistillate flowers have been documented producing abundant fruits.

Studies showing polyembryony in species based on the occurrence of multiple seedlings in germination tests or inconclusive embryology were included in Table 7.1 only when evidence for asexual embryos in the same taxa is presented elsewhere [e.g., in *Shorea ovalis* Blume, Dipterocarpaceae Blume, Kaur et al. (1978, 1986); in *Genipa americana* L., Rubiaceae Juss., Crestana (1995); Salomão and Allem (2001)] or in studies including related taxa [e.g., *Commiphora leptophloeos* (Mart.) J.B.Gillett and *C. wightii* (Arn.) Bhandari, Burseraceae Kunth, Gupta et al. (1996); Salomão and Allem (2001); in *Syzygium* P. Browne ex Gaertn., Myrtaceae Juss., Tiwary (1926); van der Pijl (1934); Merril and Perry (1938); Narayanaswami and Roy (1960); Chantaranothai and Parnell (1994); Thurlby et al. (2011); in *Miconia* Ruiz & Pav., Melastomataceae Juss., Mendes-Rodrigues and Oliveira (2012)]. For example, many cases presented by Braun (1860) have subsequently been corroborated (e.g., in Mauritzon, 1935; Subba Rao, 1940; Symon, 1956).

Embryological evidence on nucellar or integumentary embryony without mentioning relative number of studied ovules is placed in Table 7.1; for such cases, reassessments are recommended [e.g., *Withania somnifera* (L.) Dunal, Solanaceae Adans., Banerji and Bhaduri (1933); *Ptelea trifoliata* L., Rutaceae Juss., Mauritzon (1935); *Daphne papyracea* Wall. ex G. Don =*Daphne cannabina* Wall., Thymelaeaceae Juss., Venkateswarlu (1946); *Combretum coccineum* (Sonn.) Lam., Combretaceae R.Br., Venkateswarlu (1947)]. When detailed analyses were available, cases of unusual aposporous developments in sexual species or species with regular adventitious embryony [e.g., in *Rhexia mariana* L., Melastomataceae Juss., Goldenberg and Shepherd (1998); in *Syzygium cumini* (L.) Skeels, Myrtaceae Juss., van der Pijl (1934); Narayanaswami and Roy (1960); in *Zanthoxylum* spp., Rutaceae Juss., Mauritzon (1935); Naumova (1993)] were not considered unless explicitly stated (see comments in section 7.3).

As mentioned in Chapter 2, cases of polyembryony based on mechanisms other than sporophytic apomixis (adventitious embryony) like cleavage polyembryony [e.g., in *Erythronium americanum* Ker Gawl., Liliaceae Juss., Jeffrey (1985); in *Eulophia epidendraea* (J. Koenig ex Retz.) C.E.C. Fisch., Orchidaceae Juss., Swamy (1943); in *Hippobroma longiflora* (L.) G. Don ≡*Isotoma longiflora* (L.) C. Presl, Campanulaceae Juss., Kausik and Subramanyam (1946); in *Ophiopogon japonicus* (Thunb.) Ker Gawl., Asparagaceae, Fukai et al. (2000)], simultaneous development of female gametophytes in the same ovule [e.g., among Orchidaceae Juss., in *Anacamptis morio* (L.) R.M. Bateman, Pridgeon & M.W. Chase ≡*Orchis morio* L., Schacht (1850), Braun (1860); in *Lecanorchis japonica* Blume, Tohda (1971); or in *Neottia nipponica* (Makino) Szlach. ≡*Listera nipponica* Makino, Abe (1972)], or apogamy [e.g., in *Spiranthes sinensis* (Pers.) Ames =*Spiranthes australis* (R.Br.) Lindl., Orchidaceae Juss., Maheshwari and Narayanaswami (1952); in *Caryota rumphiana* Mart., Arecaceae Bercht. & J. Presl, Shirke (1963); in *Hypericum tetrapterum* Fr. or *H. perforatum* L., Hypericaceae Juss., Bougnicourt (1970, 1971); Robson (1981)] are not considered. Cleavage polyembryony and apogamy have already been reported in many taxa (e.g., Swamy, 1946; Hagerup, 1947; Chua and Rao, 1978; Nowaczyk, 1987; Martinez-Gómez and Gradziel, 2003; Nowaczyk and Nowaczyk, 2006; Sriyot et al., 2015) and are common among orchids (Batygina et al., 2003; Ghimire et al., 2020) and therefore, our Table 7.2 does not list all reported cases of polyembryony in this family, even though some may represent cases of adventitious embryony.

As apomixis refers to alternative mechanisms involved in the formation of a seed carrying maternal (clonal) embryos (see Chapter 1), this also excludes cases of vegetative multiplication like bulbils or foliar embryos that simulate ovule-like structures [e.g., in *Hammarbya paludosa* (L.) Kuntze ≡*Malaxis paludosa* (L.) Sw., Orchidaceae Juss., Dickie (1875); Taylor (1967)].

In a few cases, apomixis was suggested based on the formation of fruits in emasculated or unpollinated flowers but without apparent seed formation [e.g., in *Vochysia tucanorum* Mart., Vochysiaceae A.St.-Hil., Costa et al. (1992); or in *Calophyllum papuanum* Lauterb. and *C. pseudovitiense* P.F. Stevens, Clusiaceae Lindl., Stevens (1974)]. These cases are placed in Table 7.2. Likewise, Lughadha and Proença (1996) have discussed extensively about polyembryony in Myrtaceae Juss. These authors indicate that van der Pijl (1934) reported polyembryony in *Syzygium aqueum* (Burm.f.) Alston ≡*Eugenia aquea* Burm.f., but the author states that the species is sterile and parthenocarpic (van der Pijl, 1934, pp. 174, 185) (Table 7.2). Taxa previously cited as apomict (polyembryonic) [e.g., *Cassia australis* var. *revoluta* (F. Muell.) Benth.; *Cassia desolata* var. *involucrata* (J.M. Black) J.M. Black; *Cassia phyllodinea* R. Br., in Naumova (1993), p. 65] but found to be originally reported as monoembryonic (see e.g., Symon, 1956) are placed in our quarantine Table 7.2.

Molecular evidence for clonality was accepted when based on tests using five or more individuals (e.g., Palacios and González-Candelas, 1997a, 1997b; Hedrén et al., 2000; Martins and Oliveira, 2003; Mayo and Langridge, 2003; Huang et al., 2009; Pangsuban et al., 2009; Lu et al., 2010; Arakaki et al., 2013; Moura et al., 2020). In these cases, vegetative propagation (see Chapter 1) should be ruled out. Some studies were designed to disclose genotypic variability rather than clonality and, hence, were not considered here. In specific cases, apomixis was corroborated or inferred from the presence of markers linked to apomeiosis (e.g., by amplification of the marker allele APOLLO, or by linkage of male and female apomeiosis marker alleles across *Boechera* species, Mau et al., 2015; see Appendix 1.10).

Different new apomictic species had been detected using FCSS (e.g., Mártonfi, 2001; Matzk et al., 2003; Aliyu et al., 2010; Brožová et al., 2019). As mentioned in previous chapters, in these apomicts it is only possible to speculate about the type of female gametophyte development, and embryological analysis are needed. Even in genera with characterized apomictic developments, observed variations in type of female gametophyte and embryo developments within a species makes embryological inference from FCSS data inaccurate. For instance, in ovules of *Burmannia coelestis* D. Don (Burmanniaceae Blume), the Antennaria type of female gametophyte development predominates but Taraxacum type had also been observed (Gustafsson, 1935), as well as the occurrence of occasional polyembryony (Ernst und Bernard, 1912, p. 248).

We apologize for any missing data or errors resulting from lack of access to certain works or from our language limitations (e.g., Tiwary, 1926; Marklund, 1941; Fedortschuk, 1944; Vaclav, 1956; Okabe, 1963; Chochlov et al., 1978; Gvaladze and Krialashvili, 1988; Bogomolov, 2005).

7.2.1 Doubtful Cases of Apomixis in Different Plant Families

The occurrence of apomixis, either sporophytic or gametophytic, is pending confirmation in 120 taxa (Table 7.2). This number will undoubtedly exceed 200 if all cases with some indication of apomixis excluded in Tables 7.1 and 7.2 are considered, specially from taxa of biodiverse genera like *Limonium* spp., *Ranunculus* spp., or *Boechera* spp. For the cases in Table 7.2, current circumstantial evidence is not sufficient to demonstrate or conclude asexual seed formation and exhaustive experiments are needed. Besides the above-mentioned reasons to include species in the quarantine list, other specific cases are listed below. The most frequent occurrences in our quarantine list include diploids (25% of species with ploidy records), triploids (38%), and tetraploids (35%). In contrast to previous chapters, most taxa –including diploids – in this list are associated with evidence suggesting adventitious embryony.

Cases cited as doubtful in Naumova (1993) were reviewed whenever possible. For example, nucellar embryony in *Ardisia* spp. is validated (Table 7.1) based on the work of Kume (1959) showing the occurrence of nucellar embryos in *A. compressa* Kunth (p. 21) and polyembryony in several species along with the observations of Braun (1857). Both authors also agree that *A. crispa* (Thunb.) A. DC. is sexual, but show disparate results for *A. humilis* Vahl, which is placed here in Table 7.2.

Nucellar polyembryony was suggested for several taxa of the family Onagraceae Juss. (Naumova, 1993). It was first observed experimentally in *Oenothera lamarckiana* Ser. (Haberlandt, 1921a, 1921b, 1927, 1938) and after interspecific crossings between *Oenothera* species and *Epilobium* species (Haberlandt, 1922, 1923; Michaelis, 1925; Lakshmanan and Ambegaokar, 1984). Yet, the occurrence of multiple archespores in *Epilobium luteum* Pursh (Michaelis, 1925) and failed attempts to replicate Haberlandt's results (Beth, 1938) weaken the evidence for apomixis on these taxa and no other study on the occurrence of nucellar embryony was found despite *Oenothera* being a model plant for genetic and developmental studies (Harte, 1994). Genetic studies pointing to clonality among *Oenothera* spp. (e.g., Hollister et al., 2015, 2019) do not support apomixis *sensu stricto*, but a system of permanent translocation heterozygosity that prevents recombination during meiosis and random gamete fusion in these taxa (Golczyk et al., 2014) thus functionally mimicking asexuality. In a related genus, adventitious embryony in *Clarkia unguiculata* Lindl. (=*Clarkia elegans* Douglas) is suggested based on a mention by Johansen (1930, p. 87) of one case over many studied ovules. Thus, the occurrence of apomixis *sensu stricto* in the family Onagraceae is not conclusive (Table 7.2).

In a few species, castration or pollen exclusion tests show questionable approaches or results. For example, in *Cymbopetalum brasiliense* (Vell.) Benth. ex Baill., Annonaceae Juss., castration experiments show initial fruit development, but fruit ripening and seed formation was not confirmed (Braun et al., 2011). Parthenocarpy cannot be ruled out and, therefore, this species is quarantined (Table 7.2). In *Salix integra* Thunb. (=*Salix multinervis* Franch. & Sav.), Salicaceae Mirb., Ikeno (1922) reported apomixis based on exceptionally low production of autonomous seeds in pollen exclusion tests and production of some maternal progeny in interspecific crossing experiments (Table 7.2).

In *Trifolium pratense* L., Fabaceae Juss., a variety of mechanisms likely leading to the formation of asexual seeds in low proportion have been observed. These include the spontaneous formation of nucellar embryos (Fedortschuk, 1944, in Maheshwari, 1963), the development of multiple gametophytes (Jönsonn, 1883), as well as aposporous (Hieracium type; Hindmarsh, 1964) and diplosporous (Taraxacum and Ixeris types; Algan and Büyükkartal, 1999) female gametophytes. Yet, the occurrence of multiple archespores within an ovule and the high level of abortion of female gametophytes and low seed set despite egg-cell fertilization (Fedortschuk, 1944; Algan and Büyükkartal, 1999; Büyükkartal, 2008, 2009) do not allow conclusions.

Schürhoff and Jüssen (1925) described the occurrence of nucellar embryony in *Spathiphyllum patinii* (R. Hogg) N.E. Br., Araceae Juss. In *S. humboldtii* Schott, Hentrich et al. (2010) observed fructification in the absence of pollinators. Since the species has a clear separation of the male-female sexual phases, the authors suggest that seeds are produced by apomixis. Despite a recent study on two other *Spathiphyllum* species shows no self-pollination or apomixis (Díaz Jiménez, 2016), autogamy and self-pollination in *S. humboldtii* were not ruled out in the original study and the occurrence of apomixis requires verification (Table 7.2).

Jeffrey and Haertl (1939b) carried out embryological evaluations in ovules of four *Trillium* species, including *T. erectum* L. and *T. grandiflorum* (Michx.) Salisb., Melanthiaceae Batsch ex Borkh. These authors described a lack of fertilization, the development of a haploid, abortive embryo and posterior appearance of a diploid embryo in the micropylar region of the endosperm. However, this work received several critics and was not corroborated by later studies (Gyer, 1996) including a more recent study using castration experiments that shows a lack of seed formation in unpollinated flowers in *T. erectum* and *T. grandiflorum* (Irwin, 2000). Whether these apparently contradictory results correspond to plant materials with different ploidy and reproductive modes needs to be confirmed. In these species, genetic analysis in support of apomixis is impeded by the occurrence of vegetative propagation (Gonzales, 2005). We placed these taxa in our quarantine Table 7.2.

In *Allium roseum* L. var. *carneum* (O. Targ.Tozz.) Rchb. (sub *Allium roseum* L. var. *bulbilliferum* Kunth.), Amaryllidaceae J.St.-Hil., Messeri (1931) described a normal development of the female gametophyte and indicated the occurrence of apomixis based on the observation of inflorescences transformed in bulbils (Table 7.2).

TABLE 7.2
Table of Doubtful Apomictic Species (A) from all Angiosperm Families Except Poaceae Barnhart, Asteraceae Giseke, and Rosaceae Juss., Alphabetically Sorted (M: Monocot; D: Dicot)

Species	Ploidy	Type of Apomixis[a]	References
Adoxaceae E.Mey. (D)			
Adoxa moschatellina L.	*2x*	A; Ap	Czapik (1976)
Adoxa moschatellina subsp. *cescae* Peruzzi & N.G. Passal.	*3x*	A	Peruzzi and Passalacqua (2006)
Sambucus nigra L.	*4x*	A	Czapik and Kościńska-Pająk (2000)
Sambucus racemosa L.		A	Tock (1981)
Amaranthaceae Juss. (D)			
Beta vulgaris L.	*2x, 4x*	A	Zaikovskaja et al. (1978), Shiryaeva and Yarmoluk (1983), Seilova (1996), Dedong *et al. (*1999*)*, Szkutnik (2010)
Amaryllidaceae J.St.-Hil. (M)			
Allium oleraceum L.	*5x*	A; Ps	Åström and Haeggström (2004)
Allium roseum L. var. *carneum* (O.Targ. Tozz.) Rchb.	*6x*	A	Messeri (1931)
Anacardiaceae R.Br. (D)			
Myracrodruon urundeuva Allemão		A; AE	Salomão and Allem (2001)
Annonaceae Juss. (D)			
Cymbopetalum brasiliense (Vell.) Benth. ex Baill.		A	Braun et al. (2011)
Apocynaceae Juss. (D)			
Cascabela thevetia (L.) Lippold [=*Thevetia peruviana* (Pers.) K.Schum.]		A; AE	Rai (2014)
Hancornia speciosa Gomes		A; AE	Salomão and Allem (2001)
Asparagaceae Juss. (M)			
Ophiopogon japonicus (Thunb.) Ker-Gawl.		A; N	Fukai et al. (2000)

(Continued)

TABLE 7.2 (Continued)
Table of Doubtful Apomictic Species (A) from all Angiosperm Families Except Poaceae Barnhart, Asteraceae Giseke, and Rosaceae Juss., Alphabetically Sorted (M: Monocot; D: Dicot)

Species	Ploidy	Type of Apomixis[a]	References
Bignoniaceae Juss. (D)			
Anemopaegma chamberlaynii (Sims) Bureau & K. Schum.	*2x*	A; AE; Ps	Correia et al. (2006)
Handroanthus ochraceus (Cham.) Mattos [≡*Tabebuia ochracea* (Cham.) Standl.]		A; AE	Salomão and Allem (2001)
Dolichandra unguis-cati (L.) L.G. Lohmann		A; AE	Buru et al. (2016)
Oroxylum indicum (L.) Kurz		A; AE	Sharma et al. (2004)
Brassicaceae Burnett (D)			
Arabis hirsuta (L.) Scop.	*4x*	A; Ps	Roy (1995)
Boechera grahamii (Lehm.) Windham & Al-Shehbaz (=*Arabis boivinii* G.A. Mulligan)	*3x*	A	Dobeš et al. (2006)
Boechera caeruleamontana		A	Carman et al. (2019)
Boechera oxylobula (Greene) W.A. Weber [=*Boechera demissa* (Greene) W.A. Weber]	*3x*	A	Dobeš et al. (2006)
Boechera gracilenta (Greene) Windham & Al-Shehbaz [=*Boechera selbyi* (Rydb.) W.A. Weber]	*3x*	A	Rollins and Rüdenberg (1971)
Campanulaceae Juss. (D)			
Adenophora liliifolia (L.) A.DC.		A	Sukatschew (1940)
Caprifoliaceae Juss. (D)			
Scabiosa atropurpurea L.	*2x*	A	Braun (1860, in Naumova, 1993)
Caricaceae Dumort.			
Carica papaya L.		A; AE	Vegas et al. (2003)
Clusiaceae Lindl. (D)			
Calophyllum papuanum Lauterb.		A; Au	Stevens (1974)
Calophyllum pseudovitiense P.F.Stevens		A; Au	Stevens (1974)
Colchicaceae DC. (M)			
Colchicum speciosum Steven		A	Gvaladze and Krialashvili (1988)
Cucurbitaceae Juss. (D)			
Bryonia aspera Steven ex Ledeb. (=*Bryonia dioica* Jacq.)		A	Bitter (1906), Jones and Rayner (1916)
Cyrillaceae Lindl. (D)			
Cliftonia monophylla (Lam.) Britton ex Sarg.		A; Ap (Hi)	Vijayaraghavan and Dhar (1978)
Dioscoreaceae R.Br. (M)			
Tacca integrifolia Ker Gawl. (=*Tacca cristata* Jack)		A; Ap	Suessenguth (1921)
Tacca plantaginea (Hance) Drenth (≡*Schizocapsa plantaginea* Hance)		A	Håkansson (1921)
Dipterocarpaceae Blume (D)			
Vateria indica L.		A; AE	Kader et al. (2000)

TABLE 7.2 (Continued)

Species	Ploidy	Type of Apomixis[a]	References
Euphorbiaceae Juss. (D)			
Manihot esculenta Crantz	2*x*	A; AE	Nassar (1994, 2002, 2006), Grattapaglia et al. (1996), Freitas and Nassar (2013)
Manihot neusana Nassar		A; Ap	Nassar et al. (2000), Freitas and Nassar (2013)
Jatropha curcas L.	2*x*	A; AE	Ambrosi et al. (2010), Francis et al. (2020)
Fabaceae Juss. (D)			
Acacia polyphylla DC.		A; AE	Salomão and Allem (2001)
Copaifera langsdorffii Desf.		A; AE	Salomão and Allem (2001)
Inga affinis DC. [=*Inga vera* Willd. subsp. *affinis* (DC.) T.D. Penn.]		A	Faria et al. (2004)
Inga laurina (Sw.) Willd.		A	Mendes-Rodrigues et al. (2018)
Saraca asoca (Roxb.) Willd.		A; AE; Ps	Singh et al. (2005), Wanage et al. (2010)
Cassia australis var. *revoluta* (F. Muell.) Benth.		A; AE; Ps	Symon (1956)
Cassia desolata var. *involucrata* (J.M. Black) J.M. Black	4*x*	A; AE; Ps	Symon (1956), Randell (1970)
Cassia phyllodinea R. Br.	4*x*	A; AE; Ps	Symon (1956), Randell (1970)
Trifolium pratense L.	4*x*	A; Ap (Hi); Di (Ta, Ix)	Hindmarsh (1964), Algan and Büyükkartal (1999)
Gentianaceae Juss. (D)			
Gentiana lutea L.		A; AE	Rudenko (1961, in Naumova, 1993)
Gentiana punctata L.		A; AE	Rudenko (1961, in Naumova, 1993)
Gentianella amarella subsp. *livonica* (Eschsch. ex Griseb.) Dostál (=*Gentiana livonica* Eschsch. ex Griseb.)		A; AE	Rudenko (1961, in Naumova, 1993)
Gentianella lutescens (Velen.) Holub (≡*Gentiana lutescens* Velen.)		A; AE	Rudenko (1961, in Gregor, 2013)
Hypericaceae Juss. (D)			
Hypericum brasiliense Choisy		A; Ps	de Moraes et al. (2009)
Lecythidaceae A.Rich. (D)			
Cariniana estrellensis (Raddi) Kuntze		A; AE	Salomão and Allem (2001)
Liliaceae Juss. (M)			
Tulipa gesneriana L.		A; AE	Ernst (1901), Bambacioni-Mazzetti (1932), Naumova (1993)
Erythronium americanum Ker-Gawl.		A; AE	Jeffrey (1985)
Malpighiaceae Juss. (D)			
Byrsonima basiloba A.Juss.		A; AE	Salomão and Allem (2001)
Melanthiaceae Batsch ex Borkh. (M)			
Trillium erectum L.		A; AE	Jeffrey and Haertl (1939a, 1939b), Irwin (2000)
Trillium grandiflorum (Michx.) Salisb.		A; AE	Jeffrey and Haertl (1939a, 1939b), Irwin (2000)

(Continued)

TABLE 7.2 (Continued)
Table of Doubtful Apomictic Species (A) from all Angiosperm Families Except Poaceae Barnhart, Asteraceae Giseke, and Rosaceae Juss., Alphabetically Sorted (M: Monocot; D: Dicot)

Species	Ploidy	Type of Apomixis[a]	References
Melastomataceae Juss. (D)			
Chaetostoma armatum (Spreng.) Cogn.		A	Mendes-Rodrigues and Oliveira (2012)
Miconia biserrata (DC.) Michelang. (≡*Clidemia biserrata* DC.)		A; AE	Mendes-Rodrigues and Oliveira (2012)
Miconia cuspidata Mart. ex Naudin		A; AE	Mendes-Rodrigues and Oliveira (2012)
Microlicia helvola Triana		A; AE	Mendes-Rodrigues and Oliveira (2012)
Meliaceae Juss. (D)			
Carapa surinamensis Miq.		A; AE	Souza Ferreira et al. (2017, 2019)
Musaceae Juss. (M)			
Musa acuminata Colla		A	Okoro et al. (2011)
Oleaceae Hoffmanns. & Link (D)			
Fraxinus americana L.		A; AE	Steinbauer (1943)
Fraxinus velutina Torr. [=*Fraxinus velutina* var. *toumeyi* (Britton) Rehder]		A; AE	Steinbauer (1943)
Onagraceae Juss. (D)			
Clarkia unguiculata Lindl. (=*Clarkia elegans* Douglas)		A; AE	Johansen (1930)
Epilobium angustifolium L. [≡*Chamerion angustifolium* (L.) Holub subsp. *angustifolium*]	4*x*	A; AE	Michaelis (1925), Lakshmanan and Ambegaokar (1984)
Epilobium hirsutum L.	4*x*	A; AE	Michaelis (1925), Lakshmanan and Ambegaokar (1984)
Epilobium luteum Pursh	4*x*	A; AE	Michaelis (1925), Lakshmanan and Ambegaokar (1984)
Oenothera biennis L. (=*Oenothera muricata* L.)	2*x*	A; AE	Haberlandt (1921a, 1921b, 1922, 1923), Beth (1938)
Oenothera lamarckiana Ser.	2*x*	A; AE	Haberlandt (1921a, 1921b, 1922, 1923), Beth (1938)
Oenothera glazioviana Micheli	2*x*	A; AE	Haberlandt (1922)
Orchidaceae Juss. (M)			
Cyrtopodium flavum (Nees) Link & Otto ex Rchb. [=*Cyrtopodium polyphyllum* (Vell.) Pabst ex F.Barros]		A; AE; Ps	Pansarin et al. (2008)
Neottia borealis (Morong) Szlach. (≡*Listera borealis* Morong)	3*x*	A	Simon (1968)
Epipogium aphyllum Sw.		A	Krawczyk et al. (2016)
Geodorum densiflorum (Lam.) Schltr.	3*x*	A; AE	Ansari (1977)
Heterotaxis brasiliensis (Brieger & Illg) F.Barros (≡*Maxillaria brasiliensis* Brieger & Illg)		A; Ps	Illg (1977)
Prasophyllum goldsackii J. Weber & R.J. Bates		A; Ps	Bates (1984)
Prasophyllum macrostachyum R. Br.		A; Ps	Bates (1984)

TABLE 7.2 (Continued)

Species	Ploidy	Type of Apomixis[a]	References
Spathoglottis plicata Blume	*2x, 3x*	A; AE	Ansari (1977)
Paeoniaceae Raf. (D)			
Paeonia daurica subsp. *coriifolia* (Rupr.) D.Y.Hong (=*Paeonia caucasica* Schipcz.)	*2x*	A; AE	Naumova (1993)
Paeonia majko Ketzch.		A; AE	Naumova (1993)
Paeonia tenuifolia L.	*2x*	A; AE	Naumova (1993)
Pandanaceae R.Br. (D)			
Pandanus forsteri C.Moore & F.Muell.		A; Au	Cox (1985, 1990)
Pandanus kaida Kurz (=*Pandanus forceps* Martelli)		A; Au	Cox (1990)
Pandanus dubius Spreng.		A; Au	Kurz (1867, in Cox, 1990)
Plumbaginaceae Juss. (D)			
Limonium barceloi Gil & Llorens	*4x*	A	Gil and Lorens (1991)
Limonium bolosii Gil & Llorens	*4x*	A	Gil and Lorens (1991)
Limonium britannicum Ingrouille	*4x*	A	Ingrouille and Stace (1985)
Limonium carvalhoi Rosselló & L.Sáez	*3x*	A	Rosselló et al. (1998)
Limonium cythereum R. Artelari & Georgiou	*6x*	A	Artelari and Georgiou (1999)
Limonium emarginatum (Willd.) Kuntze	*4x*	A	Erben (1978)
Limonium graecum (Poir.) Kuntze	*4x*	A	Georgakopoulou et al. (2006)
Limonium interjectum Soler & Rosselló		A	Soler and Rosselló (1997)
Limonium ilergabonum López-Alvarado, Cobacho, Arán & L.Sáez	*3x*	A	López-Alvarado et al. (2017)
Limonium leonardi-llorensii L. Sáez, Carvalho & Rosselló	*3x*	A	Sáez et al. (1998)
Limonium marisolii L. Llorens	*3x*	A	Sáez et al. (1998)
Limonium maritimum Caperta, Cortinhas, Paes, Guara, Espírito-Santo and Erben.	*3x*	A	Cortinhas et al. (2015)
Limonium mucronulatum (H.Lindb.) Greuter & Burdet		A	Kouzali et al. (2012)
Limonium palmare (Sm.) Rech.	*4x*	A	Georgakopoulou et al. (2006)
Limonium perplexum L.Sáez & Rosselló	*3x*	A	Sáez and Rosselló (1999)
Limonium roridum (Sm.) Brullo & Guarino	*4x*	A	Georgakopoulou et al. (2006)
Limonium scopulorum M.B. Crespo & Lledó	*3x*	A	Crespo (2009)
Limonium tobarrense J.Moreno, Terrones, M.A.Alonso, Juan & M.B.Crespo		A	Moreno et al. (2016)
Limonium vigoi L.Sáez, Curcó & Rosselló	*4x*	A	Sáez et al. (1998)
Primulaceae Batsch ex Borkh. (D)			
Ardisia humilis Vahl		A; AE	Braun (1860), Kume (1959)
Rhamnaceae Juss. (D)			
Ziziphus joazeiro Mart.		A; AE	Salomão and Allem (2001)
Rubiaceae Juss. (D)			
Coprosma parviflora Hook. f.	*4x*	A; Au	Heenan et al. (2002)
Coprosma propinqua A.Cunn.	*4x*	A; Au	Molloy (2019)
Salicaceae Mirb. (D)			
Salix integra Thunb. (=*Salix multinervis* Franch. & Sav.)		A	Ikeno (1918, 1922)

(Continued)

TABLE 7.2 (Continued)
Table of Doubtful Apomictic Species (A) from all Angiosperm Families Except Poaceae Barnhart, Asteraceae Giseke, and Rosaceae Juss., Alphabetically Sorted (M: Monocot; D: Dicot)

Species	Ploidy	Type of Apomixis[a]	References
Sapindaceae Juss. (D)			
Magonia pubescens A. St.-Hil. (sub *Magonia pubescens* Benth.)		A; AE	Salomão and Allem (2001)
Sapotaceae Juss. (D)			
Madhuca longifolia var. *latifolia* (Roxb.) A.Chev. [=*Madhuca latifolia* (Roxb.) J.F.Macbr.]		A; AE	Chavan et al. (2014)
Solanaceae Adans. (D)			
Solanum lycopersicum L. (≡*Lycopersicon esculentum* Mill.)	2*x*	A; N	Banerji and Bhaduri (1933), Marshall (1958), Naumova (1993)
Urticaceae Juss. (D)			
Elatostema fengshanense W.T.Wang & Y.G.Wei	3*x*	A	Fu et al. (2017)
Elatostema coriaceifolium W.T.Wang	3*x*	A	Fu et al. (2017)
Elatostema longistipulum Hand.-Mazz.	3*x*	A	Fu et al. (2017)
Elatostema macintyrei Dunn	3*x*	A	Fu et al. (2017)
Elatostema cyrtandrifolium (Zoll. & Moritzi) Miq.	3*x*	A	Fu et al. (2017)
Elatostema xanthophyllum W.T.Wang	3*x*	A	Fu et al. (2017)
Elatostema asterocephalum W.T.Wang	3*x*	A	Fu et al. (2017)
Violaceae Batsch (D)			
Viola declinata Waldst. & Kit.	2*x*	A; AE	Naumova (1993)
Vochysiaceae A.St.-Hil. (D)			
Vochysia tucanorum Mart.		A; Ps; Au	Costa et al. (1992)

[a]For species with sporophytic apomixis, AE: adventitious embryony; N: nucellar embryony; I: integumentary embryony. For species with gametophytic apomixis, Ap ( ): apospory (type of female gametophyte development); Di ( ): diplospory (type of female gametophyte development); Ps: pseudogamous endosperm development; Au: autonomous endosperm development.

Hi: Hieracium type; Ta: Taraxacum type; An: Antennaria type; Ix: Ixeris type.

Tulipa gesneriana L., Liliaceae Juss., show a remarkably complex reproductive development (Bambacioni and Giombini, 1930) and the occurrence of nucellar embryony (Naumova, 1993) is unclear. While Ernst (1901) described the occurrence of cleavage polyembryony, Bambacioni-Mezzetti (1932) suggests the meiotic gametophyte degenerate before reaching maturity and a new one develops from cells in the inner integument. Nucellar embryony is cited by Naumova (1993) based on Bambacioni-Mezzetti (1932).

Håkansson (1921) noted the occurrence of aposporous female gametophytes (Fig. 44a,b, p. 218) while describing the reproductive development in *Tacca plantaginea* (Hance) Drenth ≡*Schizocapsa plantaginea* Hance, Dioscoreaceae R.Br.. However, no other details were provided which may indicate a spontaneous case of apospory (Table 7.2).

Studies in *Paeonia* L., Paeoniaceae Raf., remain inconclusive, mainly due to the occurrence of ovules with a multicellular archesporium and the difficulty to discern the cytological events in developing ovules (e.g., Shamrov, 1997; Vinogradova and Zhinkina, 2021). The occurrence of nucellar embryony in three species (Naumova, 1993) based on a report by Zhgonti (1978) disagree with the embryological observations made by several authors (e.g., Cave et al., 1961; Carniel, 1967 and references therein) showing the formation of a coenocyte from the dividing zygote and an embryonic mass from which multiple proembryos arise but only one develops to maturity. Thus, it seems plausible that the case relates to cleavage polyembryony and not nucellar embryony. The species are placed in quarantine (Table 7.2).

7.3 PATHWAYS TO ASEXUAL SEEDS IN OTHER PLANT FAMILIES

A total of 577 taxa, including 560 species, 8 subspecies and 14 varieties, from 69 plant families have at least one record of apomixis (Table 7.1), either sporophytic or gametophytic apomixis (see Chapter 2).

The pathways of apomictic development are known for 70.9% of species (*n*=409); in the remaining 29.1% of cases (*n*=168), apomixis has been determined without embryological details or hints to the embryological mechanism. Sporophytic apomixis is recorded in 307 taxa, while gametophytic apomixis is present in 127 taxa. In 24 cases, gametophytic and sporophytic apomixis has been recorded in the same taxa, either simultaneously or in different studies. This is mainly due to the occurrence of diplospory plus adventitious embryony (*n*=17; *n*=9 correspond to nucellar embryony) and apospory plus adventitious embryony (*n*=11; *n*=6 correspond to nucellar embryony) (Table 7.1). Among cases with gametophytic apomixis, apospory and diplospory occur simultaneously in 7 cases. Among cases with sporophytic apomixis, nucellar and integumentary embryony are concurrent in 18 taxa (Table 7.1).

Information about the development of the endosperm is collected for about 57.6% of all apomictic taxa (*n*=334; Table 7.1). Most recorded apomicts show autonomous endosperm development (*n*=210); the rest depend on fertilization for seed development and are pseudogamous (*n*=124). Autonomous endosperm development includes cases where viability of an adventive embryo hangs upon endosperm development but not on fertilization [e.g., in *Alchornea aquifolia* (Js.Sm.) Domin, Euphorbiaceae Juss., Strasburger, 1878], and cases where embryo viability does not depend on fertilization nor endosperm development (e.g., *Opuntia aurantiaca* Lindl., Cactaceae Juss., Archibald, 1939).

Ploidy is known for 61.7% of these apomictic taxa (*n*=356) and 50 cases have two or more ploidy records (Table 7.1). Most frequently found ploidies are diploids (*n*=123), triploids (*n*=107), and tetraploids (*n*=127). Other ploidies found among apomicts are pentaploids (*n*=19), hexaploids (*n*=24) and octoploids (*n*=14), and a few more cases with higher ploidy (Table 7.1).

Diploids are almost exclusively associated with cases of sporophytic apomixis (*n*=90; Table 7.1). The only exception of diploids with regular occurrence of gametophytic apomixis (apospory or diplospory) belongs to the genus *Boechera* Á. Löve & D. Löve and *Borodinia laevigata* (Muhl. ex Willd.) P.J. Alexander & Windham ≡*Boechera laevigata* (Muhl. ex Willd.) Al-Shehbaz, Brassicaceae Burnett (e.g., Mulligan, 1996; Aliyu et al., 2010; Mateo de Arias, 2015; Mau et al., 2015; Rojek et al., 2018; Carman et al., 2019). Diploid species in these genera are likely paleopolyploids (i.e., diploidized polyploids) having a complex history of hybridizations and genome duplications (Dobeš et al., 2007; Beck et al., 2012) like in other Brassicaceae genomes (Schranz et al., 2006; Barker et al., 2009). Likewise, paleopolyploidy is also suggested for *Cistanche* Hoffmanns. & Link spp., Orobanchaceae Vent., and the diploid apomict *Cistanche mongolica* Beck =*Cistanche tubulosa* (Schenk) Hook. f. (Table 7.1), in which diploidy is associated with high basic chromosome number (*x*=20) (Schneeweiss et al., 2004).

Other cases of gametophytic apomixis at the diploid level are rare and are associated with an irregular prevalence of functional apomixis. In ovules of diploid *Luffa cylindrica* (L.) M. Roem., Cucurbitaceae Juss., Deshpande et al. (1984) described a normal development of reduced female gametophytes with occasional formation of unreduced diplosporous Antennaria type female gametophytes. In *Genista salzmannii* DC., Fabaceae Juss., diplospory occurs in diploids and tetraploids, but it is functional only in the polyploids as diploids form no seeds or very few seeds reach maturity (Diana and Villa, 1991). Another example involves apospory in diploid *Skimmia japonica* Thunb., Rutaceae Juss. (Desai, 1961). The work of Chochlov et al. (1978) about the occurrence of apospory in three diploid species of *Salix* L., Salicaceae Mirb., was unavailable.

Finally, in other four species, the manifestation of gametophytic apomixis in diploids also does not seem to prevail and is associated with the dominant occurrence of sporophytic apomixis. This is the case in *Triphasia trifolia* (Burm.f.) P. Wilson =*Triphasia aurantiola* Lour., Rutaceae Juss.

(Braun, 1860; Mauritzon, 1935; Naumova, 1993) and in three species of Melastomataceae Juss., *Miconia crenata* (Vahl) Michelang. =*Clidemia hirta* (L.) D. Don (Caetano, 2014), *Miconia fallax* DC. (Caetano et al., 2018), and *Miconia albicans* (Sw.) Steud. (Caetano et al., 2013).

Polyploids are associated with cases of gametophytic apomixis (*n*=79), as expected, but also with sporophytic apomixis, in which polyploidy is observed in 122 cases (mostly triploids, *n*=30 and tetraploids, *n*=68) categorized as having adventitious embryony, nucellar embryony or integumentary embryony (Table 7.1).

The observed diversity of unreduced female gametophyte developments is similar to that found among aposporous and diplosporous daisies, whereby bipolar developments (i.e., Taraxacum type, Antennaria type, Ixeris type or Hieracium type) but no monopolar developments [except for a questionable case of a Panicum type development in *Elaeagnus rhamnoides* (L.) A. Nelson; see Appendix 1.8] are found among all screened angiosperm families (Table 7.1).

7.3.1 Sporophytic Apomixis

The formation of adventive embryos is exceedingly common among plant families other than the Poaceae Barnhart, Asteraceae Giseke and Rosaceae Juss. families, and has been observed in 307 taxa from 49 angiosperm families (Table 7.1).

7.3.1.1 Nucellar Embryony

This is the most frequent type of development of maternal embryos, observed in 54.1% of all taxa showing sporophytic apomixis (*n*=166), belonging to 43 angiosperm families (Table 7.1).

Nucellar embryony is recurrent in the Rutaceae Juss., among species of *Citrus* spp. (19 species; Braun, 1860; Strasburger, 1878; Biermann, 1896; Osawa, 1913; Ensign, 1919; Frost, 1926, 1938a; Bacchi, 1943; Furusato, 1952; Banerji, 1954; Yang, 1968; Iglesias et al., 1974; Kobayashi et al., 1978; Wakana and Uemoto, 1987; Ashari et al., 1988; Xiang and Roose, 1988; Naumova, 1993; Andrade-Rodríguez et al., 2004, 2005; Woo et al., 2019) and *Zanthoxylum* spp. (7 species; Longo, 1908; Mauritzon, 1935; Desai, 1962; Ghosh, 1974; Liu et al., 1987; Naumova, 1993; Cai et al., 1997, 2000), but it also occurs in *Aegle* spp. (1 species; Chakravarthy, 1936; Johri and Ahuja, 1956), *Esenbeckia* spp. (1 species; Braun, 1860; Naumova, 1993), *Murraya* spp. (2 species; Chakravarthy, 1935, 1936), *Ptelea* spp. (1 species; Schürhoff, 1924; Mauritzon, 1935), *Ruta* spp. (1 species; Cappelletti, 1929), *Toddalia* spp. (1 species; Narmatha Bai and Lakshmanan, 1982), and in *Triphasia* spp. (1 species; Braun, 1860; Mauritzon, 1935).

This type of development is also found in the Orchidaceae Juss. family, among species of *Gymnadenia* spp. (8 species; Strasburger, 1878; Afzelius, 1928, 1932; Chiarugi, 1929; Teppner and Klein, 1985; Teppner, 1996; Hedrén et al., 2000, 2018; Hedrén and Lorenz, 2019), *Cynorkis* spp. (3 species; Veyret, 1967, 1972), *Genoplesium* spp. (1 species; Sorensen et al., 2009), *Mormolyca* spp. (1 species; Illg, 1977), *Spiranthes* spp. (1 species; Leavitt, 1900; Swamy, 1948b; Catling, 1982; Schmidt and Antlfinger, 1992), *Zeuxine* spp. (1 species; Seshagiriah, 1941; Swamy, 1946; Sun, 1997; Sun and Wong, 2001; Kant and Verma, 2012), *Zygopetalum* spp. (1 species; Hurst, 1898; Suessenguth, 1923; Afzelius, 1959; Campacci et al., 2017; Moura et al., 2020).

In Myrtaceae Juss., it occurs in *Syzygium* spp. (7 species; Cook, 1907; van der Pijl, 1934; Merril and Perry, 1938; Gurgel and Soubihe Sobrinho, 1951; Roy, 1953; Narayanaswami and Roy, 1960; Roy and Sahai, 1962; Thurlby et al., 2011), *Eugenia* spp. (1 species; Silva et al., 2013), in *Luma* spp. (1 species; Johnson, 1936; Titow de Tschischow, 1956).

Nucellar embryony is also found in the following families:

In Amaranthaceae Juss., in *Beta* spp. (3 species; Barocka, 1966; Filutowicz and Dalke, 1972; Jassem and Jassem, 1970; Jassem, 1976, 1990).

Amaryllidaceae J.St.-Hil., in *Calostemma* spp. (1 species; Schlimbach, 1924), and in *Nothoscordum* spp. (1 species; Strasburger, 1878; Messeri, 1931; D'Amato, 1940; Håkansson, 1953).

Anacardiaceae R.Br., in *Mangifera* spp. (2 species; Strasburger, 1878; Juliano, 1934, 1937; Sachar and Chopra, 1957; Martínez Ochoa et al., 2012).

Apocynaceae Juss., in *Vincetoxicum* spp. (2 species; Seefeldner, 1912; Guignard, 1922; Kordyum, 1961; von Hausner, 1976).

Araceae Juss., in *Spathiphyllum* spp. (1 species; Schürhoff and Jüssen, 1925),

Asparagaceae Juss., in *Hosta* spp. (2 species; Braun, 1860; Strasburger, 1878; Fagerlind, 1944a, 1944b; Hu, 1963) and *Ornithogalum* spp. (1 species; Naumova, 1993).

Betulaceae Gray, in *Alnus* spp. (3 species; Woodworth, 1929, 1930; Naumova, 1993) and in *Betula* spp. (2 species; Korchagina, 1981; Naumova, 1993).

Bignoniaceae Juss., *Anemopaegma* spp. (3 species; Firetti-Leggieri et al., 2013), and in *Handroanthus* spp. (3 species; Piazzano, 1998; Salomão and Allem, 2001; Costa et al., 2004; Souza et al., 2005; Bittencourt and Moraes, 2010; Mendes et al., 2018; Alves et al., 2016; Braga et al., 2019).

Burseraceae Kunth, *Commiphora* spp. (1 species; Gupta et al., 1996; Kavane et al., 2021), and in *Garuga* spp. (1 species; Ghosh, 1970).

Buxaceae Dumort., in *Sarcococca* spp. (4 species and 1 variety; Orr, 1923; Wiger, 1930; Naumova, 1980, 1993; Naumova and Willemse, 1983a, 1983b).

Cactaceae Juss., in *Cactus* spp. (1 species; Tiagi, 1956), in *Opuntia* spp. (9 species; Braun, 1860; Ganong, 1898a, 1898b; Montemartini, 1899; Hull, 1915; Archibald, 1939; Maheshwari and Chopra, 1955; Naumova, 1993)

Calycanthaceae Lindl., *Calycanthus* spp. (2 species and 1 variety; Peter, 1920; Schürhoff, 1923).

Celastraceae R.Br., in *Euonymus* spp. (1 species; Braun, 1860; Strasburger, 1878; Andersson, 1931; Brizicky, 1964; Naumova, 1971, 1979).

Clusiaceae Lindl., in *Clusia* spp. (1 species; Goebel, 1900).

Colchicaceae DC., in *Colchicum* spp. (1 species; Braun, 1860; Naumova, 1993).

Combretaceae R.Br., in *Combretum* spp. (2 species; Venkateswarlu, 1947; Naumova, 1993).

Cucurbitaceae Juss., in *Cucumis* spp. (1 species; Singh, 1955), in *Momordica* spp. (1 species; Agrawal and Singh, 1957; Bhuskute, 1987), and in *Sicyos* spp. (1 species; Naumova, 1993).

Dipterocarpaceae Blume, in *Hopea* spp. (1 species; Chan, 1981; Kaur et al., 1986), in *Shorea* spp. (2 species and 2 subspecies; Chan, 1981; Kaur et al., 1978, 1986; Ng et al., 2004).

Elaeagnaceae Juss., in *Elaeagnus* spp. (1 species; Mangla et al., 2015).

Euphorbiaceae Juss., in *Euphorbia* spp. (Hegelmaier, 1911, 1913; Carano, 1926; Cesca, 1961; Kapil, 1961; Naumova, 1993), in *Mallotus* spp. (1 species; Ventura, 1934).

Fabaceae Juss., in *Cassia* spp. (4 species and 1 variety; Symon, 1956; Randell, 1970; Holman and Playford, 2000), in *Millettia* spp. (1 species; Pal, 1960; Naumova, 1993).

Grossulariaceae DC., in *Ribes* spp. (3 species; Kobasnidze, 1976, 1983).

Juglandaceae DC. ex Perleb, in *Carya* spp. (1 species; Wang et al., 2010; Zhang et al., 2012).

Malpighiaceae Juss., in *Aspicarpa* spp. (3 species; Ritzerow, 1907), in *Galphimia* spp. (1 species; Singh, 1959), in *Heteropterys* spp. (3 species; Braun, 1860; Subba Rao, 1937), in *Hiptage* spp. (1 species; Subba Rao, 1937, 1940), and in *Stigmaphyllon* spp. (1 species; Braun, 1860; Naumova, 1993).

Malvaceae Juss., in *Eriotheca* spp. (2 species; Oliveira et al., 1992; Salomão and Allem, 2001; Martins and Oliveira, 2003; Mendes-Rodrigues et al., 2005, 2018), and in *Pachira* spp. (2 species; Robins, 1959; Baker, 1960; Duncan, 1970; Naumova, 1993).

Melanthiaceae Batsch ex Borkh., in *Trillium* spp. (2 species; Jeffrey and Haertl, 1939b; Swamy, 1948a; Gyer, 1996; Gonzales, 2005; Heckel and Leege, 2007).

Melastomataceae Juss., in *Leandra* spp. (2 species; Borges, 1991; Goldenberg and Varassin, 2001; Caetano and Cortez, 2014; Maia et al., 2016), in *Miconia* spp. (2 species; Borges, 1991; Saraiva et al., 1996; Goldenberg and Shepherd, 1998; Mendes-Rodrigues and Oliveira, 2012;

Caetano and Cortez, 2014; Caetano et al., 2018), in *Osbeckia* spp. (1 species; Subramanyam, 1942).

Meliaceae Juss., in *Aphanamixis* spp. (1 species; Ghosh, 1962; Ghosh, 1972; Rai, 2014).

Moraceae Gaudich., in *Ficus* spp. (2 species; Naumova, 1993), and in *Streblus* spp. (1 species; Naumova, 1993).

Polygonaceae Juss., in *Atraphaxis* spp. (1 species; Edman, 1931).

Primulaceae Batsch ex Borkh., in *Ardisia* spp. (1 species; Braun, 1860; Kume, 1959).

Ranunculaceae Juss., in *Thalictrum* spp. (1 species; Daskalova, 2004).

Smilacaceae Vent., in *Maianthemum* spp. (1 species; McAllister, 1913; Gorham, 1952).

Solanaceae Adans., in *Petunia* spp. (1 species; Banerji and Bhaduri, 1933).

Symplocaceae Desf., in *Symplocos* spp. (1 species; Ravishankar and Gulati, 1985).

Tecophilaeaceae Leyb., in *Cyanella* spp. (1 species; de Vos, 1950; Maheshwari and Sachar, 1963).

Thymelaeaceae Juss., in *Daphne* spp. (1 species; Venkateswarlu, 1946).

Tropaeolaceae Juss. ex DC., in *Tropaeolum* spp. (1 species; Braun, 1860; Naumova, 1993).

Urticaceae Juss., in *Elatostema* spp. (2 species; Fagerlind, 1944a, 1944b; Naumova, 1993).

Zygophyllaceae R.Br., in *Zygophyllum* spp. (2 species; Phatak, 1971).

7.3.1.2 Integumentary Embryony

Integumentary embryony has been found in 65 taxa from 13 families (Table 7.1).

This type of embryo and seed development is recurrent in the Celastraceae R.Br., among species of *Euonymus* spp. (17 species; Braun, 1860; Strasburger, 1878; Andersson, 1931; Brizicky, 1964; Copeland, 1966; Naumova, 1970a, 1970b, 1971, 1979, 1993), and in *Celastrus* spp. (2 species; Naumova, 1993).

This type of development is also found in the Orchidaceae Juss. family, among species of *Spiranthes* spp. (5 species and 2 varieties; Leavitt, 1900; Swamy, 1948b; Catling, 1982; Schmidt and Antlfinger, 1992), *Dactylorhiza* spp. (1 species; Diana, 1997), *Epidendrum* spp. (1 species; Veyret, 1982; Stort and Dos Santos Pavanelli, 1985), *Genoplesium* spp. (1 species; Sorensen et al., 2009), *Sacoila* spp. (2 species; Catling, 1987), and *Zygopetalum* spp. (1 species; Hurst, 1898; Suessenguth, 1923; Afzelius, 1959; Campacci et al., 2017; Moura et al., 2020).

Integumentary embryony is also found in the following families:

Amaranthaceae Juss., in *Beta* spp. (3 species; Barocka, 1966; Filutowicz and Dalke, 1972; Jassem and Jassem, 1970; Jassem, 1976, 1990).

Amaryllidaceae J.St.-Hil., in *Allium* spp. (1 species; Tretjakow, 1895; Modilewsky, 1925, 1931; Håkansson, 1951; Håkansson and Levan, 1957).

Apiaceae Lindl., in *Ammi* spp. (1 species; Naumova, 1993).

Apocynaceae Juss., in *Vincetoxicum* spp. (2 species; Seefeldner, 1912; Guignard, 1922; Kordyum, 1961; von Hausner, 1976).

Bignoniaceae Juss., *Anemopaegma* spp. (3 species; Firetti-Leggieri et al., 2013), and in *Handroanthus* spp. (2 species; Salomão and Allem, 2001; Costa et al., 2004; Bittencourt and Moraes, 2010; Mendes et al., 2018; Alves et al., 2016; Braga et al., 2019).

Calycanthaceae Lindl., *Calycanthus* spp. (2 species and 1 variety; Peter, 1920; Schürhoff, 1923), and in *Chimonanthus* spp. (1 species; Naumova, 1993).

Capparaceae Juss., in *Capparidastrum* spp. (1 species; Mauritzon, 1934).

Clusiaceae Lindl., among species of *Garcinia* spp. (7 species; Treub, 1911a, 1911b; Sprecher, 1919; Puri, 1939; Horn, 1940; Kaur et al., 1978; Lim, 1974, 1984; Ha et al., 1988; Thomas, 1997; Richards, 1990).

Myrtaceae Juss., in *Myrceugenia* spp. (1 species; Naumova, 1993), in *Syzygium* spp. (7 species; Cook, 1907; van der Pijl, 1934; Merril and Perry, 1938; Gurgel and Soubihe Sobrinho, 1951; Roy, 1953, 1961), and in *Ugni* spp. (1 species; Naumova, 1993).

Ochnaceae DC., in *Ochna* spp. (3 species; Francini, 1928; Chiarugi and Francini, 1930; Ghosh, 1964; Pauzé, 1973; Naumova, 1993).
Solanaceae Adans., in *Scopolia* spp. (1 species; Glutchenko, 1968).

7.3.2 Gametophytic Apomixis

In total, 126 species and subspecies from 32 families are recorded as apomict by direct evidence of the formation of unreduced gametophytes and clonal offspring (Table 7.1).

7.3.2.1 The Aposporous Development

Apospory has been recorded in taxa from 16 families and represents 40.8% of all cases with gametophytic apomixis (*n*=52; Table 7.1). The aposporous development starts in a nucellar cell near the germline (Mateo de Arias, 2015; Carman et al., 2019) which acquires a gametogenetic fate and develops into an unreduced female gametophyte (Asker and Jerling, 1992). In these aposporous apomicts, the germline often degenerates before or during the meiotic division. In some cases, the developing meiotic gametophyte can either degenerate or reach maturity and produce a sexual seed.

7.3.2.1.1 The Hieracium Type

This type of unreduced female gametophyte development is the only type of aposporous development observed among families and is present in 35 species (Table 7.1).

The Hieracium type is frequent among *Ranunculus* spp., Ranunculaceae Juss. (12 species; Koch, 1939; Häflinger, 1943; Rutishauser and Hunziker, 1950; Rutishauser, 1953, 1954; Nogler, 1971, 1973, 1984; Vuille and Küpfer, 1985; Huber, 1988; Hojsgaard et al., 2014b).

This type of development is also recurrent among Brassicaceae Burnett, found in *Boechera* spp. (6 species; Mateo de Arias, 2015; Carman et al., 2019); in *Borodinia* spp. (1 species; Carman et al., 2019); in *Polyctenium* spp. (1 species; Mandáková et al., 2020); and in *Sandbergia* spp. (1 species; Mandáková et al., 2020).

It is also found in the following families:

Amaranthaceae Juss., in *Beta* spp. (2 species; Barocka, 1966; Jassem and Jassem, 1970; Jassem, 1976, 1990).
Araceae Juss., in *Aglaonema* spp. (2 species; Pfitzer, 1957, 1962).
Cucurbitaceae Juss., in *Cucumis* spp. (1 species; Singh and Dathann, 1972).
Erythroxylaceae Kunth, in *Erythroxylum* spp. (1 species; Berry et al., 1991).
Hypericaceae Juss., in *Hypericum* spp. (1 species; Noack, 1939; Mártonfi et al., 1996; Barcaccia et al., 2007).
Melastomataceae Juss., in *Miconia* spp. (2 species; Caetano, 2014; Caetano et al., 2018).
Ochnaceae DC., in *Ochna* spp. (2 species; Francini, 1928; Ghosh, 1964; Pauzé, 1973).
Polygonaceae Juss., in *Atraphaxis* spp. (1 species; Edman, 1931).
Urticaceae Juss., in *Elatostema* spp. (2 species; Fagerlind, 1944b).

7.3.2.2 The Diplosporous Development

Diplospory has been recorded in taxa from 22 families and represents 59.2% of all species and subspecies with gametophytic apomixis (*n*=75; Table 7.1). In these apomicts, diplospory occurs with alteration of the steps of meiosis (either the entrance into meiosis or its progression; see Chapter 2) and the subsequent formation of an unreduced megaspore that develops into a female gametophyte (Asker and Jerling, 1992). Occasionally, the germline may progress all through meiosis to generate reduced megaspores first, and later a reduced female gametophyte and a sexual seed after fertilization.

7.3.2.2.1 The Antennaria Type

This type of development is recorded for 29 species among all families (Table 7.1).

The Antennaria type is common among species of *Zephyranthes* Herb., Amaryllidaceae J.St.-Hil. (4 species; Pace, 1913; Church, 1916; Brown, 1951; Coe, 1953; Raina and Khoshoo, 1972; Crane, 2019; Gianini Aquino et al., 2019; Gianini Aquino, 2021), and among members of the Melastomataceae Juss., in *Leandra* spp. (2 species; Caetano, 2014; Maia et al. 2016); in *Miconia* spp. (2 species; Caetano, 2014); in *Ossae* spp. (2 species; Caetano, 2014), and the Urticaceae Juss., in *Boehmeria* spp. (1 species; Okabe, 1963; Tang et al., 2016, 2017); in *Elatostema* spp. (2 species; Modilewsky, 1908; Fagerlind, 1944b); and in *Procris* spp. (1 species; Fagerlind, 1944b).

Other families with this development include:

Amaranthaceae Juss., in *Aerva* spp. (1 species; Sachar and Murgai, 1958; Khan et al., 1970), and in *Beta* spp. (1 species; Barocka, 1966; Jassem and Jassem, 1970; Jassem, 1976, 1990).
Burmanniaceae Blume, in *Burmannia* spp. (1 species; Ernst and Bernard, 1912).
Cucurbitaceae Juss., in *Luffa* spp. (1 species; Deshpande et al., 1984).
Fabaceae Juss., in *Genista* spp. (1 species; Diana and Villa, 1991).
Myrtaceae Juss., in *Psidium* spp. (1 species; Souza-Pérez and Speroni, 2017; Souza-Pérez et al., 2021).
Ochnaceae DC., in *Ochna* spp. (3 species; Francini, 1928; Chiarugi and Francini, 1930; Ghosh, 1964; Pauzé, 1973).
Orchidaceae Juss., in *Epidendrum* spp. (1 species; Veyret, 1982; Stort and Dos Santos Pavanelli, 1985); and in *Genoplesium* spp. (1 species; Sorensen et al., 2009).
Rhamnaceae Juss., in *Pomaderris* spp. (3 species; Harvey and Braggins, 1985).
Saururaceae Rich. ex T. Lestib., in *Houttuynia* spp. (1 species; Shibata and Miyake, 1908; Okabe, 1930).

7.3.2.2.2 The Taraxacum Type

The Taraxacum type of female gametophyte development is recorded for 20 species and subspecies among all families (Table 7.1).

This type occurs frequently among *Boechera* spp., Brassicaceae Burnett (9 species; Böcher, 1951, 1969; Vorobik, 1985; Roy, 1995; Taşkin et al., 2004; Mau et al., 2015; Carman et al., 2019).

The Taraxacum type is also found in the following families:

Balanophoraceae Rich., in *Balanophora* spp. (2 species; Kuwada, 1928; Zweifel, 1939; Fagerlind, 1945a, 1945b).
Betulaceae Gray, in *Alnus* spp. (1 species; Woodworth, 1929, 1930).
Clusiaceae Lindl., in *Garcinia* spp. (1 species; Ha et al., 1988).
Melastomataceae Juss., in *Miconia* spp. (2 species; Caetano, 2014).
Potamogetonaceae Bercht. & J. Presl, in *Potamogeton* spp. (1 species; Teryokhin, 1994).
Thymelaeaceae Juss., in *Wikstroemia* spp. (1 species; Winkler, 1904, 1906; Fagerlind, 1940a, 1940b).
Urticaceae Juss., in *Boehmeria* spp. (1 species; Yahara, 1986, 1990; Zhang et al., 1997); in *Elatostema* spp. (2 species; Fagerlind, 1944a, 1944b).

7.3.2.2.3 The Ixeris Type

This is the less frequent type of female gametophyte development (Table 7.1), found in six species, all from *Limonium* spp., Plumbaginaceae Juss. (D'Amato, 1940, 1949; Hjelmqvist and Grazi, 1964; Erben, 1978; Ingrouille and Stace, 1985; Marras, 1985; Masia, 1985; Róis et al., 2016).

7.3.3 Variations in the Development of Asexual Seeds

Variations on the types of female gametophyte development had been recorded in species from different plant families. In species from the same or different genera, the occurrence of meiotic and apomeiotic female gametophytes in ovules from one or more individuals is recorded directly by embryological observation or indirectly by FCSS or progeny tests. For example, in diplosporous *Ochna atropurpurea* DC, aposporous development occurs at low frequency (Ghosh, 1964; Pauzé, 1973). Particularly interesting is the case of *Boechera* spp., a genus exhibiting a remarkable variation in types of female gametophyte and seed developments (Aliyu et al., 2010; Carman et al., 2019). Unlike other apomictic genera in which apospory and diplospory rarely occur together and when they are concurrent, one pathway is dominant (see Chapters 4–6), among *Boechera* species both pathways are observed at high frequency in several interspecific hybrids, likely due to a young evolutionary origin (Carman et al., 2019). Moreover, an interesting variation in the mechanisms of endosperm development is observed in these species. Although most diploid and triploid apomicts are pseudogamous, pseudogamy is achieved in different ways (involving haploid, diploid or polyploid pollen) and low rates of autonomous endosperm development are also observed (Aliyu et al., 2010).

Among species with adventitious embryony, a remarkable variation in the proportion of seeds carrying adventive embryos and in the number of extra embryos per seed is observed. For example, among Rutaceae, the proportion of apomictic seeds can go from a low percentage to 100% of seeds carrying up to >50 adventive embryos (e.g., Liu et al., 1987). Variability is also observed along with the development of the endosperm. In most cases, the successful development of adventive embryos depends upon endosperm development, either involving or not pollination and fertilization by male gametes (Lakshmanan and Ambegaokar, 1984). Yet, pollination and fertilization requirements are different among apomictic species from separate families, having variable rates of pollen tube growth and time needed for the pollen tube to reach the female gametophyte in ovules. This time can vary substantially and, together with the function of the male gametes, may explain apparently contrasting results in some apomicts. For example, among Rutaceae Juss., in *Citrus* L., several studies indicate that pseudogamy is the rule (Ikeda, 1904; Nagai and Tanikawa, 1928; Frost, 1938b; Wong, 1939; Esen and Soost, 1977). Reports of nucellar embryos developing and reaching a seed stage without pollination (e.g., Wright, 1937) could be due to the variable and long time (one to five weeks) between pollination and fertilization observed among citrus species (see e.g. Yang, 1968). In *Zanthoxylum* L., autonomous polyembryony was described for *Zanthoxylum bungei* Planch. & Linden ex Hance (Desai, 1962), but a recent study suggests fertilization is needed for fruit production (Munter et al., 2018). In the Smilacaceae Vent. family, *Maianthemum racemosum* (L.) Link ≡*Smilacina racemosa* (L.) Desf. is a pseudogamous species (McAllister, 1913), and even though pollination happens, Gorham (1953) indicates it is a case of endosperm development without fertilization.

Another related fact is that, although autonomous endosperm development is assumed when seeds are produced in emasculated and bagged flowers, this does not imply that the proportion of seeds formed is the same as in control trials. For example, emasculated and bagged flowers of *Garcinia mangostana* L., Clusiaceae Lindl., produce about 20% of seeds while non-emasculated and bagged flowers produce 53% of seeds (Lim, 1984).

In several species, multiple, unorganized, multinucleate and non-celled female gametophytes at anthesis were observed in low frequency. For example, during his embryological investigations in *Limonium oleifolium* Mill. ≡*Statice oleifolia* (Mill.) Scop., Plumbaginaceae Juss., D'Amato (1949) found low proportion of anomalous gametophytes.

The occurrence of multiple archespores in the same ovule, a common feature in many plant families (see examples above, Appendix 1.5), is a relevant feature for embryological analyses and characterization of apomictic species. In the family Rutaceae, the formation of multiple female gametophytes from megaspores was observed in sexual species (e.g., in *Ravenia spectabilis* Engl.,

Ghosh, 1974) and in apomictic species [e.g., in *Toddalia asiatica* (L.) Lam., Ghosh et al., 1976; in *Limonia acidissima* L. =*Feronia limonia* (L.) Swingle, Banerjee and Pal, 1958].

APPENDIX: APOMIXIS IN OTHER ANGIOSPERM FAMILIES

This Appendix includes details on specific observations on apomictic taxa from different plant families collected during the reassessment of their reproductive classification.

APPENDIX 1.1

Seed set analyses using emasculation/castration experiments on *Trillium reliquum* J.D. Freeman, Melanthiaceae Batsch ex Borkh. (Heckel and Leege, 2007) contradicts the observation by Waddell (2006) of absence fruits in self-pollinated plants. Since the studied materials are different, we accept the results from the emasculation test and placed the species in our Table 7.1, but other analyses are required to understand these contrasting results and the reproductive system of the species.

APPENDIX 1.2

In *Juglans regia* L., Juglandaceae DC. ex Perleb, independent emasculation experiments have showed that autonomous development occurs (e.g., Gao et al., 1999; Zhang et al., 2000; Cosmulescu et al., 2012). Yet, there is no agreement on the type of apomictic development. Nucellar embryony is reported by Schanderl (1964) and Valdeviesso (1990). Terziĭski and Stefanova (1990) report unreduced female parthenogenesis following apospory, and rarer instances of adventitious nucellar embryony. Contrary, Sartorius and Stösser (1997) state that they have not observed apospory or adventitious embryony and that apomixis is of diplosporous type, but no details or probe about altered meiosis are presented.

APPENDIX 1.3

In Magnoliaceae Juss., apomixis and apospory in *Magnolia kwangsiensis* Figlar & Noot. [=*Woonyoungia septentrionalis* (Dandy) Y.W. Law] were suggested by Zeng et al. (2003) and Wang et al. (2004) based on pollination and seed set experiments and embryology, respectively (Table 7.1). Yet, a recent comprehensive evaluation by Fu et al. (2012) showed a standard Polygonum type of female gametophyte development and formation of sexual seeds. More analysis and information on the ploidy of the studied materials would provide useful information to resolve the discrepancies.

The Antennaria type of diplospory has been reported in *Houttuynia cordata* Thunb., Saururaceae Rich. ex T. Lestib. (Okabe, 1930). Shibata and Miyake (1908) mention that in this species the development of the female gametophyte runs like that of the male gametophyte, for which they mention the occurrence of meiotic abnormalities and lack of chromosomal associations responsible for the high level of pollen sterility. However, in their description of the female gametophyte development, the megaspore mother cell goes through the normal steps of sexual reproduction and produces three or four megaspores, the chalazal one give rise to the female gametophyte. Then, the secondary nucleus of the central cell starts the divisions and the egg cell start dividing when the endosperm is advanced. From the description of the developing steps, it is not possible to deduce a possible type of apomixis for this species. Despite the fact that the species reproduces through vegetative propagation (rhizomes or stolons) and – expectedly – apomixis, recent molecular genetic studies show a lack or exceptionally low clonality, and levels of genetic variability within and among populations are like those found in sexual species (Wei and Wu, 2012; Zhong et al., 2016; Gupta and Bharalee, 2021). Unfortunately, the work by Okabe (1930) is not available, and all other studies mentioning the occurrence of apomixis in the species are based on Shibata and Miyake (1908) and Okabe (1930).

APPENDIX 1.4

Apomixis in *Calycanthus floridus* var. *glaucus* (Willd.) Torr. & A. Gray (= *Calycanthus fertilis* Walter), *C. floridus* L. and *C. occidentalis* Hook. & Arn., Calycanthaceae Lindl., is suggested based on old studies and therefore require new observations. Peter (1920) mentions that the female gametophytes develop after a normal meiosis and multiple nucellar embryos emerge once the endosperm is formed. Schürhoff (1923, p. 73) critizes such interpretation and describes the formation of unreduced female gametophytes through a modified meiosis in which the synapsis and diakinesis steps between homolog chromosomes proceed normally but homologs associations are disentangled during the formation of the spindle and the equatorial plate to finally segregate sister chromatids (Schürhoff, 1923, pp. 76–78, 83). Thus, it seems as if meiosis is condensated sequentially in one, longer division that creates unreduced nuclei carrying recombinant chromosomes. If this is the case, the development in these species would not represent a case of apomixis *sensu stricto* but a modification towards apogamy or hemigamy. The observation of an abnormal egg apparatus and egg cell carrying more than one nucleus (Schürhoff, 1923, p. 76) supports this interpretation. More specific studies are needed in this genus.

APPENDIX 1.5

Campbell (1903) studied the embryology of *Aglaonema commutatum* Schott, Araceae Juss., and described deviations from the usual type of development including variable number of nuclei, lack of polarization and the occurrence of more than one female gametophytes. Since the author also found the occurrence of a multicellular archesporium (Campbell, 1903, p. 669), a probable somatic origin for the extra female gametophytes was not corroborated. Later, Pfitzer (1957) studied the chromosome numbers in different Araceae and revealed that the hexaploid material of *A. commutatum* and the triploid *A. crispum* (Pitcher & Manda) Nicolson (=*Aglaonema roebelinii* Pitcher & Manda) evaluated were male sterile and yet, fruitified and produced seeds. Even though Pfitzer did not carry out pollen exclusion tests, he concluded apomixis in these species (Pfitzer, 1957, p. 442) considering Campbell (1903) observations and later provided embryological support (Pfitzer, 1962; based on Veyret, 1965, p. 20). Here we placed these species in Table 7.1, but more studies are needed to exclude doubts.

APPENDIX 1.6

In another species of the family, diplospory is accepted for *Psidium cattleyanum* Sabine based on the apparent lack of meiotic stages (Souza-Pérez and Speroni, 2017) and recent preliminary population genetic analysis providing partial support to the occurrence of apomixis in one out of three populations of the species (Machado et al., 2021).

Similarly, among taxa of Orobanchaceae Vent., absence of meiosis in ovules of *Orobanche uniflora* L. (Jensen, 1951) and in *Cistanche salsa* (C.A. Mey.) Beck and *C. mongolica* Beck [=*Cistanche tubulosa* (Schenk) Hook. f.] (Pazy, 1998), together with reduced male viability and mass production of seeds, was taken as evidence of apomixis (Table 7.1). However, more studies on these families will bring useful details on the patterns of apomixis.

APPENDIX 1.7

In *Vincetoxicum hirundinaria* Medik. [=*Cynanchum vincetoxicum* (L.) Pers.], Apocynaceae Juss., Seefeldner (1912) presented a description for the origin of multiple embryos from the fertilized egg-cell (see Seefeldner, 1912, pp. 287, 293) and Guignard (1922) found similar evidence of cleavage polyembryony in *Vincetoxicum hirundinaria* Medik. (=*Vincetoxicum officinale* Moench) and *Vincetoxicum nigrum* (L.) Moench [≡*Cynanchum nigrum* (L.) Pers.]. Later,

Kordyum (1961) described the occurrence of integumentary embryony in *Vincetoxicum officinale* Moench from cells near the micropyle, and von Hausner (1976) pointed out that the initial embryonic cells of the materials from *Cynanchum nigrum* (L.) Pers. and *Cynanchum medium* (Decne.) K. Schum. studied are from the nucellar tissue. Yet, Naumova (1993, p. 63) cites integumentary embryony for these species. The different interpretations are likely connected to the sharp reduction of nucellar tissue observed during the development of ovules in species of Apocynaceae (Johri et al., 1992). Here we did not make a distinction between integumentary or nucellar embryony.

APPENDIX 1.8

Apomixis in *Elaeagnus rhamnoides* (L.) A. Nelson (≡*Hippophae rhamnoides* L.) is likely of the nucellar embryony type (Mangla et al., 2015). However, the images supporting the occurrence of a Panicum type apospory and multiple female gametophytes are not convincing mostly because the imposed colours and shapes over the images do not clarify the case, and the details and interpretations are questionable. For example, an image showing four aposporous initials perfectly aligned to the side of the meiotic products might represent a case of multiple archespores growing in the ovule (Mangla et al., 2015, Figure 3, p. 6). Similarly, the anatomy of the claimed four-nucleate unreduced gametophyte looks much like that one of the reduced gametophytes, and the nucleus of the central cell is located at the antipods of the gametophyte instead of near the egg cell at the micropyle (Mangla et al., 2015, Figure 3, p. 6), as is the case in other angiosperms.

APPENDIX 1.9

Even though the observation of abundant fruit and seed formation in spatially isolated trees of dioecious or self-incompatible species suggests apomixis as a likely explanation, without experimental analysis and controls it is not possible to discard pollination. Long-distance transport of pollen and effective pollination seems to be responsible for the production of outcrossed seeds in isolated trees (Finkeldey and Hattemer, 2007).

APPENDIX 1.10

In *Boechera* spp., detailed embryological and genetic studies allowed the development of markers linked to apomeiosis in female (APOLLO) and male (UPGRADE2) gametophyte developments (Corral et al., 2013; Mau et al., 2013). Here, the term "linkage" refers to the co-occurrence [as used elsewhere in the text] or coabsence of both apomictic alleles from the UPGRADE2 and APOLLO genes per individual accession. These markers are useful both to corroborate the occurrence of apomixis and to identify apomictic genotypes in samples from natural populations. For example, partial evidence for apomixis presented earlier by Rollins (1940) or Rollins and Rüdenberg (1971) in species like *Boechera fendleri* (S. Watson) W.A. Weber, *Boechera inyoensis* (Rollins) Al-Shehbaz or *Boechera pulchra* (M.E. Jones ex S. Watson) W.A. Weber was recently corroborated by Mateo de Arias (2015) through embryological analysis, by Alexander et al. (2015) using molecular markers (SSR), and by Mau et al. (2015) through the presence of the apomixis marker genes APOLLO and UPGRADE2.

8 Apomixis in Angiosperms

8.1 INTRODUCTION

Signs or records of apomixis among angiosperm species date back to the eighteenth and nineteenth centuries. Although Smith's (1841) observation of seed production by a female plant without male contribution in the dioecious species *Alchornea aquifolia* (Js.Sm.) Domin represented direct evidence of agamospermy, later corroborated embryologically by Strasburger (1878), the occurrence of twin embryos in orange as indirect evidence of sporophytic apomixis, later corroborated by Strasburger (1878) and other different studies, was first reported by Leeuwenhoek (1719) (Chapter 7). Since then, records of apomixis have gradually accumulated along with knowledge about the different types of apomixis found in nature and alternative classifications.

In recent reviews on the occurrence of apomixis at the genus level, Hörandl and Hojsgaard (2012) and Hojsgaard et al. (2014a) found a steady increase of apomictic records with an average of 3–4 new apomictic genera published every year. This trend is expected to continue or even increase with the use of flow cytometry (see e.g. Brožová et al., 2019; Mráz and Zdvořák, 2019; Pegoraro et al., 2020), a technological advance that facilitates the recognition of apomictic seed-producing plants (Chapter 3).

However, all reports on apomictic species have so far remained scattered, and in many cases need critical re-evaluation. The present chapter summarizes the current information on apomixis at the species level in the angiosperm families (Tables 4.1, 5.1, 6.1 and 7.1) and the main phylogenetic and evolutionary characteristics.

8.2 APOMIXIS IN ANGIOSPERMS

Following a reassessment of the current literature on reported apomictic species, reliable evidence of apomixis has been found in 1,469 angiosperm species (Tables 4.1, 5.1, 6.1 and 7.1). Circumstantial evidence of apomixis has also been found in 223 other angiosperms species (Tables 4.1, 5.1, 6.1 and 7.1) placed in quarantine lists (Tables 4.2, 5.2, 6.2 and 7.2). Although this figure is much higher than previous rough estimates that considered the number of apomictic species to be <500 (see e.g. Asker and Jerling, 1992; Richards, 2003), it is clear that the number of apomictic species presented here is only a partial indication of reality. As many taxa within the apomictic genera have not yet been evaluated from a reproductive point of view, especially in diverse genera such as *Alchemilla* spp., *Hieracium* spp., *Limonium* spp., *Paspalum* spp., *Ranunculus* spp., *Rubus* spp. or *Taraxacum* spp., the number of apomictic species is expected to be much higher. Furthermore, considering that many species with polyembryony still require embryological confirmation of sporophytic apomixis (e.g., Salomão and Allem, 2001; Ferreira et al., 2019), or that a large number of species are temporarily quarantined and, for example, for *Limonium* spp. alone 19 species are listed with circumstantial evidence of apomixis, or that genera such as *Rubus* spp. or *Ranunculus* spp. are expected to contain more than 750 microspecies (Kurtto et al., 2010) and 830 microspecies (Karbstein, 2021), respectively, we can expect that in the short to medium term the number of apomictic species will double or triple the number presented here.

The proportion of angiosperm species with apomixis cannot be known with certainty, as this number depends on the number of reproductively studied species. A rough estimate suggests that about 10% of the angiosperm species studied are apomictic or have apomictic cytotypes in nature.

DOI: 10.1201/9781003088561-8

8.2.1 Phylogenetic Distribution of Apomixis

The phylogenetic origin of apomixis remains elusive. The two current alternative hypotheses suggest that apomixis is either a phylogenetically ancient and epigenetically inherited state whose expression in sexual lineages was repeatedly triggered during evolution (conserved pathways hypothesis) (Carman et al., 2011; Hojsgaard et al., 2014a; Albertini et al., 2019), or is a phylogenetically derived trait that evolved repeatedly *de novo* during angiosperm evolution (multiple *de novo* origins hypothesis) (van Dijk and Vijverberg, 2005; Carman et al., 2011; Hojsgaard et al., 2014a).

Previous phylogenetic reconstruction using character optimization and comparative genomics approaches failed to conclude an ancestry for apomixis in angiosperms (Hörandl and Hojsgaard, 2012; Hojsgaard et al., 2014a; Barcaccia et al., 2020). Apomixis is taxonomically scattered in 32 orders out of the 64, with evidence of multiple origins (Figure 8.1). Apomixis appears in basal stem groups and in clades derived within monocots and dicots, and in some major clades (commelinids, fabids and lamiids; Figure 8.1) apomixis is potentially ancient (Hörandl and Hojsgaard, 2012).

The most apomictically diverse orders are Asterales (*n*=352), Rosales (*n*=329), Poales (*n*=242), Myrtales (*n*=98), Caryophyllales (*n*=69) Asparagales (*n*=57), Malpighiales (*n*=52), Sapindales (*n*=48) and Brassicales (*n*=47). With the exception of Ranunculales (*n*=29), the Celastrales, Fabales, Fagales, Gentianales, Lamiales and Malvales have between 20 and 11 apomictic species. All other 16 orders contain <10 apomictic species each.

Adventitious embryony, diplospory and apospory occur in the seven major angiosperm clades in 27, 18 and 13 orders, respectively, with no clear phylogenetic predisposition (Figure 8.1). Adventitious embryony is the predominant mode of apomixis in fabids, and in the Myrtales, Asparagales, Malpighiales and Sapindales it is frequently found in species of certain families (see Section 8.2.2). Apospory is common in the Poales and diplospory in the Asterales. Apospory and diplospory are common in the Rosales and Brassicales.

8.2.2 Ranking of Apomictic Families and Prevalence of Apomictic Types

A total of 284 genera in 73 angiosperm families contain at least one apomictic species. The ranking of families with the most apomictic species is headed by the Asteraceae (*n*=352), Rosaceae (*n*=298) and Poaceae (*n*=242), which account for 60.8% (*n*=892) of all apomictic species (Table 8.1). Next in the ranking is the family Melastomataceae, an intensively studied family with 70 species with clear evidence of apomixis. On a similar level is the family Plumbaginaceae, with 49 apomictic species and 19 species with circumstantial evidence of apomixis (Tables 7.1, 7.2 and 8.1). A comparable case is the one of the family Ranunculaceae, which despite having 29 apomictic species this number is expected to increase in future studies (see e.g. Dunkel, 2011, 2021; Halamsky et al., 2019).

The families Brassicaceae, Orchidaceae, Rutaceae show similar numbers of apomictic species (Table 8.1), yet in the former more species are likely to be added as taxonomy resolves the complex life history and network of interspecific hybridizations in the group (Carman et al., 2019). The family Myrtaceae contain 26 apomictic species and the next 8 families in the ranking, including well-known apomictic cases within the Urticaceae, Celastraceae and Hypericaceae, contain between 20 and 11 apomictic species (Tables 7.1 and 8.1). Other 55 families show fewer than 10 apomictic species each, and 22 families contain only one species reported with apomixis (Table 7.1).

Apospory is the most frequent type of apomictic development among angiosperm species (*n*=357), followed by adventitious embryony (*n*=338) and diplospory (*n*=287) (Table 8.1). Among families, apospory is more frequent in the Poaceae (*n*=166; Table 8.1), although a few genera within the family consistently show diplospory (*Calamagrostis* spp., *Eragrostis* spp. and *Tripsacum* spp.; Table 4.1). Diplospory is more frequent within the Asteraceae (*n*=121; Table 8.1), although some genera

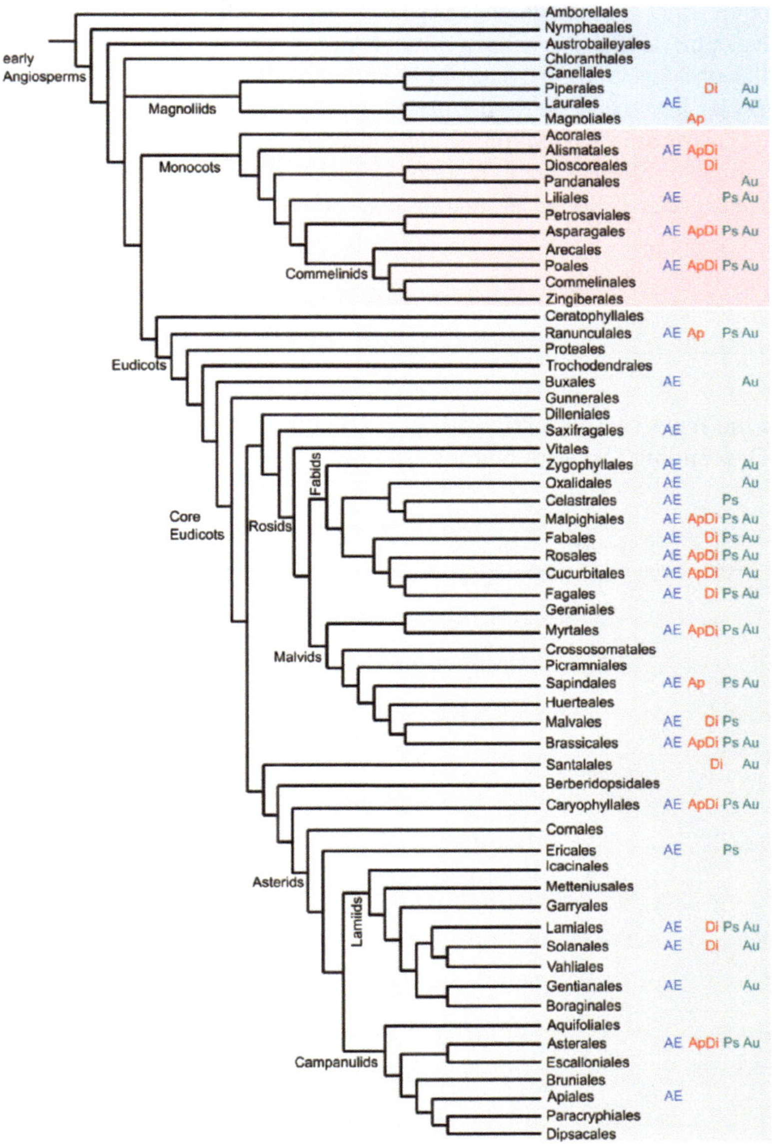

FIGURE 8.1 Distribution of apomixis among angiosperm orders. Apomixis records are superimposed on a phylogenetic tree of angiosperm orders [tree topology after APG IV (2016) and Stevens (2020)]. AE: adventitious embryony; Ap: apospory; Di: diplospory; Ps: pseudogamous endosperm development; Au: autonomous endosperm development.

are consistently aposporous (*Crepis* spp., *Pilosella* spp.; Table 5.1). Adventitious embryony, either integumentary or nucellar, is more frequent in the Rutaceae (n=37; Table 8.1).

In the Rosaceae, often considered as a family with mainly aposporic development, about one third of the embryologically assessed species show diplospory. This may probably be related to earlier studies that did not always distinguish between generative apospory (i.e., diplospory) and somatic apospory (i.e., apospory) (see Glossary), and to the occurrence in many species of a multicellular archesporium that makes interpretation and identification of developmental mechanisms difficult (Chapter 6).

Finally, for about 67% of all apomictic species there is embryological evidence for the type of apomictic development. In the remaining 33% of species, apomixis has been detected by methods that do not allow the developmental pathway to be determined (see Chapter 3). While in most families not all species have been assessed embryologically, in some families such as the Cactaceae, Celastraceae, Dipterocarpaceae, Myrtaceae or Rutaceae (but see Table 7.1), evidence for the type of apomixis has been collected for all reported species (Table 8.1). The families for which embryological data have been collected on one or few species are Hypericaceae and Plumbaginaceae (Table 8.1).

TABLE 8.1
Summary of the Types of Apomictic Pathways Occurring Among the Angiosperm Families (Ranked in Descending Order of Number of Species)

			Type of Apomixis			Endosperm		
Families	**Genera**	**Species**	**Ap**	**Di**	**AE**	**Ps**	**Au**	**Known§**
Asteraceae Giseke	33	**352**	30	**121**	3	8	**294**	0.44
Rosaceae Juss.	16	298	109	56	17	**226**	15	0.61
Poaceae Barnhart	**56**	242	**166**	35	11	131	22	0.89
Melastomataceae Juss.	17	70	3	8	19	0	65	0.43
Plumbaginaceae Juss.	1	49	0	6	0	0	8	0.12
Brassicaceae Burnett	7	45	13	9	0	13	21	0.45
Orchidaceae Juss.	15	38	0	2	28	4	26	0.79
Rutaceae Juss.	12	38	2	0	**37**	13	6	**1.0**
Ranunculaceae Juss.	3	29	13	0	1	20	2	0.48
Myrtaceae Juss.	9	26	0	1	25	10	0	**1.0**
Urticaceae Juss.	4	20	4	11	3	0	9	0.90
Celastraceae R.Br.	2	19	0	0	19	2	0	**1.0**
Hypericaceae Juss.	1	17	1	0	0	17	0	0.06
Fabaceae Juss.	4	16	0	1	13	4	1	0.88
Dipterocarpaceae Blume	3	14	0	0	14	4	0	**1.0**
Amaryllidaceae J.St.-Hil.	4	14	0	9	4	10	3	0.79
Clusiaceae Lindl.	2	13	0	1	12	1	6	0.92
Cactaceae Juss.	3	11	0	0	11	2	2	**1.0**
Other families (*n*=55)	92	158	16	27	121	24	61	0.84
TOTAL	**284**	**1,469**	**357**	**287**	**338**	**489**	**541**	**0.67**

Ap: apospory; Di: diplospory; AE: adventitious embryony; Ps: pseudogamous endosperm development; Au: autonomous endosperm development.

§ proportion of the total number of apomictic species for which the type of apomictic development is known.
In each column, the highest value is in bold.

8.2.3 Apomixis and Endosperm Development

In general, none of the alternative types of endosperm development show a tendency to occur more frequently during asexual seed formation. In total, information on the mode of endosperm development in apomictic seeds is recorded for 1,030 apomictic species (Tables 4.1, 5.1, 6.1, 7.1 and 8.1). Pseudogamous endosperm development, whereby the central cell of the female gametophyte needs to be fertilized to initiate and complete endosperm development, occurs in 489 apomictic species (Table 8.1). Autonomous endosperm development, whereby the central cell develops the endosperm independently of fertilization, occurs in 541 apomictic species.

In species with sporophytic apomixis, experimental evidence of pseudogamous endosperm development is found in 79 species (Tables 4.1, 6.1 and 7.1), and evidence of autonomous endosperm development is found in 83 species (Tables 4.1, 6.1 and 7.1).

In species with gametophytic apomixis, experimental evidence of pseudogamous endosperm development is found in 433 species (Tables 4.1, 6.1 and 7.1), and evidence of autonomous endosperm development is found in 401 species (Tables 4.1, 6.1 and 7.1).

Among families holding species with gametophytic apomixis, the overall figures for the type of gametophytic apomixis (i.e., diplospory or apospory) and the types of endosperm development (i.e., pseudogamous or autonomous) show positive correlations with strong statistical significance between apospory and pseudogamous endosperm development (r=0.87; p<0.001; df. 23) and between diplospory and autonomous endosperm development (r=0.91; p<0.001; df. 23), but not between apospory and autonomous endosperm development (r=0.15; p= 0.451; df. 23) or between diplospory and pseudogamous endosperm development (r=0.43; p=0.033; df. 23).

Apomictic species of the families Rosaceae, Poaceae, Hypericaceae, Ranunculaceae usually require pollen stimulation for endosperm development. Apomictic species of the families Asteraceae, Melastomataceae and Urticaceae usually develop the endosperm without pollen stimulus. In the families Brassicaceae, Orchidaceae and Rutaceae some species can produce apomictic seeds only after fertilization, and some species are able to produce apomictic seeds without fertilization (Table 8.1).

8.2.4 Apomixis and Ploidy

In general, apomixis as a mechanism of asexual seed formation shows a slight tendency to occur in individuals with a certain ploidy. In total, 123 apomictic species are diploid or include a diploid apomictic cytotype, and 948 apomictic species are polyploid or include one or more polyploid cytotypes (Tables 4.1, 5.1, 6.1, 7.1 and 8.2). The most frequent cytotype among apomictic species is tetraploid (n=505) followed by triploid (n=422). Pentaploid and hexaploid cytotypes are also frequent among apomictic species (n=87 and n=104, respectively; Table 8.2). Diploidy is frequent in apomictic species of the families Brassicaceae, Melastomataceae and Rutaceae (Table 8.2). Triploidy is most frequently found in apomictic species of the families Asteraceae, Brassicaceae, Plumbaginaceae, Rosaceae and Urticaceae (Table 8.2). Tetraploidy is most frequently found in apomictic species of the families Asteraceae, Hypericaceae, Poaceae, Ranunculaceae and Rosaceae (Table 8.2).

When considering the alternative types of apomixis separately (only those with evidence of the type of apomictic development), no association to a particular ploidy is found among species with sporophytic apomixis, but a clear association to ploidy classes is observed in species with gametophytic apomixis.

Of the 338 species with sporophytic apomixis, the ploidy of about half of species is known (n=167), and 21 cases have two or more cytotypes (Table 8.2). Species with adventitious embryony are mainly diploid (n=79) and tetraploid (n=60), although triploid cytotypes (n=29) are also frequent (Tables 5.1, 6.1 and 7.1). The least frequent cytotypes are pentaploids (n=3), hexaploids (n=7), octoploids (n=9) and hexadecaploids (n=1).

TABLE 8.2
Summary of Ploidies Found in Apomictic Species Among the Angiosperm Families (Ranked in Descending Order of Number of Species)

Family	Spp.	Cases with Ploidy Records		2x	3x	4x	5x	6x	7x	8x	9x	10–20x
		1	2 or >§									
Asteraceae Giseke	352	306	85	–	196	150	37	24	10	12	1	5
Rosaceae Juss.	298	245	40	–	108	133	13	18	7	8	6	5
Poaceae Barnhart	242	145	53	1	11	95	18	40	11	26	13	36
Melastomataceae Juss.	70	22	3	12	7	5	–	–	–	1	–	–
Plumbaginaceae Juss.	49	47	2	–	37	10	2	–	–	–	–	–
Brassicaceae Burnett	45	43	10	30	21	3	–	–	–	1	–	–
Orchidaceae Juss.	38	22	4	6	5	13	1	–	–	1	–	–
Rutaceae Juss.	38	31	4	25	2	6	–	–	–	3	–	1
Ranunculaceae Juss.	29	25	7	1	1	20	6	8	1	–	–	–
Myrtaceae Juss.	26	8	2	5	1	3	–	1	1	–	–	–
Urticaceae Juss.	20	15	1	–	9	4	3	1	–	–	–	–
Celastraceae R.Br.	19	8		2	–	1	1	–	–	4	–	–
Hypericaceae Juss.	17	15	3	–	–	14	3	2	–	–	–	–
Fabaceae Juss.	16	14	10	2	9	13	1	–	–	–	–	–
Dipterocarpaceae Blume	14	12		6	4	2	–	–	–	–	–	–
Amaryllidaceae J.St.-Hil.	14	11	2	–	1	3	1	1	–	4	1	2
Clusiaceae Lindl.	13	3		2	–	1	–	–	–	–	–	–
Cactaceae Juss.	11	2		–	1	–	–	1	–	–	–	–
Other families (*n*=54)	158	77	2	32	9	29	1	8	–	–	–	1
TOTAL	**1,469**	**1,052**	**228**	**124**	**422**	**505**	**87**	**104**	**30**	**60**	**21**	**50**

§ Species with multiple cytotypes are counted more than once and therefore create a bias between the number of species with at least one ploidy record and the total number of cytotypes in each family.

The 644 cases of gametophytic apomixis reported in Table 8.1 represent 571 species, with 73 species showing both apospory and diplospory (51 in the family Rosaceae). Of these 571 species with embryological evidence of gametophytic apomixis, ploidy is known in 80.9% of cases (*n*=462, including 15 cases of *Alchemilla* species with high chromosome numbers; Table 6.1), and 129 cases have two or more cytotypes (Table 8.2). Species with gametophytic apomixis are mainly tetraploid (*n*=240) and triploid (*n*=163), although hexaploid (*n*=79), pentaploid (*n*=48) and octoploid (*n*=35) cytotypes are also frequent. Species showing nonaploid (*n*=19), decaploid (*n*=18) or higher ploidy cytotypes are less frequent and are mostly from the Poaceae family. Disregarding the questioned diploid cases reported in the literature [e.g., *Elaeagnus rhamnoides* (L.) A. Nelson, see Chapter 7 and Appendix 1.8], a few diploid cases associated with sporophytic apomixis showing a tendency to apospory or diplospory (Chapter 7, Table 7.1), and three cases of apospory in diploid *Salix* species from an unavailable report by Chochlov et al. (1978), regular occurrence of apospory or diplospory at the diploid level is only known for genera of the family Brassicaceae, in *Boechera* spp. (*n*=9), *Borodinia* spp. (*n*=1) and *Phoenicaulis* spp. (*n*=1).

Genera with regular occurrence of diplospory such as *Hieracium* spp. and *Taraxacum* spp. in the Asteraceae or *Limonium* spp. in the Plumbaginaceae, show a higher frequency of triploidy. Genera with regular occurrence of apospory such as *Crepis* spp. in the Asteraceae, *Chloris* spp. and *Paspalum* spp. in the Poaceae, *Ranunculus* spp. in the Ranunculaceae or *Amelanchier* spp. in the Rosaceae, show a higher frequency of tetraploidy. Genera with regular occurrence of diplospory and apospory such as *Cotoneaster* spp., *Rubus* spp. and *Sorbus* spp. in the family Rosaceae, show similar frequencies of triploidy and tetraploidy. Yet, this is not a general trend. Genera with regular occurrence of apospory (such as *Crataegus* spp. and *Malus* spp. in the Rosaceae) or diplospory (such as *Eupatorium* spp. and *Townsendia* spp. in the Asteraceae) and similar frequencies of triploidy and tetraploidy do also occur.

Species with gametophytic apomixis seem to comply with the rule of being exclusively polyploid. This agrees with previous reviews (Asker and Jerling, 1992; Carman, 1997). An exception to this rule is three of the six genera of Brassicaceae. Diploid brassicas underwent a whole-genome triplication event that has marked their evolutionary history (Bowers et al., 2003; Barker et al., 2009; Beilstein et al., 2010). Whether ancient paleopolyploidy can explain this exception remains to be tested.

8.3 APOMIXIS AND SEXUALITY

Angiosperms use two general mechanisms to produce seeds: one exploits sexuality and the other apomixis. Both are interrelated at the cellular, physiological and genetic levels (see Chapter 9). Despite there still being no clear view on how molecular changes are involved in the sexual-to-apomixis reproductive switch and in the formation *de novo* of apomictic individuals, much useful information on the population dynamics of apomixis has been gathered from laboratory studies and from studies on natural populations in different plant systems. Apomicts arise in nature from sexual relatives through repeated polyploidization and hybridization events (Hojsgaard and Hörandl, 2019). The first individuals capable of producing apomictic seeds are likely to be odd polyploids, meiotically unstable and with low fertility, as found in experimental studies (Hojsgaard et al., 2014b; Barke et al., 2018). Different intrinsic (balanced/unbalanced chromosome segregations, genomic reorganization, gene expression networks, phenotypic instability) and extrinsic factors (disadvantages of minority cytotypes, environmental conditions) may influence the restoration of balanced meiosis and sexual fertility or maintain a functional apomictic pathway and establish a new asexual lineage recovering fertility rates (Hojsgaard, 2018). During this transition from sexuality to apomixis, meiosis is depleted to a variable degree, only to allow the nascent lineage to refine its reproductive fitness and survivability (Hojsgaard, 2018; Carman et al., 2019).

Apomictic taxa are intrinsically associated with sexuality (functionally) and sexual relatives (genetically and ecologically). Functionally, apomicts are facultative, meaning that they retain varying levels of functional sexuality. In general, studies show that functional sex in apomicts is low or residual (e.g., Aliyu et al., 2010; Cosendai and Hörandl, 2010; Mártonfiová et al. 2010; Rebozzio et al., 2011; Šarhanová et al., 2012; Krahulcová et al., 2013, 2014; Hojsgaard et al., 2014b; Molins et al., 2014; Cushman et al., 2017; Carman et al., 2019; Doležal et al., 2020; Karunarathne et al., 2020a). Studies indicating the existence of obligated apomicts are not unusual. In some cases, apomictic taxa are found to behave as obligates (Aliyu et al., 2010; Šingliarová et al., 2011; Kant and Verma, 2012; Šarhanová et al., 2012; Caetano et al., 2013; Hojsgaard et al., 2013; Lo et al., 2013; Mráz et al., 2019). However, such studies typically look at one or few individuals over a specific period of time and, in some cases, more detailed analyses have shown variable levels of functional sexuality (e.g., Dias et al., 2017, 2021). Moreover, the incidence of functional levels of residual sex in apomictic plants is influenced by abiotic variables as light and temperature (Klatt et al., 2016, 2018; Rodrigo et al., 2017; Ulum et al., 2020) and these environmental conditions modulate residual sex rates (against apomixis) in natural populations (Karunarathne et al., 2020a). Therefore, even though apomictic plants may temporarily behave as obligates, if longer time frames during the life of any apomict are considered, the occurrence of genetic exchange through intraspecific or interspecific hybridization enabled by residual sex is very likely or even a fact of life among surviving asexuals.

In this framework, the occurrence of sexual ancestors and derivative apomictic taxa that maintain gene flow, either by *de novo* formation of asexual lineages or by residual sexuality between apomictic partners, creates an association among species called agamic complex (Stebbins, 1950; Grant, 1981). Such complexes are regularly found in nature, in species with sporophytic and gametophytic apomixis. In the first case, sexually reproducing individuals and individuals with adventitious embryony may share the same ploidy level and form homoploid complexes, or have a different ploidy and form heteroploid complexes, usually of sexual diploids and apomictic auto- and allopolyploids. Examples are found in the genus *Anemopaegma* spp. (Bignoniaceae; Correia et al., 2006; Firetti-Leggieri et al., 2013; Sampaio et al., 2013a, 2013b), in *Handroanthus* spp. (Bignoniaceae; Alves et al., 2016; Mendes et al., 2018; Braga et al., 2019), in *Sarcococca* spp. (Buxaceae; Naumova, 1980, 1993; Naumova and Willemse, 1983a, 1983b; Hoskins and Contreras, 2019), in *Euonymus* spp. (Celastraceae; Brizicky, 1964; Naumova, 1971, 1979, 1993), in *Garcinia* spp. (Clusiaceae; Ha et al., 1988, Richards, 1990, Thomas, 1997; Pangsuban et al., 2009), in *Cassia* spp. (Fabaceae; Symon, 1956; Randell, 1970; Holman and Playford, 2000), in *Miconia* spp. (Melastomataceae; Goldenberg and Varassin, 2001; Mendes-Rodrigues and Oliveira, 2012; dos Santos et al., 2012; Maia et al., 2016), in *Spiranthes* spp. (Orchidaceae; Swamy, 1948a, 1948b; Catling, 1982; Schmidt and Antlfinger, 1992), or in *Gymnadenia* spp. (Orchidaceae; Teppner and Klein, 1985; Teppner, 1996; Hedrén et al., 2000, 2018; Hedrén and Lorenz, 2019).

A similar situation occurs in species with gametophytic apomixis, as they can form homoploid or heteroploid agamic complexes. However, homoploid complexes with fully sexual individuals and apomictic individuals with the same ploidy are rare and are only represented by *Boechera* spp. (Brassicaceae; Schranz et al., 2005; Voigt-Zielinski et al., 2012; Mau et al., 2015). Most species with gametophytic apomixis form heteroploid agamic complexes. Examples are common in families with high numbers of apomictic species, but also in families where apomixis is less frequent. Just to mention a few, in the Asteraceae family, agamic complexes occur in genera such as *Antennaria* spp. (Bayer, 1997), *Chondrilla* spp. (Poddubnaja-Arnoldi, 1933, 1964; van Dijk, 2003). *Crepis* spp. (Babcock and Stebbins, 1938; Stebbins and Jenkins, 1939; Sears and Whitton, 2016), *Hieracium* spp. (Krahulec et al., 2008), *Taraxacum* spp. (Lyman and Ellstrand, 1984; Verduijn et al., 2004), *Townsendia* spp. (Thompson and Whitton, 2006), *Pilosella* spp. (Krahulcová et al., 2009, 2016; Šingliarová et al., 2011; Rosenbaumová and Krahulec, 2015). In the family Rosaceae, in *Amelanchier* spp. (Campbell et al., 1987, 1997; Burgess et al., 2014), *Aronia* spp. (Hovmalm et al., 2004; Mahoney et al., 2019), *Cotoneaster* spp. (Sax, 1954; Hjelmqvist, 1962; Nybom and Bartish, 2007), *Crataegus* spp. (Talent and Dickinson, 2007a, 2007b, 2007c; Lo et al., 2009, 2013; Vašková and Kolarčik, 2019), *Potentilla* spp. (Rutishauser, 1943a, 1943b, 1943c, 1945a, 1945b; Paule et al., 2012; Dobeš et al., 2013a, 2013b, 2015), *Rubus* spp. (Christen, 1950; Dowrick, 1961; Czapik, 1981a, 1981b; Spies and du Plessis, 1985, 1986; Šarhanová et al., 2012, 2017), in *Sorbus* spp. (Liljefors, 1953, 1955; Warburg and Kárpáti, 1968; Jankun, 1993; Kuntzelnigg, 1994; Lepší et al., 2009; Hamston et al., 2018). In the family Poaceae, agamic complexes occur in *Calamagrostis* spp. (Nygren, 1946, 1954a, 1954b; Greene, 1984), *Cenchrus* spp. (Narayan, 1962; Schmelzer and Renno, 1997, 1999; Kharrat-Souissi et al., 2011), *Dichanthium* spp. and *Bothriochloa* spp. (Celarier and Harlan, 1957; de Wet and Harlan, 1968, 1970; Roy et al., 2009), *Paspalum* spp. (Norrmann et al., 1989; Hojsgaard et al., 2008; Ortiz et al., 2013), *Themeda* spp. (Birari, 1980; Liebenberg, 1990), *Thrasya* spp. (Acuña et al., 2005). Heteroploid agamic complexes are also found in species of genera in families where apomixis is less recurrent, as in the genus *Hypericum* spp. (Hypericaceae; Noack, 1939; Lihová et al., 2000; Mártonfi et al., 1996; Matzk et al., 2001, 2003), in *Limonium* spp. (Plumbaginaceae; Ingrouille and Stace, 1985; Erben, 1978; Georgakopoulou et al., 2006; Kouzali et al., 2012; Caperta et al., 2017), in *Ranunculus* spp. (Ranunculaceae; Nogler, 1971, 1995; Izmaiłow, 1996; Hörandl et al., 2000; Cosendai et al., 2011; Hojsgaard et al., 2014b), in *Zanthoxylum* spp. (Rutaceae; Naumova, 1993; Cai et al., 1997, 2000), in *Boehmeria* spp. (Urticaceae; Okabe, 1963; Yahara, 1986, 1990; Tang et al., 2016), in *Elatostema* spp. (Urticaceae; Fagerlind, 1944a, 1944b).

The morphological and genetic differentiation of sexual and apomictic taxa within these complexes varies in each case, and is often obscured by poor morphological and molecular post polyploidization differentiation (see e.g. Šingliarová et al., 2011; Zilli et al., 2014; Feulner et al., 2017; Bigl et al., 2019; Karunarathne and Hojsgaard, 2021), the formation of intermingled forms from recurrent intra- and interspecific hybridization (Christensen et al., 2014; Krahulcová et al., 2018; Cushman et al., 2017; Uhrinová et al., 2017; Nardi et al., 2018), unequal genomic parental contributions, maternal inheritance or epigenetic effects (Hodač et al., 2014; Schinkel et al., 2020). In agamic complexes, sexual and apomictic taxa usually differ in ecological attributes and geographic ranges (Thompson and Whitton, 2006; Mráz et al., 2009; Lo et al., 2013; Schinkel et al., 2016; Karunarathne et al., 2018; Paule et al., 2018; Alonso-Marcos et al., 2019).

The combination of apomixis, sexuality, polyploidy and hybridization creates complex patterns of reticulate evolution and intergradation of taxa (e.g., Lo et al., 2010; Krak et al., 2013; Hamston et al., 2018; Persson et al., 2020) that hinders species boundaries and taxonomy.

8.4 TAXONOMIC ISSUES RAISED BY APOMIXIS AND SPECIES DELIMITATIONS

Taxonomy is intricately connected to population biology. Defining the taxonomic status of a group of organisms is linked to defining their genetic cohesion, an issue that is addressed by almost all species concepts. Because of the challenge posed by the panmictic population-based species concept (Wheeler and Meier, 2000), plants that reproduce asexually – in particular through asexual seeds – defy the practical and theoretical definitions used to categorize a group of plants at the level of species.

In these cases, the use of the concepts of basic *versus* intermediate and narrow *versus* broad species to treat apomictic lineages and the use of criteria to recognize apomictic species have been widely discussed (Majeský et al., 2017; Dickinson, 2018; Hörandl, 2018; Sochor et al., 2018), but a general solution is probably out of reach. Narrow delimitations and taxonomic approaches that consider a group of organisms as a species only if they are morphologically stable over a wider geographical range and ignore local morphotypes/biotypes of presumed hybrid origin (e.g., the Weberian concept; Weber, 1996) may be a solution to partially reduce the number of accepted species compared to approaches using broader species concepts that merge stable genotypes sharing morphological similarities into one species, but may not help to obtain reliable intrageneric phylogenies (Kurtto et al., 2010; Sochor et al., 2015, 2018).

Even when, at first, ignoring hybrid morphotypes/biotypes is practically useful (Kellog, 1990; Weber, 1996), a proper understanding of the life history of any agamic complex, whether as a group of morphologically variable individuals or as an aggregate of (micro)species, should not be neglected but should be considered as an undoubted aid to standardize taxonomic treatments. This is important, as several parameters related to diversity, ecology and evolution depend on genus-level measurements. The use of standardized approaches will help to eliminate intra- and intergeneric bias.

As already mentioned in the previous chapters, a detailed discussion is beyond our expertise and the scope of this book. However, we would like to express our humble opinion and highlight two significant points towards the *utopia* of standardizing the taxonomic treatments used on apomictic taxa. First, as each plant group is bound to different biological and ecological attributes, the application of distinct species concepts within families, tribes or even between genera is envisaged. Nevertheless, at the genus level, specialists should consent on the use of a single taxonomic approach. Second, once an approach is consented, recommendations on the merging of crossbreeding species should be taken to actions and implemented, even if this means lumping species together.

If the eighteenth and nineteenth centuries focused on discovering, describing, cataloguing and mapping the distribution of plants and served as a starting point for the emergence of established twentieth-century disciplines such as physiology, genetics, ecology and biogeography, the twenty-first century will focus on uncovering and characterizing functional intraspecific variability occurring at all levels. The relevance of intraspecific variability in plants has already begun to be revealed (e.g., Anderegg et al., 2021) and will play a central role for humankind, especially in ensuring food production. Taxonomists have already embarked on this path, in many cases encouraged by the occurrence of apomixis as an amplifier of intraspecific variants.

8.5 APOMIXIS AND EVOLUTION

Apomictic species show developmental particularities in each family, either by the type of apomictic development, requirements for endosperm development during seed formation, ploidy or by other associated features as growth habit (e.g., Noyes, 2006; Talent, 2009; Ortiz et al., 2013). However, in different plant families, apomictic species or species with apomictic cytotypes are often associated with geographical parthenogenesis, a term used to denote when sexual taxa show a restricted distribution compared to apomictic taxa in agamic complexes (Bierzychudek, 1985; Hörandl, 2006), and with positive biodiversity parameters (Burgess et al., 2014; Hojsgaard et al., 2014a).

Polyploidy itself is often associated with characteristics that promote wider plant distributions (te Beest et al., 2012; Linder and Barker, 2014), increased diversity (Vamosi and Dickinson, 2006; Román-Palacios et al., 2020) and higher diversification rates (Han et al., 2020). In clades harbouring taxa with gametophytic apomixis in which polyploidy is inherently associated with apomixis, it is not possible to examine the effects of asexuality and polyploidy separately. In the exceptional case of *Boechera* spp. where gametophytic apomixis is found in diploids, Mau et al. (2015) observed no difference between sexual and apomictic *Boechera* in genus-wide geographic range and a weak pattern of geographical parthenogenesis for both ploidy and reproductive mode across *Boechera* species. However, evidence from species with sporophytic apomixis and diploid taxa indicates that, at least regarding plant distributions, clades with diploid apomicts show comparable patterns of geographical parthenogenesis (dos Santos et al., 2012). Apomictic taxa not always exhibit clear patterns of geographical parthenogenesis (e.g., Burgess et al., 2014), but reproductive assurance systems, such as selfing or apomixis, have advantages when colonizing new areas, as new populations can be founded by single individuals without requiring a mate and pollinator limitation (Baker, 1955, 1967). Apomixis in particular, creates clonal offspring and can therefore multiply and locally establish the fittest genotypes through differential seed sets and better competing abilities, contributing to the observed patterns of geographical parthenogenesis (Hojsgaard and Hörandl, 2019).

Polyploidy also alters cellular, genetic and physiological traits of adaptive potential (Comai, 2005; Weiss-Schneeweiss et al., 2013) with effects on ecological tolerance and breadth. In association with apomixis (i.e., reproductive assurance and clonality) and often hybridization, polyploidy can promote ecological niche differentiation, as observed in different auto- and allopolyploid apomicts (Kirchheimer et al., 2016, 2018; Coughlan et al., 2017; Karunarathne et al., 2018). If polyploid apomicts behave as generalists and are less competitive than specialized sexual parents in their native range, but are more competitive in peripheral areas (Karunarathne et al., 2018, 2020a, 2020b), this may prelude ecological differentiation between cytotypes and foster occupation of new habitats. By preserving favourable characters and features related to the fixed heterozygosity or hybrid vigour (Richards, 2003), apomixis can enhance range expansion of ecologically tolerant genotypes adapted to novel environments. Despite clonality may give advantages to ecologically fit genotypes, natural populations of known apomicts rarely contain single clones (Hojsgaard and Hörandl, 2019). Instead, they show different levels of genetic and genotypic variability owing to residual sex or multiple independent origins of apomictic lineages. While genetic variability in apomicts can be higher or lower than that found in the sexual parents (Paun et al., 2006a, 2006b; Šarhanová et al., 2017; Karunarathne and Hojsgaard, 2021), in all cases genotype diversity is inherently low. The observed rates of residual sexuality in apomicts can both explain the genetic variation existing in natural populations and prevent negative effects of asexuality, such as the accumulation of deleterious mutations via Muller's ratchet (van Dijk et al., 2009; Majeský et al., 2012; Pellino et al., 2013; Hojsgaard and Hörandl, 2015; Hodač et al., 2019), an expected outcome in obligated asexual lineages (Muller,

1964; Hollister et al., 2015) and observed in some apomictic plants (Lovell et al., 2017). Thus, facultative apomixis and polyploidy can generate genetically persistent evolutionary lineages.

The presence of sporophytic and gametophytic apomixis in angiosperms is positively associated with measures of biodiversity. Orders, families and subfamilies containing apomictic genera are larger than those non-containing apomictic genera and their numbers are positively correlated with the total numbers of families and genera (Hojsgaard et al., 2014a). To explain this association, the authors proposed the apomixis-to-sex reversal hypothesis, which suggests that apomixis can facilitate diversification through polyploidy cycles and sexual reversions (Carman, 1997; Hörandl and Hojsgaard, 2012, Hojsgaard et al., 2014a; Carman et al., 2019). According to this hypothesis, a new polyploid arising in a diploid sexual population may shift to apomixis and establish an apomictic population reproductively isolated from sexual parents by interploidy barriers (Ramsey and Schemske, 1998) and parthenogenetic embryo development refractory to fertilization. In the new population, facultative apomictic polyploids can serve as pioneer explorers of new habitats whereby they rapidly expand the distribution areas of their sexual parent populations by occupying new ecological niches and geographic areas (Hojsgaard et al., 2014a). Because apomixis freezes genetic variation and reduces genotype variability, new geographically distant populations are expected to evolve independently by reduced gene flow and, over time, hold differentiated gene pools (Hojsgaard and Hörandl, 2019). Alternatively, a reversion to sexuality by restating functional meiosis (see Section 8.3) would allow the establishment of new sexual populations in new habitats distant from their more recent sexual ancestors and without the long-term disadvantages of asexuality. If the new sexuals become established in an area of sympatry, they may avoid hybridization via pre- and postzygotic barriers and become subspecies or new species (e.g., Arrigo and Barker, 2012). If the new lineages are geographically isolated from the parental populations, they may then diversify by ecological and allopatric speciation. The notion that sexuality is the default reproductive mode in facultative apomicts on which apomixis is over-imposed (Koltunow et al., 2011), that examples of reversals of apomixis to sex have been reported for species of *Pennisetum* (Taliaferro and Bashaw, 1966), *Pilosella* (Chapman et al., 2003) and *Paspalum* (Ortiz et al., 2013) and that apomixis-to-sex reversals facilitate speciation in *Boechera* (Carman et al., 2019) supports the hypothesis.

In addition, apomixis may function as a temporary escape from sterility in sexual neotriploids and enhance polyploid establishment (Hojsgaard, 2018). In this case, transient activation of apomictic development during polyploidization events in a diploid sexual population may help the nascent polyploid lineage to overcome the problems imposed by sexuality in intermediate triploid individuals with unbalanced genome sets (e.g., aneuploid gametes, sterility, minority cytotype disadvantages) and thus enhance the efficiency of the triploid bridge for the establishment of genomically balanced, fertile sexual polyploids.

Apomixis is a reproductive wild card for lineage subsistence and, in connection with polyploidy, hybridization and reversion to sexuality, promotes divergence at the genetic, genomic and habitat adaptation levels and the evolution of new species and genera.

9 Molecular Basis of Apomixis and Applications

9.1 INTRODUCTION

As presented in the previous chapter, apomixis is a trait of ecological and evolutionary relevance. Another significant point of apomixis relates directly to its short-term genetic consequence, i.e., the formation of clonal embryo-bearing seeds. This elementary point has long attracted scientists in hopes of functionally exploiting apomixis in plant breeding to fix heterosis and capture gene combinations of elite phenotypes (van Dijk et al., 2016).

The use of apomixis in hybrid seeds is a game-changing innovation in plant breeding that would allow rapid mass production of new cultivars better adapted to local environments around the world. Apomixis breeding would not only reduce the time and costs involved in releasing new commercial hybrid varieties (Sharma et al., 2002; Spillane et al., 2004; Acquaah, 2012; van Dijk et al., 2016) but would also reduce problems due to pathogen or insect epidemics and massive use of damaging pesticides; prevent the spread and dissemination of diseases and viruses; reduce storage, shipping and planting costs in tuber or vegetatively propagated crops; and it would avoid fruit and cereal crop losses related to difficulties in pollination (Gröber and Zacharias, 1983; Jefferson, 1994, Hanna, 1995; Vielle-Calzada et al., 1996; Grossniklaus et al., 1998; Ramulu et al., 1999). Thus, apomixis will help plant breeders to create more resilient crops and develop more sustainable farming systems, solving phytosanitary problems and reducing pressures on biodiversity and water resources, an urgent task both to ensure access to food and to combat climate change and environmental degradation (Alkemade et al., 2009; FAO, 2009; Ray et al., 2013).

To exploit apomixis, breeders face a two-sided problem. On the one hand, if apomixis is present in the target plant, it limits the possibilities to make crosses and exploit the variability present in different genotypes. On the other hand, if apomixis is not present in the target plant, breeders face the challenge of introducing the trait through wide crosses.

In many minor crops where apomixis is present, breeders have found different strategies to overcome apomixis and exploit sexual recombination to create cultivars and varieties. These include the use of different breeding schemes or the creation of artificial sexual genotypes for use in crosses with genotypes selected from the wild to generate new genetic combinations that are fixed by apomixis (e.g., do Valle and Miles, 2001; Worthington and Miles, 2015; Zilli et al., 2018, 2019; Acuña et al., 2019).

In major crops and many other minor crops where apomixis is not present, early efforts focused on the introgression of apomixis from wild relatives (for details see Asker and Jerling, 1992; Savidan, 2001). The central concept that pollen of apomicts could transmit dominant apomixis genes in wide crosses and backcrosses between maize × *Tripsacum dactyloides* (Petrov et al., 1979, 1984), pearl millet × *Cenchrus squamulatus* (≡*Pennisetum squamulatum*, Hanna et al., 1993), wheat × *Anthosachne rectiseta* (≡*Elymus rectisetus*, Peel et al., 1997a, 1997b) encountered different problems, mostly associated with interspecific sterility and ploidy barriers (Savidan, 2000), and failed to introduce apomixis in these sexual crop plants.

Nowadays, the emphasis is placed on understanding the molecular basis of apomixis to induce or mimic apomixis in sexual crops by altering sexual development. In a recent study, Sailer et al. (2016) proved that clonal seeds can fix genotype and complex phenotypic traits in apomictic hybrids. This is the first empirical proof of principle of transgenerational phenotype fixation in hybrid plants and, in agreement with previous studies based on visual classifications of phenotypes, demonstrates that

DOI: 10.1201/9781003088561-9

apomixis can be used in plant breeding for commercial deployment. However, molecular manipulation of sexual reproduction to induce apomixis is only in its early development. Despite substantial advances (see Section 9.3), the challenge for apomixis breeding remains the poor understanding of the genomic background and molecular mechanisms that initiate apomixis and guide the formation of a clonal seed (Hojsgaard, 2020).

This chapter briefly outlines some examples of the commercial use of apomixis in plant breeding, summarizes what is known about the molecular basis of apomixis in plants and mentions recent technological advances that are facilitating gene manipulation and are likely to have an impact on the biotechnological use of apomixis in the coming years.

9.2 APOMIXIS IN PLANT BREEDING

Although efforts to introduce apomixis into major crops have so far been unsuccessful, apomixis is used in many minor crops to fix elite genotypes. A number of apomictic species of different genera and families are cultivated for various purposes and have been participated or are participating in breeding programmes. These programmes exploit the genetic variability existing in nature through direct selection and different breeding schemes (Singh et al., 2021) to generate high-yielding apomictic genotypes that are marketed as cultivars and varieties.

Within the Poaceae, a great deal of experimental work has been done on the breeding of mainly tropical and subtropical grasses for forage and turf (e.g., Busey, 1989; Miles, 2007; Hanna et al., 2013; Worthington and Miles, 2015). Examples include selected genotypes, cultivars or varieties of *Brachiaria brizantha* and related species, planted on approximately 120 million hectares in Brazil alone (Rueda da Silva et al., 2020), *Paspalum notatum* and related species, used as cultivated pasture and turf in warm, humid areas of different countries in the Americas but also in Australia and Japan (Blount and Acuña, 2009; Acuña et al., 2019), *Cenchrus ciliaris* and related species, also as forage (Griffa et al., 2010; Quiroga et al., 2013; Kumar et al., 2017a, 2017b), *Panicum maximum* and related species (Sato et al., 1992; Savidan et al., 2004; Lipka et al., 2014), to mention a few. An example of an alternative use of an apomictic grass, apart from fodder, is *Jatropha curcas*, which is used as a biofuel feedstock (Francis et al., 2020). A few species used as cereal reproduce sexually but share apomictic cytotypes or close related apomictic species [i.e., tef (*Eragrostis tef*); kodo millet (*Paspalum scrobiculatum*); pearl millet (*Cenchrus americanus*)].

Within the Rosaceae, some cultivars of species forming agamic complexes involve sexual cytotypes that are propagated by seed or vegetatively but may not directly exploit apomixis as a breeding tool. Examples in this family that include apomictic species or species with apomictic cytotypes are mostly used for berry fruit consumption, either fresh or processed in jams, jellies and beverages, such as *Fragaria* species and cultivars (Hancock et al., 2002; Luby et al., 2008; Ather-uz-Zaman and Islam, 2018), *Aronia melanocarpa* (Hovmalm et al., 2004), many *Rubus* varieties and cultivars (Thompson, 1995; Dossett et al., 2012; Clark et al., 2013; Foster et al., 2019), or in *Sorbus* cultivars and hybrids, also used as ornamental plants for their attractive habit and decorative flowers and fruits (Kampuss et al., 2009; Jurikova et al., 2014; Kaplan et al., 2015; Zymone et al., 2018; Asbaganov et al., 2020). Other species with apomixis are used as ornamental or for their pharmaceutical properties and include hybrids and cultivars of *Alchemilla* (Gehrke et al., 2008; Galambosi et al., 2010; Vitkova et al., 2013), *Cotoneaster* (Contreras et al., 2014), *Crataegus* (Nieto-Angel et al., 2013; Jablonski, 2017), *Malus* (Schmidt, 1977; Gröber and Zacharias, 1983).

Within the Asteraceae, species with apomixis involved in breeding programmes for natural rubber production belong to the genera *Taraxacum* and *Parthenium*, particularly *Taraxacum kok-saghyz* (Arias et al., 2016; Stolze et al., 2017; Iaffaldano et al., 2018; Zeisek et al., 2019) and *Parthenium argentatum* (Tysdal, 1950; Abdel-Haleem et al., 2019; Stonebloom and Scheller, 2019). In different species of *Hieracium*, *Pilosella* and *Crepis*, cultivars or pre-varieties are offered as ornamental in local markets (e.g., in Germany, the Netherlands or England), as summer fodder (Tilley et al., 2012) or for restoration (Gucker and Shaw, 2018, 2019).

Species with apomixis from other plant families used for various purposes and directly or indirectly involved in breeding programmes are or belong to *Beta* spp. (Amaranthaceae; Cleij et al., 1976; Szkutnik, 2010; Sokolova, 2020), *Citrus* spp. (Rutaceae; Asker and Jerling, 1992; Koltunow et al., 1995), *Mangifera indica* (Anacardiaceae; Ochoa et al., 2012), *Hypericum* spp. (Hypericaceae; Pank et al., 2003; Olsen et al., 2006; Trueblood et al., 2010), *Juglans regia* (Juglandaceae; Meiyong et al., 2000; Guo-Liang et al., 2007, 2010; Bujdoso et al., 2021), *Garcinia mangostana* (Clusiaceae; Ramage et al., 2004; Sobir and Poerwanto, 2007; Sobir et al., 2013), *Pachira glabra* (≡*Bombacopsis glabra*, Malvaceae; Da Silva Araújo et al., 2017).

Other cases of minor crops in genera such as *Allium* spp. and *Humulus* spp. appear to have apomictic relatives, but to our knowledge, no breeding experiments have been found that take advantage of apomixis.

9.3 GENETICS OF APOMICTIC DEVELOPMENT

Apomixis is a trait that appears in natural populations, likely in diploid hybrids and neotriploid individuals before a new asexual lineage is established (Hojsgaard and Hörandl, 2019). Understanding the genetic nature of apomixis means unravelling the molecular steps that drive the emergence and establishment of a regular pathway to form asexual seeds in the ovules of apomictic plants.

In the ovules of plants with sporophytic apomixis, somatic embryos develop from initial cells of the nucellus or integuments, in parallel or after zygotic embryogenesis in the fertilized female gametophyte (Naumova, 1993; Chapter 2). As this type of apomixis does not interfere with sexual development, the genetic control and molecular regulatory network for trait expression is expected to be simple.

In the ovules of plants with gametophytic apomixis, embryos with a somatic genetic constitution develop from unreduced female gametophytes arising from initial cells in the nucellus or germline cells with altered meiosis (Asker and Jerling, 1992; Chapter 2). As this type of apomixis partially or totally interfere with sexual development, the genetic control and molecular regulatory network for trait expression is predictably complex.

Current knowledge of the genetic basis of apomixis in angiosperms comes from a rich variety of cytological, biochemical, physiological, genetic and genomic studies. Unfortunately, we cannot cite them all here and try to include the most appropriate or recent studies. The interested reader will find further references in these studies.

9.3.1 Sporophytic Apomixis

Until recently, little was known about the genetic control of sporophytic apomixis. Most of the findings on this type of apomixis have been made in species of the family Rutaceae. Earlier studies using data from crosses between monoembryonic and polyembryonic cultivars postulated that one dominant gene (Parlevliet and Cameron, 1959; Iwamasa et al., 1967), two (Cameron and Soost, 1979) or three (Deidda and Chessa, 1982) control the occurrence of nucellar embryony in citrus fruit trees.

More recent studies using molecular markers to characterize genetic segregation ratios and linkage in complex populations and mono- × polyembryonic F_1 populations recognized a cluster of markers linked to a polyembryony locus (Nakano et al., 2008a, 2008b; Kepiro and Roose, 2010; Raga et al., 2012) mapped to a genomic region spanning approximately 380 kb (Nakano et al., 2012). This supports a monogenic inheritance for nucellar embryony in citrus, or at least the occurrence of a single locus controlling the trait.

Gene expression analyses in citrus ovules revealed many differentially expressed genes during the initiation and development of nucellar embryony, including stress-related genes and several genes associated with sexual embryogenesis (Nakano et al., 2013; Kumar et al., 2014; Long et al., 2016; Wang et al., 2017), pointing to a close interplay among genes and shared pathways of sexual and nucellar embryogenesis in citrus.

Sequence analysis of 6 genomes and 100 accessions, including primitive, wild and cultivated *Citrus* species and varieties, revealed an 80-kb interval containing 11 genes within the 380-Kb polyembryony locus, one of which was identified as a candidate for polyembryony in citrus (Wang et al., 2017). Transcriptomic analyses show that the *CitRWP* gene is expressed specifically in ovules and at higher levels in polyembryonic cultivars compared to monoembryonic cultivars, and harbours a MITE (Miniature Inverted-repeat Transposable Element) insertion mutation in its promoter region that co-segregates with the polyembryonic phenotype and is not present in monoembryonic plants (Wang et al., 2017; Xu et al., 2021). Genes from the RWP-RK domain-containing (RKD) family are found only in plants and function in the maintenance of egg cell identity (Koszegi et al., 2011). The *AtRKD4* (*GROUNDED*) gene of *Arabidopsis thaliana* promotes embryogenesis in somatic cells when transiently overexpressed (Jeong et al., 2011; Waki et al., 2011).

Another study focusing on the *CitRWP* gene (termed as *CitRKD1*) found similar overexpression of the MITE-containing *CitRKD1* gene in flowers and fruits of polyembryonic citrus varieties, while its antisense overexpression in transgenic plants abolished nucellar embryogenesis without altering zygotic embryogenesis (Shimada et al., 2018).

These studies point to a central role of the MITE-containing *CitRKD1* gene in enabling nucellar embryo development that is not essential for zygotic embryo development among *Citrus* species. However, no MITE insertion in the *CitRWP* gene promoter was found in the genome of the highly polyembryonic trifoliate orange (*Poncirus trifoliata*) (Smith et al., 2019) and the *CitRWP* gene is not expressed in ovules, suggesting that another dominant gene may regulate polyembryony in this species (Garcia et al., 1999; Hong et al., 2001; Kepiro and Roose, 2010; Xu et al., 2021).

The gain-of-function phenotype associated with a MITE insertion is absent in the homologous *CitRWP* gene in other highly polyembryonic species of the genera *Zanthoxylum* and *Murraya* (Xu et al., 2021), indicating that the genetic control of nucellar embryony is more complex and governed by genetic changes with independent origins.

Consistent with this, a recent analysis of small RNAs in *Zanthoxylum bungeanum* flowers shows an alternative route involving microRNA-target pairs interactions and expression patterns associated with apomixis (Fei et al., 2021). The study indicates that during the development of adventitious embryos a miR172 significantly reduces the expression levels of *TARGET OF EAT3* (*TOE3*) and *APETALA 2* (*AP2*) genes (Fei et al., 2021). *TOE3* and *AP2* are MADS-BOX transcription factors class A genes that control flowering time, meristem determination, floral organ and fruit development (Zumajo-Cardona et al., 2021), with antagonistic effects to *AGAMOUS*, a MADS-BOX class C gene that regulates the floral meristem activity and stamen and carpel development (Uemura et al., 2018).

Thus, by inhibiting *TOE3* and *AP2*, miR172 increases the expression of *AGAMOUS* during nucellar embryogenesis in *Zanthoxylum bungeanum*. The study also shows that the expression level of auxin response factor (ARF) gradually increases during fruit development (Fei et al., 2021). Similarly, auxin content in ovaries of polyembryonic citrus varieties is higher than that of monoembryonic varieties, and regulatory changes in the expression of auxin transporter genes and genes regulating auxin responses were detected at the onset of nucellar embryo initiation (Xu et al., 2021). Collectively, these results suggest that auxin is an active player in diverting resources and regulating the development of nucellar embryos in Rutaceae.

Adventitious embryony in species with sporophytic apomixis is likely to be under the monogenic control of genes and regulatory sequences with independent origins in different species, and shared molecular networks and hormonal regulation for sexual and asexual embryogenesis.

9.3.2 Gametophytic Apomixis

Much research has been done to elucidate the genetic control of gametophytic apomixis. Yet, as this type of apomixis interferes with the sexual developmental programme and alters sporogenesis, gametogenesis and the interaction of gametes during seed formation, the genetics behind it is essentially complex. Moreover, as gametophytic apomixis involves the formation of an unreduced female gametophyte, which usually replaces the reduced (meiotic) one, both embryo and endosperm formation must occur to produce a functional seed. Thus, gametophytic apomixis is linked to two developmental steps necessary for seed formation, i.e., the development of an unreduced female gametophyte, and the development by parthenogenesis of the embryo from the unreduced egg cell (which may or may not require fertilization of the central cell). Surprisingly, even though findings are slightly different according to species and type of apomixis development, genetic control shows consistent results across species (Ozias-Akins and van Dijk, 2007).

In general, apomixis (i.e., apospory or diplospory plus parthenogenesis) is dominant over sexuality and its occurrence is under the control of genes physically located at one or two loci. Species in which the genetic control of apomeiosis and parthenogenesis are inherited together as a single Mendelian locus belong to the tribe Paniceae within the Poaceae, and the locus is often denoted as the "apospory-specific genomic region" (ASGR) or simply the "apomixis-controlling locus" (ACL). Examples include *Brachiaria brizantha* (Pessino et al., 1997, 1998), *B. decumbens* (Worthington et al., 2016), *Cenchrus ciliaris* (Roche et al., 1999), *C. squamulatus* (Akiyama et al., 2004; Goel et al., 2006, Ozias-Akins et al., 1998), *Panicum maximum* (Savidan 1981), *Paspalum notatum* (Martínez et al., 2001, 2003; Stein et al., 2004, 2007), *P. simplex* (Pupilli et al., 2001, 2004; Labombarda et al., 2002), *P. procurrens* (Hojsgaard et al., 2011) and *Urochloa humidicola* (Worthington et al., 2019).

In species where apomixis is controlled by a single locus, genetic mapping studies localized the apomixis locus to large chromosomal regions of up to 50 Mb with suppressed recombination (Goel et al., 2003; Pupilli et al., 2004), containing several genes. Recombination is impaired by the presence of chromosomal rearrangements, not only inversions and translocations (Pupilli et al., 2004; Okada et al., 2011; Worthington et al., 2019) but also gene duplications, insertions or deletions of transposable elements that disrupt gene sequences (Akiyama et al., 2005; Calderini et al., 2006, 2011; Conner et al., 2008; Siena et al., 2016), enhance heterochromatin formation and further promote allelic sequence divergence, retrotransposon activity and mutational degradation.

However, Conner et al. (2013) found in *Cenchrus ciliaris* a rare, near-sterile recombinant plant that lost a portion of the ASGR locus and the ability to express parthenogenesis, yet it produces unreduced female gametophytes. The possible occurrence of rare recombination events in other species (e.g., Jessup et al., 2002) suggests that repression of recombination may not be obligatory, but affects the functionality of apomixis and fertility, and underlines the fact that apomeiosis and parthenogenesis are under independent genetic control, as in species with two loci.

Species in which the genetic control of apomeiosis and parthenogenesis are inherited separately as independent factors include *Allium ramosum* (Yamashita et al., 2012), *Erigeron annuus* (Noyes and Rieseberg, 2000; Noyes, 2006; Noyes et al., 2007), *Cenchrus ciliaris* (Conner et al., 2013), *Hieracium aurantiacum* (Koltunow et al., 2011; Okada et al., 2011), *H. caespitosum* (Koltunow et al., 2011), *H. praealtum* (Catanach et al., 2006; Koltunow et al., 2011; Okada et al., 2011), *H. piloselloides* (Koltunow et al., 2011; Okada et al., 2011), *Hypericum perforatum* (Schallau et al., 2010), *Tripsacum dactyloides* (Blakey et al., 2001), *Taraxacum officinale* (van Dijk et al., 1999, 2020; van Dijk and Bakx-Schotman, 2004; Vijverberg et al., 2004) and *Poa pratensis* (Albertini et al., 2001a, 2001b; but see Matzk et al., 2005 for a more complex model of genetic control with differences in gene expressivity and penetrance).

An additional locus associated with the control of endosperm development is likely tightly linked to parthenogenesis. Whereas in species with pseudogamous endosperm development, completion of parthenogenetic embryo development requires the central cell of the unreduced female gametophyte to be fertilized, evidence in *Taraxacum* spp. and *Hieracium* spp. suggests that fertilization-independent endosperm formation segregates independently of apomeiosis and parthenogenesis with complex underlying genetics (Ogawa et al., 2013; Henderson et al., 2017; van Dijk et al., 2020).

The above-mentioned features and the impaired intra-locus recombination in species with one or two loci have reduced the efficiency of map-based cloning strategies and genetic dissection of apomixis (but see below the case of the *DIP* locus in *Taraxacum officinale*). Still, by using different approaches, including deletion mutants, chromosome walking and transcriptomics, different authors have successfully pursued a functional dissection of the trait by pinpointing genes with different relevant roles in the emergence of apomixis.

In *Boechera*, the *APOLLO* gene (apomixis linked locus) encodes an Aspartate Glutamate Aspartate Aspartate histidine exonuclease whose transcripts are upregulated in apomeiotic ovules and downregulated in sexual ovules entering meiosis (Sharbel et al., 2009, 2010; Corral et al., 2013). Highly polymorphic apo- and sex-specific alleles have been identified among *Boechera* spp. accessions, with all apomicts being heterozygous for the *APOLLO* gene (having at least one apo-allele and one sex-allele), while sexual genotypes are homozygous for sex-alleles. The dominant apo-alleles share different apomixis-specific polymorphisms, the most relevant being a 20 nucleotide polymorphism in the 5′ untranslated region (UTR) containing binding sites for specific transcription factors (e.g., ARABIDOPSIS THALIANA HOMEOBOX PROTEIN5; Corral et al., 2013). The expression of a deregulated apo-allele is hypothesized to trigger the cascade of events leading to the formation of unreduced female gametophytes in apomictic *Boechera* plants.

In *Cenchrus* species, the *PsASGR-BBML* gene (*Pennisetum squamulatum ASGR-BABY BOOM-like*), a member of the BBM-like clade of conserved AP2-like transcription factors in the *AINTEGUMENTA-LIKE* gene family (Zumajo-Cardona et al., 2021), induces the onset of embryonic development in unfertilized egg cells (Conner et al., 2015, 2017). AINTEGUMENTA-LIKE proteins are regulators of cell division and embryo and stem cell niche specification expressed in all dividing tissues (Horstman et al., 2014). *BBM* or *BBM-like* genes are known to induce ectopic embryogenesis in sexual model systems such as *Brassica* and *Arabidopsis* (Boutilier et al., 2002). *PsASGR-BBML* transgenic plants induce parthenogenesis in haploid egg cells in sexual pearl millet, maize, rice and sorghum (Conner et al., 2015, 2017; Khanday et al., 2019; Ozias-Akins and Conner, 2020). The conservation of the ASGR region and *PsASGR-BBML* gene sequences in *Cenchrus*, *Brachiaria* and *Panicum* species, despite having different chromosomal backgrounds, suggests this gene likely have a key role for parthenogenesis in apomictic species of the Paniceae (Worthington et al., 2016, 2019).

In *Poa pratensis*, the *APOSTART* gene (apomixis START domain) member with flower-specific expression shows delayed expression in the nucellar cells of developing ovules in apomictic but not in sexual individuals, possibly causing the switch from a somatic to a reproductive cell fate observed during the emergence of aposporous initial cells (Albertini et al., 2004; Marconi et al., 2020). START domains are conserved lipid/sterol-binding domains that in plants are primarily found within homeodomain transcription factors involved in the transcriptional regulation of key developmental processes, including embryogenesis (Schrick et al., 2004). Together with *APOSTART*, another gene differentially expressed in apomictic and sexual ovules identified as *PpSERK* (*Poa pratensis SOMATIC EMBRYOGENESIS RECEPTOR KINASE*) may also be a key component of the embryogenesis-signalling pathway (Albertini et al., 2004, 2005). *SERK* genes encode Leucine-rich repeats receptor-like kinases responsible for the acquisition of embryogenic competence and during zygotic embryogenesis after fertilization (Schmidt et al., 1997). Screens of *SERK* family members in *Paspalum notatum* and *Brachiaria brizantha* identified paralogous genes showing strong differential spatial expression patterns in ovules of apomict and sexual genotypes (Podio et al., 2014; Koehler et al., 2020).

In *Taraxacum officinale*, the *ToPAR* gene (*Taraxacum officinale PARTHENOGENESIS*) encodes a K2-2 zinc finger domain protein with an EAR (ethylene-responsive element-binding factor-associated amphiphilic repression) motif (Underwood et al., 2022). In apomict dandelions, the dominant *PAR* allele has a MITE (Miniature Inverted-repeat Transposable Element) insertion in its promoter that is absent in sexuals and appears to act as a *cis*-regulatory element responsible for its expression in egg cells and for triggering embryo development without fertilization. Loss of parthenogenesis *Taraxacum* mutants transformed with a construct carrying the MITE-containing promoter and a homologous gene from sexual lettuce (*Lssex*) were able to recover parthenogenesis, demonstrating that the *ToPAR* promoter can invoke a lettuce gene to induce parthenogenesis (Underwood et al., 2022). A *PAR* ortholog of *Pilosella piloselloides* (*PpPAR*) containing a similar MITE absent in sexual biotypes is also specifically expressed in ovules of apomicts during late gametogenesis. In addition, heterologous *PAR* expression using an egg-cell-specific promoter in the sexual crop lettuce (*Lactuca sativa*) induced ectopic cell divisions and haploid embryo-like structures in the absence of fertilization (Underwood et al., 2022).

Many other genetic studies have identified genes with apparent roles in different steps of apomictic reproduction. These studies uncovered a variety of genes and proteins, small RNAs and long non-coding RNAs participating in the regulatory cascade of apomixis, but little is known about their function and interconnection. We briefly mention a few here, and the reader is also referred to recent detailed reviews (Vijverberg et al., 2019; Hojsgaard, 2020; Schmidt, 2020; Ozias-Akins and Conner, 2020).

In addition to the *APOLLO* gene in *Boechera*, Mau et al. 2013 characterized the *BspUPG2* gene (*Boechera spp. UPGRADE2*), associated with unreduced pollen formation. Furthermore, F-box and E3 ligases genes were identified as likely regulators of germline development in sexual and apomictic ovular tissues (Schmidt et al., 2014; Zühl et al., 2019), along with several microRNAs, differentially upregulated at the megaspore mother cell stage in apomictic ovules, and small RNAs targeting exonic regions likely having roles in post-transcriptional gene regulation (Amiteye et al., 2011, 2013).

In *Hieracium*, apomixis is controlled by two independent and dominant loci: *LOA* (*LOSS OF APOMEIOSIS*) and *LOP* (*LOSS OF PARTHENOGENESIS*) (Catanach et al., 2006; Koltunow et al., 2011). Mutants lacking *LOA* fail to initiate unreduced female gametophytes, whereas mutants lacking *LOP* generate unreduced gametophytes but require fertilization to form a seed. Different genes associated to *LOA* and *LOP* loci have been identified, including ovary-expressed ARGONAUTE genes and genes involved in small RNA biogenesis, chromatin silencing and candidates for the specification and initiation of aposporous initial cells (Kotani et al., 2014; Rabiger et al., 2016; Bräuning et al., 2018). Moreover, early expression of Arabinogalactan-proteins (AGPs) in aposporous initial and functional megaspore cells likely has a role in communication and signal transduction during megaspore death and progression to gametogenesis (Juranić et al., 2018).

In *Hypericum*, the *HAPPY* locus (*Hypericum APOSPORY*) contains a truncated version of the *AtARI7* gene (*Arabidopsis thaliana ARIADNE7*) encoding a ring finger E3 ligase involved in ubiquitin-mediated protein degradation that could have a role in protein populations that alter gametophyte development (Schallau et al., 2010). In addition, the differentiation of aposporous initial cells and progression of apomixis is likely a consequence of misregulation of RNA-directed DNA methylation (RdDM) pathways, chromatin-remodeling proteins and hormonal homeostasis (Galla et al., 2019; Basso et al., 2019).

In *Paspalum*, three reproductive candidate genes have been functionally characterized (Ortiz et al., 2020). The *PN_QGJ* gene (*Paspalum notatum QUI-GON JINN*) shows homology with mitogen-activated protein kinase kinase kinases (MAP3K) and RNAi inhibition of *PN_QGJ* in the ovule of transformed plants resulted in significantly reduced formation of aposporic female gametophytes, indicating a likely role in the progression of aposporous gametophytes (Mancini et al., 2018). The *PN_TGS1*-like gene (*Paspalum notatum TRIMETHYLGUANOSINE SYNTHASE 1*), positively correlated with sexuality rates in facultative apomictic plants, influences ovule, gametophyte and possibly embryo developments (Siena et al., 2014; Colono et al., 2019). The pseudogene *PsORC3* (*Paspalum simplex ORIGIN RECOGNITION COMPLEX subunit 3*), located in the ACL, might regulate endospermogenesis in apomictic plants by trans-silencing the functional homolog *ORC3* required for endosperm formation in sexual plants (Siena et al., 2016). Other regulatory sequences, such as long non-coding RNAs or small RNAs differentially represented in sexual and apomictic flowers, are likely participating in the regulatory cascade of apomixis (Ochogavía et al., 2018; Ortiz et al., 2019).

In *Taraxacum*, the *PAR* locus and the *DIP* locus (*DIPLOSPORY*) control apomixis, both showing dominant and monogenic inheritance (van Dijk et al., 1999, 2020; van Dijk and Bakx-Schotman, 2004). While the *PAR* locus shows some suppression of recombination, the *DIP*-chromosomal region does not and lacks allelic sequence divergence, being so far the only case among all other map-based cloning efforts in apomictic species (Vijverberg et al., 2004, 2010). Hence, fine mapping defined the *DIP* locus to 0.6 cM (centi Morgan or mapping units), covering ~200–300 Kb and containing at least two *DIP* genes (Vijverberg et al., 2010), with the *Vps13* gene (Vacuolar Protein Sorting 13-like protein) being a candidate whose function has been tested and is currently the subject of a patent application (WO/2017/039452).

In other apomictic systems such as *Brachiaria brizantha* (Rodrigues et al., 2003; Silveira et al., 2012; Koehler et al., 2020), *Eragrostis curvula* (Selva et al., 2017, 2020; Garbus et al., 2019), *Panicum maximum* (Yamada-Akiyama et al., 2009) and *Ranunculus auricomus* (Pellino et al., 2013, 2020) current research has identified tissue-specific expression differences for several homologous genes encoding ARGONAUTE proteins, DNA methyltransferases, helicases, transcription factors, among others, as well as tissue-specific differences in small RNA populations, all relevant for cell fate specificity and cell maintenance and control of early germline differentiation during megasporogenesis and megagametogenesis (Olmedo-Monfil et al., 2010; Petrella et al., 2021).

9.4 APOMIXIS TECHNOLOGY: THE QUEST FOR THE MOLECULAR BASIS OF APOMIXIS

The genetic basis of apomixis is of great importance for any research aiming to induce apomixis in non-apomictic plants (Hanna, 1995; Ronceret and Vielle-Calzada, 2015).

Many genes have been characterized that are likely to play relevant roles in the activation and/or progression of apomeiosis and/or parthenogenesis in different apomictic plant systems. However, despite the enormous amount of data gathered over the last three decades, no gene or genes have been identified that can trigger a stable apomixis. A central problem in unravelling the molecular control of apomixis is that no single hypothesis can explain all experimental observations related to apomixis, its occurrence in nature, development and molecular behaviour (Hojsgaard, 2020; Scheben and Hojsgaard, 2020). Instead, three separate hypotheses are needed to explain the data, each covering a part of the facts and characteristics of asexual seed formation.

The first hypothesis, which can be traced back to the original hybridization theory of apomixis by Ernst (Ernst, 1918) based on various contemporary studies of the time (e.g., Ostenfeld, 1910; Rosenberg, 1917), suggests that apomixis is caused by developmental asynchronies and genetic deregulation of sexual developmental programmes resulting from divergent parental regulatory sequences during hybridization and/or from dosage effects and stoichiometric disbalances of macromolecular complexes during polyploidization (Koltunow, 1993; Carman, 1997; Grimanelli et al., 2001; Grossniklaus, 2001). Therefore, there is no apomixis gene but rather temporal-spatial differences in the expression of genes controlling megasporogenesis, embryony and endosperm formation.

The second hypothesis suggests that apomixis is a mutation-based anomaly involving one or a few master genes often modulated by genetic modifiers, segregation distortion, epistasis, polyploidy or aneuploidy (Mogie, 1992; Richards, 2003; Ozias-Akins and van Dijk, 2007). Mutations in at least two or three genes controlling key steps activate the developmental cascade leading to apomixis.

The third and most recent hypothesis considers apomixis to be an ancient, polyphenic and epigenetically regulated state whereby all eukaryotes retain remnant capacities for reproductive transitions (Carman et al., 2011; Albertini et al., 2019; Mateo de Arias et al., 2020). According to the polyphenism hypothesis, environmental conditions can alter the metabolic status in ovules and canalize reproductive sex-apomixis switches.

Thus, as pointed out in a recent article, each hypothesis implies a dissimilar molecular basis for apomixis and different regulatory frames for the manipulation of sexuality and the induction of apomixis in plants, either through *heterochrony*, *de novo* by mutations or by *restitutio* of an ancient polyphenic switch (Scheben and Hojsgaard, 2020). The lack of a single hypothesis on the genetic basis of apomixis relies on the lack of a proper understanding of its genomic architecture. Even when this may blur the prospect for genetically engineered apomictic crops, recent advances in the genomic basis of apomixis (e.g., Xu et al., 2021; Underwood et al., 2022) hold on the dream of apomixis breeding.

Scheben and Hojsgaard (2020) discussed the use of alternative gene editing methods, depending on the (hypothesis or) type of regulatory control and gene targets, and considered possible obstacles during the induction of apomixis in a sexual plant. Distinct methods can be used, either to modulate gene expression, to target specific gene sequences or to use allelic series for trait optimization. Several sexual genes with key roles in megasporogenesis and megagametogenesis have been targeted through various genetic modification techniques [e.g., Transgene-directed Mutagenesis, RNA-directed DNA-Methylation, or new restriction enzyme techniques (gene-editing)] aimed at inducing apomixis in sexual plants. However, the number of genes from sexual plants whose mutants display apomixis-like phenotypes is large, and a description of all cases is beyond the scope of this book. A classic example is *MiMe* (Mitosis instead of Meiosis), an *Arabidopsis* triple mutant designed to eliminate recombination and homolog segregation during the first division and omit the second division of meiosis to create non-recombined, unreduced gametes (d'Erfurth et al., 2009). *MiMe* combines knockout versions of the genes *spo11* (eliminates pairing and recombination), *rec8* (alters sister chromatid cohesion allowing chromatids to separate during the first division) and *osd1* (eliminates the second meiotic division), and thus generates apomeiotic-like female gametophytes carrying diploid clonal gametes that are then fertilized to form seeds and polyploid offspring (d'Erfurth et al., 2009). A *MiMe* phenotype has been generated in rice using a similar combination of mutants (Mieulet et al., 2016). Another recent example of sexual mutants, in this case mimicking parthenogenesis, regards haploid induction genes such as *MTL* or *MATL* (*MATRILINEAL*), a gene encoding a patatin-like phospholipase A causing sperm chromosome fragmentation and paternal genome elimination in fertilized egg cells (Gilles et al., 2017; Kelliher et al., 2017; Liu et al., 2017). Haploid induction phenotypes have been generated by knocking out the homologue in rice *OsMATL* (Yao et al., 2018), and in wheat *TaMTL* (Liu et al., 2020).

The variety of sexual mutants used to manipulate meiosis, gamete formation and parthenogenesis in attempts to synthesize apomixis in sexual plants has been reviewed elsewhere (Vijverberg et al., 2019; Hojsgaard, 2020; Ozias-Akins and Conner, 2020; Scheben and Hojsgaard, 2020; Schmidt, 2020). Current strategies to obtain a synthetic apomictic plant start from the assumption that apomixis is a mutation-based phenomenon, perhaps the only methodologically and economically affordable molecular framework at the moment, and experiment with the use of genetic engineering and gene editing methods to introduce or modify genes that are expected to have key functions in the apomixis cascade (Figure 9.1). In general, most of these mutants (Lambing et al., 2017), as well as the synthetic apomictic plants, show feeble phenotypes and low expressivity (Ozias-Akins and Conner, 2020; Scheben and Hojsgaard, 2020). For instance, the combination of *MiMe* (non-recombinant unreduced gamete formation) and *MATL* (haploid induction) mutants in rice was successful in the formation of clonal seeds, but with exceptionally low frequencies (Xie et al., 2019; Wang et al., 2019). So far, the few cases where mutants have been exploited to induce clonal seed formation by synthesizing an apomixis-like phenotype have resulted in plants with low expressivity and reduced fertility (Scheben and Hojsgaard, 2020). Induction of synthetic apomixis is still limited by its low efficiency, caused by ineffective molecular coordination between the mutated genes and the other plant genes involved in reproductive development (i.e., the products of the modified genes must interact at different regulatory levels with gene networks involved in sporogenesis and gametogenesis to form unreduced gametes, and for embryo and endosperm progression). However, the fact that distinct mutations can be combined to enhance trait expressivity by five- to sixfold, as has been observed with *MTL* mutants (e.g., Zhong et al., 2019), opens new avenues of research.

Although the results of engineering apomixis *de novo* have shown little success, the enormous progress and advances made towards inducing apomixis in sexual plants are encouraging. Relevant examples mentioned in the previous section include cases in *Cenchrus squamulatus* (Conner et al., 2017), in *Taraxacum officinale* (Underwood et al., 2022) or in *Hieracium piloselloides* (Henderson et al., 2020). While the first two cases have proven the function of apomeiosis- and

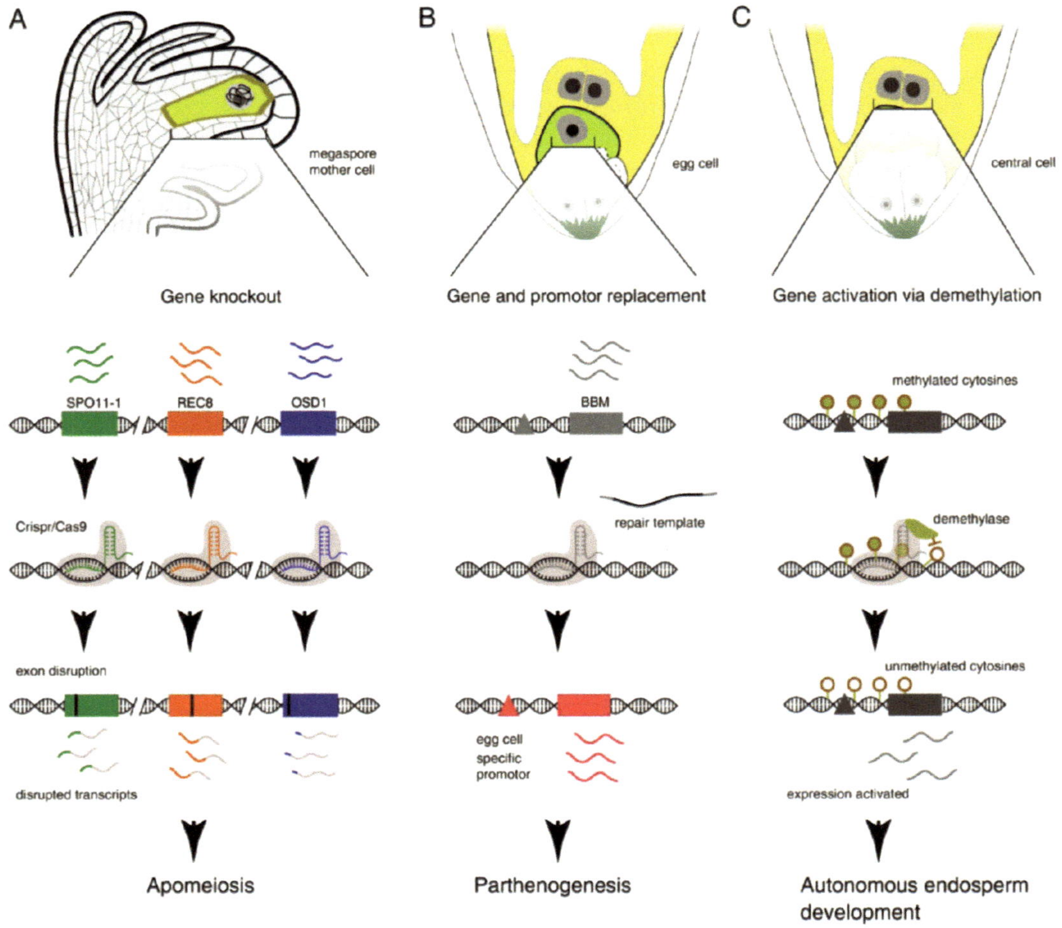

FIGURE 9.1 Alternative gene editing approaches to induce apomixis *de novo* by targeting known apomixis genes or genes having a master-like effect on the development. (A) Gene knockout can be used to skip meiosis or convert it to mitosis, as with the *MiMe* mutant. (B). Tissue-specific promoters and gene replacement can be used to induce embryo development without fertilization, as with the *BBM* gene. (C). Cas9 can be used to demethylate repressed female genes, erase genomic imprinting and trigger autonomous endosperm development. Adapted from Scheben and Hojsgaard (2020).

parthenogenesis-inducing genes in target plants, including both apomictic and sexual species, and using different genetic engineering techniques, the last two cases have also demonstrated that gene editing technology – so far only tested in sexual plants – can be used in apomictic plants obtaining high rates of gene editing.

In apomixis there are clearly no "silver bullet" inducer genes capable of mimicking natural apomixis in terms of high trait penetrance and fertility (as is regularly observed in apomictic systems). Exploitation of *de novo* apomixis will therefore rely on understanding the regulatory network and interaction of the different molecular players in natural apomixis and using it to fine-tune molecular networks in synthetic apomicts.

Upcoming genomic analyses in different plant systems will give scientists a much better view of the molecular network underlying apomixis and the observed gene expression patterns to integrate more complex models of apomixis regulation (e.g., Matzk et al., 2005) and environmental modulation (Fehér, 2015; Karunarathne et al., 2020a; Selva et al., 2020), and will expose ways to tweak specific points to increase trait expressivity and unfold a penetrant apomictic plant.

9.5 APOMIXIS FOR BACKING UP GENETIC DIVERSITY

Different studies have analysed the consequence of releasing an apomictic cultivar and the likelihood and environmental threat that an apomictic gene may spread to wild relatives and landraces of that crop (e.g., van Dijk and van Damme, 2000; Berthaud, 2001). In this case, as apomixis reduces genetic recombination rates, it may pose a risk to the conservation of genetic diversity of wild relatives and landraces, which can be avoided to some extent by manipulating residual sexuality rates and using landraces with high genetic diversity (Berthaud, 2001).

However, in nature, apomictic species may have been using apomixis to back up genetic variability. In agamic complexes, as apomixis prevents recombination and reduces gene flow to a minimum, allele evolution is expected to freeze, and asexual lineages may maintain unchanged the alleles present in the sexual population at the time of their origin. Of course, the occurrence of residual sexuality and the possible accumulation of mutations will affect the conservatism and functionality of alleles (Birky, 1996), as discussed in Chapter 8. Still, similar to the way apomixis genes are transmitted over generations and are older than asexual lineages by moving to a sexual genomic background through introgression (van Dijk, 2003; van Dijk et al., 2009), alleles for certain traits or metabolic pathways may remain conserved in long-lived asexual lineages and be available even after they have diverged or disappeared from sexual parental populations.

From a biotechnological point of view, with the advancement of molecular knowledge on apomixis and assuming the possibility of inducing apomixis in a sexual plant, another use of apomixis emerges that scientists could exploit. If a set of changes in genes that regulate meiosis and parthenogenesis could be used to induce apomixis – as discussed above – in any target species, then apomixis could function as a tool in gene banks for the multiplication and conservation of rare genotypes and genes present in natural populations that might otherwise be lost through sexual reproduction. Alternatively, it could complement or replace *in vitro* plant regeneration systems in recalcitrant species of commercial interest, such as fruit trees. Thus, by inducing apomixis in a given genotype or using a breeding scheme to capture genes of a genotype of interest, scientists could generate asexual seeds to multiply the copy number of that genotype/gene and its genetic background to back up such genetic diversity. This may be especially important for relatives of crop species and individuals from low-density or endangered plant populations.

Glossary

Brief list of common terms used in the present book.

Adventitious embryony: a synonym for sporophytic apomixis. The embryo in the seed develops from initial embryonic cells in nucellar or integumentary tissues.

Agamic complex: assemblage of sexual and apomictic cytotypes or agamospecies formed via repeated hybridization and polyploidization. An agamic complex is a polyploid complex in which some or most members reproduce by apomixis.

Agamospecies: a group of organisms that reproduce asexually and are to be considered as having a common origin.

Amphimixis: a synonym for sexual reproduction. A reproductive process in which megaspore and microspore mother cells undergo meiosis, form a reduced female gametophyte and male gametophyte, respectively, whose gametes fuse during double fertilization and give rise to a seed with both maternal and paternal contributions.

Apomixis: a reproductive process in which the seed embryo is of asexual origin, developed solely with the maternal contribution (clonal). See Chapters 1 and 2.

Apospory: a type of gametophytic apomixis where the (unreduced) female gametophyte arises from a somatic nucellar cell that acquires a developmental programme alike that of a functional megaspore.

Asynapsis: the failure of homologous chromosomes to pair or synapse during the prophase of the first meiotic division.

Autonomous apomixis: a reproductive process that develops the embryo and endosperm spontaneously, without requiring pollination and fertilization of the female gametophyte.

Desynapsis: a condition in which homologous chromosomes pair or synapse at the beginning of prophase of the first meiotic division, but then fail to maintain this association and prematurely separate in later stages of meiosis.

Dicots: a paraphyletic group of angiosperms characterized by the presence of two embryonic leaves or cotyledons in the seed.

Diploidy: a genetic constitution where a cell, organ or organism has one pair of homologous chromosomes (two sets of chromosomes).

Diplospory: a type of gametophytic apomixis where the (unreduced) female gametophyte arises from a megaspore mother cell that suppresses or modifies its meiosis.

Facultative apomixis: the occurrence of both sexual and apomictic reproductive pathways within an individual and the formation of sexual and apomictic seeds with varying frequencies.

Female gametophyte: also called embryo sac or megagametophyte. A multicellular organism that develops in the ovule embedded in the ovary of the carpel and contains the female gametes, the egg cell and central cell.

Gametogenesis: the biological process by which a reduced or unreduced (somatic) precursor (spore or spore-like) cell divides by mitosis and differentiates to form gametes (see female and male gametophytes).

Gametophytic apomixis: a reproductive process that involves the formation of a maternal seed embryo from an egg cell in an unreduced female gametophyte developed by apospory or diplospory. Fertilization may be required for the development of the endosperm.

Geographic parthenogenesis: a term that describes a phenomenon that sexual and related asexual organisms differ in their geographic distribution. Sexuals have often smaller ranges than their asexual relatives.

Haploidy: a genetic constitution where a cell, organ or organism has one set of chromosomes.
Hemizygous: a condition in an individual that has only one member of a gene, chromosome segment or chromosome pair instead of the usual two.

Integumentary embryony: a type of sporophytic apomixis in which the embryo in the seed develops from initial embryonic cells in somatic integumentary tissues.

Male gametophyte: also called pollen grain or microgametophyte. A multicellular organism that develops in the anther of the stamen and contains the male gametes or sperms.
Meiosis: a type of cell division that generates four daughter cells, each with half (reduced) the number of chromosomes and a different genetic make-up to the parent cell. Is involved in the production of spores and plant gametes.
Microspecies: used as a synonym for agamospecies.
Mitosis: a type of cell division that generates two daughter cells, each with the same number of chromosomes and genetic make-up as the parent cell. Is involved in the ordinary growth of tissues.
Monocots: a monophyletic group of angiosperms characterized by the presence of one embryonic leaf or cotyledons in the seed.

Nothotaxon: the taxonomic name for any organism that is a hybrid of natural taxa.
Nucellar embryony: a type of sporophytic apomixis in which the embryo in the seed develops from initial embryonic cells in somatic nucellar tissue.

Obligate apomixis: the occurrence of (only) apomictic reproductive pathways within an individual and the exclusive formation of apomictic seeds.

Parthenogenesis: the development of the seed embryo from an unfertilized egg cell.
Polyphenism: a phenomenon in which two or more distinct phenotypes are produced by the same genotype depending on environmental conditions.
Polyploid complex: a group of distinct and interrelated sexual cytotypes, partially associated or connected by the recurrent exchange of genetic material, while their respective gene pools retain some level of genetic cohesion. When the different cytotypes belong to one species, they form autopolyploid complexes. When cytotypes of more than one species are involved, they form allopolyploid complexes.
Polyploidy: a genetic constitution where a cell, organ or organism has more than one pair of homologous chromosomes (more than two sets of chromosomes).
Pseudogamous apomixis or pseudogamy: a reproductive process that requires pollination and fertilization of the central cell of the female gametophyte to form the endosperm and trigger parthenogenesis in the egg cell during the formation of a seed.

Sporogenesis: the biological process by which a somatic (often diploid) precursor (archespore) cell divides by meiosis to form reduced (often haploid) spores.
Sporophytic apomixis: a reproductive process that involves the formation of an embryo directly from one or more somatic cells of the ovule, which attaches to the developing meiotic female gametophyte and gives rise to a seed carrying maternal embryos.

References

Abdel-Haleem, H., Z. Luo and D. Ray. 2019. Genetic improvement of guayule (*Parthenium argentatum* A. Gray): An alternative rubber crop. In: Al-Khayri J., Jain S., Johnson D. (eds.), *Advances in Plant Breeding Strategies: Industrial and Food Crops*. Springer, Cham. doi:10.1007/978-3-030-23265-8_6

Abe, K. 1972. Contributions to the embryology of the family Orchidaceae, VI. Development of the embryo sac in 15 species of orchids. *Sci. Rep. Tûhoku Univer.* 4, *Biol.* 36: 135–178.

Acquaah, G. 2012. *Principles of plant genetics and breeding*. 2nd ed.; Wiley-Blackwell, Chichester, UK.

Acuña, C.A., E.J. Martínez and C.L. Quarin. 2005. Sexual diploid and apomictic tetraploid races in *Thrasya petrosa* (Gramineae). *Aust. J. Bot.* 53: 479–484. doi:10.1071/BT04190

Acuña, C.A., E.J. Martínez, A.L. Zilli, E.A. Brugnoli, F. Espinoza, F. Marcón, M.H. Urbani and C.L. Quarin. 2019. Reproductive systems in *Paspalum*: Relevance for germplasm collection and conservation, breeding techniques, and adoption of released cultivars. *Front. Plant Sci.* 10: 1377. doi:10.3389/fpls.2019.01377

Afzelius, K. 1928. Die Embryobildung bei *Nigritella nigra*. *Svensk Bot. Tidskr.* 22: 82–91.

Afzelius, K. 1932. Zur Kenntnis der Fortpflanzungsverhältnisse und Chromosomezahlen bei *Nigritella nigra*. *Svensk. Bot. Tidskr.* 26: 365–369.

Afzelius, K. 1936. Apomixis in der Gattung *Arnica*. *Svensk Bot. Tidskr.* 30: 572–679.

Afzelius, K. 1959. Apomixis and polyembryony in *Zygopetalum mackayi* Hook. *Acta Horti Berg.* 19: 7–13.

Agafonov, A.V., N.B. Sukhareva, S.O. Baturin and O.A. Struzhkina. 2004. Identification of apomixis in the Kentucky bluegrass (*Poa pratensis* L.) using SDS-PAGE of endosperm storage proteins. *Biol. Bull. Russian Acad. Sci.* 31: 36–41.

Agrawal, J.S. and S.P. Singh. 1957. Nucellar polyembryony in *Momordica charantia* Linn. *Sci. Cult.* 22(ll): 630–631.

Aimé, S. and P. Roiron. 1985. Perennial species of the genus *Limonium* in Oran (Algeria). I. Taxonomy and distribution. *Candollea* 40: 409–424.

Åkerberg, E. 1941. Cytogenetic studies in *Poa pratensis* and its hybrid with *Poa alpina*. *Hereditas* 28: 1–126.

Akhter, S., T. Morita and Y. Yoshida. 1993. Clonal diversity in the agamospermous polyploids of *Taraxacum hondoense* in northern Honshu, Japan. *J. Plant Res.* 106, 167–179. doi:10.1007/BF02344420

Akiyama, Y., J.A. Conner, S. Goel, D.T. Morishige, J.E. Mullet, W.W. Hanna and P. Ozias-Akins. 2004. High-resolution physical mapping in *Pennisetum squamulatum* reveals extensive chromosomal heteromorphism of the genomic region associated with apomixis. *Plant Physiol.* 134: 1733–1741.

Akiyama, Y., W.W. Hanna and P. Ozias-Akins. 2005. High-resolution physical mapping reveals that the apospory-specific genomic region (ASGR) in *Cenchrus ciliaris* is located on a heterochromatic and hemizygous region of a single chromosome. *Theor. Appl. Genet.* 111: 1042–1051.

Albertini, E., G. Barcaccia, J.G. Carman and F. Pupilli. 2019. Did apomixis evolve from sex or was it the other way around? *J. Exp. Bot.* 70: 2951–2964. doi:10.1093/jxb/erz109

Albertini, E., G. Barcaccia, A. Porceddu, S. Sorbolini and M. Falcinelli. 2001b. Mode of reproduction is detected by Parth1 and Sex1 SCAR markers in a wide range of facultative apomictic Kentucky bluegrass varieties. *Mol. Breeding* 7: 293–300.

Albertini, E., G. Marconi, G. Barcaccia, L. Raggi and M. Falcinelli. 2004. Isolation of candidate genes for apomixis in *Poa pratensis* L. *Plant Mol. Biol.* 56: 879–894.

Albertini, E., G. Marconi, L. Reale, G. Barcaccia, A. Porceddu, F. Ferranti and M. Falcinelli. 2005. *SERK* and *APOSTART*. Candidate genes for apomixis in *Poa pratensis*. *Plant Physiol.* 138: 2185–2199.

Albertini, E., A. Porceddu, F. Ferranti, L. Reale, G. Barcaccia, B. Romano and M. Falcinelli. 2001a. Apospory and parthenogenesis may be uncoupled in *Poa pratensis*: a cytological investigation. *Sex. Plant Reprod.* 14: 213–217.

Alexander, P.J., M.D. Windham, J.B. Beck, I.A. Al-Shehbaz, L. Allphin and C.D. Bailey. 2015. Weaving a tangled web: Divergent and reticulate speciation in *Boechera fendleri* sensu lato (Brassicaceae: Boechereae). *Syst. Bot.* 40(2): 572–596.

Aleza, P., J. Juárez, P. Ollitrault and L. Navarro. 2010. Polyembryony in non-apomictic citrus genotypes. *Ann. Bot.* 106: 533–545.

Algan, G. and H.N.B. Büyükkartal. 1999. Apomictic development in tetraploid Red Clover (*Trifolium pratense* L.). *Turk. J. Agric. For.* 23: 519–525.

Alice, L.A., D.H. Goldman, J.A. Macklin and G. Moore. 2015. *Rubus*. In: *Flora of North America. North of Mexico*, Vol. 9, Oxford University Press, New York and Oxford.

Aliyu, O.M., M.E. Schranz and T.F. Sharbel. 2010. Quantitative variation for apomictic reproduction in the genus *Boechera* (Brassicaceae). *Am. J. Bot.* 97(10):1719–1731. doi:10.3732/ajb.10001888

Alkemade, R., M. van Oorschot, L. Miles, C. Nellemann, M. Bakkenes and B. ten Brink. 2009. A framework to investigate options for reducing global terrestrial biodiversity loss. *Ecosystems*, 12: 374–390.

Allem, A.C. 1991. Estudo da biologia reprodutiva de duas espécies florestais (aroeira e gonçalo-alves) da região do cerrado. *Cenargen* 2: 1–5 (Pesquisa em andamento=ongoing research).

Almeda, F. and T.I. Chuang. 1992. Chromosome numbers and their systematic significance in some Mexican Melastomataceae. *Syst. Bot.* 17: 583–593.

Alonso-Marcos, H., F.D. Nardi, S. Scheffknecht, A. Tribsch, K. Hülber and C. Dobeš. 2019. Difference in reproductive mode rather than ploidy explains niche differentiation in sympatric sexual and apomictic populations of *Potentilla puberula*. *Ecol. Evol.* 9(6): 3588–3598. doi:10.1002/ece3.4992

Alves, E.R., V.T.C. Carneiro and A.C.G. Araujo. 2001. Direct evidence of pseudogamy in an apomictic *Brachiaria brizantha* (Poaceae). *Sex. Plant Reprod.* 14: 207–212.

Alves, M.F., M.O. Duarte, N.S. Bittencourt, P.E. Oliveira and D.S. Sampaio. 2016. Sporophytic apomixis in polyembryonic *Handroanthus serratifolius* (Vahl) S.O. Grose (Bignoniaceae) characterizes the species as an agamic polyploid complex. *Plant Syst. Evol.* 302(6): 651–659.

Ambrosi, D.G., G. Galla, P. Marina and B. Tommaso. 2010. DNA markers and FCSS analyses shed light on the genetic diversity and reproductive strategy of *Jatropha curcas* L. *Diversity* 2(5): doi:10.3390/d2050810

Amiteye, S., J.M. Corral, H. Vogel, M. Kuhlmann, M.F. Mette and T.F. Sharbel. 2013. Novel microRNAs and microsatellite-like small RNAs in sexual and apomictic *Boechera* species. *MicroRNA* 2: 45–62.

Amiteye, S., J.M. Corral, H. Vogel and T.F. Sharbel. 2011. Analysis of conserved microRNAs in floral tissues of sexual and apomictic *Boechera* species. *BMC Genomics* 12: 500.

Amsellem, L., J.L. Noyer and M. Hossaert-Mckey. 2001. Evidence for a switch in the reproductive biology of *Rubus alceifolius* (Rosaceae) towards apomixis, between its native range and its area of introduction. *Am. J. Bot.* 88: 2243–2251.

Amsellem, L., T. Pailler, J-L. Noyer and M. Hossaert-McKey. 2002. Characterization of pseudogamous apospory in the reproductive biology of the invasive weed *Rubus alceifolius* (Rosaceae) in its area of introduction. *Acta Bot. Gall.* 149: 217–224.

Anderegg, L.D.L., X. Loy, I.P. Markham, C.M. Elmer, M.J. Hovenden, J. HilleRisLambers and M.M. Mayfield. 2021. Aridity drives coordinated trait shifts but not decreased trait variance across the geographic range of eight Australian trees. *New Phytol.* 229: 1375–1387. doi:10.1111/nph.16795

Andersen, A.M. 1927. Development of the female gametophyte and caryopsis of *Poa pratensis* and *Poa compressa*. *J. Agric. Res.* 34(11): 1001–1018.

Andersson, A. 1931. Studien über die Embryonalentwicklung der Familien Celastraceae, Oleaceae und Apocynaceae. *Acta Univ. Lund.* 27: 1–110.

Andrade-Rodríguez, M., A. Villegas-Monter, G. Carrillo-Castañeda and A. García-Velázquez. 2004. Polyembryony and identification of Volkamerian lemon zygotic and nucellar seedlings using RAPD. *Pesqui. Agropecu. Bras.* 39(6): 551–559. doi:10.1590/S0100-204X2004000600006

Andrade-Rodríguez, M., Á. Villegas-Monter, A. Gutiérrez Espinoza, G. Carrillo-Castañeda and A. García-Velázquez. 2005. Polyembryony and RAPD markers for identification of zygotic and nucellar seedlings in *Citrus*. *Agrociencia* 39: 371–383.

Andrew, R.L., J.T. Miller, R. Peakall, M.D. Crisp and R.J. Bayer. 2003. Genetic, cytogenetic and morphological patterns in a mixed mulga population: Evidence for apomixis. *Aust. Syst. Bot.* 16: 69–80. doi:10.1071/SB01043

The Angiosperm Phylogeny Group, M.W. Chase, M.J.M. Christenhusz, M.F. Fay, J.W. Byng, W.S. Judd, D.E. Soltis, D.J. Mabberley, A.N. Sennikov, P.S. Soltis, P.F. Stevens. 2016. An update of the Angiosperm Phylogeny Group classification for the orders and families of flowering plants: APG IV. *Bot. J. Lin. Soc.* 181(1): 1–20. doi:10.1111/boj.12385

Ansari, R. 1977. Observations on the occurrence of polyembryony in two species of orchids. *Curr. Sci.* 46(17): 607–607. http://www.jstor.org/stable/24098414

Aparicio, A. 2005. *Limonium silvestrei* (Plumbaginaceae), a new agamospecies from southern Spain. *Ann. Bot. Fenn.* 42(5): 371–377. http://www.jstor.org/stable/23726732

Arakaki, M., P. Speranza, P.S. Soltis and D.E. Soltis. 2013. Genetic variability of an unusual apomictic triploid cactus—*Haageocereus tenuis* Ritter—from the coast of central Peru. *J. Heredity* 104(1): 127–133. doi:10.1093/jhered/ess072

Araujo, A.C.G., S. Mukhambetzhanov, M.T. Pozzobon, E.F. Santana and V.T.C. Carniero. 2000. Female gametophyte development in apomictic and sexual *Brachiaria brizantha* (Poaceae). *Rev. Cytol. Biol. Veg. Bot.* 23: 13–28.

Araujo, A.C.G., J.M. Nobrega, M.T. Pozzobon and V.T.C. Carneiro. 2005. Evidence of sexuality in *Brachiaria brizantha* (Poaceae) artificial tetraploid plants and progeny. *Euphytica* 144: 39–50.
Archibald, E.E.A. 1939. The development of ovule and seed of jointed cactus (*Opuntia aurantiaca* Lindley). *S. Afr. J. Sci.* 36: 195–211.
Arias, M., J. Herrero, M. Ricobaraza, M. Hernández and E. Ritter. 2016. Evaluation of root biomass, rubber and inulin contents in nine *Taraxacum koksaghyz* Rodin populations. *Ind. Crops Prod.* 83: 316–321. doi:10.1016/j.indcrop.2016.01.023
Armenta-Medina, A. and C.S. Gilmor. 2019. Genetic, molecular and parent-of-origin regulation of early embryogenesis in flowering plants. *Curr. Top. Dev. Biol.* 131: 497–543.
Armstrong, J.M. 1937. A cytological study of the genus *Poa*. *Canad. J. Res. C* 15: 281–295.
Arrigo, N. and M.S. Barker. 2012. Rarely successful polyploids and their legacy in plant genomes. *Curr. Opin. Plant Biol.* 15: 140–146. doi:10.1016/j.pbi.2012.03.010
Artelari, R. 1989a. Biosystematic study of the genus *Limonium* (Plumbaginaceae) in the Aegean area (Greece). I. Some *Limonium* species from the Kikladhes islands. *Willdenowia* 18: 399–408.
Artelari, R. 1989b. *Limonium creticum* (Plumbaginaceae), a new species from Kriti island (Aegean sea) Greece.*Candollea* 44: 415–420.
Artelari, R. 1989c. Biosystematic study of the genus *Limonium* (Plumbaginaceae) in the Aegean area (Greece). II. *Limonium hierapterae* RECH. fil. from Kriti island. *Webbia* 43: 33–40.
Artelari, R. and O. Georgiou. 1999. Two new species of *Limonium* (Plumbaginaceae) from the island of Kithira (Greece). *Bot. J. Linn. Soc.* 131: 399–415.
Asadian, G. and K. Pieber. 2005. Morphological variations in walnut varieties of the Mediterranean regions. *Int. J. Agr. Biol.* 7: 71–73.
Asbaganov, S., A. Lokteva and Z. Rupasova. 2020. New cultivar of mountain ash (*Sorbus*) Siberian breeding. *BIO Web Conf.* 24: 00003. doi:10.1051/bioconf/20202400003
Asbaganov, S.V., E.V. Kobozeva and A.V. Agafonov. 2015. Application of the electrophoresis of cotyledon storage protein and ISSR-markers to the identification of hybrids between *Sorbus sibirica* Hedl. and *Sorbocotoneaster pozdnjakovii* Pojark. *Russ. J. Genet. Appl. Res.* 5: 33–40. doi:10.1134/S2079059715010013
Ashari, S., D. Aspinall and M. Sedgley. 1988. Discrimination of zygotic and nucellar seedlings of five polyembryonic citrus rootstocks by isozyme analysis and seedling morphology. *J. Am. Soc. Hortic. Sci.* 63: 695–703
Asker, S. 1967. Induced sexuality after chromosome doubling in an apomictic *Potentilla argentea* biotype. *Hereditas* 57: 339–342.
Asker, S. 1970a. Apomictic biotypes in *Potentilla intermedia* and *P. norvegica*. *Hereditas* 66: 101–108.
Asker, S. 1970b. Apomixis and sexuality in the *Potentilla argentea* complex: I. Crosses within other species. *Hereditas* 66: 127–144.
Asker, S. 1970c. Apomixis and sexuality in the *Potentilla argentea* complex: II. Crosses within the complex. *Hereditas* 66: 189–204.
Asker, S. 1971. Apomixis and sexuality in the *Potentilla argentea* complex: III. Euploid and aneuploid derivatives (including trisomics) of some apomictic biotypes. *Hereditas* 67: 111–142.
Asker, S. 1985. Polymorphism of *Potentilla tabernaemontani* and related taxa on Gotland. *Hereditas* 102: 39–45.
Asker, S. 1986. Variation in some apomictic Nordic *Potentilla* species. *Symb. Bot. Upsal.* 27: 199–205.
Asker, S.E. 1980. Gametophytic apomixis: Elements and genetic regulation. *Hereditas* 93: 277–293.
Asker, S.E. and L. Jerling. 1992. *Apomixis in Plants*. Boca Raton, FL: CRC Press, pp. 298.
Asplund, E. 1928. Eine neue Balanophoreen-Gattung aus Bolivien. *Svensk Bot. Tidskr.* 22: 261–277.
Astegiano, M.E., A.M. Anton and H.E. Connor 1995. Sinopsis del género *Cortaderia* (Poaceae) en Argentina. *Darwiniana* 33(1/4): 43–51. http://www.jstor.org/stable/23222990
Åström, H. and C-A. Haeggström. 2004. Generative reproduction in *Allium oleraceum* (Alliaceae). *Ann. Bot. Fenn.* 41: 1–14.
Ather-uz-Zaman, J.M. Al-Khayri and Islam R. 2018. Genetic improvement of strawberry (*Fragaria × ananassa* Duchesne). In: Al-Khayri J., Jain S., Johnson D. (eds.), *Advances in Plant Breeding Strategies: Fruits*. Springer, Cham. doi:10.1007/978-3-319-91944-7_6
Azevedo, L.B. 2014. *Developmed and application of stressor – response relationships of nutrients*. PhD thesis, Radboud University Nijmegen, Nijmegen, the Netherlands.
van Baarlen, P., H.J. de Jong and P.J. van Dijk. 2002. Comparative cyto-embryological investigations of sexual and apomictic dandelions (*Taraxacum*) and their apomictic hybrids. *Sex. Plant Reprod.* 15: 31–38.

Babcock, E.B. and G.L. Stebbins. 1938. The American species of *Crepis*: Their interrelationships and distribution as affected by polyploidy and apomixis. Carnegie Institution of Washington Publication, no. 504, Washington, DC.

Bacchi, O. 1943. Cytological observations in *Citrus*: III. Megasporogenesis, fertilization, and polyembryony. *Bot. Gaz.* 105(2): 221–225.

Bailey, L.H. 1941–1945. Species Batorum I-VIII. *Gentes Herbarum* 5: 1–918.

Baker, H.G. 1953. Dimorphism and monomorphism in the Plumbaginaceae II. Pollen and stigmata in the genus *Limonium*. *Ann. Bot.* (n.s.) 17: 433–445.

Baker, H.G. 1955. Self-Compatibility and establishment after 'Long-Distance' dispersal. *Evolution* 9: 347–349. doi:10.2307/2405656

Baker, H.G. 1960. Apomixis and polyembryony in *Pachira oleaginea* (Bombacaceae). *Am. J. Bot.* 47: 296–302.

Baker, H.G. 1966. The evolution, functioning and breakdown of heteromorphic incompatibility systems. I. the Plumbaginaceae. *Evolution* 20: 349–368. doi:10.2307/2406635

Baker, H.G. 1967. Support for Baker's law-as a rule. *Evolution* 21: 853–856. doi:10.1111/j.1558-5646.1967.tb03440.x

Baldjiev, G. and P. Robeva-Davidova, 2008. Female gametophyte, embryo- and endospermogenesis in *Alchemilla glabra* (Rosaceae: Rosoideae). *Compt. Rend. Acad. Bulg. Sci., Biologie, Botany* 61(8): 1007–1012.

Bambacioni, V. and A. Giombini. 1930. Sullo sviluppo del gametofito femminile in *Tulipa gesneriana* L. *Ann. Bot.* (Rome) 18: 373–386.

Bambacioni-Mezzetti, V. 1932. Sullo sviluppo dell'embrione in *Tulipa gesneriana* L. *Ann. Bot.* (Rome) 19(1): 145.

Banerjee, I. and S. Pal. 1958. Studies in the embryology of *Feronia limonia* Swingle. *Bull. Bol. Soc. Bengal* 12: 18–28.

Banerji, I. 1954. Morphological and cytological studies on *Citrus grandis*. *Phytomorphology* 4: 390–396.

Banerji, E.A. and P.N. Bhaduri. 1933. Polyembryony in Solanaceae. *Curr. Sci.* 1: 310–311.

Bantin, J., F. Matzk and T. Dresselhaus. 2001. *Tripsacum dactyloides* (Poaceae): A natural model system to study parthenogenesis. *Sex. Plant Reprod.* 14: 219–226. doi:10.1007/s00497-001-0119-z

Baranec, T. 1992. *Cotoneaster* Medicus. In: Bertová L (ed.), *Flóra Slovenska* IV/3. Veda, Bratislava, pp. 452–462.

Barcaccia, G., F. Arzenton, T.F. Sharbel, S. Varotto, P. Parrini and M. Lucchin. 2006. Genetic diversity and reproductive biology in ecotypes of the facultative apomict *Hypericum perforatum* L. *Heredity* 96: 322–334.

Barcaccia, G., H. Bäumlein and T. Sharbel. 2007. Apomixis in St. John's wort (*Hypericum perforatum* L.): an overview and glimpse towards the future. In: Hörandl, E., U. Grossniklaus, P. van Dijk, T. Sharbel (eds.), *Apomixis: Evolution, mechanisms and perspectives*. Gantner Verlag, Ruggell, Liechtenstein, pp. 259–280.

Barcaccia, G., A. Mazzucato, E. Albertini, J. Zethof, A. Gerats, M. Pezzotti and M. Falcinelli. 1998. Inheritance of parthenogenesis in *Poa pratensis* L: Auxin test and AFLP linkage analyses support monogenic control. *Theor. Appl. Genet.* 97: 74–82.

Barcaccia, G., A. Mazzucato, A. Belardinelli, M. Pezzotti, S. Lucretti and M. Falcinelli. 1997. Inheritance of parental genomes in progenies of *Poa pratensis* L. from sexual and apomictic genotypes as assessed by RAPD markers and flow cytometry. *Theor. Appl. Genet.* 95: 516–524.

Barcaccia, G., A. Mazzucato, M. Falcinelli and F. Veronesi. 1996. Callose localization in cell wall during meiotic and apomeiotic megasporogenesis in diploid alfalfa (*Medicago* sp.). *Caryologia* 49: 45–56.

Barcaccia, G., A. Mazzucato, M. Pezzotti and M. Falcinelli. 1994. Comparison between isozyme and RAPD analyses to screen aberrant plants in *Poa pratensis* L. progenies. *Apomixis Newsl.* 7: 29–30.

Barcaccia, G., F. Palumbo, S. Sgorbati, E. Albertini and F. Pupilli. 2020. A reappraisal of the evolutionary and developmental pathway of apomixis and its genetic control in angiosperms. *Genes* 11(8): 859. doi:10.3390/genes11080859

Barke, H., M. Daubert and E. Hörandl. 2018. Establishment of apomixis in diploid F2 hybrids and inheritance of apospory from F1 to F2 hybrids of the *Ranunculus auricomus* complex. *Front. Plant Sci.* 9:1111. doi:10.3389/fpls.2018.01111

Barker, M.S., H. Vogel and M.E. Schranz. 2009. Paleopolyploidy in the Brassicales: analyses of the *Cleome* transcriptome elucidate the history of genome duplications in *Arabidopsis* and other Brassicales. *Genome Biol. Evol.* 1: 391–399. doi:10.1093/gbe/evp040

Barlow, B.A. 1958a. A stage in the evolution of apomixis in *Casuarina*. *Aust. J. Sci.* 21: 108.

Barlow, B.A. 1958b. Heteroploid twins and apomixis in *Casuarina nana* Sieb. *Aust. J. Bot.* 6: 204–219.

Barlow, B.A. 1959. Polyploidy and apomixis in the *Casuarina distyla* species group. *Aust. J. Bot.* 7: 238–251.

Barocka, K.H. 1966. Die Sektion *Corollinae* der Gattung *Beta* (Tournef.) L. *Zeitschrift für Pflanzenzüchtung* 56: 379–388.

Baroux, C. and U. Grossniklaus. 2019. Seeds-An evolutionary innovation underlying reproductive success in flowering plants. *Curr. Top. Dev. Biol.* 131: 606–642.

Bartish, I.V., B. Hylmö and H. Nybom. 2001. RAPD analysis of interspecific relationships in presumably apomictic *Cotoneaster* species. *Euphytica* 120: 273–280. doi:10.1023/A:1017585600386

Basappa, G.P. and M. Muniyamma. 1981. Reproduction in two species of *Arundinella* Raddi (Poaceae). *Proc. Indian Acad. Sci. (Plant Sci.)* 90: 477–483.

Bashaw, E.C. and E.C. Holt. 1958. Megasporogenesis, embryo sac development and embryogenesis in Dallis grass, *Paspalum dilatatum. Agron. J.* (USA) 50: 753–756.

Bashaw, E.C., A.W. Howin and E.C. Holt. 1970. Apomixis, its evolutionary significance and utilization in plant breeding. *Proc.11th Int. Grasslnd Congress*, Queensland, Australia, Queensland University Press, pp. 245–248.

Bashaw, E.C., M.A. Hussey and K.W. Hignight. 1992. Hybridization (n+n and 2n+n) of facultative apomictic species in the *Pennisetum* agamic complex. *Int. J. Plant Sci.* 153: 466–470.

Basso, A., G. Barcaccia and G. Galla. 2019. Annotation and expression of IDN2-like and FDM-like genes in sexual and aposporous *Hypericum perforatum* L. accessions. *Plants* 8(6): 158. doi:10.3390/plants8060158

Bates, R. 1983. Apomixis in two South Australian orchids (*Microtis orbicularis* R. Rogers and *Paracaleana minor* (R.Br.) Blaxell). *J. Nat. Orch. Soc. South Australia* 7: 35–36.

Bates, R. 1984. The genus *Microtis* R.Br. (Orchidaceae): A taxonomic revision with notes on Biology. *J. Adelide Bot. Gard.* 7(1): 45–89.

Battaglia, E. 1945. Fenomini citologici nuovi nella embriogenesi (Semigamia) e nella microsporogenesis (doppio nucleo di retituzioni) di *Rudbeckia laciniata. N.G. Bot. Ital.* 52: 34–38.

Battaglia, E. 1946a. Ricerche cariologiche ed embriologiche sul genere *Rudbeckia*. 1-5. Il gametofito femminile e maschile di *R. bicolour, R. hirta, R. hirta* var. *meine feude, R. amplexicaule* e *R. purpurea* (*Echinacea purpurea*). *N.G. Bot. Ital.* 53: 1–26.

Battaglia, E. 1946b. Richerche cariologiche ed embriologiche sul genere *Rudbeckia*. 6. Apomissia in *R. speciosa. N.G. Bot. Ital.* 53: 27–69.

Battaglia, E. 1946c. Richerche cariologiche ed embriologiche sul genere *Rudbeckia*. 7. Apomissia in *R. laciniata* ed amfimissia nella sua varieta a fiori doppi. *N.G. Bot. Ital.* 53: 437–482.

Battaglia, E. 1946d. Richerche cariologiche ed embriologiche sul genere *Rudbeckia*. 8. Semigamia in *R. laciniata. N.G. Bot. Ital.* 53: 483–511.

Battaglia, E. 1947a. Richerche cariologiche ed embriologiche sul genere *Rudbeckia*. 9. Le anomalie del gametofito femminile cellularizzato di *R. laciniata. N.G. Bot. Ital.* 54: 377–405.

Battaglia, E. 1947b. Richerche cariologiche ed embriologiche sul genere *Rudbeckia*. 11. Semigamia in *R. speciosa. N.G. Bot. Ital.* 54: 531–559.

Battaglia, E. 1947c. La "Semigamia" singolare compartamento denucleo spermatico nella uova diploidi delle specie apomittiche del genere *Rudbeckia* (Asteraceae) e consequente embriogenesi di tipi chimerico. *C.R. Accad. Naz. Lincei* 8: 63–67.

Battaglia, E. 1947d. Apomissia *Hieracium ramosum. Atti Soc. Tosc. Sci. Nat.* (Pisa) B 54: 50–69.

Battaglia, E. 1948. Ricerche sulla parameiosi restituzionale nel genere *Taraxacum. Caryologia* 1: 1–47.

Battaglia, E. 1949. L'alterazione della meiosi nella riproduzione apomittica di *Chondrilla juncea. Caryologia* 2: 23–30.

Battaglia, E. 1950. L'alterazione della meiosi nella riproduzione apomittica di *Erigeron karwinskianus* var. *mucronatus. Caryologia* 2: 165–204.

Battaglia, E. 1952. Nuovo reperti di apomissia nel genere *Rudbeckia* (Compositae). *Atti Soc. Tosc. Sci. Nat.* (Pisa) B 59: 205–209.

Battaglia, E. 1955. Unusual cytological features in the apomictic *Rudbeckia sullivantii. Caryologia* 8: 1–32.

Battaglia, E. 1963. *Apomixis*. In Maheshwari, P. (ed.), *Recent Advances in the Embryology of Angiosperms*. International Society of Plant Morphologists, University of Delhi, Delhi, pp. 221–264.

Battaglia, E. 1981. Embryological questions: 3. Semigamy, hemigamy and gynandroembryony. *Ann. Bot.* (Rome) 39: 173–175.

Battaglia, E. 1989. Embryological questions. 14. The evolution of the female gametophyte of angiosperms: an interpretive key. *Ann. Bot.* (Rome) 47: 7–144.

Battaglia, E. 1991. Embryological questions: 16. Unreduced embryo sacs and related problems in Angiosperms (apomixis, cyclosis, cellularization…). *Atti Soc. Tosc. Sci. Nat. Mem. Serie B* 98: 1–134.

Battjes J., S.B.J. Menken and H.C.M. den Nijs. 1992. Clonal diversity in some microspecies of *Taraxacum* sect. *Palustria* (Lindb. fil.) Dahlst. from Czechoslovakia. *Bot. Jahrb. Syst.* 114: 315–328.

Baturin, S.O. 1997. Experimental apomixis in garden strawberry (*Fragaria* × *ananassa* Duch.). (Eksperimental'nyj apomiksis u sadovoj zemlyaniki (*Fragaria* × *ananassa* Duch.)): Extended Abstract of Thesis of cand. Dissertation, Novosibirsk, Russia, 16 pages. (in Russian).

Baturin, S.O. 2007. Diploid parthenogenesis induction in *Fragaria vesca* L. *Sibirskii Ekol. Zh.* 6: 987–991.

Baturin S.O. 2009. Agamospermy in outbred crosses of *Fragaria vesca* L. (Rosaceae). *Russ. J. Genet.* 45: 322–326.

Batygina, B.T., E.A. Bragina and V.E. Vasilyeva. 2003. The reproductive system and germination in orchids. *Acta Biol. Crac. Series Bot.* 45(2): 21–34.

Batygina, T.B. 1991. Nucellar embryoidogeny in *Poa pratensis* (Poaceae), *Polish. Bot. Stud.* 2: 121–125.

Batygina, T.B. 2002. (ed.), *Embryology of flowering plants: Terminology and concepts, Vol. 1. Generative organs of flower.* CRC Press, Boca Raton, Florida, 442 pages.

Batygina, T.B. 2005. (ed.), *Embryology of flowering plants: Terminology and concepts, Vol. 2. The seed.* CRC Press, Boca Raton, Florida, 810 pages.

Batygina, T.B. 2009. (ed.), *Embryology of flowering plants: Terminology and concepts, Vol. 3. Reproductive systems.* CRC Press, Boca Raton, Florida, 598 pages

Batygina, T.B. and T.E. Freiberg. 1979. Polyembryony in *Poa pratensis* L. *Bot. Zhu.* 6: 793–804.

Batygina, T.B. and G.Y. Vinogradova 2007. Phenomenon of polyembryony. Genetic heterogeneity of seeds. *Russ. J. Dev. Biol.* 38(3): 126–151. doi:10.1134/S1062360407030022

Bayer, R.J. 1991. Allozymic and morphological variation in *Antennaria*. Asteraceae: Inuleae. From the low arctic of north-western North America. *Syst. Bot.* 16: 492–506.

Bayer, R.J. 1997. *Antennaria rosea* (Asteraceae) – a model group for the study of the evolution of polyploid agamic complexes. *Opera Bot.* 132: 53–65.

Bayer, R.J. and T.M. Minish. 1993. Isozyme variation, ecology and phytogeography of *Antennaria soliceps* (Asteraceae: Inuleae), an alpine apomict from the Spring mountains, Nevada. *Madroño* 40(2): 75–89.

Bayer, R.J., K. Ritland and B.G. Purdy. 1990. Evidence of partial apomixis in *Antennaria media* (Asteraceae: Inuleae) detected by the segregation of genetic markers. *Am. J. Bot.* 77(8): 1078–1083.

Bayer, R.J. and G.L. Stebbins.1983. Distribution of sexual and apomictic populations of *Antennaria parlinii. Evolution* 37: 555–561.

Bayer, R.J. and G.L. Stebbins. 1987. Chromosome numbers, patterns of distribution, and apomixis in *Antennaria* (Asteraceae: Inuleae). *Syst. Bot.* 12: 305–319.

Beals, G.R. 1970. A study of apomixis in *Erigeron karwinskianums* (Compositae: Astereae). PhD thesis, University of Cincinatti, OH, USA, 60 pages.

Beaman, J.H. 1957. The sytematics and evolution of *Townsendia* (Compositae). *Contr. Gray Herb.* 183: 87–134.

Beck, J.B., P.J. Alexander, L. Allphin, I.A. Al-Shehbaz, C. Rushworth, C.D. Bailey and M.D. Windham. 2012. Does hybridization drive the transition to asexuality in diploid *Boechera*? *Evolution* 66: 985–995. doi:10.1111/j.1558-5646.2011.01507.x

te Beest, M., J.J. Le Roux, D.M. Richardson, A.K. Brysting, J. Suda, M. Kubesova, and P. Pysek. 2012. The more the better? The role of polyploidy in facilitating plant invasions. *Ann. Bot.* 109: 19–45.

Beilstein M.A., N.S. Nagalingum, M.D. Clements, S.R. Manchester and S. Mathews. 2010. Dated molecular phylogenies indicate a Miocene origin for *Arabidopsis thaliana. Proc. Natl. Acad. Sci. U.S.A.* 107: 18724–18728.

Bennett, H.W. 1944. Embryology of *Paspalum dilatatum. Bot. Gaz.* 106: 40–45.

Bennett, H.W. and E.C. Bashaw. 1966. Interspecific hybridization with *Paspalum* spp. *Crop Sci.* 6: 52–58.

Berger, X. 1953. Untersuchungen über die Embryologie partiell apomiktischer *Rubus*-Arten. *Ber. Schweiz. Bot. Ges.* 63: 224–266.

Bergman, B. 1932. Aposporie bei *Picris hieracioides* and *Leontodon hispidus. Svensk Bot. Tidskr.* 26: 453–457.

Bergman, B. 1935a. Zytologische Studien über sexualle und asexualle *Hieracium umbellatum. Hereditas* 20: 47–64.

Bergman, B. 1935b. Zytologische Studien über die Fortflanzung bei den Gattungen *Leontodon* und *Picris. Svensk Bot. Tidskr.* 29: 155–301.

Bergman, B. 1935c. Zur Kenntnis der Zytologie der Skandinavischen *Antennaria*-Arten. *Hereditas* 20: 214–226. doi:10.1111/j.1601-5223.1935.tb03187.x

Bergman, B. 1937. Eine neue apomiktische *Antennaria. Svensk Bot. Tidskr.* 31: 391–394.

Bergman, B. 1941. Studies on the embryo sac mother cell and its development in *Hieracium* subg. *Archieracium.Svensk Bot. Tidskr.* 35: 1–42.

Bergman, B. 1942. Zur Embryologie der Gattung *Erigeron. Svensk Bot. Tidskr.* 36: 429–443.

Bergman, B. 1944. A contribution to the knowledge of the embryo sac mother cell and its development in two apomicts. *Svensk Bot. Tidskr.* 38: 249–259.

Bergman, B. 1950. Meiosis in two different clones of apomictic *Chondrilla juncea*. *Hereditas* 36: 297–320.

Bergman, B. 1951. On the formation of reduced and unreduced gametophytes in the females of *Antennaria carpatica*. *Hereditas* 38: 501–518.

Beridze, K. 1964. Kizuchenyu embriologi roda *Fragaria* L. *Trans. Tbilisi, Bot. nst.* 23: 235–239.

Bernardo, F.A., C.C. Jessena and C.C. Ramirez. 1961. Parthenocarpy and apomixis in *Lansium domesticum*. *Philipp. Agric. Rev*. 44: 415–421.

Berry, P.E., H. Tobe and J.A. Gómez. 1991. Agamospermy and the loss of distyly in *Erythroxylum undulatum* (Erythroxylaceae) from Northern Venezuela. *Am. J. Bot.* 78(5): 595–600.

Bertasso-Borges, M.S. and J.R. Coleman. 1998. Embryology and cytogenetics of *Eupatorium pauciflorum* and *E. intermedium* (Compositae). *Genet. Mol. Biol.* 21: 507–514.

Bertasso-Borges, M.S. and J.R. Coleman. 2005. Cytogenetics and embryology of *Eupatorium laevigatum* (Compositae). *Genet. Mol. Biol.* 28: 123–128.

Berthaud, J. 2001. Apomixis and the management of genetic diversity. In: Savidan, Y., J.G. Carman, T. Dresselhaus (eds.), *The flowering of apomixis: From mechanisms to genetic engineering*. CIMMYT, IRD, European Commission OC VI (FAIR): Mexico City, Mexico; pp. 8–23.

Berzenko, N.P. 1979. Abnormal phenomena in the development of ovules, female gametophyte and embryo i *Fragaria grandiflora* Ehrh. *Ukr. Bot. Zhurn.* 36: 16–20.

Beth, K. 1938. Untersuchungen über de Auflösung von Adventivembryonie durch Wundreiz. *Planta* 28: 296–343.

Bhanwra, R.K. 1988. Embryology in relation to systematic of Gramineae. *Ann. Bot.* 62: 215–233.

Bhanwra, R.K. and S.P. Choda. 1981. Apomixis in *Eremopogon foveolatus* (Gramineae). *Nordic J. Bot.* 1: 97–101.

Bhanwra, R.K., S.P. Choda and S. Kumar. 1982. Comparative embryology of some grasses. *Proc. Indian Natl. Sci. Acad. B*. 48: 152–162.

Bhanwra, R.K., J. Kaur and A. Garg. 1991. Embryological studies in some grasses and their taxonomic significance. *Bot. J. Linn. Soc.* 107: 405–419.

Bhuskute, S.M. 1987. Generative apospory in *Momordica charantia* var. *muricata* (Willd.) Chakrav. *Indian Bot. Rep*. 6: 131–133.

Bicknell, R.A. 1994. *Hieracium*: A model system for studying the molecular genetics of apomixis. *Apomixis Newsl.* 7: 8–10.

Bicknell, R.A. 1997. Isolation of a diploid, apomictic plant of *Hieracium aurantiacum*. *Sex. Plant Reprod.* 10: 168–172.

Bicknell, R.A., S.C. Lambie and R.C. Butler. 2003. Quantification of progeny classes in two facultatively apomictic accessions of *Hieracium*. *Hereditas* 138: 11–20. doi:10.1034/j.1601-5223.2003.01624.x

Biermann, M. 1896. Beiträge zur Kenntnis der Entwicklungsgeschichte der Früchte von *Citrus vulgaris* Risso und anderer Citrusarten. PhD Dissertation, Universität Bern, Switzerland, 52 pages.

Bierzychudek, P. 1985. Patterns in plant parthenogenesis. *Experientia* 41:1255–1264.

Bigl, K., J. Paule and C. Dobeš. 2019. The morphometrics of autopolyploidy: insignificant differentiation among sexual–apomictic cytotypes. *AoB Plants* 11(3): plz028. doi:10.1093/aobpla/plz028

Birari, S.P. 1980. Apomixis and sexuality in *Themeda* Forssk. at different ploidy levels (Gramineae). *Genetica* 54: 133–139.

Birari, S.P. 1981. Mechanism of Apomixis in *Pennisetum polystachyon* Schultz. *J. Maharashtra Agric. Univ. Pune* 6: 206–212.

Birky Jr, C.W. 1996 Heterozygosity, heteromorphy, and phylogenetic trees in asexual eukaryotes. *Genetics* 144: 427–437.

Bisognin, C., E. Seemüller, S. Citterio, R. Velasco, M.S. Grando and W. Jarausch. 2009. Use of SSR markers to assess sexual vs. apomictic origin and ploidy level of breeding progeny derived from crosses of apple proliferation-resistant *Malus sieboldii* and its hybrids with *Malus* × *domestica* cultivars. *Plant Breed.* 128(5): 507–513.

Bittencourt Jr, N.S. and C.I.G. Moraes. 2010. Self-fertility and polyembryony in South American yellow trumpet trees (*Handroanthus chrysotrichus* and *H. ochraceous*, Bignoniaceae): A histological study of post-pollination events. *Plant Syst. Evol.* 288: 59–76. doi:10.1007/s00606-010-0313-2

Bitter, G. 1906. Parthenogenesis und Variabilität der *Bryonia dioica*. *Abh. Naturwiss. Ver. Bremen* 138: 99–107.

Blakey, C.A., S.L. Goldman and C.L. Dewald. 2001. Apomixis in *Tripsacum*: Comparative mapping of a multigene phenomenon. *Genome* 44: 222–230. doi:10.1139/gen-44-2-222

Blount, A.R. and C.A. Acuña. 2009. Bahiagrass. In: *Genetic resources, chromosome engineering, and crop improvement series: Forage crops*, Vol. 5. Singh, R.J. (Ed.); Boca Raton, FL: CRC Press, pp. 81–101. doi:10.1201/9781420047400.ch4

Böcher, T.W. 1951. Cytological and embryological studies in the amphi-apomictic *Arabis holboellii* complex. *K. Dansk. Vidensk. Selsk.* 6(7):1–59.

Böcher, T.W. 1969. Further studies in *Arabis holboellii* and allied species. *Bot. Tidsskr.* 64: 141–161.

Bogomolov, M.A. 2005. Apomiksis i ego rol' v selektsii sakharnoy svekly. *Sakharnaya svekla* 8: 19–22.

Bogunić, F., S. Siljak-Yakovlev, I. Mahmutović-Dizdarević, A. Hajrudinović-Bogunić, M. Bourge, S.C. Brown and E. Muratović. 2021. Genome size, cytotype diversity and reproductive mode variation of *Cotoneaster integerrimus* (Rosaceae) from the Balkans. *Plants* 10: 2798. doi:10.3390/plants10122798

Bolòs, O. and J. Vigo. 1996. *Flora dels pai'sos Catalans 3*. Editorial Barcino, Barcelona.

Bonilla, J.R. and C.L. Quarin. 1997. Diplosporous and aposporous apomixis in a pentaploid race of *Paspalum minus*. *Plant Sci.* 127: 97–104.

Böös, G. 1917. Über Parthenogenesis in der Gruppe *Aphanes* der Gattung *Alchemilla* nebst einigen im Zusammenhang damit stehenden Fragen. *Lunds Univ. Årsskr.* 2(13): 1–37

Böös, G. 1920. Der experimentelle Nachweis der Parthenogenesis in der Gruppe *Aphanes* der Gattung *Alchemilla*. *Bot. Not.* 1920: 145–150.

Böös, G. 1924. Neue embryologische Studien über *Alchemilla arvensis* (L.) Scop. *Bot. Not.* 1924: 209–250.

Borg, M., L. Brownfield and D. Twell. 2009. Male gametophyte development: a molecular perspective. *J. Exp. Bot.* 60(5): 1465–1478.

Borges, H.B.N. 1991. Biologia reprodutiva de quatro espécies de Melastomataceae. Dissertação de Mestrado, Universidade Estadual de Campinas, Campinas, Brazil.

Bougnicourt, M. 1970. Polyembryonie chez l'*Hypericum tetrapterum* Fries (=*Hypericum acutum* Moench). *Bull. Soc. Bot. France* 116: 495–504.

Bougnicourt, M. 1971. Embryogénie des Hypericacées. Developpement de l'embryon chez l'*Hypericum tetrapterum* Fries. *Bull. Soc. Bot. France* 118: 319–334.

Bouquet, A. 1980. Differences observed in graft compatibility between some cultivars of muscadine grape (*Vitis rotundifolia* Michx.) and European grape (*Vitis vinifera* L. cv. Cabemet Sauvignon). *Vitis* 19: 99–104.

Boutilier, K., R. Offringa, V.K. Sharma, H. Kieft, T. Ouellet, L. Zhang, J. Hattori, C.M. Liu, A.A. van Lammeren, B.L. Miki, J.B. Custers and M.M. van Lookeren Campagne. 2002. Ectopic expression of *BABY BOOM* triggers a conversion from vegetative to embryonic growth. *Plant Cell* 14(8): 1737–1749. doi:10.1105/tpc.001941

Bowers, J.E., B.A. Chapman, J. Rong and A.H. Paterson. 2003. Unravelling angiosperm genome evolution by phylogenetic analysis of chromosomal duplication events. *Nature* 422: 433–438.

Bradley, J., J. Carman, M. Jamison and T. Naumova. 2007. Heterochronic features of the female germline among several sexual diploid *Tripsacum* L. (Andropogoneae, Poaceae). *Sex. Plant Reprod.* 20: 9–17.

Braga, A.C., G. Sousa da Silva, J. de Melo Nascimento, G. da Silva Gomes, G. M. da Conceição. 2019. Um, dois, três, quatro! Um estudo de caso de poliembrionia em *Handroanthus serratifolius* (Vahl) S. Grose, (Bignoniaceae), Maranhão, Brasil. *Revista Arquivos Científicos* (IMMES) 2: 119–124. doi:10.5935/2595-4407/rac.immes.v2n2p119-124

Braun, A. 1857. *Über Parthenogenesis bei Pflanzen*. Akademie der Wissenschaften, Berlin.

Braun, A. 1860. Über Polyembryonie und Keimung von *Coelobogyne*. Ein Nachtrag zu der Abhandlung über Parthenogenesis der Pflanzen. *Abh. Kon Akad. Wiss. Berlin Phys. Abt.* 1860: 107–263.

Braun, M., S. Dötterl and G. Gottsberger. 2011. Absence of pollinators and apomictic fruit production in an Atlantic rainforest population of *Cymbopetalum Brasiliense* (Annonaceae). *Plant Syst. Evol.* 296: 265–273.

Bräuning, S., A. Catanach, J.M. Lord, R. Bicknell and R.C. Macknight. 2018. Comparative transcriptome analysis of the wild-type model apomict *Hieracium praealtum* and its loss of parthenogenesis (lop) mutant. *BMC Plant Biol.* 18(1): 206. doi:10.1186/s12870-018-1423-1

Bray, R.A. 1978. Evidence for facultative apomixis in *Cenchrus ciliaris*. *Euphytica* 27: 801–804.

Bremer, G. 1923. A cytological investigation of some species and species-hybrids of the genus *Saccharum*. *Genetica* 5: 273–326. doi:10.1007/BF01508814

Bremer, G. 1946. De cytologie van soortbastaarden bij *Saccharum*. *Vakblad voor Biologen* 26: 3–10.

Brix, K. 1974. Sexual reproduction in *Eragrostis curvula*. *Z. Pflanzenzuecht.* 71: 25–32.

Brizicky, G.K. 1964. Polyembryony in *Euonymus* (Celastraceae). *J. Arnold Arbor.* 45(2): 251–259. http://www.jstor.org/stable/43781500

Brown, A.G., 1975. Breeding for specific characters. In: *Advances in fruit breeding*. J. Janick and J.N. Moore (Eds.). Purdue University Press, West Lafayette, pp. 3–57.

Brown, W.V. 1951. Apomixis in *Zephyranthes texana*. *Am. J. Bot.* 38: 697–702.

Brown, W.V. and W.H.P. Emery. 1957. Apomixis in the Gramineae, tribe Andropogoneae: *Themeda triandra* and *Bothriochloa ischaemum*. *Bot. Gaz.* 118. 246–253.

Brown, W.V. and W.H.P. Emery. 1958. Apomixis in the Gramineae-Panicoideae. *Am. J. Bot.* 45: 253–263. doi:10.2307/2439258

Brožová, V., P. Koutecký and J. Doležal. 2019. Plant apomixis is rare in Himalayan high-alpine flora. *Sci. Rep.* 9: 14386. doi:10.1038/s41598-019-50907-5

Brullo, S. 1978. Il génère *"Limonium"* MILLER in Cirenaica (The genus *"Limonium"* in Cirenaica). *Webbia* 33: 137–158.

Brullo, S. and M. Erben. 1989. The genus *Limonium* (Plumbaginaceae) in Tunisia. *Mitt. Bot. Staatssamml. München* 28: 419–500.

Bryant, G.D. 1952. The apomictic mechanism in *Bouteloua curtipendula*. PhD thesis, University of Oklahoma, Oklahoma city, OK, USA.

Bujdoso, G., B. Illes, V. Varjas and K. Cseke. 2021. Is "Esterhazy II", an old walnut variety in the Hungarian Gene Bank, the original genotype? *Plants* 10(5): 854. doi:10.3390/plants10050854

Burgess, M.B., K.R. Cushman, E.T. Doucette, N. Talent, C.T. Frye and C.S. Campbell. 2014. Effects of apomixis and polyploidy on diversification and geographic distribution in *Amelanchier* (Rosaceae). *Am. J. Bot.* 101: 1375–1387 doi:10.3732/ajb.1400113

Burke, J.L. and J.F. Bain. 2008.The origins of *Erigeron trifidus* (Asteraceae), an agamospermous rare species in Alberta. *Botany* 86(8): 866–876.

Burnier, J., S. Buerki, N. Arrigo, P. Küpfer and N. Alvarez. 2009. Genetic structure and evolution of alpine polyploid complexes: *Ranunculus kuepferi* (Ranunculaceae) as a case study. *Mol. Ecol.* 18: 3730–3744.

Burson, B.L. 1975. Cytology of some apomictic *Paspalum* species. *Crop Sci.* 16: 229–232.

Burson, B.L. 1985. Cytology of *Paspalum chacoense* and *P. durifolium* and their relationship to *P. dilatatum*. *Bot. Gaz.* 146: 124–129.

Burson, B.L. 1997. Apomixis and sexuality in some *Paspalum* species. *Crop Sci.* 37: 1347–1351.

Burson, B.L. and H.W. Bennett. 1971a. Chromosome numbers, microsporogenesis and mode of reproduction of seven *Paspalum* species. *Crop Sci.* 11: 292–294.

Burson, B.L. and H.W. Bennett. 1971b. Meiotic and reproductive behaviour of some introduced *Paspalum* species. *J. Miss. Acad. Sci.* 17: 5–8.

Burson, B.L. and M.A. Hussey. 1998. Cytology of *Paspalum malacophyllum* and its relationship to *P. juergensii* and *P. dilatatum*. *Int. J. Plant sci.* 159: 153–159.

Burson, B.L. and P.W. Voigt. 1996. Cytogenetic relationships between the *Eragrostis curvula* and *E. lehmanniana* complexes. *Int. J. Plant Sci.* 157: 632–637.

Burson, B.L., P.W. Voigt and W. Evers. 1991. Cytology, reproductive behaviour and forage potential of hexaploid dallis grass biotypes. *Crop Sci.* 31: 636–641.

Burson, B.L., P.W. Voigt, R.A. Sherman and C.L. Dewald. 1990. Apomixis and sexuality in Eastern gammagrass. *Crop Sci.* 30: 86–89.

Burton, G.W. 1948. The method of reproduction in common Bahia grass *Paspalum notatum*. *J. Am. Soc. Agron.* 40: 443–452.

Burton, G.W., J.C. Millot and W.G. Monson. 1973. Breeding procedures for *Panicum maximum* Jacq. suggested by plant variability and mode of reproduction. *Crop Sci.* 13: 717–720.

Buru, J.C., K. Dhileepan, O.O. Osunkoya and J. Firn. 2016. Comparison of growth traits between abundant and uncommon forms of a non-native vine, *Dolichandra unguis-cati* (Bignoniaceae) in Australia. *NeoBiota* 30: 91–109.

Busey, P. 1989. Progress and benefits to humanity from breeding warm-season grasses for turf. In: Sleper D.A., K.H. Asay and J.F. Pedersen (eds.), *Contributions from breeding forage and turf grasses, CSSA Special Publication no. 15*, Crop Science Society of America, Wisconsin: 49–70.

Büyükkartal, H.N. 2009. Ultrastructural changes of the egg apparatus associated with fertilisation of natural tetraploid *Trifolium pratense* L. (Fabaceae). *Biol. Res.* 42(1): 25–30.

Büyükkartal, H.N.B. 2008. Causes of low seed set in the natural tetraploid *Trifolium pratense* L. (Fabaceae). *Afr. J. Biotechnol.* 7(9): 1240–1249. doi:10.5897/AJB07.771

Cáceres, M.E., F. Matzk, A. Busti, F. Pupilli and S. Arcioni. 2001. Apomixis and sexuality in *Paspalum simplex*: Characterization of the mode of reproduction in segregating progenies by different methods. *Sex. Plant Reprod.* 14: 201–206.

Caetano, A.P.S. 2014. Contribuição da embriologia na sistemática e na elucidaão da apomixia em Melastomataceae Juss. Doctoral thesis. Universidade Stadual de Campinas, Campinas, Brazil. http://repositorio.unicamp.br/jspui/handle/REPOSIP/3155575

Caetano, A.P.S., J.P. Basso-Alves, P.A. Cortez, V.L.G. De Brito, F.A. Michelangeli, M. Reginato, R. Goldenberg, S.M. Carmello-Guerreiro and S.P. Teixera. 2018. Evolution of the outer ovule integument and its systematic significance in Melastomataceae. *Bot. J. Linn. Soc.* 186(2): 224–246. doi:10.1093/botlinnean/box093

Caetano, A.P.S. and Cortez, P.A. 2014. Reprodução assexuada. In: Rech, A.R.., K. Agostini, P.E. Oliveira, I.C. Machado (eds.), *Biologia da Polinização*. Editora Projeto Cultural. pp. 93–111.

Caetano, A.P.S., D.G. Simão, R. Carmo-Oliveira and P.E. Oliveira. 2013. Diplospory and obligate apomixis in *Miconia albicans* (Miconieae, Melastomataceae) and an embryological comparison with its sexual congener *M. chamissois*. *Plant Syst. Evol.* 299: 1253–1262.

Cai, X., X-J. Mu, Z-Q. Zhu, Z-M. Hau, Y-H. Huang and Q.X. Sun. 1997. Polyembryony and multiple seedlings in the apomictic plants. *Acta Bot. Sin.* 39: 590–595.

Cai, X., D. Sun, X. Guan and H. Wang. 2000. Origin and development of the nucellar embryo and endosperm in apomictic *Xanthoxylum bungeanum* Maxim. *Acta Bot. Boreal-Occid. Sin.* 20(4): 533–538.

Calderini, O., S.B. Chang, H. de Jong, A. Busti, F. Paolocci, S. Arcioni, S.C. de Vries, M.H.C. Abma-Henkens, R.M.K. Lankhorst, I.S. Donnison and F. Pupilli. 2006. Molecular cytogenetics and DNA sequence analysis of an apomixis-linked BAC in *Paspalum simplex* reveal a non pericentromere location and partial microcolinearity with rice. *Theor. Appl. Genet.* 112: 1179–1191.

Calderini, O., I. Donnison, L. Polegri, F. Panara, A. Thomas, S. Arcioni and F. Pupilli. 2011. Partial isolation of the genomic region linked with apomixis in *Paspalum simplex*. *Mol. Breed.* 28: 265–276.

Cameron, J.W. and M.J. Garber. 1968. Identical-twin hybrids of *Citrus* × *Poncirus* from strictly sexual seed parents. *Am. J. Bot.* 55(2): 199–205. doi:10.2307/2440453

Cameron, J.W. and R.K. Soost. 1979. Sexual and nucellar embryony in F1 hybrids and advanced crosses of *Citrus* with *Poncirus*. *J. Am. Soc. Hortic. Sci.* 104: 408–410.

Campacci, T.V.S., C.T. Castanho, R.L.F. Oliveira, R.M. Suzuki, E.L.M. Catharino and S. Koehler. 2017. Effects of pollen origin on apomixis in *Zygopetalum mackayi* orchids. *Flora* 226: 96–103. doi:10.1016/j.flora.2016.11.013

Campbell, A.I. 1960. *Apomictic seedling rootstocks for apples*. Progress report II. Annual Report for 1959. Long Aston Research Station, pp. 50–56.

Campbell, A.I. and D. Wilson. 1962. *Apomictic seedling rootstocks for apples*. Progress report III. Annu. Rep. for 1961. Long Ashton Research Station, pp. 68–70.

Campbell, C.S., L.A. Alice and W.A. Wright. 1999. Comparisons of within-population genetic variation in sexual and agamospermous *Amelanchier* (Rosaceae) using RAPD markers. *Plant Syst. Evol.* 215: 157–167.

Campbell, C.S. and T.A. Dickinson. 1990. Apomixis patterns of morphological variation and species concepts in subfam. Maloideae (Rosaceae). *Syst. Bot.* 15(1): 124–135. doi:10.2307/2419022

Campbell, C.S., C.W. Greene and S.E. Bergquist. 1987. Apomixis and sexuality in three species of *Amelanchier* (Rosaceae, Maloideae). *Am. J. Bot.* 74: 321–328.

Campbell, C.S., C.W. Greene and T.A. Dickinson. 1991. Reproductive biology in the *Maloideae* (Rosaceae). *Syst. Bot.* 16: 333–349.

Campbell, C.S., C.W. Greene, B.F. Neubauer and J.M. Higgins. 1985. Apomixis in *Amelanchier laevis*, shadbush (Rosaceae, Maloideae). *Am. J. Bot.* 72: 1397–1403.

Campbell, C.S., M.F. Wojciechowski, B.G. Baldwin, L.A. Alice and M.J. Donoghue. 1997. Persistent nuclear ribosomal DNA sequence polymorphism in the *Amelanchier* agamic complex (Rosaceae). *Mol. Biol. Evol.* 14(1): 81–90. doi:10.1093/oxfordjournals.molbev.a025705

Campbell, C.S. and W.A. Wright. 1996. Apomixis, hybridization, and taxonomic complexity in eastern North American *Amelanchier* (Rosaceae). *Folia Geobot. Phytotax.* 31: 345–354.

Campbell, D.H. 1903. Studies on the Araceae. 2. The embryo sac and embryo of *Aglaonema* and *Spathicarpa*. *Ann. Bot.* 17: 665–687.

Caperta, A.D., S. Castro, J. Loureiro, A.S. Róis, S. Conceição, J. Costa, L. Rhazi, D. Espírito Santo and P. Arsénio. 2017. Biogeographical, ecological and ploidy variation in related asexual and sexual *Limonium* taxa (Plumbaginaceae). *Bot. J. Linn. Soc.* 183: 75–93. doi:10.1111/boj.12498

Caponio, I. and C.L. Quarin. 1987. El sistema genético de *Paspalum simplex* y de un híbrido interespecífico con *P. dilatatum*. *Kurtziana* 19: 35–45.

Caponio, I., S.C. Torres, G. Andersson and G.A. Norrmann. 2016. Aspects of reproductive biology of *Stevia rebaudiana* (Asteraceae). *Bonplandia* 25(1): 59–71 (in Spanish).

Cappelletti, C. 1929. Sterilità di origine micotica nella *Ruta patavina* L. *Ann. Bot.* (Rome) 18: 145–166.

Carano, E. 1913. Su particolari anomalie del sacco embrionali di *Bellis perennis*. *Ann. Bot.* (Rome) 11: 435–439.

Carano, E. 1921. Nuovo recherché sulla embriologie delle Asteraceae. *Ann. Bot.* (Rome) 15: 1–100.

Carano, E. 1924. Osservazioni sul mecanismo di divisione delle cellula madre del sacco embrionale nella piantee apogame. *C.R. Acad. Lincei Roma* 33: 150–155.

Carano, E. 1925. Sul particolare sviluppa del gametofito O di *Euphorbia dulcis*. *C.R. Acad. Lincei Roma* (6). 1: 633–635.
Carano, E. 1926. Ulteriori osservazioni su *Euphorbia dulcis* L. in rapport col suo compartment spomittico. *Ann. Bot.* (Rome) 17: 50–79.
Carman, J.G. 1995. Gametophytic angiosperm apomicts and the occurrence of polyspory and polyembryony among their relatives. *Apomixis Newsl.* 8: 39–53.
Carman, J.G. 1997. Asynchronous expression of duplicate genes in angiosperms may cause apomixis, bispory, tetraspory, and polyembryony. *Biol. J. Linn. Soc.* 61: 51–94.
Carman, J.G. 2007. Do duplicate genes cause apomixis. In: Hörandl, E., U. Grossniklaus, P. Van Dijk, T. Sharbel (eds.), *Apomixis: Evolution, mechanisms and perspectives*. Gantner Verlag: Ruggell, Liechtenstein, pp. 63–91.
Carman, J.G., C.F. Crane and O. Reir-Lizarazu. 1991. Comparative histology of cell walls during meiotic and apomeiotic megasporogenesis in two hexaploid Australian *Elymus* species. *Crop Sci.* 31: 1527–1532.
Carman, J.G. and S.L. Hatch 1982. Aposporous apomixis in *Schizachyrium* (Poaceae, Andropogoneae). *Crop Sci.* 32: 1252–1254.
Carman, J.G., M. Jamison, E. Elliott, K.K. Dwivedi and T.N. Naumova. 2011. Apospory appears to accelerate onset of meiosis and sexual embryo sac formation in sorghum ovules. *BMC Plant Biol.* 11: 9. doi:10.1186/1471-2229-11-9
Carman, J.G., M. Mateo de Arias, L. Gao, X. Zhao, B.M. Kowallis, D.A. Sherwood, M.K. Srivastava, K.K. Dwivedi, B. Price, L. Watts and M.D. Windham. 2019. Apospory and diplospory in diploid *Boechera* (Brassicaceae) may facilitate speciation by recombination-driven apomixis-to-sex reversals. *Front. Plant Sci.* 10: 724. doi:10.3389/fpls.2019.00724
Carmichael, J.S. and S.M. Selbo. 1999. Ovule, embryo sac, embryo, and endosperm development in leafy spurge (*Euphorbia esula*). *Can. J. Bot.* 77(4): 599–610. doi:10.1139/b99-020
Carniel, K. 1967. Über die Embryobildung in der Gattung *Paeonia*. *Osterr. Bot Z.* 114: 4–19. doi:10.1007/BF01373930
Castro, D.C., M.C. Cerino, N. Gariglio and S. Radice. 2016. Study of reproductive behaviour in low-chill apples in warmer zones of Argentina. *Sci. Hortic.* 199: 124–132. doi:10.1016/j.scienta.2015.12.018
Catanach, A.S., S.K. Erasmuson, E. Podivinsky, B.R. Jordan and R. Bicknell. 2006. Deletion mapping of genetic regions associated with apomixis in *Hieracium*. *Proc. Natl. Acad. Sci. U.S.A.* 103: 18650–18655. doi:10.1073/pnas.0605588103
Catling, P. 1987. Notes on the breeding systems of *Sacoila lanceolata* (Aublet) Garay (Orchidaceae). *Ann. Mo. Bot. Gard.* 74(1): 58–68. doi:10.2307/2399262
Catling, P.M. 1982. Breeding systems of northeastern North American *Spiranthes* (Orchidaceae). *Can. J. Bot.* 60: 3017–3039.
Catling, P.M. and V.R. Catling. 1991. A Synopsis of breeding systems and pollination in North American orchids. *Lindleyana* 6: 187–210.
Catling, P.M. and K.L. McIntosh. 1979. Rediscovery of *Spiranthes parksii* Correll. *Sida, Contrib. Bot.* 8(2): 188–193.
Cave, M., H. Arnott and S. Cook. 1961. Embryogeny in the California Peonies with reference to their taxonomic position. *Am. J. Bot.* 48(5): 397–404. doi:10.2307/2439332
Celarier, P. and J.R. Harlan. 1957. Apomixis in *Bothriochloa*, *Dichanthium*, and *Capillipedium*. *Phytomorphology* 7: 93–102.
Cesca, G. 1961. Ricerche cariologiche ed embriologiche sulle Euphorbiaceae. I — Su alcuni biotipi di *Euphorbia dulcis* L. della Toscana. *Caryologia* 14(1): 79–96. doi:10.1080/00087114.1961.10796013
Chaboudez, P. 1994. Patterns of clonal cariation in skeleton weed (*Chondrilla juncea*), an apomictic species. *Aust. J. Bot.* 42: 283–295.
Chakravarthy, R.S. 1935. Polyembryony in *Murraya koenigii*. *Curr. Sci.* 3(8): 361.
Chakravarthy, R.S. 1936. Nucellar polyembryony in Rutaceae. *Curr. Sci.* 5: 202–203.
Chan H.T. 1981. Reproductive biology of some Malaysian dipterocarps III. Breeding systems. *Malays. For.* 44: 28–36.
Chantaranothai, P. and J.A.N. Parnell 1994. The breeding biology of of some Thai *Syzygium* species. *Tropic. Ecol.* 35(2): 199–208.
Chao, C.Y. 1964. Megasporogenesis, megagametogenesis and embryogeny in *Paspalum orbiculare*. *New Asia College Acad. Ann.* 6: 15–25.
Chao, C.Y. 1974. Megasporogenesis and megagametogenesis in *Paspalum commersonii* and *P. longifolium* at two polyploidy levels. *Bot. Not.* 127: 267–275.
Chao, C.Y. 1980. Autonomous development of embryo in *Paspalum conjugatum*. *Bot. Not.* 133: 215–222.

Chapman, H., G.J. Houliston, B. Robson and I. Iline. 2003. A case of reversal: The evolution and maintenance of sexuals from parthenogenetic clones in *Hieracium pilosella. Int. J. Plant Sci.* 164: 719–728.
Chatterji, A.K. 1969. Apomixis and tetraploidy in *Pennisetum orientale. Crop Sci.* 9: 796–799.
Chatterji, A.K. and G.K. Pillai. 1970. Apomixis in *Pennisetum pedicellatum. Sci. & Cult.* 36: 667–669.
Chatterji, A.K. and D.H. Timothy. 1969. Microsporogenesis and embryogenesis in *Pennisetum flaccidum* Griseb. *Crop Sci.* 9: 219–222.
Chavan, S., A. Keerthika, A. Jha, A.K. Handa and R. Newaj. 2014. Abnormal seedlings in *Madhuca latifolia* – An important biodiesel tree. *Life Sci. Leaflets* 58: 18–21.
Cheek, M.D. and R. Lalla. 2017. A description of the naturalised *Clusia rosea* Jacq. (Clusiaceae) populations in South Africa. *Bothalia* 47(1): a2229. doi:10.4102/abc.v47i1.2229
Chemisquy, M.A., L.M. Giussani, M.A. Scataglini, E.A. Kellogg and O. Morrone. 2010. Phylogenetic studies favour the unification of *Pennisetum, Cenchrus* and *Odontelytrum* (Poaceae): a combined nuclear, plastid and morphological analysis, and nomenclatural combinations in *Cenchrus. Ann. Bot.* 106(1): 107–130. doi:10.1093/aob/mcq090
Chen, L., L. Guan, A. Kojima and T. Adachi. 2000. The mechanisms of appearance of aposporous initial cell and apomictic embryo sac formation in *Paspalum notatum. Cytologia* 65(3): 333–341. doi:10.1508/cytologia.65.333
Chen, M. and X-Y. Zhao. 2017. Effect of pollen and resource limitation on reproduction of *Zygophyllum xanthoxylum* in fragmented habitats. *Ecol Evol.* 7: 9076–9084. doi:10.1002/ece3.3465
Chen, X., H-G. Lai, Q. Sun, J-P. Liu, S-B. Chen and W-L. Zhu. 2018. Induction of apomixis by dimethyl sulfoxide (DMSO) and genetic identification of apomictic plants in cassava. *Breed. Sci.* 68: 227–232. doi:10.1270/jsbbs.17089
Chen X.Y. and J.Y. Kim. 2009. Callose synthesis in higher plants. *Plant Signal Behav.* 4(6): 489–492. doi:10.4161/psb.4.6.8359
Cheng, M.H., X.H. Yang and W.G. Zeng. 1984. A preliminary report on investigation and study of *M. xiaojinensis*-an apple stock resource. *J. Southwest University* 6:38–43.
Chennaveeraiah, M.S. and R.M. Patil. 1971. Apomixis in *Blumea. Phytomorphology* 21: 71–76.
Chiarugi, A. 1926. Aposporia e apogamia in *Artemisia nitida. N.G. Bot. Ital.* 33: 501–626.
Chiarugi, A. 1929. Diploidismo con anfimissia e tetraploidismo con apomissia in una medesima specie *Nigritella nigra* Rchb. *Boll. Soc. Ital. Biol. Sper.* 4(6): 659–661.
Chiarugi, A. and E. Francini. 1930. Apomissia in *Ochna serrulata. N.G. Bot. Ital.* 7: 1–250.
Chochlov, S.S., M.I. Zajceva and P.G. Kutrijanov. 1978. Vyjavlenie apomiktichiykh form vo flore cvetkovykh rastenij SSSR. Programma, metodika, rezul'taty (*Revelation of the apomictic forms in the flora of the angiosperms of USSR. Programs, methods and results*) [in Russian]. Saratovskogo Universiteta, Saratov.
Choda, S.P. and R.K. Bhanwra. 1977. The mode of reproduction in *Bothriochloa odorata* and *Paspalum distichum. Proc. Indian Natl. Sci. Acad.* 43: 175–183.
Choda, S.P. and R.K. Bhanwra. 1980. Apomixis in *Capillipedium huegelli* (Graminae). *Proc. Indian Natl. Sci. Acad.* B. 46: 572–578.
Choda, S.P. and S.K. Sharma. 1981. Apomixis in *Pennisetum orientale* (Graminae). In: S.C. Verma (ed.), *Contemporary Trends in Plant sciences.* Kalyani Publishers, New Delhi.
Christen, H.R. 1950. Untersuchungen über die Embryologie pseudogamer und sexueller *Rubus*-Arten. *Ber. Schweiz. Bot.Ges.* 60: 153–198. doi:10.3929/ethz-a-000098769
Christen, H.R. 1952. Die Embryologie von *Rubus idaeus* und von Bastarden zwischen *Rubus caesius* und *Rubus idaeus. Zeitschr. f. Indukt. Abstamm. Vererbungsl.* 84: 454–461.
Christensen, C.A., E.J. King, J.R. Jordan and G.N. Drews. 1997. Megagametogenesis in *Arabidopsis* wild type and the Gf mutant. *Sex. Plant Reprod.* 10: 49–64.
Christensen, K.I., M. Zarrei, M. Kuzmina, N. Talent, C. Lin and T.A. Dickinson. 2014. *Crataegus* × *ninaecelottiae* and *C.* × *cogswellii* (Rosaceae, Maleae), two spontaneously formed intersectional nothospecies. *PhytoKeys* 36: 1–26.
Christoff, M. 1942. Die genetische Grundlage der apomiktischen Fortpflanzung bei *Hieracium aurantiacum* L. *Zeitschr. f. Indukt. Abstamm. Vererbungsl.* 80: 103–125.
Christoff, M. and M.A. Christoff. 1948. Meiosis in the somatic tissue responsible for the reduction of chromosome number in the progeny of *Hieracium hoppeanum* Schult. *Genetics* 33: 36–42.
Christoff, M. and G. Papasova. 1943. Die genetischen Grundlagen der apomiktischen Fortpflanzung in der Gattung *Potentilla. Zeitschr. f. Indukt. Abstamm. Vererbungsl.* 81: 1–27.
Christoff, M.A. and R. Moskova. 1972. The apomixis and polyembryony in *Bothriochloa ischaemum* (=*Andropogon ischaemum*). *Genet. Sel.* 5: 71–86.

Chrtek, J. and K. Marhold. 1998. Taxonomy of the *Hieracium fritzei* group (Asteraceae) in the Sudeten Mts. and the West Carpathians. (Studies in *Hieracium* sect. *Alpina* II.). *Phyton* (Horn, Austria) 37: 181–217.

Chrtek, J., P. Mráz, J. Zahradníček, G. Mateo and Z. Szeląg. 2007. Chromosome numbers and DNA-ploidy levels of selected species of *Hieracium* s.str. (Asteraceae). *Folia Geobot*. 42:411–430.

Chrtek J. Jr., I. Plačková, Z. Dočkalová, A. Krahulcová and P. Trávníček. 2014. Patterns of genetic variation in *Pilosella echioides* and its selected relatives: results of variation in ploidy level, facultative apomixis and past and present hybridization. *Plant Syst. Evol*. 300: 2091–2104.

Chua, L.G. and A.N. Rao. 1978. Polyembryony and suspensor characteristics in *Spathoglottis*. *Flora* 167(5): 399–402.

Chung, S. 1974. Meiosis and embryo sac development in *Paspalum thunbergii*. M.Phil. Dissertation, Chinese University of Hong Kong, Hong Kong, China.

Church, M. 1916. The Development of the embryo sac and embryo of *Cooperia drummondii*. *Bull. Torrey Bot. Cl*. 43(8): 397–405. doi:10.2307/2479510

Clark, L.V., K.J. Evans and M. Jasieniuk. 2013. Origins and distribution of invasive *Rubus fruticosus* L. agg. (Rosaceae) clones in the Western United States. *Biol. Invasions* 15: 1331–1342.

Cleij, G., T.S.M. De Bock and B. Lekkerkerker. 1976. Crosses between *Beta vulgaris* L. and *Beta lomatogona* F. et M. *Euphytica* 25: 539–547. doi:10.1007/BF00041589

Coe, G. 1953. Cytology of reproduction in *Cooperia pedunculata*. *Am. J. Bot*. 40(5): 335–343. doi:10.2307/2438757

Coleman, J.R. 1989. Embryology and cytogenetics of apomictic hexaploid *Eupatorium odoratum* L. (Compositae). *Rev. Bras. Genet*. 12: 803–817.

Coleman, J.R. and M.A. Coleman 1984. Apomixis in two triploid Brazilian species of *Eupatorium: E. bupleurifolium* and *E. callilepis*. *Rev. Bras. Genet*. 7: 549–567.

Coleman, J.R. and M.A. Coleman 1988. Embryology and cytogenetics of apomictic triploid *Eupatorium squalidum* DC. (Compositae). *Rev. Bras. Genet*. 11: 129–184.

Colono, C., J.P.A. Ortiz, H.R. Permingeat, E.D. Souza Canada, L.A. Siena, N. Spoto, F. Galdeano, F. Espinoza, O. Leblanc and S.C. Pessino. 2019. A plant-specific *TGS1* homolog influences gametophyte development in sexual tetraploid *Paspalum notatum* ovules. *Front. Plant Sci*. 10: 1566. doi:10.3389/fpls.2019.01566

Comai, L. 2005. The advantages and disadvantages of being polyploid. *Nat. Rev. Genet*. 6: 836–846.

Conceição S.I.R., Fernandes, J., E.B. da Silva and A.D. Caperta. 2021. Reproductive output and insect behavior in hybrids and apomicts from *Limonium ovalifolium* and *L. binervosum* complexes (Plumbaginaceae) in an open cross-pollination experiment. *Plants* 10(1): 169. doi:10.3390/plants10010169

Conner, H.E. 1979. Breeding systems in the grasses: A survey. *N. Z. J. Bot*. 17: 547–574.

Conner, J.A., S. Goel, G. Gunawan, M.M. Cordonnier-Pratt, V.E. Johnson, C. Liang, H. Wang, L.H. Pratt, J.E. Mullet, J. DeBarry, L. Yang, J.L. Bennetzen, P.E. Klein and P. Ozias-Akins. 2008. Sequence analysis of bacterial artificial chromosome clones from the apospory-specific genomic region of *Pennisetum* and *Cenchrus*. *Plant Physiol*. 147: 1396–1411.

Conner, J.A., G. Gunawan and P. Ozias-Akins. 2013. Recombination within the apospory specific genomic region leads to the uncoupling of apomixis components in *Cenchrus ciliaris*. *Planta* 238(1): 51–63. doi:10.1007/s00425-013-1873-5

Conner, J.A., M. Mookkan, H.Q. Huo, K. Chae and P. Ozias-Akins. 2015. A parthenogenesis gene of apomict origin elicits embryo formation from unfertilized eggs in a sexual plant. *Proc. Nat. Acad. Sci. U.S.A.* 112(36): 11205–11210. doi:10.1073/pnas.1505856112

Conner, J.A., M. Podio and P. Ozias-Akins. 2017. Haploid embryo production in rice and maize induced by *PsASGR-BBML* transgenes. *Plant Reprod*. 30: 41–52.

Connor, H.E. and M.I. Dawson. 1993. Evolution of reproduction in *Lamprothyrsus* (Arundineae: Gramineae). *Ann. Mo. Bot. Gard*. 80(2): 512–517.

Contreras, R., J. Rothleutner and V.O. Stockwell. 2014. Breeding for fire blight resistance and sterility in *Cotoneaster*. *Acta Hortic*. 1056: 221–224. doi:10.17660/ActaHortic.2014.1056.35

Cook, M.T. 1907. Notes on polyembryony. *Torreya* 7(6): 113–117. http://www.jstor.org/stable/40594604

Cooper, D.C. and R.A. Brink 1949. The endosperm-embryo relationship in the autonomous apomict, *Taraxacum officinale*. *Bot. Gaz*. 111: 139–152.

Copeland, H.F. 1966. Morphology and embryology of *Euonymus japonica*. *Phytomorphology* 16: 326–334.

Cornier, B. 2009. *Sorbus legrei* (spec. nov.) et *Sorbus remensis* (spec. nov.) (Rosaceae), deux nouvelles espèces françaises. *Bulletin Mensuel de la Société Linnéenne de Lyon* 78(1): 27–46.

Corral, J.M., H. Vogel, O.M. Aliyu, G. Hensel, T. Thiel, T. Kumlehn and T.F. Sharbel. 2013. A conserved apomixis-specific polymorphism is correlated with exclusive exonucleate expression in premeiotic ovules of apomictic *Boechera* species. *Plant Physiol*. 163(4): 1660–1672.

Correia, M.C.R., M.C.B. Pinheiro and H.A. Lima. 2006. Biologia floral e polinização de *Anemopaegma chamberlaynii* Bur. and K. Schum. (Bignoniaceae). *Lundiana* 7: 39–46.
Cortinhas, A.L., M. Erben, A.P. Paes, D. Espírito-Santo, M. Guara-Requena and A.D. Caperta. 2015. Taxonomic complexity in the halophyte *Limonium vulgare* and related taxa (Plumbaginaceae): insights from analysis of morphological, reproductive and karyological data. Special issue on Halophytes and Saline Adaptations. *Ann. Bot.* 115: 369–383.
Cosendai, A.C. and E Hörandl. 2010. Cytotype stability, facultative apomixis and geographical parthenogenesis in *Ranunculus kuepferi* (Ranunculaceae). *Ann. Bot.* 105: 457–470. doi:10.1093/aob/mcp304
Cosendai, A-C., J. Rodewald and E. Hörandl. 2011. Origin and distribution of autopolyploids via apomixis in the alpine species *Ranunculus kuepferi* (Ranunculaceae). *Taxon* 60(2): 355–364.
Cosmulescu, S.N., M. Botu and G. Achim. 2012. Determination of apomictic fruit set ratio in several Romanian Walnut (*Juglans regia* L.) cultivars. *Notulae Botanicae Horti Agrobotanici Cluj-Napoca* 40(1): 229–233. doi:10.15835/nbha4016407
Costa, M.E., D.S. Sampaio, A.A.S. Paoli and S.C.A.L. Leite. 2004. Poliembrionia e aspectos da embriogênese em *Tabebuia ochracea* (Chamisso) Standley (Bignoniaceae). *Rev. Bras. Bot.* 27: 395–406.
Costa, R.B., P.Y. Kageyama and G. Mariano. 1992. Estudo do sistema de cruzamento de *Anadenanthera falcata* Benth., *Vochysia tucanorum* Mart. e *Xylopia aromatica* Baill. em área de cerrado. *Rev. Bras. Sementes* 14: 93–96.
Coughlan, J.M., S. Han, S. Stefanović and T.A. Dickinson. 2017. Widespread generalist clones are associated with range and niche expansion in allopolyploids of Pacific Northwest Hawthorns (*Crataegus* L.). *Mol. Ecol.* 26: 5484–5499. doi:10.1111/mec.14331
Coulon, J.De 1923. *Nardus stricta.* Etude physiologique, anatomique et embryologique. *Mém. Soc. Vaud. Sci. Nat. (Lausanne)* 6: 246–332.
Cowan, R., M.J. Ingrouille and M.D. Lledó. 1998. The taxonomic treatment of agamosperms in the genus *Limonium* Mill. (Plumbaginaceae). *Folia Geobot.* 33(3): 353–366. http://www.jstor.org/stable/4201312
Cowan, R.S., R.J. Smith and M.F. Fay. 2008. Genetic variation in Irish Whitebeam, *Sorbus hibernica* E. F. Warb. (Rosaceae) and its relationship to a *Sorbus* from the Menai Strait, North Wales. *Watsonia* 27: 99–108.
Cox, P. 1990. Pollination and the evolution of breeding systems in Pandanaceae. *Ann. Mo. Bot. Gard.* 77(4): 816–840. doi:10.2307/2399673
Cox, P.A. 1985. Islands and dioecism: insights from the reproductive ecology of *Pandanus tectorius* in Polynesia. In: J. White (ed.), *Studies on plant demography: A festschrift for John L. Harper*. Academic Press, London. pp. 355–368.
Crane, C.F. 1989. A proposed classification of apomictic mechanisms in flowering plants. *Apomixis Newsl.* 1: 11–19.
Crane, C.F. 2001. Classification of apomixis mechanisms. In: Y. Savidan, T. Dresselhaus and J.G. Carman (eds.), *The Flowering of Apomixis: From Mechanisms to Genetic Engineering*. CIMMYT, IRD, European Commission OC VI (FAIR), Mexico City, Mexico. pp. 24–43.
Crane, C.F. 2019. Megagametophyte differentiation in *Zephyranthes drummondii* D. Don and *Zephyranthes chlorosolen* (Herb.) D. Dietr. (Amaryllidaceae). *Caryologia* 72(4): 105–119. doi:10.13128/caryologia-670
Crane, C.F. and J.G. Carman. 1987. Mechanism of apomixis in *Elymus retisetus* from Australia and New Zealand. *Am. J. Bot.* 74: 477–496.
Crane, M.B. 1940. Reproductive versatility in *Rubus*. *J. Genetics* 40: 109–118. doi:10.1007/BF02982484
Crane, M.B. and P.T. Thomas. 1939. Segregation in asexual (apomictic) offspring in *Rubus*. *Nature* 143: 684. doi:10.1038/143684a0
Crespel, L. and S. Gudin. 2003. Evidence for the production of unreduced gametes by tetraploid *Rosa hybrida* L. *Euphytica* 133:65–69.
Crespo, M.B. 2009. A new coastal species of *Limonium* (Plumbaginaceae) from southeastern Spain. *Folia Geobot.* 44: 177–190. doi:10.1007/s12224-009-9035-z
Crestana, C.M.S. 1995. Ecologia da polinização de *Genipa americana* L. (Rubiaceae) na estação ecológica de Moji-Guaçu, estado de São Paulo. *Rev. Inst. Florest.* 7: 169–195.
Cruz, V.V., A. Pantel, D.T. Ray, W. Niaura, P.H. Purdy and D.A. Dierig. 2017. Analysis of mode of reproduction of guayule (*Parthenium argentatum* A. Gray) using flow cytometry and identification of polyhaploids for breeding. *Ind. Crops Prod.* 107(6): 618–623. doi:10.1016/j.indcrop.2017.07.004
Cushman, K.R., M.B. Burgess, E.T. Doucette, G.A. Nelson and C.S. Campbell. 2017. Species delimitation in tetraploid, apomictic *Amelanchier* (Rosaceae). *Syst. Bot.* 42(2): 234–256. doi:10.1600/036364417X695529

Czapik, R. 1954. Cytological and embryological studies in *Centaurea scabiosa* L. *Acta Soc. Bot. Pol.* 23: 175–194.

Czapik, R. 1961. Embryological studies in the genus *Potentilla* L. I. *P. crantzii*. *Acta Biol. Cracov., Ser. Bot.* 4: 97–119.

Czapik, R. 1962. Embryological studies in the genus *Potentilla* L. II. *Potentilla arenaria*. *Acta Biol. Cracov. Ser. Bot.* 5: 29–42.

Czapik, R. 1976. Variability in the embryological cycle of *Adoxa moschatellina* L. *Acta Biol. Cracov. Ser. Bot.* 19: 1–13.

Czapik, R. 1981a. Elementary apomictic processes in *Rubus* L. *Acta Soc. Bot. Pol.* 50: 201–204.

Czapik, R. 1981b. Embryology of *Rubus saxatilis* L. *Acta Biol. Cracov. Ser. Bot.* 23: 7–13.

Czapik, R. 1985. Apomictic embryo sacs in diploid *Waldsteinia geoides* Willd. (Rosaceae). *Acta Biol. Cracov. Ser. Bot.* 27: 29–37.

Czapik, R. 1987. Apomixis in *Rubus bellardi* Weihe. *Acta Biol. Cracov. Ser. Bot.* 29: 45–54.

Czapik, R. 1990. Haploid parthenogenesis and polyembryony in *Sibbaldia procumbens* L. (Rosaceae). In: J. Pare and M. Bugnicort (eds.), *Recent aspects and actual orientations in plant embryology*, Dedicatory volume to Professor A. Lebègue, Moulet, Amiems, pp. 137–143.

Czapik, R. 2000. Apomixis in monocotyledons. In: Jacobs, S.W.L., Everett, J. (eds.), *Grasses: Systematics and Evolution*. CSIRO, Kerenhaven, pp. 316–321.

Czapik, R. and M. Kościńska-Pająk. 2000. Apomictic studies and agamospermy in Polish flora. *Botanical Guidebooks* 24: 129–149.

D'Amato. 1940. Apomissia in *Statice oleifolia* Scop. var. *confusa* Godr. (Plumbaginaceae). *N.G. Bot. Ital.* 47: 504.

D'Amato. 1949. Triploidia e apomissia *Statice oleifolia* Scop. var. *confusa* Godr. *Caryologia* 2: 71–84.

d'Erfurth, I., S. Jolivet, N. Froger, O. Catrice, M. Novatchkova and R. Mercier. 2009. Turning meiosis into mitosis. *PLoS Biol.* 7: e1000124. doi:10.1371/journal.pbio.1000124

Dahlgren, K.V.O. 1920. Zur Embryologie der Kompositen mit besonderer Berücksichtigung der Endospermbildung. *Z. Bot.* 12: 481–516.

Das, A. and A.S. Islam. 1977. Embryological studies of three species of *Cenchrus* and their relationships. *Bangladesh J. Bot.* 6: 123–132.

Daskalova, C. 2004. Nucellar embryony in *Thalictrum aquilegifolium* L. ssp. *aquillegifolium*. *Dokl. Bulg. Akad. Nauk* 57(3): 71–74.

Daskalova, C. 2005. On the cytoembryology of *Chaenomeles japonica* (Thunb.) Lindl. *Dokl. Bolg. Akad. Nauk.* 58: 711–716.

Davis, G.L. 1963. Generative apospory and diploid parthenogenesis in *Brachyscome ciliaris* var. *lanuginosus*. *Aust. J. Sci.* 26: 90.

Davis, G.L. 1964a. Somatic apospory and polyembryony in *Minuria integerrima*. *Nature* 204: 94.

Davis, G.L. 1964b. Embryological studies in the Compositae. 4. Sporogenesis gametogenesis and embryogeny in *Brachyscome ciliaris*. *Aust. J. Bot.* 12: 142–151.

Davis, G.L. 1966. *Systematic embryology of the angiosperms*. John Wiley & sons, New York.

Davis, G.L. 1967. Apomixis in Compositae. *Phytomorphology* 17: 270–277.

Davis, G.L. 1968. Apomixis and abnormal anther development in *Calotis lappulacea* (Compositae). *Aust. J. Bot.* 16: 1–17.

Davis, H.A., A.M. Fuller and T. Davis. 1967. Contributions toward the revision of the *Eubati* of eastern North America. *Castanea* 32: 20–37.

Day, D.F. 1896. Parthenogenesis in *Thalictrum jendleri*. *Bot. Gaz.* 22: 241.

D'Cruz, R. and P.S. Reddy. 1968. Apomixis in *Pennisetum massaicum* Stapf. *Sci. Cult.* 34: 255–257.

De Candolle, C. 1893. Contribution à l'étude du genre *Alchimilla*. *Bull. Herb. Boissier* 1: 485–495.

Decker, K. 2006. *Draba weberi* Price & Rollins (Weber's draba): a technical conservation assessment. [Online]. USDA Forest Service, Rocky Mountain Region. Available: http://www.fs.fed.us/r2/projects/scp/assessments/drabaweberi.pdf [September 28, 2021].

Dedong, G., K. Chuanhong, L. Liping, L. Yong. 1999. Study of apomixis in the allotriploid Beet (VVC). *Sci. Agric. Sin.* 32(4): 1–5.

Deidda, P. and I. Chessa. 1982. A three-gene hypothesis for the inheritance of nucellar embryony in *Citrus*. *Riv. Ortoflorofrutt. It.* 66(6): 431–436. http://www.jstor.org/stable/42878849

Delgado, L., F. Galdeano, M.E. Sartor, C.L. Quarin, F. Espinoza and J.P.A. Ortiz. 2014. Analysis of variation for apomictic reproduction in diploid *Paspalum rufum*. *Ann. Bot.* 113(7): 1211–1218. doi:10.1093/aob/mcu056

Dent-Acosta S.R. and G.J. Breckon. 1991. Reproductive biology of six species of Melastomataceae in Western Porto Rico. In: *First International Melastomataceae Symposium* - Washington: Smithsonian Institution.
Dermen, H. 1936. Aposporic parthenogenesis in a triploid apple, *Malus hupehensis*. *J. Arnold Arbor.* 17: 90–105.
Desai, S. 1961. Studies on apospory in *Skimmia*. *Phyton* (Argentina) 16: 147–152.
Desai, S.1962. Polyembryony in *Xanthoxylum*. *Phytomorphology* 12: 184–189.
Deshpande, P.K., S.M. Bhuskute and K.H. Makde. 1984. A case of diplospory in *Luffa cylindrica*. *Curr. Sci.* 53: 820–822.
Diana, S. 1992. Alcuni aspetti della biologia della riproduzione del genere *Limonium*. *G. Bot. Ital.* 126 (2): 187–195. doi:10.1080/11263509209430276
Diana, S. 1997. Sulla presenza di poliembrionia in *Dactylorhiza insularis* (Sommier) Landw. (Orchidaceae). *Boll. Soc. Sarda Sci. Nat.* 31: 201–205. http://eprints.uniss.it/3167/
Diana, S. and R.S. Villa. 1991. Apomixis in *Genista salzmannii* DC. (Fabaceae). *G. Bot. Ital.* 125 (1-2): 29–37. http://eprints.uniss.it/7857/
Dias, A.C.C., R.C. Marinho, D.S. Sampaio, A.M. Bonetti and P.E. Oliveira. 2021. Clone worth? Genetic diversity in obligate apomictic *Miconia albicans* (Melastomataceae). *Plant Biol.* 23(5): 743–748. doi:10.1111/plb.13273
Dias, A.C.C., A.C. Serra, D.S. Sampaio, E.L. Borba, A.M. Bonetti and P.E. Oliveira. 2017. Unexpectedly high genetic diversity and divergence among populations of the apomictic Neotropical tree *Miconia albicans*. *Plant Biol.* 20: 244–251.
Díaz Jiménez, P. 2016. Biología floral y reproductiva de *Spathiphyllum cochlearispathum* (Liebm.) Engl. y *Spathiphyllum ortgiesii* Regel (Araceae) en la región de Los Tuxtlas, Veracruz, México. MSc thesis, Universidad Veracruzana, Centro de Investigaciones Tropicales, Veracruz, Mexico, 103 pages.
Dickie, G. 1875. Note on the buds developed on leaves of *Malaxis*. *J. Linnean Soc. London, Botany* 14: 1–3.
Dickinson, T.A. 2018. Sex and Rosaceae apomicts. *Taxon* 67(6): 1093–1107.
Dickinson, T.A., S. Belaoussoff, R.M. Lov and M. Muniyamma. 1996. North American black-fruited hawthorns. I. Variation in floral construction, breeding system correlates, and their possible evolutionary significance in *Crataegus* sect. *Douglasii* Loudon. *Folia Geobot. Phytotax.* 31: 355–371.
Dickinson, T.A., E. Lo and N. Talent. 2007. Polyploidy, reproductive biology and Rosaceae: Understanding evolution and classifications. *Plant Syst. Evol.* 266: 59–78.
Dickinson, T.A. and J.B. Phipps. 1986. Studies in *Crataegus* (Rosaceae: Maloideae). XIV. The breeding system of *Crataegus crus-galli sensu lato* in Ontario (Canada). *Am. J. Bot.* 73: 116–130.
Dickore, B. and G. Kasparek. 2010. Species of *Cotoneaster* (Rosaceae, Maloideae) indigenous to, naturalising or commonly cultivated in Central Europe. *Willdenowia* 40(1): 13–45.
van Dijk, P. and J. van Damme. 2000. Apomixis technology and the paradox of sex. *Trends Plant Sci.* 5(2): 81–84.
van Dijk, P.J. 2003. Ecological and evolutionary opportunities of apomixis: insights from *Taraxacum* and *Chondrilla*. *Phil. Trans. R. Soc. Lond.* B 358: 1113–1121. doi:10.1098/rstb.2003.1302
van Dijk, P.J. and J.M.T. Bakx-Schotman. 2004. Formation of unreduced megaspores (diplospory) in apomictic dandelions (*Taraxacum*) is controlled by a sex-specific dominant gene. *Genetics* 166: 483–492.
van Dijk, P.J., H. de Jong, K. Vijverberg and A. Biere. 2009. An apomixis gene's view on dandelions. In: Schön, I., P. van Dijk, K. Martens (eds.), *Lost Sex*. Springer Netherlands, Dordrecht, pp. 475–493. doi:10.1007/978-90-481-2770-2_22
van Dijk, P.J., R. Op den Camp and S.E. Schauer. 2020. Genetic dissection of apomixis in dandelions identifies a dominant parthenogenesis locus and highlights the complexity of autonomous endosperm formation. *Genes* 11(9): 961. doi:10.3390/genes11090961
van Dijk, P.J., D. Rigola and S.E. Schauer. 2016. Plant breeding: surprisingly, less sex is better. *Curr. Biol.* 26: R122–R124.
van Dijk, P.J., I.C.Q. Tas, M. Falque and J.M.T. Bakx-Schotman. 1999. Crosses between sexual and apomictic dandelions (*Taraxacum*). II. The breakdown of apomixis. *Heredity* 83: 715–721.
van Dijk, P.J. and K. Vijverberg. 2005. The significance of apomixis in the evolution of the angiosperms: a reappraisal. In: Bakker, F.T., L.W. Chatrou, B. Gravendeel and P.B. Pelser (eds.), *Plant Species-level Systematics. New Perspectives on Pattern & Process*, A.R.G. Gantner Verlag, Ruggell, Liechtenstein, pp. 101–116.
Dikshit, B.K. and G. Panigrahi. 1998. The family Rosaceae in India (revisionary studies on *Potentilla* L., *Sibbaldia* L., *Brachycaulos* Dikshit & Panigr.). Bishen Singh Mahendra Pal Singh, Dehra Dun.
do Valle, C.B. and C. Glienke. 1991. New sexual accessions in *Brachiaria*. *Apomixis Newsl.* 13: 11–13.

do Valle, C.B. and J.W. Miles. 2001. Breeding of apomictic species. In: Savidan, Y., J.G. Carman, T. Dresselhaus (eds.), *The flowering of apomixis: From mechanisms to genetic engineering*. CIMMYT, IRD, European Commission OC VI (FAIR), Mexico City, Mexico, pp. 137–152.

Dobeš, C., M. Koch and T.F. Sharbel. 2006. Embryology, karyology, and modes of reproduction in the North American genus *Boechera* (Brassicaceae): A compilation of seven decades of research. *Ann. Mo. Bot. Gard.* 93(3): 517–534.

Dobeš, C., A. Lückl, K. Hülber and J. Paule. 2013a. Prospects and limits of the flow cytometric seed screen - Insights from *Potentilla* sensu lato (Potentilleae, Rosaceae). *New Phytol.* 198: 605–616. doi:10.1111/nph.12149

Dobeš, C., A. Lückl, L. Kausche, S. Scheffknecht, D. Prohaska, C. Sykora and J. Paule. 2015. Parallel origins of apomixis in two diverged evolutionary lineages in tribe Potentilleae (Rosaceae). *Bot. J. Linn. Soc.* 177 (2): 214–229. doi:10.1111/boj.12239

Dobeš, C., A. Milosevic, D. Prohaska, S. Scheffknecht, T.F. Sharbel and K. Hülber. 2013b. Reproductive differentiation into sexual and apomictic polyploid cytotypes in *Potentilla puberula* (Potentilleae, Rosaceae). *Ann. Bot.* 112(6): 1159–1168. doi:10.1093/aob/mct167

Dobeš, C., T. Mitchell-Olds and M. Koch. 2004a. Intraspecific diversification in North American *Boechera stricta* (= *Arabis drummondii*), *Boechera* × *divaricarpa*, and *Boechera holboellii* (Brassicaceae) inferred from nuclear and chloroplast molecular markers—an integrative approach. *Am. J. Bot.* 98: 2087–2101.

Dobeš, C., T. Mitchell-Olds and M. Koch. 2004b. Phylogeographic analysis of extensively sympatric and highly diverse chloroplast haplotypes (trnL intron-trnF IGS) in North American *Arabis drummondii*, *A.* × *divaricarpa*, and *A. holboellii* (Brassicaceae). *Molec. Ecol.* 13: 349–370.

Dobeš, C., S. Scheffknecht, Y. Fenko, D. Prohaska, C. Sykora and K. Hülber. 2018. Asymmetric reproductive interference: The consequences of cross-pollination on reproductive success in sexual–apomictic populations of *Potentilla puberula* (Rosaceae). *Ecol. Evol.* 8: 365–381. doi:10.1002/ece3.3684

Dobeš, C., T.F. Sharbel and M. Koch. 2007. Towards understanding the dynamics of hybridization and apomixis in the evolution of the genus *Boechera* (Brassicaceae). *Syst. Biodivers.* 5(3):321–331. doi:10.1017/S1477200007002423

Doležal, J., A. Krahulcová, T. Urfus and F. Krahulec. 2020. Residual sexuality of the apomict *Pilosella rubra* under natural conditions in the Krkonoše Mts. *Preslia* 92: 403–428.

dos Santos, A.P.M., C. Magioni Fracasso, M.L. dos Santos, R. Romero, M. Sazima and P.E. Oliveira. 2012. Reproductive biology and species geographical distribution in the Melastomataceae: a survey based on New World taxa. *Ann. Bot.* 110(3): 667–679. doi:10.1093/aob/mcs125

Dossett, M., N.V. Bassil, K.S. Lewers and C.E. Finn. 2012. Genetic diversity in wild and cultivated black raspberry (*Rubus occidentalis* L.) evaluated by simple sequence repeat markers. Publications from USDA-ARS / UNL Faculty. 1243.

Dowrick, G.J. 1961. Biology and reproduction in *Rubus*. *Nature* (Lindl) 191: 680–682.

Dowrick, G.J. 1966. Breeding systems in tetraploid *Rubus* species. *Genet. Res.* 7(2): 245–253. doi:10.1017/S0016672300009654

Dresselhaus, T. and S. Coimbra. 2016. Plant reproduction: AMOR enables males to respond to female signals. *Curr. Biol.* 26(8): R321–323. doi:10.1016/j.cub.2016.03.019

Drews, G.N. and A.M. Koltunow. 2011. The female gametophyte. *Arabidopsis Book* 9: e0155. doi:10.1199/tab.0155

Drews, G.N., D. Lee and G.A. Christensen. 1998. Genetic analysis of female gametophyte development and function. *Plant Cell* 10: 5–17.

Dujardin, M. and W.W. Hanna. 1984a. Microsporogenesis, reproductive behaviour and fertility in five *Pennisetum* species. *Theor. Appl. Genet.* 67: 197–202.

Dujardin, M. and W.W. Hanna. 1984b. Cytogenetics of double cross hybrids between *Pennisetum americanum* – *P. purpureum* amphiploids and *P. americanum* × *Pennisetum squamulatum* interspecific hybrids. *Theor. Appl. Genet.* 69: 97–100.

Dujardin, M. and W.W. Hanna. 1989. Developing apomictic pearl millet: characterisation of a BC3 plant. *J. Genet. Breed.* 43: 145–150.

Duncan, E.J. 1970. Ovule and embryo ontogenesis in *Bombacopsis glabra* (Pasq.) A. Robins. *Ann. Bot.* 34: 32–46.

Dunkel, F.G. 2011. Neue Arten aus dem Schweizer *Ranunculus auriomus*-Komplex. *Bauhinia* 23: 33–55.

Dunkel, F.G. 2021. Sieben neue Arten aus dem Schweizer *Ranunculus auricomus*-Komplex. *Forum Geobot.* 10: 1–27. doi:10.3264/FG.2021.0825

Dupont, Y.L. 2002. Evolution of apomixis as a strategy of colonization in the dioecious species *Lindera glauca* (Lauraceae). *Popul. Ecol.* 44: 293–297.

Durand, J., L. Garnier, I. Dajoz, S. Mousset and M. Veuille. 2000. Gene flow in a facultative apomictic Poaceae, the savana grass *Hyparrhenia diplandra*. *Genetics* 156: 823–831.
Dusi, D.M.A. 2001. Apomixis in *Brachiaria decumbens* Stapf. PhD thesis, Wageningen University, Brasilia, Brazil, 167 pages.
Dusi, D.M.A. and M.T.M. Willemse. 1999. Apomixis in *Brachiaria decumbens* Stapf.: Gametophytic development and reproductive calendar. *Acta Biol. Cracov. Ser. Bot.* 41: 151–162.
Duthie, A.V. 1915. Note on apparent apogamy in *Pterygodium newdigatae*. *Trans. Roy. Soc. South Africa* 5(1): 593–598. doi:10.1080/00359191509519730
Dwivedi, K.K., S.R. Bhat, V. Bhat, B.V. Bhat and M.G. Gupta. 2007. Identification of a SCAR marker linked to apomixis in buffelgrass (*Cenchrus ciliaris* L.). *Plant Sci.* 172(4):788–795.
Dwivedi, K.K., A. Radhakishna, S. Kumar, M.K. Srivastava, M.G. Gupta and P. Kaushal. 2015. Development of an ISSR– derived SCAR marker linked to apospory in buffel grass (*Cenchrus ciliaris* L.). *Indian J. Gen.* 75(2): 271–273.
Dziadczyk, E., M. Domaciuk, M. Nowak, E. Szczuka and J. Bednara. 2011. The development of feale gametophyte in *Fragaria* × *ananassa* Duch. cv. Selva. *Acta Biol. Cracov. Ser. Bot.* 53(2): 104–112.
Ebina, M., H. Nakagawa, T. Yamamoto, H. Araya, S. Tsuruta, M. Takahara, and K. Nakajima. 2005. Co-segregation of AFLP and RAPD markers to apospory in guineagrass (*Panicum maximum* Jacq.). *Grassl. Sci.* 51: 71–78.
Edman, G. 1929. Zur Entwicklungsgeschichte der Gattung *Oxyria* Hill, nebst zytologischen, embryologischen und systematischen Bemerkungen über einige andere Polygonaceen. *Acta Horti Berg.* 9: 165–291.
Edman, G. 1931. Apomeiosis und Apomixis bei *Atraphaxis frutescens* C.Koch. *Acta Horti Berg.* 11: 13–66.
Ehtheridge, A.L. and J.M. Herr Jr. 1968. The development of the ovule and megagametophyte in *Rhexia mariana*. *Can. J. Bot.* 46: 133–139.
Einset, J. 1951. Apomixis in American polyploid blackberries. *Amer. J. Bot.* 38: 768–772.
El Mokadem, H., L. Crepel, J. Meynet, S. Gudin and Y. Jacob. 2001. Gametic behaviour of parthenogenetic plants of *Rosa canina* L. and *R. hybrida* L. *Acta Hortic.* 547: 289–296.
El'konin, L.A., E.V. Beliaeva and I. Fadeeva. 2012. Expression of the apomictic potential and selection for apomixis in *Sorghum* line AS-1a. *Genetika* 48(1): 40–49.
Eliott, F.G., M. Shepherd, M. Rossetto, P. Bundock, N. Rice and R.J. Henry 2014. Contrasting breeding systems revealed in the rainforest genus *Davidsonia* (Cunoniaceae): can polyembryony turn the tables on rarity? *Austral. J. Bot.* 62, 451–464. doi:10.1071/BT14063
Ellerstrom, S. 1983. Apomictic progeny from *Raphanobrassica*. *Hereditas* 99: 315.
Ellerstrom, S. and L. Zagorcheva. 1977. Sterility and apomictic embryo-sac formation in *Raphanobrassica*. *Hereditas* 87: 107–120.
Emery, W.H.P. 1957. A study of reproduction in *Setaria macrostachya* and its relatives in the Southwestern United States and Northern Mexico. *Bull. Torrey Bot. Club* 84(2) 106–121. http://www.jstor.org/stable/2482786
Emery, W.H.P. and W.V. Brown. 1958. Apomixis in Gramineae. Tribe Andropogoneae: *Heteropogon contortus*. *Madroño* 14: 238–246.
Engelbert, V. 1941. Development of twin embryo sacs, embryos and endosperm in *Poa arctica*. *Canad. J. Res.* 19: 135–144.
Engell, K. 1970. Embryological investigations in *Arnica alpina* from Greenland. *Bot. Tidskr.* 71: 237–244.
Engle, J. 1988. The spread and effect of the vine *Bryonia alba* in Whitman Co., Washington. MSc thesis, Washington State University. Pullman, WA, USA.
Ensign, M.R. 1919. Venation and senescence of polyembryonic *Citrus* plants. *Am. J. Bot.* 6(8): 311. doi:10.1002/j.1537-2197.1919.tb05546.x
Erben, M. 1978. Die Gattung *Limonium* im südwest-mediterranen Raum. *Mitt. Bot. Staatssamml. Münch.* 14: 361–631.
Erben, M. 1979. Karyotype differentation and its consequences in Mediterranean *Limonium*. *Webbia* 34: 409–417.
Erben, M. 1981. Bemerkungen zur Taxonomie der Gattung *Limonium* II. *Mitt. Bot. Staatssamml. Münch.* 17: 485–510.
Erben, M. 1988. Bemerkungen zur Taxonomie der Gattung *Limonium* IV. *Mitt. Bot. Staatssamml. Münch.* 27: 381–406.
Erben, M. 1989. Bemerkungen zur Taxonomie der Gattung *Limonium* V. *Mitt. Bot. Staatssamml. Münch.* 28: 313–417.
Erben, M. 1993. *Limonium* Miller. *In*: *Flora Iberica* 3. C.S.I.C., Madrid, pp. 2–143.

Eriksen, B. 1996. Mating systems in two species of *Potentilla* from Alaska. *Folia Geobot.* 31, 333–344. doi:10.1007/BF02815378
Eriksen, B. and M. Fredrikson. 2000. Megagametophyte development in *Potentilla nivea* (Rosaceae) from Northern Swedish Lapland. *Am. J. Bot.* 87: 642–651.
Erilova, A., L. Brownfield, V. Exner, M. Rosa, D. Twell, O. Mittelsten Scheid, L. Hennig and C. Köhler. 2009. Imprinting of the polycomb group gene *MEDEA* serves as a ploidy sensor in Arabidopsis. *PLoS Genet.* 5: e1000663. doi:10.1371/journal.pgen.1000663
Ernst, A. 1901. Beitrage zur Kenntnis der Entwicklung des Embryosackes und des Embryo (Polyembryonie) von *Tulipa gesneriana* L. *Flora* 88: 37–77.
Ernst, A. 1918. *Bastardierung als Ursache der Apogamie im Pflanzenreich.* Verlag von Gustav Fischer, Jena, Germany.
Ernst, A. and C. Bernard. 1912. Beitrage zur Kenntniss der Saprophyten Javas. 9. Entwicklungsgeschichte des Embryosackes, des Embryos und des Endosperm von *Burmannia coelestis. Ann. Jard. Bot. Buitenz.* 2. Ser. 11: 234–257.
Esau, K. 1944. Apomixis in Guyayule. *Proc. Nat. Acad. Sci. U.S.A.* 30: 352–355.
Esau, K. 1946. Morphology of reproduction in guyayule and certain other species of *Parthenium. Hilgardia* 77: 61–101.
Esen, A. and K. Soost. 1977. Adventive embryogenesis in *Citrus* and its relation to pollination and fertilization. *Am. J. Bot.*, 64: 607–614.
Espinoza, F., M.H. Urbani, E.J. Martínez and C.L. Quarin. 2001. The breeding system of three *Paspalum* species. *Trop. Grassl.* 35: 211–217.
Evans, L.T. and R.B. Knox. 1969. Environmental control of reproduction in *Themeda australis. Aust. J. Bot.* 17: 375–389.
Fagerlind, F. 1940a. Die Terminologie der Apomixis-Prozesse. *Hereditas* 26: 1–22.
Fagerlind, F. 1940b. Zytologie und Gametophytenbildung in der Gattung *Wikstroemia. Hereditas* 30: 179–200.
Fagerlind, F. 1944a. Der Zusammenhang zwischen Perennität, Apomixis und Polyploidie. *Hereditas* 30: 179–200.
Fagerlind, F. 1944b. Die samenbildung und die Zytologie bei agamospermischen und sexuallen Arten von *Elatostema* und einigen nehestehenden Gattungen nebst Beleuchtung einiger damit zusammenhängender Probleme. *Kgl. Svensk. Vet.-Akad. Handl.* 21(4): 1–130.
Fagerlind, F. 1945a. Bildung und Entwicklung des Embryosacks bei sexuellen und agamospermischen *Balanophora*-Arten. *Svensk Bot. Tidskr.* 39: 65–82.
Fagerlind, F. 1945b. Die Bastarde der *Canina*-Rosen, ihre Syndese- und Formbildungsverhältnisse. *Acta Horti Berg.* 14(2): 9–32.
Fagerlind, F. 1946. Sporogenesis, Embryosackentwicklung und pseudogame Samenbildung bei *Rudbeckia laciniata* L. *Acta Horti Berg.* 14: 39–90.
Fagerlind, F. 1947. Macrogametophyte formation in two agamospermous *Erigeron* spp. *Acta Horti Berg.* 14: 221–247.
Faltermeier, M., N. Meyer, T. Gregor, J. Paule, M. Lepší, B. Fussi, P. Koutecký, H. Schäfer and J. Ewald. 2016. Genetische und blattmorphologische Variabilität von *Sorbus ratisbonensis. Ber. Bayer. Bot. Ges.* 86: 37–56.
FAO. 2009. *The state of food and agriculture*, FAO, Rome.
Farco, G.E., M.M. Sosa, M. Dematteis and A. Fernández. 2012. Cytology and embryology of the pompom weed, *Campuloclinium macrocephalum* (Eupatorieae, Asteraceae). *South Afr. J. Bot.* 78: 21–29.
Faria, J.M.R., A.A.M. van Lamneren and H.W.M. Hilhorst. 2004. Desiccation sensitivity and cell cycle aspects in seed of *Inga vera* subsp. *affins. Seed Sci. Res.* 14: 165–178.
Farquharson, L.I. 1955. Apomixis and polyembryony in *Tripsacum dactyloides. Am. J. Bot.* 42: 737–743. doi:10.2307/2438651
Faruqi, S.A., S. Afridi and S.M.H. Usmani. 1975. Apomixis and polyembryony in some Andropogoneae from Pakistan. *Pak. J. Bot.* 7: 135–138.
Fayeun, L.S., L.A. Hammed, O.A. Oduwaye, J.U. Madike and E.U. Ushie. 2016. Estimates of Genetic variability for seedling traits in fluted pumpkin (*Telfairia occidentalis* Hook. f.). *Plant Breed. Biotechnol.* 4: 262–270.
Febulaus, G.N.V. 1992. Apomixis in *Panicum notatum* Retz. (Poaceae). *J. Indian Bot. Soc.* 71: 147–151.
Febulaus, G.N.V. and T. Pullaiah. 1992. Embryology of *Eragrostiella bifaria* (Vahl) Bor (Poaceae). *Ann. Bot.* (Rome) 50: 5–15.
Febulaus, G.N.V. and T. Pullaiah. 1995. Apomixis in *Cenchrus ciliaris* L (Poaceae). In Chauhan, Y.S. and A.K. Pandey (eds.), *Advances in plant reproductive biology.* Narendra Publishing House, India, pp. 153–161.
Fedortschuk, V.R. 1944. Development and organization of the embryo and seed of red clover (*Trifolium pratense* L). *Moscow State Publ. Agric. Lit.* 25: 1–39.

Fehér, A. 2015. Somatic embryogenesis — Stress-induced remodelling of plant cell fate. *Biochim. Biophys. Acta Gene Regul. Mech.* 1849(4): 385–402. doi:10.1016/j.bbagrm.2014.07.005

Fehrer, J., A. Krahulcová, F. Karahulec, J. Chrtek and S. Bräutigam. 2007. Evolutionary aspects in *Hieracium* subgenus *Pilosella*. In: Hörandl, E., U. Grossniklaus, P. van Dijk and T. Sharbel (eds.), *Apomixis: Evolution, Mechanisms and Perspectives*. Gantner Verlag, Ruggell, Liechtenstein, pp. 359–390.

Fei, X., Y. Lei, Y. Qi, S. Wang, H. Hu and A. Wei. 2021. Small RNA sequencing provides candidate miRNA-target pairs for revealing the mechanism of apomixis in *Zanthoxylum bungeanum*. *BMC Plant Biol.* 21: 178. doi:10.1186/s12870-021-02935-5

Ferreira, D.N.S., J.L.C. Camargo and I.D.K. Ferraz. 2019. Do polyembryonic seeds of *Carapa surinamensis* (Meliaceae) have advantages for seedling development? *Acta Amazonica* 49(2): 97–104. doi:10.1590/1809-4392201801211

Ferri S. 1961. Ricerche sullo sviluppo embriologico di *Erigeron speciosus* DC. Compositae. *G. Bot. Ital.* 68(3–4): 269–284. doi:10.1080/11263506109433009

Feulner, M., A. Weig, J. Paule, T. Gregor, L.F. Schott and G. Aas. 2017. Genetic variability and morphology of tri- and tetraploid members of the *Sorbus aria* complex in northern Bavaria. *Preslia* 89: 275–290.

Figueiredo, D.D., R.A. Batista, P.J. Roszak and C. Köhler. 2015. Auxin production couples endosperm development to fertilization. *Nat. Plants* 1: 15184. doi:10.1038/nplants.2015.184

Figueiredo, D.D. and C. Köhler. 2018. Auxin: a molecular trigger of seed development. *Genes & Dev.* 32: 479–490. doi:10.1101/gad.312546.118

Filutowicz, A and L. Dalke. 1972. Mieszańce buraka cukrowego (*Beta vulgaris* L.) z *Beta lomatogona* F. et M. *Hodowla Roślin Aklimatyzacja i Nasiennictwo* 16(2): 85–93.

Finkeldey, R. and H.H. Hattemer. 2007. Sexual and asexual reproduction in tropical forests. In: Tropical Forest Genetics. Tropical Forestry. Springer, Berlin, Heidelberg. doi:10.1007/978-3-540-37398-8_4

Firetti-Leggieri, F., L.G. Lohmann, S. Alcantara, I.R. Costa and J. Semir 2013. Polyploidy and polyembryony of *Anemopaegma* (Bignonieae, Bignoniaceae). *Plant Reprod.* 26: 43–53.

Fisher, W.D., E.C. Bashaw and E.C. Holt. 1954. Evidence for apomixis in *Pennisetum ciliare* and *Cenchrus setigerus*. *Agron. J.* 46: 401–404.

Flagg, R.O. and W.S. Flory. 1976. Origins of three Texas species of *Zephyranthes*. *Pl. Life* 32: 67–80.

Flinck, K.E. and B. Hylmö. 1962. *Cotoneaster harrysmithii*, a new species from Western China. *Bot. Not.* 115: 29–34.

Flora of North America Editorial Committee, ed. 2006. Magnoliophyta: Asteridae, part 6: Asteraceae, part 1. 19: i–xxiv, 1–579. In Fl. N. Amer. Oxford University Press, New York.

Ford, H. and A. Richards. 1985. Isozyme variation within and between *Taraxacum* agamospecies in a single locality. *Heredity* 55: 289–291.

Forenbacher, A. 1913. Rasplodne prilike u roda *Potentilla*. (Die Fortpflanzungsverhältnisse bei der Gattung *Potentilla*.) [in Serbo-Croation]. *Rad Jugoslav. Akad. Znan.* 55: 132–160.

Foster, T.M., N.V. Bassil, M. Dossett, M.L. Worthington and J. Graham. 2019. Genetic and genomic resources for *Rubus* breeding: a roadmap for the future. *Hortic. Res.* 6: 116. doi:10.1038/s41438-019-0199-2

Foxworthy, F.W. 1932. Dipterocarpaceae of the Malaysian Peninsula. *Malaysian Forest Records* 10: 169.

Francini, B. 1928. Fenomeni di aposporia somatica, di aposporia goniale e di embrionia avventzia in *Ochna multiflora*. *C.R. Acad. Lincei (Roma)*. 7: 92.

Francis, G., O. John, S. Piergiorgio and S. Mulpuri. 2020. Apomixis as a tool for development of high yielding clones and selections in *Jatropha curcas* L. *Genet. Resour. Crop. Evol.* 67: 727–743. doi: 10.1007/s10722-019-00851-0

Freire-Fierro, A. 2002. Monograph of *Aciotis* (Melastomataceae). *Syst. Bot. Monogr.* 62: 1–99.

Freitas, D.Y.H. and N.M.A. Nassar. 2013. Apomixis in cassava: advances and challenges. *Genet. Mol. Res.* 12(2): 988–994.

Friedman, W.E. 1999. Expression of the cell cycle in sperm of *Arabidopsis*: implications for understanding patterns of gametogenesis and fertilization in plants and other eukaryotes. *Development* 126: 1065–1075.

Friedman, W.E. 2001. Developmental and evolutionary hypotheses for the origin of double fertilization and endosperm. *C.R. Acad. Sci. III* 324(6): 559–567. doi:10.1016/s0764-4469(01)01326-9

Friedman, W.E. 2006. Embryological evidence for developmental lability during early angiosperm evolution. *Nature* 441:337–340.

Frost, H.B. 1926. Polyembryony, heterozygosis and chimeras in *Citrus*. *Hilgardia* 1: 365–402.

Frost, H.B. 1938a. Nucellar embryony and juvenile character in clonal variaties of *Citrus*. *J. Hered.* 29(11): 423–492. doi:10.1093/oxfordjournals.jhered.a104431

Frost, H.B. 1938b. The genetics and cytology of *Citrus*. *Current Sci.* 1938 (special no., Genetics): 24–27.

Fryer, J. and B. Hylmö. 2001. Captivating *Cotoneaster*. *New Plantsman* 8: 227–238.

Fu, L., F-X. Xu and Q-W. Zeng. 2012. Embryology of the dioecious *Woonyoungia septentrionalis* (Magnoliaceae). *Nord. J. Bot.* 30: 215–225. doi:10.1111/j.1756-1051.2011.01151.x

Fu, L-F., L-Y. Su, A. Mallik, F. Wen and Y-G. Wei. 2017. Cytology and sexuality of 11 species of *Elatostema* (Urticaceae) in limestone karsts suggests that apomixis is a recurring phenomenon. *Nordic J. Bot.* 35(2): 251–256.

Fukai, S., T. Shimomura and T. Kondo. 2000. Morphological observation of polyembryony in *Ophiopogon japonicus* (Liliaceae). *J. Japanese Soc. Hort. Sci.* 69 (5): 614–616. doi:10.2503/jjshs.69.614

Furusato, K. 1952. Genetics and breeding of *Citrus*. *Nat. Inst. Genet. (Japan) Ann. Rep.* 41.

Furusato, K., Y. Ohta and K. Ishibashi. 1957. Studies on polyembryony in *Citrus*. *Rep. Kihara Inst. Biol. Res.* 8: 40–48.

Gao, S-T., C-B. Liu, D-L. Liu, Y.N. Lin and Y.F. Chen. 1999. Apomixis characteristics of *Juglans regia* L. *J. Northwest Forest. Coll.* 14(1): 45–47.

Gadella, T.W.J. 1982. Cytology and reproduction of *Hieracium pilosella* L. and some related diploid species. *Acta Bot. Neerl.* 31: 140–141

Gadella, T.W.J. 1984. Cytology and the mode of reproduction of some taxa of *Hieracium* subgenus *Pilosella*. *Proc. Kon. Ned. Acad. Wetensch* 87: 387–399.

Gadella, T.W.J. 1992. Notes on some triple and inter-sectional hybrids in *Hieracium* L. subgenus *Pilosella* (Hill) S. F. Gray. *Proc. Kon. Ned. Akad. Wetensch., Biol.* 95: 51–63.

Galambosi, B., C. Rey and J.F. Vouillamoz. 2010. Suitability of Swiss herb cultivars under Finnish climatic conditions. *Acta Hortic.* 860: 173–180. doi:10.17660/ActaHortic.2010.860.25

Galdeano, F., M.H. Urbani, M.E. Sartor, A.I. Honfi, F. Espinoza and C.L. Quarin. 2016. Relative DNA content in diploid, polyploid, and multiploid species of *Paspalum* (Poaceae) with relation to reproductive mode and taxonomy. *J. Plant Res.* 129(4): 697–710. doi:10.1007/s10265-016-0813-4

Galla, G., G. Barcaccia, A. Schallau, M.P. Molins, H. Bäumlein and T.F. Sharbel. 2011. The cytohistological basis of apospory in *Hypericum perforatum* L. *Sex. Plant Reprod.* 24: 47–61.

Galla, G., A. Basso, S. Grisan, M. Bellucci, F. Pupilli and G. Barcaccia. 2019. Ovule gene expression analysis in sexual and aposporous apomictic *Hypericum perforatum* L. (Hypericaceae) accessions. *Front. Plant Sci.* 10: 654. doi:10.3389/fpls.2019.00654

Galla, G., H. Vogel, T.F. Sharbel and G. Barcaccia. 2015. *De novo* sequencing of the *Hypericum perforatum* L. flower transcriptome to identify potential genes that are related to plant reproduction sensu lato. *BMC Genom.* 16: 254. doi:10.1186/s12864-015-1439-y

Ganong, W.F. 1898a. Upon polyembryony and its morphology in *Opuntia vulgaris*. *Bot. Gaz.* 25: 221–228.

Ganong, W.F. 1898b. Contributions to a knowledge of the morphology and ecology of the Cactaceae. II. The comparative morphology of the embryos and seedlings. *Ann. Bot.* 12: 423–474.

Garbus, I., J.P. Selva, M.C. Pasten, A.M. Bellido, J. Carballo, E. Albertini and V. Echenique. 2019. Characterization and discovery of miRNA and miRNA targets from apomictic and sexual genotypes of *Eragrostis curvula*. *BMC Genom.* 20: 839. doi:10.1186/s12864-019-6169-0

Garcia, R., M. Asins, J. Forner and E. Carbonell. 1999. Genetic analysis of apomixis in *Citrus* and *Poncirus* by molecular markers. *Theor. Appl. Genet.* 99: 511–518.

Gasser, C.S. and D.J. Skinner. 2019. Development and evolution of the unique ovules of flowering plants. *Curr. Top. Dev. Biol.* 131: 373–399.

Gehrke, B., C. Bräuchler, K. Romoleroux, M. Lundberg, G. Heubl and T. Eriksson. 2008. Molecular phylogenetics of *Alchemilla, Aphanes* and *Lachemilla* (Rosaceae) inferred from plastid and nu-clear intron and spacer DNA sequences, with comments on genetic classification. *Molec. Phylog. Evol.* 47: 1030–1044.

Gelin, O.E.V. 1934. Embryologische und cytologische studien in Heliantheae-Coreopsidineae. *Acta Hort. Berg.* 11: 99–128.

Gentscheff, G. 1937. Zytologische und embryologische Studien über einige *Hieracium* Arten. *Planta* 27: 165–195.

Gentscheff, G. and A. Gustafsson. 1940. The balance system of meiosis in *Hieracium*. *Hereditas* 26: 209–249. doi:10.1111/j.1601-5223.1940.tb03233.x

Gentscheff, G.J. 1938. Über die pseudogame Fortpflanzung bei *Potentilla*. *Genetica* 20: 398–408.

Georgakopoulou, A., S. Manousou, R. Artelari and O. Georgiou. 2006. Breeding systems and cytology in Greek populations of five *Limonium* species (Plumbaginaceae). *Willdenowia* 36(2): 741–750. http://www.jstor.org/stable/3997748

Gerlach, D. 1965. Befruchtung und Autogamie bei *Rubus caesius*. *Biol. Zentralbl.* 84: 611–633.

Gerstel, D.U., B.L. Hammond and C. Kidd. 1953. An additional note on the inheritance of apomixis in guayule. *Bot. Gaz.* 115: 89–93.

Ghimire, B., S. Son, J.H. Kim and M.J. Jeong. 2020. Gametophyte and embryonic ontogeny: understanding the reproductive calendar of *Cypripedium japonicum* Thunb. (Cypripedoideae, Orchidaceae), a lady's slipper orchid endemic to East Asia. *BMC Plant Biol.* 20: 426. https://doi.org/10.1186/s12870-020-02589-9

Ghosh, R.B. 1962. A note on the nucellar polyembryony in *Aphanamixis polystachya* (Wall.) Parker. *Curr. Sci.* 31(4): 165.
Ghosh, R.B. 1964. Supernumerary archesporial cells in *Ochna atropurpurea* DC. *Curr. Sci.* 33: 116–117.
Ghosh, R.B. 1970 Studies on the embryology of Burseraceae - embryogeny of *Garuga pinnata* Roxb. *Broteria* (Lisboa) 39: 17–23.
Ghosh, R.B. 1972. Studies in the embryology of the family Meliaceae. IV. Fertilisation, endosperm and embryogeny of *Aphanamixis polystachya* (Wall.) Parker (syn. *Amoora rohituka* W. & A.) a medicinal plant with a discussion on its taxonomic status and horticulture. *An. Estac. Exp. Aula Dei* 11 (3/4): 396–403.
Ghosh, R.B. 1974. Embryological studies in the family Rutaceae. Megasporogenesis, female gametophyte and embryo development in *Zanthoxylum oxyphyllum* Edgew. *An. Estac. Exp. Aula Dei* 12(3/4): 113–121.
Ghosh, R.B., B. Ghosh and S. Das. 1976. The development of the female gametophyte in *Toddalia asiatica* (Linn.) Lamk (Syn.: *T. aculeata* Pers.). *An. Estac. Exp. Aula Dei* 13(3/4): 217–222.
Gianini Aquino, A.C. 2021. Niveles de ploidía y modos de reproducción en especies del género *Habranthus* (Amaryllidaceae). Doctoral thesis, Universidad Nacional de Córdoba, Córdoba, Argentina, 206 pages.
Gianini Aquino, A.C., A.I. Honfi and J.R. Daviña. 2019. Modo de reproducción y origen de la progenie de *Habranthus tubispathus* (L. Hér.) Traub. (Amaryllidaceae). Argentina. Ciudad Autónoma de Buenos Aires. *BAG - J. Basic Appl. Genetics* 30(1): 249.
Gifford, E.M. and A.M. Foster. 1989. *Morphology and evolution of vascular plants.* W.H.Freeman and Company, New York.
Gil, L. and L. Lorens. 1991. *Limonium barceloi* and *L. bolosii* Gil & Llorens, new species from the island of Mallorca (Balearic Islands, Spain). *An. Jard. Bot. Madr.* 49(1): 51–56.
Gildenhuys, P.S. and K. Brix. 1959. Apomixis in *Pennisetum dubium. S. Afr. J. Agr. Sci.* 2: 231–245.
Gilles, L.M., A. Khaled, J. Laffaire, S. Chaignon, G. Gendrot, J. Laplaige, H. Bergès, G. Beydon, V. Bayle, P. Barret, J. Comadran, J-P. Martinant, P.M. Rogowsky and T. Widiez. 2017. Loss of pollen-specific phospholipase NOT LIKE DAD triggers gynogenesis in maize. *EMBO J.* 36: 707–717. doi:10.15252/embj.201796603
Glazunova, K.P. 1983. Mikrosprorogenez i mikrogametofitogenez u *Alchemilla filicaulis* Buser (Microsporogenesis and microgametophytogenesis in *Alchemilla filicaulis* Buser). *Naucl. Dokl. Vysshei Shkoly Biol. Nauki* 7:79–83.
Glazunova, K.P. 1984. Apomixis the Eastern European members of the genus *Alchemilla* L. PhD thesis, Moscow, Russia. [In Russian]
Glazunova, K.P. 1987. Formation of embryo sacs in agamic species of lady's mantle (*Alchemilla* L.). *Byull. Mosk. Obshestva Prirodi. Biol.* 92(5): 96–110.
Glazunova, K.P. and V.D. Myatlev. 1990. Korrelyatsionnaya struktura i izmenchivost priznakov pri regulyarnom apomiksise (na primere agamnogo vida *Alchemilla monticola* Opiz) [Correlation structure and variability of characters in the case of regular apomixis (*Alchemilla monticola* Opiz as an example of agamic species)]. *Byull. Mosk. Obshch. Ispyt. Prir., Otd. Biol.* 95: 96–109.
Glutchenko, G.I. 1968. Early manifestation of adventive embryony in *Scopolia carniolica.* In: *Apomixis in Animals and Plants.* Nauka, Novosibirsk, pp. 10.
Goebel, K. 1900. Bemerkung zu vorstehende Mitteilung. *Biol. Ztbl.* 20: 571–572.
Goel, S., Z. Chen, Y. Akiyama, J.A. Conner, M. Basu, G. Gualtieri, W.W. Hanna and P. Ozias-Akins. 2006. Comparative physical mapping of the apospory-specific genomic region in two apomictic grasses: *Pennisetum squamulatum* and *Cenchrus ciliaris. Genetics* 173(1): 389–400. doi:10.1534/genetics.105.054429
Goel, S., Z. Chen, J.A. Conner, Y. Akiyama, W.W. Hanna and P. Ozias-Akins. 2003. Delineation by fluorescence in situ hybridization of a single hemizygous chromosomal region associated with aposporous embryo sac formation in *Pennisetum squamulatum* and *Cenchrus ciliaris. Genetics* 163: 1069–1082.
Goel, S., H.D. Singh and S.N. Raina. 2009. *Cenchrus*: a potential forage crop. Springer WCR (Wild Crop Relatives) Series Volume 2. *Wild Relatives of Millets and Forage Grasses.*
Gohil, R.N. and R. Kaul. 1981. Studies on male and female meiosis in Indian *Allium. Chromosoma* 82: 735–739.
Golczyk, H., A. Massouh and S. Greiner. 2014. Translocations of chromosome end-segments and facultative heterochromatin promote meiotic ring formation in evening primroses. *Plant Cell* 26: 1280–1293.
Goldenberg, R. 1994. Estudos sobre a biologia reprodutiva de espécies de Melastomataceae de Cerrado em Itirapina, SP. Dissertação de Mestrado, Universidade Estadual de Campinas, Campinas, Brazil, 88 pages.
Goldenberg, R. and G.J. Shepherd. 1998. Studies on the reproductive biology of Melastomataceae in 'cerrado' vegetation. *Plant Syst. Evol.* 211: 13–29.
Goldenberg, R. and I.G. Varassin. 2001. Sistemas reprodutivos de espécies de Melastomataceae da Serra do Japi. *Rev. Bras. Bot.* 24: 283–288.

Goldflus, M. 1898–1899. Sur la structure et les fonctions et ales fonctions de l'assie epitheliale et des antipodes chez les Composees. *J. de Botanique* 12(13): 1–38.

Gonzales, E.B. 2005. Multi-scale genetics analysis of two *Trillium* species (*Trillium reliquum* and *Trillium cuneatum*) in space and time. PhD thesis, University of Georgia, Greece, 127 pages.

Gorham, A.L. 1952. Development ot the female gametophyte and embryo in *Smilacina racemosa* (L) Dest. MSc thesis, McGill University, Montreal, Canada, 65 pages.

Gorham, A.L. 1953. The question of fertilization in *Smilacina racemosa*(L.) Desf. *Phytomorphology* 3(1,2): 44–50.

Grant, V. 1981. *Plant Speciation*. Columbia University Press, New York Chichester, West Sussex. doi:10.7312/gran92318

Grashoff, J.L., M.W. Bierner and D.K. Northington. 1972. Chromosome numbers in North and Central American Compositae. *Brittonia* 24: 379–394.

Grattapaglia, D., C. Costa e Silva and N.M.A. Nassar. 1996. Strict maternal inheritance of RAPD fingerprints confirms apomixis in cassava (*Manihot esculenta* Crantz). *Can. J. Plant Sci.* 76: 379–382.

Grazi, F., M. Umaerus and E. Akerberg. 1961. Observations on the mode of reproduction and embryology of *Poa pratensis*. *Hereditas* 46: 489–541.

Greene, C.W. 1984. Sexual and apomictic reproduction in *Calamagrostis* (Gramineae) from Eastern North America. *Am. J. Bot.* 71: 285–293.

Gregor, T. 2004a. Apomicts in the Vegetation of Central Europe. Habilitationsschrift, Univ. Vechta, Vechta (http://www.flora-deutschlands.de → Publikationen).

Gregor, T. 2004b. *Potentilla asiatica* T. Gregor, ein Fingerkraut der *Potentilla collina*-Gruppe aus der südlichen oberheinebene. *Bauhinia* 18: 5–20.

Gregor, T. 2013. Apomicts in the vegetation of Central Europe. *Tuexenia* 33: 233–257.

Greilhuber, J., M.A. Lysák, J. Doležel and M.D. Bennett. 2005. The origin, evolution and proposed stabilisation of the terms 'Genome Size' and 'C-Value' to describe nuclear DNA contents. *Ann. Bot.* 94:255–260.

Greilhuber J., E.M. Temsch and J. Loureiro. 2007. Nuclear DNA content measurement. In: Doležel J., J. Greilhuber, J. Suda (eds.), *Flow cytometry with plant cells – Analysis of Genes, Chromosomes and Genomes*. Wiley-VCH Verlag GmbH & Co. KGaA, pp. 67–101. doi:10.1002/9783527610921.ch4

Griffa, S., A. Ribotta, E. López Colomba, E. Tommasino, E. Carloni, C. Luna and K. Grunberg. 2010. Evaluation seedling biomass and its components as selection criteria for improving salt tolerance in Buffel grass genotypes. *Grass Forage Sci.* 65: 358–361.

Grimanelli, D., M. García, E. Kaszas, E. Perotti and O. Leblanc. 2003. Heterochronic expression of sexual reproductive programs during apomictic development in *Tripsacum*. *Genetics* 165: 1521–1531.

Grimanelli, D., O. Leblanc, E. Espinosa, E. Perotti, D. González-de-León and Y. Savidan. 1998. Mapping diplosporous apomixis in tetraploid *Tripsacum*: One gene or several genes? *Heredity* 80: 33–39.

Grimanelli, D., O. Leblanc, E. Perotti and U. Grossniklaus. 2001. Developmental genetics of gametophytic apomixis. *Trends Genet.* 17: 597–604.

Gröber, K. and M. Zacharias. 1983. Über Nutzungsmöglichkeiten der apomiktischen Samenausbildung bei ausgewählten Kulturpflanzen. *Kulturpflanze* 31: 145–163.

Grossniklaus, U. 2001. From sexuality to apomixis: Molecular and genetic approaches. In: Savidan, Y., J.G. Carman and T. Dresselhaus (eds.), *The flowering of Apomixis: From Mechanisms to Genetic Engineering*. CIMMYT, IRD, European Commission OC VI (FAIR), Mexico City, Mexico, pp. 168–211.

Grossniklaus, U. (ed.) 2019. *Current topics in developmental biology*. Vol. 131. Academic Press, Cambridge.

Grossniklaus, U., A.M. Koltunow and M.M. van Lookeren Campagne. 1998. A bright future for apomixis. *Trends Plant Sci.* 3:415–416.

Grover, A. and P.C. Sharma. 2016. Development and use of molecular markers: past and present. *Crit. Rev. Biotechnol.* 36(2): 290–302. doi:10.3109/07388551.2014.959891

Grun, P. 1955. Cytogenetic studies of *Poa*-III. Variation within *Poa nervosa*, an obligate apomict. *Am. J. Bot.* 42: 778–784.

Gucker, C.L. and N.L. Shaw. 2018. Tapertip hawksbeard (*Crepis acuminata*). In: Gucker, C.L. and N.L. Shaw (eds.), *Western forbs: Biology, ecology, and use in restoration*. Reno, NV: Great Basin Fire Science Exchange. 20 pages. Online: http://greatbasinfirescience.org/western-forbs-restoration

Gucker, C.L. and N.L. Shaw. 2019. Limestone hawksbeard (*Crepis intermedia* A. Gray). In: Gucker, C.L. and N.L. Shaw (eds.), *Western forbs: Biology, ecology, and use in restoration*. Great Basin Fire Science Exchange, Reno, NV, 14 pages. Online: http://greatbasinfirescience.org/western-forbs-restoration

Gudjonsson, G. 1941. Om *Aphanes arvensis* L. og *A. microcarpa* (Boiss. et Reut.) Rothm. og deres Udbredelse i Danmark. *Svensk Bot. Tidskr.* 45: 671–678.

Guignard, L. 1922. Sur l'existence de corps protéiques dans le pollen de diverses Asclepiadacées. *C.R.Acad. Sci. Paris* 175: 1015–1020.

Gunaga, R.P., Doddabasav and R. Vasudeva. 2011. Enhancement of seed germination through proper pre-sowing treatment in *Calophyllum inophyllum*. *Karnataka J. Agric. Sci.* 24(3): 413–414.

Guo-Liang, W., L. He, L. Qun-Long, W. Yong and Z. Peng-Fei. 2010. 'Qinquan 1', a new apomixis walnut cultivar. *Fruits* 65: 39–42. doi:10.1051/fruits/2009040

Guo-Liang, W., Yan-Hui, C., Peng-Fei, Z., Yang Jun-Qiang, Y. and Yu-Qin, S. 2007. Apomixis and new selections of walnut. *Acta Hortic.* 760: 541–548. doi:10.17660/ActaHortic.2007.760.77

Gupta, P., K.R. Shivanna and H.Y. Mohan Ram. 1996. Apomixis and polyembryony in the guggul plant, *Commiphora wightii*. *Ann. Bot.* 78: 67–72.

Gupta, P.K. 1968. Observations on degree of apospory in three members of Andropogoneae. *Curr. Sci.* 37: 295–296.

Gupta, P.K. 1969–1970. Apomixis in *Bothriochloa pertusa* (L.) A. Camus *Port. Acta Biol. ser.* A. 11: 279–287.

Gupta, P.K., R.P. Roy and A.P. Singh. 1969. Aposporous apomixis: Seasonal variation in tetraploid *Dichanthium annulatum* (Forssk) Stapf. *Port. Acta Biol. Sci. A* 11: 253–260.

Gupta, P.K. and Yashvir. 1971. Apomixis in *Cenchrus ciliaris* – A preliminary study. *Curr. Sci.* 40: 444–446.

Gupta, S. and R. Bharalee. 2021. Genetic diversity and population structure of a medicinal herb *Houttuynia cordata* Thunb. of North-East India. *Plant Mol. Biol. Rep.* 39: 434–442. doi:10.1007/s11105-020-01260-9

Gupta, S., B.V. Bhat, V. Bhat, M.G. Gupta and S.T. Ahmad. 1999. Estimation of frequency of apomixis by auxin induced parthenocarpy technique in *Dichanthium annulatum* (Forssk.) Stapf. *Range Manag. Agrofor.* 19: 53–58.

Gurgel, J.T.A. and J. Soubihe Sobrinho. 1951. Poliembrionia em mirtáceas frutíferas. *Bragantia* 11(4–6): 141–163.

Gurr, E. 1965. *The Rational use of dyes in biology and general staining methods*. Leonard Hill, London.

Gustafsson, Å. 1934a. Primary and secondary association in *Taraxacum*. *Hereditas* 21: 1–31.

Gustafsson, Å. 1934b. Entwicklungswege der parthenogenetischen Embryosackmutterzelle. *Bot. Not.* 333–338.

Gustafsson, Å. 1935. Studies on the mechanism of parthenogenesis. *Hereditas* 21: 1–112. doi:10.1111/j.1601-5223.1935.tb02641.x

Gustafsson, Å. 1937. Över förekomsten av en sexuell population inom *Taraxacum vulgare*-gruppen. *Bot. Not.* 1937: 332–336.

Gustafsson, Å. 1943. The genesis of the European blackberry flora. *Lunds Univ. Årsskr.* 39(6): 1–200.

Gustafsson, Å. 1946. Apomixis in higher plants. I. The mechanism of apomixis. *Lunds Univ. Årsskr.* 42(3): 1–67.

Gustafsson, Å. 1947a. Apomixis in higher plants II. *Lunds Univ. Årsskr.* 43(2): 69–180.

Gustafsson, Å. 1947b. Apomixis in higher plants III. Biotype and species formation. *Lunds Univ. Årksskr.* 43(12): 181–370.

Gustafsson, Å. and A. Håkansson, 1942. Meiosis in some *Rosa*-hybrids. *Bot. Not.* 1942: 331–343.

Gustine, D.L., R.T. Sherwood, Y. Gounaris and D. Huff. 1996. Isozyme, protein and RAPD markers within a half-sib family of buffelgrass segregating for apospory. *Crop Sci.* 36: 723–737.

Gustine, D.L., R.T. Sherwood and D.R. Huff. 1997. Apospory-linked molecular markers in buffelgrass. *Crop Sci.* 37: 947–951.

Gvaladze, G.E. 1976. Forms of apomixis in the genus *Allium* L. In: Khokhlov, S.S. (ed.), *Apomixis and breeding*. Amerind, New Delhi. pp. 160–164.

Gvaladze, G.E. and L.G. Krialashvili. 1988. Polyembryony in *Colchicum speciosum* Stev. *Soobshch Akad. Nauk. Gruz. SSR*, 132, 117.

Gyer, J.F. 1996. Observation of polyembryony in *Trillium* and *Smilacina*. *Bartonia* 59: 119–121.

Ha, C.O., V.E. Sands, E. Soepadmo and K. Jong. 1988. Reproductive patterns of selected understory trees in Malaysian rain forest: the apomictic species. *Bot. J. Linn. Soc.* 97: 317–332.

Haacke, W. 1893. Gestaltung und Vererbung. Eine Entwicklungsmechanik der Organismen. Weigel Nachfolger Leipzig, 337 pages.

Haberlandt, G. 1921a. Die Entwicklungserregung der Eizellen einiger parthenogenetischer Kompositen. *Sitzb. K. Preuss. Akad. Wiss.* 51: 861–881.

Haberlandt, G. 1921b. Über experimentelle Erzeugung von Adventivembryonen bei *Oenothera lamarckiana*. *Sitzb. Preuss. Akad. Wiss.* Berlin 40: 695–725.

Haberlandt, G. 1922. Über Zellteilungshormone und ihre Beziehungen zur Wundheilung, Befruchtung, Parthenogenesis und Adventivembryonie. *Biol. Zentralbl.* 42: 145–172.

Haberlandt, G. 1923. Über die Ursache des Ausbleibens der Reduktionsteilung in den Samenanlagen einiger parthenogenetischen Angiospermen. *Sitzb. K. Preuss. Akad. Wiss.* 1923: 283–294.

Haberlandt, G. 1927. Zur Zytologie und Physiologie des weiblichen Gametophyten von *Oenothera*. *Sitzb. Preuss. Akad. Wiss.* 1927: 33–47.
Haberlandt, G. 1938. Über experimentelle Adventivembryonie. *Sitzb. Preuss. Akad. Wiss.* 1938: 243–248.
Hackenberg, D. and D. Twell. 2019. The evolution and patterning of male gametophyte development. *Curr. Top. Dev. Biol.* 131: 257–298.
Häflinger, E. 1943. Zytologische-embryologische Untersuchungen pseudogamer Ranunkeln der *Auricomus*-Gruppe. *Ber. der Schweiz. Bot. Ges.* 53: 318–382.
Hagerup, O. 1947. The spontaneous formation of haploid, polyploid, and aneuploid embryos in some orchids. *Biol. Meddel. Kongel. Danske Vidensk. Selsk.* 20: 1–22.
Haig, D. 1990. New perspectives on the angiosperm female gametophyte. *Bot. Rev.* 56: 236–274.
Hair, J.B. 1956. Subsexual reproduction in *Agropyron*. *Heredity* 19: 129–160.
Hajrudinović, A., S. Siljak-Yakovlev, S.C. Brown, F. Pustahija, M. Bourge, D. Ballian and F. Bogunič. 2015. When sexual meets apomict: genome size, ploidy level and reproductive mode variation of *Sorbus aria s.l.* and *S. austriaca* (Rosaceae) in Bosnia and Herzegovina. *Ann. Bot.* 116: 301–312.
Håkansson, A. 1921. Beiträge zur Entwicklungsgeschichte der Taccaceen. *Bot. Not.* 1921: 189–268.
Håkansson, A. 1943. Die Entwicklung des Embryosacks und die Befruchtung bei *Poa alpina*. *Hereditas* 29: 25–61.
Håkansson, A. 1944. Erganzende Beitrage zur Embryologie von *Poa alpina*. *Bot. Not.* 1944: 299–311.
Håkansson, A. 1946. Untersuchungen über die Embryologie einiger *Potentilla*-Formen. *Lunds Univ. Årsskr.* 42: 1–70.
Håkansson, A. 1951. Parthenogenesis in *Allium*. *Bot. Not.* (Lund.) 104: 143–179.
Håkansson, A. 1953. Die Samenbildung bei *Nothoscordum fragrans*. *Bot. Not.* 1953: 129–139.
Håkansson, A. and A. Levan. 1957. Endoduplicational meiosis in *Allium odorum*. *Hereditas* 43: 179–200.
Halamsky, A.T., F.G. Dunkel and E.M. Temsch. 2019. *Ranunculus kobendzae*, a new species of the *R. auricomus* complex (Ranunculaceae) from Poland. *Ann. Bot. Fenn.* 56(1–3): 123–134. doi:10.5735/085.056.0118
Halperin, W. and D. Wetherell. 1964. Adventive embryony in tissue cultures of the wild carrot, *Daucus carota*. *Am. J. Bot.* 51(3): 274–283. doi:10.1002/j.1537-2197.1964.tb06630.x
Hammel, B. 1986. New species of Clusiaceae from Central America with notes on *Clusia* and synonymy in the tribe clusieae. *Selbyana* 9(1): 112–120. http://www.jstor.org/stable/41888793
Hamston, T.J., N. de Vere, R.A. King, J. Pellicer, M.F. Fay, J.E. Cresswell and J.R. Stevens. 2018. Apomixis and hybridization drives reticulate evolution and phyletic differentiation in *Sorbus* L.: Implications for conservation. *Front. Plant Sci.* 9:1796. doi:10.3389/fpls.2018.01796
Hamston, T.J., R.J. Wilson, N. de Vere, T.C.G. Rich, J.R. Stevens and J.E. Creswell. 2017. Breeding system and spatial isolation from congeners strongly constrain seed set in an insect-pollinated apomictic tree: *Sorbus subcuneata* (Rosaceae). *Sci. Rep.* 7: 45122.
Han, T-S., Q-J. Zheng, R.E. Onstein, B.M. Rojas-Andrés, F. Hauenschild, A.N. Muellner-Riehl and Y-W. Xing. 2020. Polyploidy promotes species diversification of *Allium* through ecological shifts. *New Phytol.* 225: 571–583. doi:10.1111/nph.16098
Hancock, J.F., J.J. Luby, A. Dale, P.W. Callow, S. Serçe and A. El-Shiek. 2002. Utilizing wild *Fragaria virginiana* in strawberry cultivar development: Inheritance of photoperiod sensitivity, fruit size, gender, female fertility and disease resistance. *Euphytica* 126: 177–184. doi:10.1023/A:1016309724998
Hand, M.L., P. Vít, A. Krahulcová, S.D. Johnson, K. Oelkers, H. Siddons, J. Chrtek Jr., J. Fehrer and A.M.G. Koltunow. 2015. Evolution of apomixis loci in *Pilosella* and *Hieracium* (Asteraceae) inferred from the conservation of apomixis-linked markers in natural and experimental populations. *Heredity* 114: 17–26. doi:10.1038/hdy.2014.61
Hanna, W., M. Dujardin, P. Ozias-Akins, E. Lubbers and L. Arthur. 1993. Reproduction, cytology, and fertility of pearl millet × *Pennisetum squamulatum* BC, plants. *J. Hered.* 84: 213–216.
Hanna, W., P. Raymer and B. Schwartz. 2013. Warm-season grasses: Biology and breeding. In: Stier, J.C., Horgan, B.P. and Bonos, S.A. (eds.), *Turfgrass: Biology, use, and management*. Agronomy Monograph No. 56. American Society of Agronomy, Madison, WI, pp. 543–590. doi:10.2134/agronmonogr56.c16
Hanna, W.W. 1995. Use of apomixis in cultivar development. *Adv. Agron.* 54: 333–350.
Hanna, W.W. and E.C. Bashaw. 1987. Apomixis: Its identification and use in Plant breeding. *Crop Sci.* 27: 1136–1139.
Hanna, W.W., K.F. Schertz and E.C. Bashaw. 1970. Apospory in *Sorghum bicolor* (L.) Moench. *Science* 170: 338–339.
Haran, N. 1952. Apomixis in *Taraxacum cyprinum*. *Palestine J. Bot.* 5: 237–247.
Hardin, J.W. 1973. The enigmatic Chokeberries (*Aronia*, Rosaceae). *Bull. Torrey Bot. Cl.* 100(3): 178–184.

Harlan, J.R. 1949. Apomixis inside Oats gramma (*Bouteloua curtipendula* [Michx.] Torr.). *Am. J. Bot.* 36: 495–499.

Harlan, J.R., M.H. Brooks, D.S. Borgaonkar and J.M.J. de Wet. 1964. Nature and inheritance of apomixis in *Bothriochloa* and *Dichanthium*. *Bot. Gaz.* 125: 41–46. doi:10.1086/336242

Harte, C. 1994. Developmental biology of *Oenothera*. In: *Oenothera*. Monographs on Theoretical and Applied Genetics, Vol. 20. Springer, Berlin, Heidelberg. doi:10.1007/978-3-642-84286-3_6

Harvey, C.F. and J.E. Braggins. 1985. Reproduction in the New Zealand taxa of *Pomaderris* Labill. (Rhamnaceae). *N. Z. J. Bot.* 23: 151–156.

Haskell, G. 1960. Role of the male parent in apomictic *Rubus* species. *Heredity* 14: 101–113.

von Hausner, G. 1976. Embryogenesis und nucellar-polyembryonie bei *Cynanchum-vincetoxicum*-Arten. *Beitr. Biol. Pflanz.* 52: 101–126.

Haveman, R. 2013. Freakish patterns – species and species concepts in apomicts. *Nord. J. Bot.* 31: 257–269. doi:10.1111/j.1756-1051.2013.00158.x

Hayman, D.L. 1956. Cytological evidence of apomixis in Australian *Paspalum dilatatum*. *J. Aust. Inst. Agric. Sci.* 22: 292–293.

Heckel, C.D. and L.M. Leege. 2007. Life history and reproductive biology of the endangered *Trillium reliquum*. *Plant Ecol.* 189: 49–57.

Hedrén, M., E. Klein and H. Teppner. 2000. Evolution of polyploids in the European orchid genus *Nigritella:* Evidence from allozyme data. *Phyt. Ann. Rei Bot.* 40: 239–275.

Hedrén, M. and R. Lorenz. 2019. Seed dispersal and fine-scale genetic structuring in the asexual *Nigritella miniata* (Orchidaceae) in the Alps. *Bot. J. Linn. Soc.* 190(1): 83–100. doi:10.1093/botlinnean/boz005

Hedrén, M., R. Lorenz and D. Ståhlberg. 2018. Evidence for bidirectional hybridization between *Gymnadenia* and *Nigritella*. *J. Eur. Orch.* 50(1): 43–60.

Heenan, P.B., M.I. Dawson and R.A. Bicknell. 2002. Evidence for apomictic seed formation in *Coprosma waima* (Rubiaceae). *N. Z. J. Bot.* 40: 347–355.

Heenan, P.B., B.P.J. Molloy, R.A. Bicknell and C. Luo. 2003. Levels of apomictic and amphimictic seed formation in a natural population of *Coprosma robusta* (Rubiaceae) in Riccarton Bush, Christchurch, New Zealand. *N. Z. J. Bot.* 41(2): 287–291.

Hegelmaier, F. 1903. Zur Kenntnis der Polyembryonie von *Euphorbia dulcis* (*purpurata*). *Ber. Deutsch. Bot. Ges.* 21: 6–19.

Hegelmaier, F. 1911. Über einen neuen Fall von habitueller Polyembryonie. *Ber. Deutsch. Bot. Ges.* 19: 488–499.

Hegelmaier, F. 1913. Zur Kenntnis der Polyembryonie von *Euphorbia dulcis*. *Ber. Deutsch. Bot. Ges.* 21: 6–19.

Heller, F.O. 1973. DNS-Bestimmung an Keimwurzeln von *Vicia faba* L. mit Hilfe der Impulscytophotometrie. *Ber. Deutsch. Bot. Ges.* 86: 437–441.

Henderson, S.T., S.D. Johnson, J. Eichmann and A.M.G. Koltunow. 2017. Genetic analyses of the inheritance and expressivity of autonomous endosperm formation in *Hieracium* with different modes of embryo sac and seed formation. *Ann. Bot.* 119(6): 1001–1010. doi:10.1093/aob/mcw262

Henderson, S.W., S.T. Henderson, M. Goetz and A.M.G. Koltunow. 2020. Efficient CRISPR/Cas9-mediated knockout of an endogenous *PHYTOENE DESATURASE* gene in T1 progeny of apomictic *Hieracium* enables new strategies for apomixis gene identification. *Genes* 11: 1064. doi:10.3390/genes11091064

Hentrich, H., R. Kaiser and G. Gottsberger. 2010. Floral biology and reproductive isolation by floral scent in three sympatric Aroid species in French Guyana. *Plant Biol.* 12(4): 587–596.

Herr, J.M. 1971. A new clearing squash technique for the study of ovule development in angiosperms. *Am. J. Bot.* 58: 785–790.

Hersh, E., J. Grimm and J. Whitton. 2016. Attack of the clones: Reproductive interference between sexuals and asexuals in the *Crepis* agamic complex. *Ecol. Evol.* 6: 6473–6483. doi:10.1002/ece3.2353

Heslop-Harrison, J. 1959. Apomictic potentialities in *Dactylorchis*. *Proc. Linnean Soc. Lond.* 170(2): 174–178. doi:10.1111/j.1095-8312.1959.tb00851.x

Hetterscheid, W.L.A. and C-I. Peng. 1995. Notes on the genus *Amorphophallus* (Araceae) IV. Revision of the species in Taiwan. *Bot. Bull. Acad. Sin.* 36: 101–112.

Hindmarsh, G.J. 1964. Gametophyte development in *Trifolium pratense* L. *Aust. J. Bot.* 12: 1–14.

Hjelmqvist, H. 1951. The embryo sac development of *Tridax trilobata*. *Bot. Not.* 1951: 180–187.

Hjelmqvist, H. 1957. The apomictic development in *Malus sieboldii*. *Bot. Not.* 110: 455–467.

Hjelmqvist, H. 1959. Studien über Embryologie und Variabilität bei einigen *Aphanes*-Arten. *Bot. Not.* 1959: 17–64.

Hjelmqvist, H. 1962. The embryo sac development of some *Cotoneaster* species. *Bot. Not.* 15: 208–236.

Hjelmqvist, H. and F. Grazi. 1964. Studies on variation in embryo sac development. *Bot. Not.* 117: 141–166.

Hochwallner, H., S. Vogel, W. Huber, B.E. Hammel and A. Weber. 2012. Aspects of reproductive ecology of *Clusia valerioi* Standl. and *Clusia peninsulae* Hammel (sp. nov.), two Central American species of Clusiaceae with resin flowers. *Plant Biol.* 14: 198–207. doi:10.1111/j.1438-8677.2011.00471.x

Hodač, L., S. Klatt, D. Hojsgaard, T.F. Sharbel and E. Hörandl. 2019. A little bit of sex prevents mutation accumulation even in apomictic polyploid plants. *BMC Evol. Biol.* 19: 170 doi:10.1186/s12862-019-1495-z

Hodač, L., A.P. Scheben, D. Hojsgaard, O. Paun and E. Hörandl. 2014. ITS polymorphisms shed light on hybrid evolution in apomictic plants: a case study on the *Ranunculus auricomus* complex. *PLoS One* 9: e103003. doi:10.1371/journal.pone.0103003

Hofmeister, W. 1858. Neure Beobachtungen über die Embryobildung der Phanerogamen. *Jahrb. Wiss. Bot.* 1: 82–190.

Hojsgaard, D. 2018. Transient activation of apomixis in sexual neotriploids may retain genomically altered states and enhance polyploid establishment. *Front. Plant Sci.* 9: 230. doi:10.3389/fpls.2018.00230

Hojsgaard, D. 2020. Apomixis technology: Separating the wheat from the chaff. *Genes* 11: 411. doi:10.3390/genes11040411

Hojsgaard, D. 2021. Molecular basis of apomixis in plants. *Genes* 12: 576. doi:10.3390/genes12040576

Hojsgaard, D., J. Greilhuber, M. Pellino, O. Paun, T.F. Sharbel and E. Hörandl. 2014b. Emergence of apospory and bypass of meiosis via apomixis after sexual hybridisation and polyploidisation. *New Phytol.* 204: 1000–1012.

Hojsgaard, D. and E. Hörandl. 2015. A little bit of sex matters for genome evolution in asexual plants. *Front. Plant Sci.* 6: 82. doi:10.3389/fpls.2015.00082

Hojsgaard, D. and E. Hörandl. 2019. The rise of apomixis in natural plant populations. *Front. Plant Sci.* 10: 358. doi:10.3389/fpls.2019.00358

Hojsgaard, D., S. Klatt, R. Baier, J.G. Carman and E. Hörandl. 2014a. Taxonomy and biogeography of apomixis in Angiosperms and associated biodiversity characteristics. *Crit. Rev. Plant Sci.* 33: 414–427.

Hojsgaard, D. and M. Schartl. 2021. Skipping sex: A non-recombinant genomic assemblage of complementary reproductive modules. *BioEssays* 43: e2000111. doi:10.1002/bies.202000111

Hojsgaard, D.H., E.J. Martínez, C.A. Acuña, C.L. Quarin and F. Pupilli. 2011. A molecular map of the apomixis-control locus in *Paspalum procurrens* and its comparative analysis with other species of *Paspalum*. *Theor. Appl. Genet.* 123: 959–971. doi:10.1007/s00122-011-1639-z

Hojsgaard, D.H., E.J. Martinez and C.L. Quarin 2013. Competition between meiotic and apomictic pathways during ovule and seed development results in clonality. *New Phytol.* 197: 336–347. doi:10.1111/j.1469-8137.2012.04381.x

Hojsgaard, D.H., E. Schegg, J.F.M. Valls, E.J. Martínez and C.L. Quarin. 2008. Sexuality, apomixis, ploidy levels, and genomic relationships among four *Paspalum* species of the subgenus *Anachyris* (Poaceae). *Flora* 203: 535–547. doi:10.1016/j.flora.2007.09.005

Hokche, D.O. and N. Ramírez. 2008. Sistemas reproductivos en especies de Melastomataceae en la gran sabana (Estado Bolívar, Venezuela). *Acta Bot. Venez.* 31(2): 387–408. http://ve.scielo.org/scielo.php?script=sci_arttext&pid=S0084-59062008000200005&lng=es&tlng=es

Hollister, J.D., S. Greiner, M.T.J. Johnson and S.I. Wright. 2019. Hybridization and a loss of sex shape genome-wide diversity and the origin of species in the evening primroses (*Oenothera*, Onagraceae). *New Phytol.* 224: 1372–1380. doi:10.1111/nph.16053

Hollister, J.D., S. Greiner, W. Wang, J. Wang, Y. Zhang, G.K. Wong, S.I. Wright and M.T. Johnson. 2015. Recurrent loss of sex is associated with accumulation of deleterious mutations in *Oenothera*. *Mol. Biol. Evol.* 32(4): 896–905. doi:10.1093/molbev/msu345

Holm, S. and L. Ghatnekar. 1996a. Apomixis and sexuality in hexaploid *Potentilla argentea*. *Hereditas* 125: 53–60.

Holm, S. and L. Ghatnekar. 1996b. Sexuality and no apomixis found in crossing experiments with diploid *Potentilla argentea*. *Hereditas* 125: 77–82.

Holm, S., L. Ghatnekar and B.O. Bengtsson. 1997. Selfing and outcrossing but no apomixis in two natural populations of diploid *Potentilla argentea*. *J. Evol. Biol.* 10: 343–352.

Holman, J.E. and J. Playford. 2000. Molecular and morphological variation in the *Senna artemisioides* complex. *Aust. J. Bot.* 48(5): 569–579.

Holmgren, I. 1916. Apogamie in der Gattung *Eupatorium*. *Svensk Bot. Tiddsk.* 10: 263–269.

Holmgren, I. 1919. Zytologische Studien über die Fortpflanzung bei den Gattungen *Erigeron* und *Eupatorium*. *Kgl. Svensk. Vet. Akad. Hendl.* 59: 1–118. Dissertation, Stockholm, Sweden.

Hombe Gowda, H.C., P. Georgi, P. Mathachen and R. Vasudeva. 2004. Twin seedlings in *Nothapodytes nimmoniana* Graham (Icacinaceae): An important medicinal plant. *J. Non-Timber For. Prod.* 11(2): 108–109.

Hong, Q.B., S.Q. Xiang, K.L. Chen and L.G. Chen. 2001. Two complementary dominant genes controlling apomixis in genus *Citrus* and *Poncirus*. *Acta Genet. Sin.* 28: 1062–1067.
Honys, D., D. Reňák and D. Twell. 2006. Male gametophyte development and function. In: da Silva J.T. (ed.), *Floriculture, ornamental and plant biotechnology: Advances and topical issues*, Vol. 1. Global Science Books, London, pp. 76–87.
Hörandl, E. 1998. Species concepts in agamic complexes: applications in the *Ranunculus auricomus* complex and general perspectives. *Folia Geobot.* 33: 335–348.
Hörandl, E. 2006. The complex causality of geographical parthenogenesis. *New Phytol.* 171: 525–538. doi:10.1111/j.1469-8137.2006.01769.x
Hörandl, E. 2018. The classification of asexual organisms: Old myths, new facts, and a novel pluralistic approach. *Taxon* 67: 1066–1081. doi:10.12705/676.5
Hörandl, E. and D. Hojsgaard. 2012. The evolution of apomixis in angiosperms: a reappraisal. *Plant Biosyst.* 146: 681–693.
Hörandl, E. and J. Greilhuber. 2002. Diploid and autotetraploid sexuals and their relationships to apomicts in the *Ranunculus cassubicus* group: insights from DNA content and isozyme variation. *Plant Syst. Evol.* 234: 85–100.
Hörandl, E. and W. Gutermann. 1998. Zur Kenntnis des *Ranunculus auricomus*-Komplexes in Österreich: Die Arten der *R. phragmiteti-* und *R. indecorus*-Gruppe. *Phyton* (Horn, Austria) 37: 263–320.
Hörandl, E., A-C. Cosendai and E. Temsch. 2008. Understanding the geographic distributions of apomictic plants: a case for a pluralistic approach. *Plant Ecol. Div.* 1: 309–320.
Hörandl, E., C. Dobeš, J. Suda, P. Vít, T. Urfus, E.M. Temsch, A-C. Cosendai, J. Wagner and U. Ladinig. 2011. Apomixis is not prevalent in subnival to nival plants of the European Alps. *Ann. Bot.* 108: 381–390. doi:10.1093/aob/mcr142
Hörandl, E., G. Jakubowsky and C. Dobeš. 2001. Isozyme and morphological diversity within apomictic and sexual taxa of the *Ranunculus auricomus* complex. *Plant Syst. Evol.* 226: 165–185. doi:10.1007/s006060170064
Hörandl, E., J. Greilhuber and C. Dobeš. 2000. Isozyme variation and ploidy levels within the apomictic *Ranunculus auricomus* complex: evidence for a sexual progenitor species in southeastern Austria. *Plant Biol.* 2: 53–62.
Hörandl, E., J. Greilhuber, K. Klímová, O. Paun, E. Temsch, K. Emadzade and I. Hodálová. 2009. Reticulate evolution and taxonomic concepts in the *Ranunculus auricomus* complex (Ranunculaceae): insights from analysis of morphological, karyological and molecular data. *Taxon* 58(4): 1194–1216. doi:10.1002/tax.584012
Horn, C.L. 1940. Existence of only one variety of cultivated mangosteen explained by asexually formed 'seed'. *Science* 92: 237–238.
Hornslien, K.S., J.R. Miller and P.E. Grini. 2019. Regulation of parent-of-origin allelic expression in the endosperm. *Plant Physiol.* 180(3): 1498–1519. doi:10.1104/pp.19.00320
Horstman, A., V. Willemsen, K. Boutilier and R. Heidstra. 2014. AINTEGUMENTA-LIKE proteins: hubs in a plethora of networks. *Trends Plant Sci.* 19: 146–157.
Hoskins, T. and R.N. Contreras. 2019. Exposing seeds of *Sarcococca confusa* to increased concentrations and durations of ethyl methanesulfonate reduced seed germination, twinning, and plant size. *HortScience* 54(11): 1902–1906. doi:10.21273/HORTSCI14428-19
Hoták, Z., J. Štěpánek, I. Plačková and V. Jarolímová. 2013. *Poa riphaea*, an endangered stenoendemic species in the Hrubý Jeseník Mts (Eastern Sudetes). *Preslia* 85: 81–96.
Hotchkiss, E.E., A. Di Tomassi, D.C. Brainard and C.L. Mohler. 2008. Survival and performace of the invasive vine *Vincetoxicum rossicum* (Apocynaceae) from seeds of different embryo number under two light environments. *Am. J. Bot.* 95(4): 447–453.
Hovmalm, H.A.P., N. Jeppsson, I.V. Bartish and H. Nybom. 2004. RAPD analysis of diploid and tetraploid populations of *Aronia* points to different reproductive strategies within the genus. *Hereditas* 141: 301–312.
Hrabětová-Uhrová, A. 1961. Skalinik černoplodý (*Cotoneaster nigra* (Wahlb.) Fries) a jeho rozšiřeni na Moravě. *Sborn Klubu Přir Brno* 33: 5–24.
Hrabětová-Uhrová, A. 1962. Beitrag zur Taxonomie und Verbreitung der Gattung *Cotoneaster* in der Tschechoslowakei. *Práce Brenske Zákl Ceskoslov Akad Věd* 34: 197–248.
Hu, S.Y. 1963. Studies in the polyembryony of *Hosta caerulea*. 2. observations on the development of adventives embryos under hormone treatments. *Acta Bot. Sin.* 1: 21–25.
Huang, B-Q. and S.D. Russell. 1992. Female germ unit: organization, isolation, and function. *Int. Rev. Cytol.* 140: 233–292.

Huang, S.-Q., Y. Lu, Y.-Z. Chen, Y.-B. Luo and L.F. Delph. 2009. Parthenogenesis maintains male sterility in a gynodioecious orchid. *Am. Nat.* 174(4): 578–584.
Huber, W. 1988. Natürliche Bastardierungen zwischen weißblühenden *Ranunculus*-Arten in den Alpen. *Veröff. Geobot. Inst. ETH, Stiftg. Rübel* 100: 1–160.
Hughes J. and Richards A.J. 1988. The genetic structure of populations of sexual and asexual *Taraxacum*. *Heredity* 60: 161–171.
Hughes, J. 1987. Variability in sexual and apomictic *Taraxacum*. PhD thesis, University of Newcastle upon Tyne, Newcastle upon Tyne, England.
Hughes, J. and A.J. Richards. 1989. Isozymes, and the status of *Taraxacum* (Asteraceae) agamospecies. *Bot. J. Linn. Soc.* 99(4): 365–376.
Hull, E.D. 1915. Polyembryony in *Opuntia rafinesquii*. *Am. J. Bot.* 21: 56–57.
Hunziker, H.R. 1954. Beitrag zur Aposporie und ihrer Genetik bei *Potentilla*. *Arch Jul. Klaus-Stift.* 29: 135–222.
Hurst, C.C. 1898. Notes on some curiosities of orchid breeding. *J. Roy. Hort. Soc.* 21: 442–486.
Hussain, S., F. Curk, P. Ollitrault, R. Morillon and F. Luro. 2011. Facultative apomixis and chromosome doubling are sources of heterogeneity in citrus rootstock trials: impact on clementine production and breeding selection. *Sci. Hortic.* 130(4): 815–819. doi:10.1016/j.scienta.2011.09.009
Iaffaldano, B., J. Cardina and K. Cornish. 2018. Hybridization potential between the rubber dandelion *Taraxacum kok-saghyz* and common dandelion *Taraxacum officinale*. *Ecosphere* 9(2): e02115. doi:10.1002/ecs2.2115
Iglesias, L., H. Lima and J.P. Simon. 1974. Isozyme identification of zygotic and nucellar seedlings in *Citrus*. *J. Hered.* 65: 81–84.
Ikeda, T. 1904. On the parthenocarpy of *Citrus* fruits. *Sci. Agr. Soc. Tokyo Jour.* 63.
Ikeno, S. 1918. On hybridization of some species of *Salix*. *J. Genet.* 8: 33–58.
Ikeno, S. 1910. Sind alle Arten der Gattung *Taraxacum* parthenogenetisch? *Ber. Deutsch. Bot. Ges.* 28: 394– 397. doi:10.1111/j.1438-8677.1910.tb06908.x
Ikeno, S. 1922. On hybridization of some species of *Salix*. II. *Ann. Bot.* 36(142): 175–191.
Illg, R.D. 1977. Sobre a reprodução em *Maxillaria brasiliensis* Brieg. et Illg e *M. cleistogama* Brieg. et Illg (Orchidaceae). *Rev. Bras. Biol.* 37: 267–279.
Inamuddin, M. and K.M. Maghboub. 1990. Embryology and apomixis investigations in *Stipagrostis ciliata* (Desf.) de Winter. *Ann. Bot.* (Rome) 48: 57–60.
Inamuddin, M., B. Were and M. Saquib. 2009. A contribution to the embryology of *Rhynchelytrum repens* (Willd) C.E. Hubbard. *Scientific World* 7(7): 37–40. doi:10.3126/sw.v7i7.3822
Ingrouille, M. 1982. The colony structure and taxonomie characterisation of agamospermous *Limonium* (*Plumbaginaceae*) in the British Isles. PhD thesis, University of Leicester, Leicester, England.
Ingrouille, M. 1984. A taxometric analysis of *Limonium* (Plumbaginaceae) in Western Europe. *Plant Syst. Evol.* 147: 103–118.
Ingrouille, M. 1985. The *Limonium auriculae-ursifolium* (Pourret) Druce group (Plumbaginaceae) in the Channel Isles. *Watsonia* 15: 221–229.
Ingrouille, M. and C.A. Stace. 1985. Pattern of variation of agamospermous *Limonium* (Plumbaginaceae) in the British Isles. *Nord J. Bot.* 5: 113–125.
Ingrouille, M.J. and C.A. Stace. 1986. The *Limonium binervosum* aggregate (Plumbaginaceae) in the British Isles. *Bot. J. Linn. Soc.* 92: 177–217.
Inoue, E., F. Sakuma, M. Kasumi, H. Hara and T. Tsukihashi. 2002. Production of apomict-like seedling through intergeneric cross between Japanese pear and apple by means of pollen irradiation. *Acta Hortic.* 587: 211–215.
Irwin, R.E. 2000. Morphological variation and female reproductive success in two sympatric *Trillium* species: evidence for phenotypic selection in *Trillium erectum* and *Trillium grandiflorum* (Liliaceae). *Am. J. Bot.* 87(2): 205–214.
Islam, A.S. and A. Das. 1970. Apomixis in *Cenchrus*. *Sci. Conf.* Rajshahi, Part III. pp. 21–22.
Iwamasa, M., I. Ueno and M. Nishiura. 1967. Inheritance of nucellar embryony in citrus. *Bull. Hort. Res. Sta. Japan* 7: 1–8.
Izmaiłow, R. 1965. Macrosporogenesis in the apomictic species *Ranunculus cassubicus* L. *Acta Biol. Cracov. Ser. Bot.* 8: 183–195.
Izmaiłow, R. 1986. Cyto-embryological studies on *Alchemilla* L. (series *Calycinae* Buser). II. Apomictic processes in ovules. *Acta Biol. Cracov. Ser. Bot.* 28: 39–63.
Izmaiłow, R. 1994a. Embryo and endosperm relations at early stages of their development in *Alchemilla* (subsect). *Heliodrosium* (Rosaceae). *Polish Bot. Stud.* 8:61–67.

Izmaiłow, R. 1994b. Further observations in embryo sac development in *Alchemilla* L. (subsection *Heliodrosium* Rothm.). *Acta Biol. Cracov. Ser. Bot.* 36:37–41.
Izmaiłow, R. 1996. Reproductive strategy in the *Ranunculus auricomus* complex (Ranunculaceae). *Acta Soc. Bot. Poloniae* 65: 167–170.
Izmaiłow, R. 1999. Embryological analysis of *Aphanes arvensis* L. (Rosaceae). *Acta Biol. Cracov. Ser. Bot.* 41: 197–203.
Jablonski, E. 2017. Cultivars of European *Crataegus* – past and present. *Belgische Dendrologie Belge* 2017: 83–95.
Jafarova, V.E., L.V. Golishkin, Є.O. Dolmatov and L.V. Tashmatova. 2013. Peculiaritis of apomictic plants development *Pyrus communis* L. × *Chaenomeles japonica* (Thunb.) Lindl. *in vitro* conditions. *Plant Varieties Studying and Protection* 2(19): 46–49. doi:10.21498/2518-1017.2(19).2013.58547
Janas, A., K. Musiał, M. Kościńska-Pająk and P. Marciniuk. 2016. Insights into developmental processes in anthers, ovaries, and ovules of *Taraxacum belorussicum* (Asteraceae-Cichorioideae) using DIC optics. *Plant Syst. Evol.* 302(6): 617–628.
Jankun A. 1993. Evolutionary significance of apomixis in the genus *Sorbus* (Rosaceae). *Fragm. Flor. Geobot.* 38: 627–686.
Jankun A. 1994. Embryological studies in *Sorbus intermedia* (Rosaceae). *Polish Bot. Stud.* 8: 69–74.
Jankun, A. and M. Kovanda. 1986. Apomixis in *Sorbus sudetica* (Embryological studies in *Sorbus* 1). *Preslia* 58: 7–19.
Jankun, A. and M. Kovanda. 1987. Apomixis and origin of *Sorbus bohemica* (Embryological studies in *Sorbus* 2). *Preslia* 59: 97–116.
Jankun, A. and M. Kovanda. 1988. Apomixis at the diploid level in *Sorbus eximia* (Embryological studies in *Sorbus* 3). *Preslia* 60: 193–213.
Jassem, B. 1976. Embryology and genetics of apomixis in the section Corollinae of the genus *Beta*. *Acta Biol. Cracov. Bot.* 19: 149–172.
Jassem, B. 1990. Apomixis in the genus *Beta*. *Apomixis Newsl.* 2: 7–23.
Jassem, B. and M. Jassem. 1970. Embryological investigation into the polyploidy *Beta* species of Corolline section. *Genet. Pol.* 11: 367–377.
Jefferson, R.A. 1994. Apomixis: a social revolution for agriculture? *Biotech. Dev. Monitor* 19: 14–16.
Jeffrey, E. 1985. Polyembryony in *Erythronium americanum*. *Ann. Bot.* 9(36): 537–541. http://www.jstor.org/stable/43234228
Jeffrey, E.C. and E.J. Haertl. 1939a. Apomixis in *Trillium*. *Cellule* 47(1): 79–88.
Jeffrey, E.C. and E.J. Haertl. 1939b. The production of unfertilized seeds in *Trillium*. *Science* 90: 81–82.
Jensen, W. 1951. The normal and parthenogenetic forms of *Orobanche uniflora* L. *La Cellule* 54: 135–141.
Jeong, S., T.M. Palmer and W. Lukowitz. 2011. The RWP-RK factor *GROUNDED* promotes embryonic polarity by facilitating YODA MAP kinase signaling. *Curr. Biol.* 21: 1268–1276.
Jessup, R.W., B.L. Burson, G.B. Burow, Y-W. Wang, C. Chang, Z. Li, A.H. Paterson and M.A. Hussey. 2002. Disomic inheritance, suppressed recombination, and allelic interactions govern apospory in buffelgrass as revealed by genome mapping. *Crop Sci.* 42: 1688–1694.
Johansen, D.A. 1930. Studies on the morphology of the Onagraceae. II. Embryonal manifestations of fasciation in *Clarkia elegans*. *Bot. Gaz.* 90(1): 75–91.
Johansen, D.A. 1940. *Plant Microtechnique*. Mc-Graw-Hill, New York.
Johnson, A.M. 1936. Polyembryony in *Eugenia hookeri*. *Am. J. Bot.* 23: 83–88.
Johri, B.M. and M.R. Ahuja. 1956. Development of endosperm and nucellar polyembryony in *Aegle marmelos* Correa. *Curr. Sci.* 25: 162–164.
Johri, B.M. (ed.) 1984. *Embryology of Angiosperms*. Springer-Verlag: Berlin.
Johri, B.M. 1936. The life history of *Butomopsis lanceolata*. *Proc. Indian Acad. Sci.* B 4: 139–162.
Johri, B.M. 1965. Chemical induction of polyembryony. In: Ramakrishnan, C.V. (ed.), *Proc. "Tissue culture" Seminar* Baroda. W. Junk, The Hague, pp. 330–337.
Johri, B.M., K.B. Ambegaokar and P.S. Srivastava. 1992. *Comparative Embryology of Angiosperms*. Springer, Berlin Heidelberg New York.
Johri, H.M., P.S. Srivastava and N. Singh. 2001. Reproductive biology of angiosperms. In: Johri, B.M. and P.S. Srivastava (eds.), *Reproductive biology of plants*, Springer-Verlag: Berlin Heidelberg, pp. 237–272.
Jones, W.N. and M.C. Rayner. 1916. Mendelian inheritance in varietal crosses of *Bryonia dioica*. *J. Genet.* 5(3): 203–224.
Jönsonn, B. 1883. Polyembryoni hos *Trifolium pratense*. *Bot. Not.* 1883: 134–137.

Jordon-Thaden, I. and M.A. Koch. 2012. Detection of apomixis in an octoploid, alpine species, *Draba oligosperma* (Brassicaceae). In: *Plant and Animal Genome XX Conference* (January 14—18, 2012), W027. https://pag.confex.com/pag/xx/webprogram/Paper1684.html

Jørgensen, C.A., T.H. Sørensen and M. Westergaard. 1958. The flowering plants of Greenland. A taxonomical and cytological survey. *Biol. Skr. Dan. Vid. Selsk.* 9: 1–172.

Juel, H.O. 1898. Parthenogenesis bei *Antennaria alpina* (L.). R. Br. *Bot. Ztbl.* 74 (13): 369–372.

Juel, H.O. 1900. Beitrage zur Kenntnis der Tetradenbildung. *Jahrb. f. Wiss. Bot.* 35: 626–659.

Juel, H.O. 1904. Die Tetradenteilung in der Samenanlagen von *Taraxacum*. *Arkiv. f. Bot.* 2: 1–9.

Juel, H.O. 1905. Die Tetradenteilung bei *Taraxacum* und anderen Cichorieen. *Kgl. Svensk. Akad. Handl* 39(4): 1–21.

Juhl-Noodt, H. 1955. Apomixis und umweltbedingte Variation bei *Poa pratensis* L. *Züchter* 25 (3): 80–86.

Juliano, J.B. 1934. Origin of embryos in strawberry and mango. *Philip. J. Sci.* 54: 533–561.

Juliano, J.B. 1937. Embryos of Carabao mango, *Mangifera indica* L. *Philip. J. Agric.* 25: 949–960.

Juranić, M., M.R. Tucker, C.J. Schultz, N.J. Shirley, J.M. Taylor, A. Spriggs, S.D. Johnson, V. Bulone and A.M. Koltunow. 2018. Asexual female gametogenesis involves contact with a sexually-fated megaspore in apomictic *Hieracium*. *Plant Physiol.* 177(3): 1027–1049. doi:10.1104/pp.18.00342

Jurikova, T., J. Sochor, J. Mlcek, S. Balla, B. Klejdus, M. Baron, S. Ercisli and S.O. Yilmaz. 2014. Polyphenolic profile of interspecific crosses of rowan (*Sorbus aucuparia* L.). *Ital. J. Food Sci.* 26: 317–324.

Kader, A., K.R. Bindu and K.C. Chacko. 2000. Polyembryony in *Syzygium cumini* (L.) and *Vateria indica* L. *Indian For.* 126: 1353–1356.

Kajala, K., P. Ramakrishna, A. Fisher, D.C. Bergmann, I. De Smet and R. Sozzani. 2014. Omics and modelling approaches for understanding regulation of asymmetric cell divisions in *Arabidopsis* and other angiosperm plants. *Ann. Bot.* 113(7): 1083–1105. doi:10.1093/aob/mcu065

Kampuss, K., S. Kampuse, E. Berna, Z. Kruma, I. Krasnova and I. Drudze. 2009. Biochemical composition and antiradical activity of rowanberry (*Sorbus* L.) cultivars and hybrids with different Rosaceae L. cultivars. *Latvian J. Agron.* 12: 59–65. (In Polish)

Kant, R. and J. Verma. 2012. Obligate apomixis in *Zeuxine strateumatica* (Lindl.) Schltr. (Orchidaceae). *Vegetos* 25(1): 274–277.

Kao, R.H. 2007. Asexuality and the coexistence of cytotypes. *New Phytol.* 175: 764–772.

Kao, R.H. 2008. Origins and widespread distribution of co-existing polyploids in *Arnica cordifolia* (Asteraceae). *Ann. Bot.* 101(1): 145–152.

Kapil, R.N. 1961. Some embryological aspects of *Euphorbia dulcis*. *Phytomorphology* 11: 24–36.

Kaplan, J., J. Matejickova, A. Matejicek and F. Paprstein. 2015. Comparison of fruit cultivar 'Moravsky Sladkopl'dy' belonging to the species *Sorbus aucuparia* L. with other rowan cultivars, interspecific hybrids and wild rowan. *Acta Hortic.* 1074: 83–86. doi:10.17660/ActaHortic.2015.1074.12

Karbstein, K. 2021. Untying Gordian knots: The evolution and biogeography of the large European apomictic polyploid *Ranunculus auricomus* plant complex. PhD Thesis, University of Goettingen, Germany, 475 pages.

Karbstein, K., S. Tomasello, L. Hodač, F.G. Dunkel, M. Daubert and E. Hörandl. 2020. Phylogenomics supported by geometric morphometrics reveals delimitation of sexual species within the polyploid apomictic *Ranunculus auricomus* complex (Ranunculaceae). *Taxon* 69: 1191–1220. doi:10.1002/tax.12365

Karunarathne, P., C. Feduzka and D. Hojsgaard. 2020b. Ecological setup, ploidy diversity, and reproductive biology of *Paspalum modestum*, a promising wetland forage grass from South America. *Genet. Mol. Biol.* 43 (1 suppl 2): e20190101. doi:10.1590/1678-4685-GMB-2019-0101

Karunarathne, P. and D. Hojsgaard. 2021. Single independent autopolyploidization events from distinct diploid gene pools and residual sexuality support range expansion of locally adapted tetraploid genotypes in a South American grass. *Front. Genet.* 12:736088. doi:10.3389/fgene.2021.736088

Karunarathne, P., A.V. Reutemann, M. Schedler, A. Glücksberg, E.J. Martinez, A.I. Honfi and D.H. Hojsgaard. 2020a. Sexual modulation in a polyploid grass: a reproductive contest between environmentally inducible sexual and genetically dominant apomictic pathways. *Sci. Rep.* 10: 8319. doi:10.1038/s41598-020-64982-6

Karunarathne, P., M. Schedler, E.J. Martínez, A.I. Honfi, A. Novichkova and D. Hojsgaard. 2018. Intraspecific ecological niche divergence and reproductive shifts foster cytotype displacement and provide ecological opportunity to polyploids. *Ann. Bot.* 121(6): 1183–1196. doi:10.1093/aob/mcy004

Kashyap K., S. Banu, M.N. Shrivastava and N. Ramchiary. 2018. Study of polyembryony and development of molecular markers for identification of zygotic and nucellar seedlings in Khasi mandarin (*Citrus reticulata* Blanco). *Int. J. Environ. Agric. Biotech.* (*IJEAB*) 3(2): 363–372. doi:10.22161/ijeab/3.2.8

Kaul, V. and D. Singh. 1980. Embryological studies on Indian *Taraxacum* Weber. Embryo sac, endosperm and embryo. *G. Bot. Ital.* 114(6): 277–294. doi:10.1080/11263508009426424

Kaur, A., C.O. Ha, V.E. Sands, H.T. Chan, E. Soepadmo and P.S. Ashton. 1978. Apomixis may be widespread among trees of the climax rain forest. *Nature* 271: 440–442.

Kaur, A., K. Jong, V.E. Sands and E. Soepadmo. 1986. Cytoembryology of some Malaysian *Dipterocarpus* with some evidence of apomixis. *Bot. J. Linn. Soc.* 92: 75–88.

Kausche, L.E. 2010. Zellkern-DNA-Gehalte und Anatomie der generativen Fortpflanzungseinheiten von *Potentilla sensu lato* (Rosaceae). MSc thesis, University of Vienna, Vienna, Austria, 116 pages.

Kausik, S.B. and K. Subramanyam. 1946. A case of polyembryony in *Isotoma longiflora* Presl. *Curr. Sci.* 15: 257–258.

Kavane, A., A.K. Bishoyi and K.A. Geetha. 2021. Assessment of facultative apomixis in *Commiphora wightii* (Burseraceae): A detailed study by controlled pollination followed by histology and flow cytometry. *Flora* 281: 151854. doi:10.1016/j.flora.2021.151854

Kawahara, T., T. Yahara and K. Watanabe. 1989. Distribution of sexual and agamospermous populations of *Eupatorium* (Compositae) in Asia. *Plant Species Biol.* 4(1): 37–46.

Kawashima, T. and R.B. Goldberg. 2010. The suspensor: not just suspending the embryo. *Trends Plant Sci.* 15(1): 23–30.

Keil, D.J., M.A. Luckow and D.J. Pinkava. 1988. Chromosome studies in Asteraceae from the United States, Mexico, the West Indies, and South America. *Amer. J. Bot.* 75: 652–658.

Kelley, A.M., P.G. Johnson, B.L. Waldron and M.D. Peel. 2009. A survey of apomixis and ploidy levels among *Poa* L. (Poaceae) using flow cytometry. *Crop Sci.* 49: 1395–1402. doi:10.2135/cropsci2008.09.0553

Kelliher, T., D. Starr, L. Richbourg, S. Chintamanani, B. Delzer, M.L. Nuccio, J. Green, Z. Chen, J. McCuiston, W. Wang, T. Liebler, P. Bullock and B. Martin. 2017. Matrilineal, a sperm-specific phospholipase, triggers maize haploid induction. *Nature* 542: 105–109. doi:10.1038/nature20827

Kellog, E.A. 1987. Apomixis in the *Poa secunda* complex. *Am. J. Bot.* 74: 1431–1437.

Kellog, E.A. 1990. Variation and species limits in agamospermous grasses. *Syst. Bot.* 15: 112–123.

Kepiro, J.L. and M.L. Roose. 2010. AFLP markers closely linked to a major gene essential for nucellar embryony (apomixis) in *Citrus maxima* × *Poncirus trifoliata*. *Tree Genet. Genome* 6: 1–11. doi:10.1007/s11295-009-0223-z

Kerner von Marilaun, A. 1898. Pflanzenleben. II: Die Geschichte der Pflanzen. 2. Bibliographisches Institut Leipzig, 778 pages (p. 419).

Kerstetter, T.A. 1994. Taxonomic investigation of *Erigeron lackschewitzii*. MSc. thesis, Montana State University, Bozeman, MT, USA.

Khan, A.S. 1990. Apomixis in *Urochloa panicoides* Beauv. and *Cymbopogon winterianus* Fowitt. *12th All India Bot. Conf. Abstract.*

Khan, I.A. and M.L. Roose. 1988. Frequency and characteristics of nucellar and zygotic seedlings in three cultivars of trifoliate orange. *J. Am. Soc. Hort. Sci.* 1988(113): 105–110.

Khan, M.A., I.S. Ali and A.S. Faruqi. 1970. Breeding system and population structure in *Aerva javanica* complex L. *Phyton* (Horn, Austria) 14:135–145.

Khanday, I., D. Skinner, B. Yang, R. Mercier and V. Sundaresan. 2019. A male-expressed rice embryogenic trigger redirected for asexual propagation through seeds. *Nature* 565: 91–95. doi:10.1038/s41586-018-0785-8

Kharrat-Souissi, A., A. Baumel, F. Torre, M. Juin, S. Siljak-Yakovlev, A. Roig and M. Chaieb. 2011. New insights into the polyploid complex *Cenchrus ciliaris* L. (Poaceae) show its capacity for gene flow and recombination processes despite its apomictic nature. *Aust. J. Bot.* 59: 543–553. doi:10.1071/BT10312

Kiellander, C.L. 1935. Apomixis bei *Poa serotina*. *Bot. Not.* 35: 87–95.

Kiellander, C.L. 1937. On the embryological basis of apomixis in *Poa palustris*. *Svensk Bot. Tidskr.* 31: 425–429.

Kiellander, C.L. 1941. Studies on apospory in *Poa pratensis* var. *alpina*. *Svensk Bot. Tidskr.* 35: 321–332.

Kim, B.J., Y.C. Kwon, Y.H. Kwack, M.S. Lim and E.H. Park. 1999. The mode of seed formation in *Allium senescens* and two Korean *Allium* species. *Plant Breed.* 118: 435–438.

Kim, K-J. and B.L. Turner. 1992. Systematic overview of *Krigia* (Asteraceae-Lactuceae). *Brittonia* 44(2): 173–198.

Kimata, Y., T. Higaki, T. Kawashima, D. Kurihara, Y. Sato, T. Yamada, S. Hasezawa, F. Berger and T. Higshiyama. 2016. Cytoskeleton dynamics control the first asymmetric cell division in *Arabidopsis* zygote. *Proc. Nat. Acad. Sci. U.S.A.* 113(49): 14157–14162.

Kindiger, B. and C. Dewald. 1994. Genome accumulation in Eastern gammagrass [*Tripsacum dactyloides* (L.) L.] Poaceae. *Genetics* 92: 197–201.

King, R.M. and H. Robinson 1967. Multiple pollen forms in two species of the genus *Stevia* (Compositae). *Sida* 3: 165–169.

Kirchheimer, B., C.C.F. Schinkel, A.S. Dellinger, S. Klatt, D. Moser, M. Winkler, J. Lenoir, M. Caccianiga, A. Guisan, D. Nieto-Lugilde, J-C. Svenning, W. Thuiller, P. Vittoz, W. Willner, N.E. Zimmermann, E. Hörandl and S. Dullinger. 2016. A matter of scale: apparent niche differentiation of diploid and tetraploid plants may depend on extent and grain of analysis. *J. Biogeogr.* 43: 716–726. doi:10.1111/jbi.12663

Kirchheimer, B., J. Wessely, A. Gattringer, K. Hülber, D. Moser, C.C.F. Schinkel, M. Appelhans, S. Klatt, M. Caccianiga, A. Dellinger, A. Guisan, M. Kuttner, J. Lenoir, L. Maiorano, D. Nieto-Lugilde, C. Plutzar, J.C. Svenning, W. Willner, E. Hörandl and S. Dullinger. 2018. Reconstructing geographical parthenogenesis: effects of niche differentiation and reproductive mode on Holocene range expansion of an alpine plant. *Ecol. Lett.* 21(3): 392–401. doi:10.1111/ele.12908

Kirchner, O. 1905. Parthenogenesis bei Blütenpflanzen. *Jahresh. d. Vereins. f. vaterl. Naturk. in Württemberg* 61: 53–54.

Kirschner, J. and J. Štěpánek. 1994. Clonality as part of the evolution process in *Taraxacum*. *Folia Geobot. Phytotax.* 29: 265–275.

Kirschner, J., J. Štěpánek and L. Klimeš. 2006. Dandelions in Central Asia: A taxonomic revision of *Taraxacum* section *Leucantha*. *Preslia* 78: 27–65.

Kirschner, J., C. Oplaat, J.F. Vehoeven, V. Zeisek, I. Uhlemann, B. Trávníček, J. Räsänen, R.A. Wilschut and J. Štěpánek. 2016. Identification of oligoclonal agamospermous microspecies: Taxonomic specialists versus microsatellites. *Preslia* 88(1): 1–17.

Kirschner, J., J. Štěpánek, T. Černý, P. De Heer and P.J. van Dijk. 2013. Available *ex situ* germplasm of the potential rubber crop *Taraxacum koksaghyz* belongs to a poor rubber producer, *T. brevicorniculatum* (Compositae-Crepidinae). *Genet. Resour. Crop Evol.* 60: 455–471.

Kirschner, J., L. Záveská Drábková, J. Štěpánek and I. Uhlemann. 2015. Towards a better understanding of the *Taraxacum* evolution (Compositae-Cichorieae) on the basis of nrDNA of sexually reproducing species. *Plant Syst. Evol.* 301: 1135–1156. https://doi.org/10.1007/s00606-014-1139-0

Kissling, W.D., J.M. Lord and M. Schnittler. 2006. Agamospermous seed production of the invasive tussock grass *Nardus stricta* L. (Poaceae) in New Zealand – evidence from pollination experiments. *Flora* 201(2): 144–151. doi:10.1016/j.flora.2005.05.003

Klatt, S., F. Hadacek, L. Hodac, G. Brinkmann, M. Eilerts, D. Hojsgaard and E. Hörandl. 2016. Photoperiod extension enhances sexual megaspore formation and triggers metabolic reprogramming in facultative apomictic *Ranunculus auricomus*. *Front. Plant Sci.* 7:278. doi: 10.3389/fpls.2016.00278

Klatt, S., C.C.F. Schinkel, B. Kirchheimer, S. Dullinger and E. Hörandl. 2018. Effects of cold treatments on fitness and mode of reproduction in the diploid and polyploid alpine plant *Ranunculus kuepferi* (Ranunculaceae). *Ann. Bot.* 121: 1287–1298. doi: 10.1093/aob/mcy017

Klopper, K.C., J.J. Spies and B. Visser. 1998. Cytogenetic studies in the genus *Pentaschistis* (Poaceae: Arundinoideae). *Bothalia* 28(2): 231–238.

Kobasnidze, E. Y. 1976. The phenomenon of polyembryony in Grossulariaceae. In: *Sbornik Trudov Molodykh Nauchnukh Rahofnikov Instituta Botaniki Akademii Nauk Gruzinskoi SR* 7: 70.

Kobasnidze, E. Y. 1983. Embryology of Grossulariaceae in the high-altitude conditions of Georgia. PhD Dissertation, Tbilisi, Georgia.

Kobayashi, S. 1979. Studies on nucellar embryogenesis in *Citrus*. II. Formation of the primordium cell of the nucellar embryo in the ovule of the flower bud and its meristematic activity. *Japan Soc. Hort. Sci.* 48: 179–185.

Kobayashi, S., I. Ikeda and M. Nakatani. 1978. Studies on the nucellar embryogenesis in *Citrus*. I. Formation of nucellar embryo and development of ovule. *Bull. Fruit Tree Res. Sta.* E2: 9–24. (In Japanese, with English summary).

Koch, W. 1933. Schweizerische Arten aus del' Verwandtschaft des *Ranunculus auricomus* L. *Ber. Schweiz. Bot. Ges.* 42: 746–753.

Koch, W. 1939. Zweiter Beitrag zur Kenntnis des Formenkreises von *Ranunculus auricomus* L. *Ber. Schweiz. Bot. Ges.* 49: 541–554.

Koehler, A.D., A.S.T. Irsigler, V.T.C. Carneiro, G.B. Cabral, J.C.M. Rodrigues, A.C.M.M. Gomes, R.C. Togawa, M.M.C. Costa, A. Pinheiro Martinelli and D.M.d.A. Dusi. 2020. SERK genes identification and expression analysis during somatic embryogenesis and sporogenesis of sexual and apomictic *Brachiaria brizantha* (Syn. *Urochloa brizantha*). *Planta* 252: 39. doi:10.1007/s00425-020-03443-w

Köhler, C., O. Mittelsten-Scheid and A. Erilova. 2010. The impact of the triploid block on the origin and evolution of polyploid plants. *Trends Genet.* 26: 142–148.

Kojima, A. and T. Kawaguchi. 1989. Apomictic nature of Chinese Chive (*Allium tuberosum* Rottl.) detected in unpollinated ovule culture. *Jpn. J. Breed.* 39: 449–456.

Kojima, A. and Y. Nagato. 1992a. Diploporous embryo sac formation and the degree of diplospory in *Allium tuberosum. Sex. Plant Reprod.* 5: 72–78.

Kojima, A. and Y. Nagato. 1992b. Pseudogamous embryogenesis and the degree of parthenogenesis in *Allium tuberosum. Sex. Plant Reprod.* 5: 79–85.

Kojima, A. and Y. Nagato. 1997. Discovery of highly apomictic and highly amphimictic dihaploids in *Allium tuberosum. Sex. Plant Reprod.* 10: 8–12.

Kojima, A., Y. Nagato and K. Hinata. 1991. Degree of apomixis in Chinese chive (*Allium tuberosum*) estimated by esterase isozyme analysis. *Jpn. J. Breed.* 41: 73–83.

Kolarčik, V., V. Kocová and D. Vašková. 2018. Flow cytometric seed screen data are consistent with models of chromosome inheritance in asymmetrically compensating allopolyploids. *Cytometry* 93: 737–748. doi:10.1002/cyto.a.23511

Kolesnik, O.B. 1994. Embriolohichne doslidzhennya *Agrimonia eupatoria* L. (Rosaceae) [Embryology of *Agrimonia eupatoria* L. (Rosaceae)]. *Ukrayins'k. Bot. Zhurn. (Ukrain. Bot. J.)* 51: 99–104. [in Ukrainian].

Koloteva, N.I., A.A. Zubov and O.S. Zhukov. 1978. Cytoembryological data on the stimulation of apomixis in strawberry. *Bull. Gl. Bot. Sada AN-SSSR* 109: 65–68.

Koltunow, A.M. 1993. Apomixis: Embryo sacs and embryos formed without meiosis or fertilization in ovules. *Plant Cell* 5: 1425–1437.

Koltunow, A.M., K. Soltys, N. Nito and S. McClure. 1995. Anther, ovule, seed and nucellar embryo development in *Citrus sinensis* cv. Valencia. *Can. J. Bot.* 73: 1567–1582.

Koltunow, A.M., N.S. Scott and A.M. Chaudhury. 2001. The use of apomixis in cloning horticultural plants: Current applications and molecular prospects. *Acta Hortic.* 560: 333–343. doi:10.17660/ActaHortic.2001.560.64

Koltunow, A.M., S.D. Johnson and R.A. Bicknell. 1998. Sexual and apomictic development in *Hieracium. Sex. Plant Reprod.* 11: 213–230.

Koltunow, A.M.G., S.D. Johnson, J.C.M. Rodrigues, T. Okada, Y.K. Hu, T. Tsuchiya, S. Wilson, P. Fletcher, K. Ito, G. Suzuki, Y. Mukai, J. Fehrer and R.A. Bicknell. 2011. Sexual reproduction is the default mode in apomictic *Hieracium* subgenus *Pilosella*, in which two dominant loci function to enable apomixis. *Plant J.* 66: 890–902.

Konstantinov, A.V. 1966. Razvitye dopolnitelnykh zarodysche vykh meskov i yavlenye poliembrioni u *Fragaria ananassa*. In: *Eksperimentalnya Botanica*. Kolos, Minsk. pp. 43–48.

Korchagina, I.A. 1981. Family Betulaceae. In: *Comparative Embryology of Flowering Plants* (Winteraceae-Juglandaceae), Nauka, Leningrad, 201 pages.

Kordyum, E.L. 1961. Do pitanija pro poliembrioniju členstvija likars'kogo (*Vincetoxicum officinale* Moench) (On polyembryony in *Vincetoxicum officinale* Moench) (in Ukrainian). *Ukr. Bot. J.* 18: 48–54.

Kościńska-Pająk, M. 1980. Embryo and endosperm in *Rudbeckia bicolor*. Nutt. *Acta Biol. Cracov. Ser. Bot.* 22(1): 71–80.

Kościńska-Pająk, M., K. Musiał and K. Janiszewska. 2014. Embryological processes in ovules of *Rudbeckia laciniata* L. (Asteraceae) from Poland. *Modern Phytomorphol.* 5: 19–23.

Koszegi, D., A.J. Johnston, T. Rutten, A. Czihal, L. Altschmied, J. Kumlehn, S.E.J. Wust, O. Kirioukhova, J. Gheyselinck, U. Grossniklaus and H. Bäumlein. 2011. Members of the RKD transcription factor family induce an egg cell-like gene expression program. *Plant J.* 67: 280–291.

Kotani, Y., S.T. Henderson, G. Suzuki, S.D. Johnson, T. Okada, H. Siddons, Y. Mukai and A.M.G. Koltunow. 2014. The *LOSS OF APOMEIOSIS* (*LOA*) locus in *Hieracium praealtum* can function independently of the associated large-scale repetitive chromosomal structure. *New Phytol.* 201(3): 973–981. doi:10.1111/nph.12574

Koul, A.K. 1970. Cytoembryological studies in Oriental Maydeae I: *Coix aquatica* Roxb. *Proc. Nat. Acad. Sci.* 40: 163–178.

Kouzali, I., R. Artelari and O. Georgiou. 2012. Breeding systems and cytology in Cyprian populations of six *Limonium* species (Plumbaginaceae). *Willdenowia* 42(2): 307–313.

Kraft, T. and H. Nybom, 1995. DNA fingerprinting and biometry can solve some taxonomic problems in apomictic blackberries (*Rubus* subgen. *Rubus*). *Watsonia* 20: 329–343.

Kraft, T., H. Nybom and G. Werlemark, 1996. DNA fingerprint variation in some apomictic blackberry species (*Rubus* subg. *Rubus*, Rosaceae). *Plant Syst. Evol.* 199: 93–108.

Krahulcová, A., V. Vladimirov, F. Krahulec and S. Bräutigam. 2009. The agamic complex of *Pilosella* (Asteraceae) in Bulgaria and SW Romania: variation in ploidy level and breeding systems. *Phytol. Balc.* 15(3): 377–384.

Krahulcová, A., V. Vladimirov, F. Krahulec and S. Bräutigam. 2016. The agamic complex of *Pilosella* (Asteraceae) in Bulgaria and SW Romania: variation in ploidy levels and breeding systems. Part 2. *Phytol. Balc.* 22(1): 39–62.

Krahulcová, A. and F. Krahulec. 1999. Chromosome numbers and reproductive systems in selected representatives of *Hieracium* subgen. *Pilosella* in the Krkonoše Mts (the Sudeten Mts). *Preslia* 71: 217–234.

Krahulcová, A. and F. Krahulec. 2000. Offspring diversity in *Hieracium* subgen. *Pilosella* (Asteraceae): new cytotypes from hybridization experiments and from open pollination. *Fragm. Flor. Geobot.* 45: 239–255.

Krahulcová, A. and J. Suda. 2006. A modified method of flow cytometric seed screen simplifies the quantification of progeny classes with different ploidy levels. *Biol. Plant.* 50:457–460.

Krahulcová, A., F. Krahulec and J. Chrtek. 2001. Chromosome numbers and reproductive systems in selected representatives of *Hieracium* subgen. *Pilosella* in the Krkonoše Mts (the Sudeten Mts) – 2. *Preslia* 73: 193–211.

Krahulcová, A., F. Krahulec and V. Vladimirov. 2018. Impact of interspecific hybridization within a polyploid agamic complex of *Pilosella* (Asteraceae, Cichorieae) in Bulgaria compared with Central Europe. *Willdenowia* 48(3): 345–362. doi:10.3372/wi.48.48302

Krahulcová, A., F. Krahulec, S. Bräutigam and J. Chrtek Jr. 2013. Chromosome numbers and reproductive systems of selected representatives of *Pilosella* from the Krkonoše Mts (the Sudetes Mts). Part 3. *Preslia* 85: 179–192.

Krahulcová, A., J. Chrtek and F. Krahulec. 1999. Autogamy in *Hieracium* subgen. *Pilosella. Folia Geobot. Phytotax.* 34: 373–376.

Krahulcová, A., O. Rotreklová and F. Krahulec. 2014. The detection, rate and manifestation of residual sexuality in apomictic populations of *Pilosella* (Asteraceae, Lactuceae). *Folia Geobot.* 49(2): 239–258. http://www.jstor.org/stable/24571967

Krahulcová, A., S. Papoušková and F. Krahulec. 2004. Reproduction mode in the allopolyploid facultatively apomictic hawkweed *Hieracium rubrum* (Asteraceae, *H.* subgen. *Pilosella*). *Hereditas* 141: 19–30.

Krahulec, F., A. Krahulcová, J. Fehrer, S. Bräutigam and F. Schuhwerk. 2008. The structure of the agamic complex of *Hieracium* subgen. *Pilosella* in the Šumava Mts and its comparison with other regions in Central Europe. *Preslia* 80: 1–26.

Krahulec, F., A. Krahulcová, J. Fehrer, S. Bräutigam, I. Plačkova and J. Chrtek. 2004. The Sudetic group of *Hieracium* subgen. *Pilosella* from the Krkonoše Mts: a synthetic view. *Preslia* 76: 223–243.

Krak, K., P. Caklová, J. Chrtek and J. Fehrer. 2013. Reconstruction of phylogenetic relationships in a highly reticulate group with deep coalescence and recent speciation (*Hieracium*, Asteraceae). *Heredity* 110: 138–151. doi:10.1038/hdy.2012.100

Krawczyk, E., J. Rojek, A.K. Kowalkowska, M. Kapusta, J. Znaniecka and J. Minasiewicz. 2016. Evidence for mixed sexual and asexual reproduction in the rare European mycoheterotrophic orchid *Epipogium aphyllum*, Orchidaceae (ghost orchid). *Ann. Bot.* 118(1): 159–172. doi:10.1093/aob/mcw084

Krch, K.L. 2002. Embriologiya *Parageum montanum* Nakai et Hara [Embryology of *Parageum montanum* Nakai et Hara]. *Ukrayins'k. Bot. Zhurn. (Ukrain. Bot. J.)* 59: 464–469. [in Ukrainian]

Křišťálová, V., J. Chrtek, A. Krahulcová, S. Bräutigam and F. Krahulec. 2010. Populations of species of *Pilosella* in ruderal habitats in the city of Prague: frequency, chromosome numbers and mode of reproduction. *Preslia* 82(4): 437–464.

Kron, P. and B.C. Husband. 2009. Hybridization and the reproductive pathways mediating gene flow between native *Malus coronaria* and domestic apple, *M. domestica. Botany* 87(9): 864–874.

Kumar, S., S. Saxena and M.G. Gupta. 2017a. Marker-assisted screening of breeding populations of an apomictic grass *Cenchrus ciliaris* L. segregating for the mode of reproduction. *Crop Breed. Appl. Biotechnol.* 17(1): 10–17. doi:10.1590/1984-70332017v17n1a2

Kumar, S., S. Saxena, A. Rai, A. Radhakrishna and P. Kaushal. 2017b. Ecological, genetic, and reproductive features of *Cenchrus* species indicate evolutionary superiority of apomixis under environmental stresses. *Ecol. Indic.* 105: 126–136. doi:10.1016/j.ecolind.2019.05.036

Kumar, V., S.K. Malik, D. Pal, R. Srinivasan and S.R. Bhat. 2014. Comparative transcriptome analysis of ovules reveals stress related genes associated with nucellar polyembryony in citrus. *Tree Genet. Genomes* 10: 449–464.

Kume, N. 1959. Morphologische-Physiologische Untersuchungen über die Entwicklung von *Ardisia. Contributions from the Biological Laboratory, Kyoto University* 8: 1–37 (i–xii).

Kuntzelnigg, H. 1994. *Sorbus.* In: Hegi, G. (ed.), *Illustrierte Flora von Mitteleuropa*, ed. 3, 4/B, Blackwell Wissenschafts-Verlag, Berlin, pp. 328–385.

Kunz, E. and K. Gröber. 1988. Über das Vorkommen meiotischer und apomeiotischer Embryosacke in der Gattung *Fragaria* (L.). *Kulturpflanze* 36: 317–330.

Kurtto, A., A.N. Sennikov and R. Lampinen. 2013. *Cotoneaster* Medicus. In: *Atlas Florae Europaeae. Distribution of vascular plants in Europe*, Vol. 16, Rosaceae (*Cydonia* to *Prunus*, exl. *Sorbus*). The Committee for Mapping the Flora of Europe & Societas Biologica Fennica Vanamo, Helsinki, pp. 49–56.

Kurtto, A., H.E. Weber, R. Lampinen and A.N. Sennikov. 2010. Atlas Florae Europaeae 15. - The Committee for Mapping the Flora of Europe & Societas Biologica Fennica Vanamo, Helsinki.

Kurup, S. 1973. Cytoembryological studies in *Pennisetum hohenackeri* Hochst. *J. Mysore Univ*. 25: 1–7.

Kurz, S.1867. Revision of Indian screwpines and their allies. *J. Bot*. 5: 93–107.

Kuwada, Y. 1928. An occurrence of restitution nuclei in the formation of the embryo sacs in *Balanophora japonica* Mak. *Bot. Mag*. Tokyo 42: 117–129.

Labombarda, P., A. Busti, M.E. Cáceres, F. Pupilli and S. Arcioni. 2002. An AFLP marker tightly linked to apomixis reveals hemizygosity in a portion of the apomixis-controlling locus in *Paspalum simplex*. *Genome* 45: 513–519.

Lafon-Placette, C. and C. Köhler. 2014. Embryo and endosperm, partners in seed development. *Curr. Opin. Plant Biol*. 17(1): 64–69. doi: 10.1016/j.pbi.2013.11.008

Lakshmanan, K.K., Ambegaokar, K.B. (1984). Polyembryony. In: Johri, B.M. (eds.), *Embryology of Angiosperms*. Springer, Berlin, Heidelberg. https://doi.org/10.1007/978-3-642-69302-1_9

Lambing, C., F.C.H. Franklin and C.J.R. Wang. 2017. Understanding and manipulating meiotic recombination in plants. *Plant Physiol*. 173: 1530–1542. doi:10.1104/pp.16.01530

Leavitt, R.G. 1901. Notes on the embryology of some New England orchids. *Rhodora* 3(31): 202–205. http://www.jstor.org/stable/23293196

Leavitt, R.G. 1900. Polyembryony in *Spiranthes cernua*. *Rhodora* 2: 227–228.

Lebègue, A. 1950. Polyembryonie chez le *Potentilla aurea* L. *Bull. Soc. Bot. Fr*. 97(7–9): 199–201.

Lebègue, A. 1952a. La polyembryonie chez les Angiospermes. *Bull. Soc. Bot. France* 99(7–9): 329–367. doi:10.1080/00378941.1952.10836500

Lebègue, A. 1952b. Recherches sur le polyembryonies chez les Rosacees (g. *Fragaria* et *Alchemilla*). *Bull. Soc. Bot. Fr*. 99: 273–275.

Leblanc, O. and Y. Savidan. 1994. Timing of megasporogenesis in *Tripsacum* species (Poaceae) as related to the control of apomixis and sexuality. *Polish Bot. Studies* 8: 75–81.

Leblanc, O. and A. Mazzucato. 2001. Screening procedures to identify and quantify apomixis. In: Savidan, Y., J.G. Carman and T. Dresselhaus (eds.), *The flowering of Apomixis: From mechanisms to genetic engineering*, CIMMYT, IRD, European Commission OC VI (FAIR), Mexico City, Mexico, pp. 121–136.

Leblanc, O., D. Grimanelli, D. Gonzalez-de-Leon and Y. Savidan. 1995a. Detection of the apomictic mode of reproduction in maize–*Tripsacum* hybrids using maize RFLP markers. *Theor. Appl. Genet*. 90: 1198–1203.

Leblanc, O., D. Grimanelli, N. Islam-Faridi, J. Berthand and Y. Savidan. 1995b. Reproductive behaviour in maize – *Tripsacum* polyploidy plants. Implications for the transfer of apomixis into maize. *J. Heredity* 87: 108–111.

Leblanc, O., M.O. Peel, J.G. Carman and Y. Savidan. 1995c. Megasporogenesis and megagametogenesis in several *Tripsacum* species (Poaceae). *Amer. J. Bot*. 82: 57–63.

Leeuwenhoek, A. 1719. *Epistolae physiologicae super compluribus naturae arcanis*. Apud Adrianum Beman, Delphis. https://bibdigital.rjb.csic.es/idurl/1/13610

Lenza, E. and P.E. Oliveira. 2005. Biologia reprodutiva de *Tapirira guianensis* Aubl. (Anacardiaceae), uma espécie dióica em mata de galeria do Triângulo Mineiro, Brasil. *Rev. Bras. Bot*. 28: 179–190.

Leonard, P.J. 2011. *Aronia mitschurinii*: Solving a horticultural Enigma. MSc. thesis, University of Connecticut, Storrs, CT, USA, 183 pages.

Lepší, M., P. Lepší, J. Sádlo, P. Koutecký, P. Vít and P. Petřík. 2013. *Sorbus pauca* species nova, the first endemic species of the *Sorbus* hybrid group for the Czech Republic. *Preslia* 85: 63–80.

Lepši, M., P. Koutecký, J. Nosková, P. Lepši, T. Urfus and T.G. Rich. 2019. Versatility of reproductive modes and ploidy level interactions in *Sorbus* s.l. (Malinae, Rosaceae). *Bot. J. Linn. Soc*. 191(4): 502–522. doi:10.1093/botlinnean/boz054

Lepší, M., P. Vít, P. Lepší, K. Boublik and F. Kolář. 2009. *Sorbus portae-bohemicae* and *Sorbus albensis*, two new apomictic endemic species recognized based on revision of *Sorbus bohemica*. *Preslia* 81: 63–89.

Leszczuk, A., M. Domaciuk and E. Szczuka. 2018. Unique features of the female gametophyte development of Strawberry *Fragaria* × *ananassa* Duch. *Sci. Hortic*. 234: 201–209.

Levin, J., M.F. Fay, J. Pellicer and M. Hedrén. 2019. Multiple independent origins of intermediate species between *Sorbus aucuparia* and *S. hybrida* (Rosaceae) in the Baltic region. *Nord. J. Bot*. 36: 02035. doi:10.1111/njb.02035

Li, Y.N. 2001. Research of germplasm resources of *Malus* Mill. China Agriculture, Beijing, pp. 207–210.

Liao, F. and Y. Li. 1997. An embryological observation of apospory in *Malus toringoides* (Rehd.) Hughes. *Xi nan Nong ye da xue xue bao = J. Southwest Agricultural University* 19(2): 98–104 (in Chinese).

Lidforss, B. 1905. Studier över artbildningen inom släktet *Rubus*. Del I. *Ark. Bot.* 4(6): 1–41. http://www.biodiversitylibrary.org/item/91288#page/247/mode/1up

Lidforss, B. 1914. Resumé seiner Arbeiten über *Rubus*. *Zeitschr. Abst. Vererb.* 12: 1–13. doi:10.1007/BF01837266

Lidforss, B. 1907a. Studien öfver artbildningen inom släktet *Rubus*. Del II. *Ark. Bot.* 6(16): 1–43. http://www.biodiversitylibrary.org/item/91215#page/567/mode/1up

Lidforss, B. 1907b. Über das Stadium polymorpher Gattungen. *Bot. Not.* 60: 241–261.

Liebenberg, H. 1961. Apomiksie by *Eragrostis* en *Themeda*. MSc verhandeling, Universiteit van Stellenbosch.

Liebenberg, H. 1990. Cytotaxonomic studies in *Themeda triandra* Forssk.Part III. Sexual and apomictic embryo sac development in 53 collections. *South Afr. J. Bot.* 56(5): 554–559.

Lihová, J., P. Mártinofi and L. Mártonfiová. 2000. Experimental study on reproduction of *Hypericum* × *desetangsii* nothosubsp. *carinthiacum* (A. Frohl). N. Robson (Hypericaceae). *Caryologia* 53: 127–132.

Liljefors, A. 1934. Über normale und apospore Embryosackentwicklung in der Gattung *Sorbus*, nebst einigen Bemerkungen über die Chromosomenzahlen. *Svensk Bot. Tidskr.* 28: 290–299.

Liljefors, A. 1953. Studies on propagation, embryology and pollination in *Sorbus*. *Acta Horti Berg.* 16: 277–329.

Liljefors, A. 1955. Cytological Studies in *Sorbus*. *Acta Horti Berg.* 17: 47–113.

Lim, A.L. 1974. Embryological studies of some Malayan fruit tree species: *Garcinia mangostana, Nephelium lappaceum* and *Lansium domesticum*. MSc thesis, University of Malaya, Kuala Lumpur, Malysia.

Lim, A.L. 1984. The embryology of *Garcinia mangostana* L. (Clusiaceae). *The Gardens' Bulletin, Singapore* 37: 93–103.

Linder, H.P. and N.P. Barker. 2014. Does polyploidy facilitate long-distance dispersal? *Ann. Bot.* 113(7): 1175–1183. doi:10.1093/aob/mcu047

Lipka, A.E., F. Lu, J.H. Cherney, E.S. Buckler, M.D. Casler and D.E. Costich. 2014. Accelerating the switchgrass (*Panicum virgatum* L.) breeding cycle using genomic selection approaches. *PLoS One* 9(11): e112227. doi:10.1371/journal.pone.0112227

Liu, C., X. Li, D. Meng, Y. Zhong, C. Chen, X. Dong, X. Xu, B. Chen, W. Li, L. Li, X. Tian, H. Zhao, W. Song, H. Luo, Q. Zhang, J. Lai, W. Jin, J. Yan and S. Chen. 2017. A 4-bp insertion at ZmPLA1 encoding a putative phospholipase A generates haploid induction in maize. *Mol. Plant* 10: 520–522. doi:10.1016/j.molp.2017.01.011

Liu, C., Y. Zhong, X. Qi, M. Chen, Z. Liu, C. Chen, X. Tian, J. Li, Y. Jiao, D. Wang, Y. Wang, M. Li, M. Xin, W. Liu, W. Jin and S. Chen. 2020. Extension of the in vivo haploid induction system from diploid maize to hexaploid wheat. *Plant Biotechnol. J.* 18: 316–318. doi:10.1111/pbi.13218

Liu, Y., W.X. Dong, L. Zhang and X.D. Cheng. 2006. Study on the development characteristics of different embryo sac types of Pingyi Tiancha (*Malus hupehensis*). *J. Fruit Sci.* 23: 330–334.

Liu, Y-H., F-X. Wang and N-F. Qian. 1987. Apomixis in *Zanthoxylum bungeanum* and *Z. simulans*. *Acta Genet. Sin.* 14: 107–113 (in Chinese).

Liu, Z.W., R.R.C. Wang and J.G. Carman. 1995. Hybrids and back cross progenies between wheat (*Triticum aestivum*) and apomictic Australian wheat grass (*Elymus retisetus*) Karyotypic and genomic analysis. *Theor. Appl. Genet.* 89: 599–605.

Lo, E.Y.Y., S. Stefanović and T.A. Dickinson 2009. Population genetic structure of diploid sexual and polyploid apomicts of *Crataegus* (hawthorns; Rosaceae) in the Pacific Northwest. *Mol. Ecol.* 18: 1145–1160.

Lo, E.Y.Y., S. Stefanović and T.A. Dickinson. 2010. Reconstructing reticulation history in a phylogenetic framework and the potential of allopatric speciation driven by polyploidy in an agamic complex in *Crataegus* (Rosaceae). *Evolution* 64: 3593–3608. doi:10.1111/j.1558-5646.2010.01063.x

Lo, E.Y.Y., S. Stefanović and T.A. Dickinson. 2013. Geographical parthenogenesis in Pacific Northwest hawthorns (*Crataegus*; Rosaceae). *Botany* 91(2): 107–116. doi:10.1139/cjb-2012-0073

Loiko, R.E. 1990. Apomixis in walnut. *Acta Hortic.* 284: 233–236.

Long, J.M., Z. Liu, X.M. Wu, Y.N. Fang, H.H. Jia, Z.Z. Xie, X.X. Deng and W.W. Guo. 2016. Genome-scale mRNA and small RNA transcriptomic insights into initiation of citrus apomixis. *J. Exp. Bot.* 67: 5743–5756.

Longo, B. 1908. La poliembrionia nello *Zanthoxylum bungei* senza fecondation. *Bull. Soc. Bot. Ital.* 11: 113–115.

López-Alvarado, J., I. Cobacho, V.J. Arán, M. Rosato, J.A. Rosselló and L. Sáez. 2017. *Limonium ilergabonum* Plumbaginaceae, a new species from Northeastern Iberian Peninsula. *Phytotaxa* 331(2): 199–212.

Löve, A. 1952. Preparatory studies for breeding Icelandic *Poa irrigata*. *Hereditas* 38: 11–32. doi:10.1111/j.1601-5223.1952.tb02912.x

Lovell, J., K. Grogan, T.F. Sharbel and J.C. McKay. 2014. Mating system and environmental variation drive patterns of adaptation in *Boechera spatifolia* (Brassicaceae). *Mol. Ecol.* 23: 4486–4497.

Lovell, J., O. Aliyu, M. Mau, M. Schranz, M. Koch, C. Kiefer, B. Song, T. Mitchell-Olds and T.F. Sharbel. 2013. On the origin and evolution of apomixis in *Boechera*. *Plant Reprod.* 26(4): 309–315.

Lovell, J., R.J. Williamson, S.I. Wright, J.K. McKay and T.F. Sharbel. 2017. Mutation accumulation in an asexual relative of *Arabidopsis*. *PLos Genet.* 13(1): e1006550. doi:10.1371/journal.pgen.1006550

Lovett Doust, L., J. Lovett Doust and P.B. Cavers. 1981. Fertility relationships in closely related taxa of *Oxalis*, section Corniculatae. *Can. J. Bot.* 59: 2603–2609.

Lu, Y., Y.B. Luo and S.Q. Huang. 2010. Reduced recombination in gynodioecious populations of a facultative apomictic orchid. *Plant Biol* (Stuttg). 12(5): 814–819. doi:10.1111/j.1438-8677.2009.00283.x

Lubbers, E.L., L. Arthur, W.W. Hanna and P. Ozias-Akins. 1994. Molecular markers shared by diverse apomictic *Pennisetum* species. *Theor. Appl. Genet.* 89: 636–642.

Luby, J.J., Hancock, J.F., Dale, A. and S. Serçe. 2008. Reconstructing *Fragaria* × *ananassa* utilizing wild *F. virginiana* and *F. chiloensis*: inheritance of winter injury, photoperiod sensitivity, fruit size, female fertility and disease resistance in hybrid progenies. *Euphytica* 163: 57–65. doi:10.1007/s10681-007-9575-3

Ludwig, S., A. Robertson, T.C.G. Rich, M. Djordjevic, R. Cerovic, L. Houston, S.A. Harris and S.J. Hiscock. 2013. Breeding systems, hybridization and continuing evolution in Avon Gorge *Sorbus*. *Ann. Bot.* 111(4): 563–575. doi:10.1093/aob/mct013

Lughadha, E.N. and C. Proença. 1996. A survey of the reproductive biology of the Myrtoideae (Myrtaceae). *Ann. Mo. Bot. Gard.* 83: 480–503.

Lundh-Almestrand, A. 1958. Studies on some microspecies of *Alchemilla vulgaris* L. *Bot. Not.* 111: 587–607.

Lunerová, J., V. Herklotz, M. Laudien, R. Vozárová, M. Groth, A. Kovařík and C.M. Ritz. 2020. Asymmetrical canina meiosis is accompanied by the expansion of a pericentromeric satellite in non-recombining univalent chromosomes in the genus *Rosa*. *Ann. Bot.* 125(7): 1025–1038. doi:10.1093/aob/mca

Lutts, S., J. Ndikumana and B.P. Louant. 1991. Fertility of *Brachiaria ruziziensis* in interspecific crosses with *Brachiaria decumbens* and *Brachiaria brizantha*: Meiotic behaviour, pollen viability and seed set. *Euphytica* 57: 267–274.

Lutts, S., J. Ndikumana and B.P. Louant. 1994. Male and female sporogenesis in apomictic *Brachiaria brizantha, Brachiaria decumbens* and F1 hybrids with sexual colchicines induced tetraploid *Brachiaria ruziziensis*. *Euphytica* 78: 19–25.

Lyman, J.C. and N.C. Ellstrand. 1984. Clonal diversity in *Taraxacum officinale* (compositae), an apomict. *Heredity* 53(1): 1–10.

Ma, G., X. Huang, N. Zhao and Q. Xu. 2004. Apospory in *Paspalum thunbergii*. *Aust. J. Bot.* 52: 81–86.

Ma, G., X.L. Huang, Q.S. Xu and E. Bunn. 2009. Multiporate pollen and apomixis in Panicoideae. *Pakistan J. Bot.* 41: 2073–2082.

Machado, R.M., F.A. de Oliveira, F. de Matos Alves, A.P. de Souza and E.R. Forni-Martins. 2021. Population genetics of polyploid complex *Psidium cattleyanum* Sabine (Myrtaceae): Preliminary analyses based on new species-specific microsatellite loci and extension to other species of the genus. *Biochem. Genet.* 59: 219–234. doi:10.1007/s10528-020-10002-1

Macková, L., J. Nosková, Ľ. Ďurišová and T. Urfus. 2020. Insights into the cytotype and reproductive puzzle of *Cotoneaster integerrimus* in the Western Carpathians. *Plant Syst. Evol.* 306(3): 58. doi:10.1007/s00606-020-01684-6

MacPhail, V.J. and P.G. Kevan. 2009. Review of the breeding systems of wild roses (*Rosa* spp.) *Floriculture Ornamental Biotech.* 3(1): 1–13.

Maguire, B. 1976. Apomixis in the genus *Clusia* (Clusiaceae). A preliminary report. *Taxon* 25: 241–244.

Maheshwari, P. 1950. *An Introduction to the Embryology of Angiosperms*. McGraw Hill, New York.

Maheshwari, P. (ed.) 1963. *Recent Advances in the Embryology of Angiosperms*. International Society of Plant Morphologists, New Delhi.

Maheshwari, P. and R.C. Sachar. 1963. Polyembryony In: Maheshwari, P. (ed.), *Recent Advance in the Embryology Angiosperms*. University of Delhi, India. pp. 265–296.

Maheshwari, P. and R.N. Chopra. 1955. The structure and development of the ovule and seed of *Opuntia dilleni* Haw. *Phytomorphology* 5: 112–122.

Maheshwari, P. and S. Narayanaswami. 1952. Embryological studies on *Spiranthes australis* Lindl. *J. Linn. Soc. Bot.* 53(355): 474–486.

Maheswari Devi, H. and T. Pullaiah. 1976a. Embryology of Safflower (*Carthamus tinctorius*). *The Botanique* 7: 63–70.

Maheswari Devi, H. and T. Pullaiah. 1976b. Embryological investigations in the Melampodinae – I. *Melampodium divaricatum*. *Phytomorphology* 26: 77–86.

Maheswari Devi, H. and T. Pullaiah. 1977a. Embryology of *Eupatorium* - an apomict. In: Dnyansagar, V.R. et al. (eds.), *Recent Trends and contacts between Cytogenetics, embryology and morphology*. Today and Tomorrows, New Delhi, pp. 307–314.

Maheswari Devi, H. and T. Pullaiah. 1977b. Embryological abnormalities in *Carthamus tinctorius*. *Acta Bot. Indica* 5: 1–8.

Mahmutović, I., A. Hajrudinović, S. Šiljak-Yakovlev, E. Silajdzic, M. Bourge, S.C. Brown, E. Muratović and F. Bogunić. 2015. Cytotypes and reproductive modes of *Cotoneaster integerrimus* (Rosaceae) from Bosnia and Herzegovina: preliminary results. *19e Congrès annuel de l'association Française de Cytométrie*, Antibes, France.

Mahoney, J.D., T.M. Hau, B.A. Connolly and M.H. Brand. 2019. Sexual and apomictic seed reproduction in *Aronia* species with different ploidy levels. *HortScience* 54(4): 642–646. doi:10.21273/HORTSCI13772-18

Maia, F.R., I.G. Varassin and R. Goldenberg. 2016. Apomixis does not affect visitation to flowers of Melastomataceae, but pollen sterility does. *Plant Biol.* 18(1): 132–138. doi:10.1111/plb.12364

Majeský, Ľ., F. Krahulec and R.J. Vašut. 2017. How apomictic taxa are treated in current taxonomy: A review. *Taxon* 66(5): 1017–1040. doi:10.12705/665.3

Majeský, Ľ., R. Vašut and M. Kittner. 2015. Genotypic diversity of apomictic microspecies of the *Taraxacum scanicum* group (*Taraxacum* sect. *Erythrosperma*). *Plant Syst. Evol.* 301: 2105–2124.

Majeský, Ľ., R.J. Vašut, M. Kiner and B. Trávníček. 2012. The pattern of genetic variability in apomictic clones of *Taraxacum officinale* indicates the alternation of asexual and sexual histories of apomicts. *PLoS One* 7(8): e41868. doi:10.1371/journal.pone.0041868

Małecka, J. 1967. Cyto-embryological studies in *Taraxacum scanicum* DT. *Acta Biol. Cracov. Ser Bot.* 10: 195–206.

Małecka, J. 1973. Cyto-embryological studies in the genus *Taraxacum* section *Palustre*. *Genet. Pol.* 14: 185–188.

Maletskii, S.I. and E.I. Maletskaya. 1996. Samofertil'nost' i agamospermiya u sakharnoy svekly (*Beta vulgaris* L.). *Genetika* 32(12): 1643–1650.

Maliro, M.F.A. and M.B. Kwapata. 2000. Apomictic embryo development and survival in *Uapaca kirkiana* under in vitro and in vivo seed germination. *Sci. Hortic.* 83(2): 139–147.

Mancini, M., H. Permingeat, C. Colono, L. Siena, F. Pupilli, C. Azzaro, D.M.A. Dusi, V.T.C. Carneiro, M. Podio, J.G. Seijo, A.M. González, S.A. Felitti, J.P.A. Ortiz, O. Leblanc and S.C. Pessino. 2018. The MAP3K-coding *QUI-GON JINN* (*QGJ*) gene is essential to the formation of unreduced embryo sacs in *Paspalum*. *Front. Plant Sci.* 9: 1547. doi:10.3389/fpls.2018.01547

Mandáková, T., K. Ashby, B.J. Price, M.D. Windham, J.G. Carman and M.A. Lysák. 2021. Genome structure and apomixis in *Phoenicaulis* (Brassicaceae; Boechereae). *J. Syst. Evol.* 59(1): 83–92. doi:10.1111/jse.12555

Mandáková, T., P. Hloušková, M.D. Windham, T. Mitchell-Olds, K. Ashby, B. Price, J. Carman and M.A. Lysák. 2020. Chromosomal evolution and apomixis in the Cruciferous tribe Boechereae. *Front. Plant Sci.* 11: 514. doi:10.3389/fpls.2020.00514

Mangla, Y., M. Chaudhary, H. Gupta, R. Thakur, S. Goel, S.N. Raina and R. Tandon. 2015. Facultative apomixis and development of fruit in a deciduous shrub with medicinal and nutritional uses. *AoB Plants* 7. doi:10.1093/aobpla/plv098

Mao, B., M. Chen, C. Liu, B. Zheng and M. Fan. 1995. A research on the ability of apomixis in *Malus* plants. *J. South West Agric. Univ.* 17(3): 220–223. https://europepmc.org/abstract/cba/283430

Marconi, G., D. Aiello, B. Kindiger, L. Storchi, A. Marrone, L. Reale, N. Terzaroli and E. Albertini. 2020. The role of *APOSTART* in switching between sexuality and apomixis in *Poa pratensis*. *Genes* 11(8): 941. doi:10.3390/genes11080941

Marklund, G. 1941. Kasvistomme apomiktisista pikkulajeista, erityisesti *Ranunculus auricomus*-ryhmästä. *Luonnon Ystävä* 45: 78–86.

Marras, M.M. 1985. Ricerche sulla riproduzione di *Limonium tibulatium* (Pign.) Pign. Bachelor's thesis.

Marshall, H.H. 1958. Polyembryony in tomatoes. *Can. J. Plant Sci.* 38(1): 67–72. doi:10.4141/cjps58-010

Martin, F.M. 1959. Staining and observing pollen tubes in the style by means of fluorescence. *Stain Technol.* 34: 436–437.

Martínez Ochoa, E.C., M. Andrade-Rodríguez, M.R. Rodríguez and A.V. Monter. 2012. Identification of zygotic and nucellar seedlings in polyembryonic mango cultivars. *Pesq. Agropec. Bras. Brasília*, 47(11): 1629–1636.

Martínez, E.J., C.L. Quarin and M.D. Hayward. 1999. Genetic control of apospory in apomictic *Paspalum* species. *Cytologia* 64: 425–433.

Martínez, E.J., E.H. Hopp, J. Stein, J.P.A. Ortiz and C.L. Quarin. 2003. Genetic characterization of apospory in tetraploid *Paspalum notatum* based on the identification of linked molecular markers. *Mol. Breed.* 12: 319–327.
Martínez, E.J., M.H. Urbani, C.L. Quarin and J.P.A. Ortiz. 2001. Inheritance of apospory in bahiagrass, *Paspalum notatum. Hereditas* 135:19–25.
Martinez-Gómez, P. and T.M. Gradziel. 2003. Sexual polyembryony in almond. *Sex. Plant Reprod.* 16: 135–139.
Martins, R.L. and P.E. Oliveira. 2003. RAPD evidence for apomixis and clonal populations in *Eriotheca* (Bombacaceae). *Plant Biol.* 5(3): 338–340.
Mártonfi, P. 2001. New species of the genus *Hypericum* sect. *Hypericum* (Guttiferae) from Slovakia. *Folia Geobot.* 36: 371–384.
Mártonfi, P. 2008. *Hypericum dubium* Leers–new data on taxonomy and biology. *Folia Geobot.* 43: 69–82.
Mártonfi, P., R. Brutovská, E. Ĉellárová and M. Repĉák. 1996. Apomixis and hybridity in *Hypericum perforatum. Folia Geobot. Phytotaxon.* 31: 389–396. doi:10.1007/BF02815383
Mártonfiová, L. 2006. Possible pathways of the gene flow in *Taraxacum* sect. *Ruderalia. Folia Geobot.* 41: 183–201. doi:10.1007/BF02806478
Mártonfiová, L., P. Mártonfi and R. Šuvada. 2010. Breeding behavior and its possible consequences for gene flow in *Taraxacum* sect. *Erythrosperma* (H. Lindb.) Dahlst. *Plant Species Biol.* 25: 93–102. doi:10.1111/j.1442-1984.2010.00270.x
Mascarenhas, J.P. 1989. The male gametophyte of flowering plants. *Plant Cell* 1(7): 657–664.
Masci, S., A. Miho and P. Marchi. 1994. *Ranunculus auricomus* L. aggr. (Ranunculaceae) in Italy. I. Sexual tetraploids on the Apennines. *Caryologia* 47(2): 97–108.
Masia, D. 1985. Ricerche sulla riproduzione di *Limonium articulatum* (Loisel.) O. Kuntze. Bachelor's thesis.
Mateo de Arias, M., L. Gao, D.A. Sherwood, K.K. Dwivedi, B.J. Price, M. Jamison, B.M. Kowallis and J.G. Carman. 2020. Whether gametophytes are reduced or unreduced in angiosperms might be determined metabolically. *Genes* 11: 1449. doi:10.3390/genes11121449
Mateo de Arias, M. 2015. Effects of Plant stress on facultative apomixis in *Boechera* (Brassicaceae). All Graduate Theses and Dissertations. 4451. https://digitalcommons.usu.edu/etd/4451
Matsson, L.P.R. 1912. Till fragan om rosornas befruktning. *Svensk Bot. Tidskr.* 3: 588–606.
Matzk, F. 1989. Genetic studies on parthenogenesis in *Poa pratensis* L. *Apomixis Newsl.* 1: 32–34.
Matzk, F. 1991. A novel approach to differentiated embryos in the absence of endosperm. *Sex. Plant Reprod.* 4: 88–94.
Matzk, F. 2007. Reproduction mode screening. In: Doležel, J., J. Greilhuber and J. Suda (eds.), *Flow cytometry with plant cells. Analysis of genes, chromosomes, and genomes.* Wiley-VCH, Weinheim, pp. 131–152.
Matzk, F., A. Meister and I. Schubert. 2000. An efficient screen for reproductive pathways using mature seeds of monocots and dicots. *Plant J.* 21: 97–108.
Matzk, F., A. Meister, R. Brutovska and I. Schubert. 2001. Reconstruction of reproductive diversity in *Hypericum perforatum* L. opens novel strategies to manage apomixis. *Plant J.* 26: 275–282. doi:10.1046/j.1365-313X.2001.01026.x
Matzk, F., K. Hammer and I. Schubert. 2003. Co-evolution of apomixis and genome size within the genus *Hypericum. Sex. Plant Reprod.* 16: 51–58.
Matzk, F., S. Prodanovic, H. Bäumlein and I. Schubert. 2005. The Inheritance of apomixis in *Poa pratensis* confirms a five locus model with differences in gene expressivity and penetrance. *Plant Cell* 17(1): 13–24. doi:10.1105/tpc.104.027359
Mau, M., J.M. Corral, H. Vogel, M. Melzer, J. Fuchs, M. Kuhlmann, N. de Storme, D. Geelen and T.F. Sharbel. 2013. The conserved chimeric transcript *UPGRADE2* is associated with unreduced pollen formation and is exclusively found in apomictic *Boechera* species. *Plant Physiol.* 163(4): 1640–1659. doi:10.1104/pp.113.222448
Mau, M., J.T. Lovell, J.M. Corral, C. Kiefer, M.A. Koch, O.M. Aliyu and T.F. Sharbel. 2015. Hybrid apomicts trapped in the ecological niches of their sexual ancestors. *Proc. Natl. Acad. Sci. U.S.A.* 112(18): E2357–E2365. doi:10.1073/pnas.1423447112
Maugini, E. 1962. Morfologia fiorale, embriologia ed embryo in *Leontopodium alpinum* Cass. var. α *typicum* Fiori e Paoletti. *G. Bot. Ital.* 69: 1–18.
Maule, A., C. Faulkner and Y. Benitez-Alfonso. 2012. Plasmodesmata "in Communicado". *Front. Plant Sci.* 3:30. doi:10.3389/fpls.2012.00030
Mauritzon, J. 1934. Die Embryologie einiger Capparidaceen sowie von *Tovaria pendula. Arkiv. Bot.* 26A(15): 1–14.
Mauritzon, J. 1935. Über die Embryologie der Familie Rutaceae. *Svensk Bot. Tidskr.* 29: 319–347.

Maury, G. 1968. Germinations anormales chez les Dipterocarpacees die Malaisie. *Bull. De la Soc. D'Histoire de Naturella de Toulouse* 106: 282–288.

Mayo, G.M. and P. Langridge. 2003. Modes of reproduction in Australian populations of *Hypericum perforatum* (St. John's wort) revealed by DNA fingerprinting and cytological methods. *Genome* 46: 573–579.

Mazzucato, A., F. Veronesi and M. Falcinelli. 1996. Evolution and adaptedness in facultatively apomictic grass *Poa pratensis* L. *Euphytica* 92: 13–19.

Mazzucato, A., G. Barcaccia, M. Pezzotti and M. Falcinelli. 1995. Biochemical and molecular markers for investigating the mode of reproduction in the facultative apomict *Poa pratensis*. *Sex. Plant Reprod.* 8: 133–138.

Mazzucato, A., M. Wagenvoort and A.P.M. den Nijs. 1994. Flow cytometric analyses to estimate the mode of reproduction in *Poa pratensis* L. *Apomixis Newsl.* 7: 22–24.

McAllister and C.M. Gillham. 1980. Tab. 792. *Sorbus forrestii*. *Curtis's Bot. Mag.* 83: 1–4.

McAllister, F. 1913. On the cytology and embryology of *Smilacina racemosa*. *Trans. Wis. Acad. Sci.* 17: 599–660.

McAllister, H.A. 1986. The Rowan and its relatives (*Sorbus* spp.). Ness Series I. Ness Gardens, University of Liverpool Botanic Gardens.

McCormick, S. 2004. Control of male gametophyte development. *Plant Cell* 16: 142–153.

McCrea, K.D. 1981. Ultraviolet floral patterning, reproductive isolation and character displacement in the genus *Rudbeckia* (Compositae). PhD Dissertation, Purdue University, West Lafayette, IN, USA.

McDonald, C.C. 1927. A study of seed development in three species of *Erigeron* with special reference to apogamy. *Bull. Torrey Bot. Cl.* 54: 479–497.

McWilliam, J.R., K. Shankar and R.B. Knox. 1970. Effects of temperature and photoperiod on growth and reproductive development in *Hyparrhenia hirta*. *Aust. J. Agric. Res.* 21: 557–569.

Mehra, P.N. and B.S. Gill. 1972. Apomixis in *Erigeron mucronatus* DC. Biology of land plants: symposium held at Botany Department, Meerut University, Meerut, Uttar Pradesh, India, pp. 271–275.

Meier, M., D. Zappacosta, J.P. Selva, S. Pessino and V. Echenique. 2011. Evaluation of different methods for assessing the reproductive mode of weeping lovegrass plants, *Eragrostis curvula* (Schrad.) Nees. *Aust. J. Bot.* 59: 253–261. doi:10.1071/BT10267

Meiyong, Z., X. Ying and M. Fengxian. 2000. Study on the apomixis ability of walnut (*Juglans regia*) varieties. *J. Fruit Sci.* 17: 314–316.

Mello, L.V., W.J. Silva, H.P. Medina Filho and R. Balve.1995. Breeding systems in *Coix lacryma-jobi* populations. *Euphytica* 81: 217–221.

Melo, G.F., I.C. Machado and M. Luceño. 1999. Reprodución de tres especies de *Clidemia* (Melastomataceae) en Brasil. *Rev. Biol. Trop.* 47(3): 359–363.

Melo, G.F.A. and I.C. Machado. 1996. Biologia da reprodução de *Henriettea succosa* DC. (Melastomataceae). *Rev. Bras. Biol.* 56: 383–389.

Mendel, G. 1870. Über einige aus künstlicher Befruchtung gewonnenen *Hieracium*-Bastarde. *Verh. Naturf. Ver. Brünn* 8 (Abhandlungen): 26–31. (English translation: On *Hieracium* hybrids obtained by artificial fertilisation)

Mendes, M.G., A.P.G.S. de Oliveira, P.E. Oliveira, A.M. Bonetti and D.S. Sampaio. 2018. Sexual, apomictic and mixed populations in *Handroanthus ochraceus* (Bignoniaceae) polyploid complex. *Plant Syst. Evol.* 304(7): 817–829. doi:10.1007/s00606-018-1512-5

Mendes-Rodrigues, C. and P.E. Oliveira. 2012. Polyembryony in Melastomataceae from Brazilian Cerrado: multiple embryos in a small world. *Plant Biol.* 14(5): 845–853.

Mendes-Rodrigues, C., R. Carmo-Oliveira, S. Talavera, M. Arista, P.L. Ortiz and P.E. Oliveira. 2005. Polyembryony and apomixis in *Eriotheca pubescens* (Malvaceae-Bombacoideae). *Plant Biol.* 7: 533–540. doi:10.1055/s-2005-865852

Mendes-Rodrigues, C., R. Morinho and M. Arista. 2018. Reproductive diversity, polyploidy, and geographical parthenogenesis in two *Eriotheca* (Malvaceae) species from Brazilian Cerrado. *Perspect. Plant Ecol. Evol. Syst.* 36: 1–12. doi:10.1016/j.ppees.2018.11.001

Menken, S.B.J. and T. Morita. 1989. Uniclonal population structure in the pentaploid obligate agamosperm *Taraxacum albidum* Dahlst. *Plant Species Biol.* 4: 29–36. doi:10.1111/j.1442-1984.1989.tb00045.x

Menken, S.J., E. Smit and J.C.M. den Nijs. 1995. Genetical population structure in plants: Gene flow between diploid sexual and triploid asexual dandelions (*Taraxacum* sect. *Ruderalia*). *Evolution* 49: 1108–1118.

Merril, E.D. and L.M. Perry. 1938. The Myrtaceae of China. *J. Arnold Arbor.* 19(3): 191–247.

Mes, T.H.M., P. Kuperus, J. Kirschner J. Štěpánek, H. Štorchová and J.C.M. den Nijs. 2002. Detection of genetically divergent clone mates in apomictic dandelions. *Molec. Ecol.* 11(2): 253–265.

Messeri, A. 1931. Ricerche embriologiche e cariologiche sopra i generi *Allium* e *Nothoscordum*. *N.G. Bot. Ital.* 38(3): 409–441. doi:10.1080/11263503109440397

Michaelis, P. 1925. Zur Cytologie und Embryoentwicklung von *Epilobium*. *Ber. Deutsch. Bot. Ges.* 43: 61–67.

Michel, M.R., M. Cruz-Requena, M.C. Avendaño-Sanchez, V.M. González-Vazquez, A.C. Flores-Gallegos, C.N. Aguilar, J. Espinoza-Velázquez and R. Rodríguez- Herrera. 2018. Polyembryony in maize: A complex, elusive, and potentially agronomical useful trait. In M.A. El-Esawi (ed.), *Maize germplasm—Characterization and genetic approaches for crop improvement*. IntechOpen. Chapter 2. London UK: IntechOpen Unlimited. doi:10.5772/intechopen.70549

Mieulet, D., S. Jolivet, M. Rivard, L. Cromer, A. Vernet, P. Mayonove, L. Pereira, G. Droc, B. Courtois, E. Guiderdoni and R. Mercier. 2016. Turning rice meiosis into mitosis. *Cell Res.* 26(11): 1242–1254. doi:10.1038/cr.2016.117

Mikoláš, V. 1996. *Sorbus dolomiticola* Mikoláš, a new hybridogenous species of the genus *Sorbus* s.l. from eastern Slovakia. *Thaiszia J. Bot., Košice* 6(1-2): 1–12.

Miles, J.W. 2007. Apomixis for cultivar development in tropical forage grasses. *Crop Sci.* 47: 238–249. doi:10.2135/cropsci2007.04.0016IPBS

Miller, J., R.L. Andrew and B.R. Maslin. 2002. Towards and understanding of variation in the Mulga complex. *Conservation Science W. Aust.* 4(3): 19–35.

Miroshnichenko, E.Y. 1970. Apomictic bluegrass forms in Siberia and outlooks of their introduction. *Apomiksis i selektsiya* (*Apomixis and Selection*), Nauka, Moscow, pp. 75–80.

Miroshnichenko, E.Y. 1978. Facultative pseudogamous apomixis and karyological polymorphism in the genus *Poa*. In: *Apomiksis u rastenii i zhivotnykh* (*Apomixis in Plants and Animals*). Nauka, Novosibirsk, pp. 224–236.

Mirzaghaderi, G. and E. Hörandl. 2016. The evolution of meiotic sex and its alternatives. *Proc. R. Soc. B* 283: 20161221. doi:10.1098/rspb.2016.1221

Miwa, T. 1952. Pollination, fertilization and fruit drop in *Citrus tamurana* Hort. *Miyazaki Univ. Bull.* 2: 1–67 (in Japanese).

Modilewsky, J. 1908. Zur Samenenwicklung einiger Urticifloren. *Flora* 98: 423–470.

Modilewsky, J. 1925. Zur Kenntnis der Polyembryonie von *Allium odorum* L. *Bull. Jard. Bot. Kiew* 2: 9–19.

Modilewsky, J. 1931. Die Embryobildung bei *Allium odorum* L. *Bull. Jard. Bot. Kiew* 12–13: 27–48.

Mogie, M. 1985. Morphological, developmental and electrophoretic variation within and between obligately apomictic *Taraxacum* agamospecies. *Biol. J. Linn. Soc.* 24: 207–216.

Mogie, M. 1992. *The Evolution of Asexual Reproduction in Plants*. Chapman & Hall, London.

Mohamed, A.H. and F.W. Gould. 1966. Biosystematic studies in the *Bouteloua curtipendula* V. Megasporogenesis and embryo sac development. *Am. J. Bot.* 53: 166–169.

Molins, M.P., J.M. Corral, O.M. Aliyu, M.A. Koch, A. Betzin, J.L. Maron and T.F. Sharbel. 2014. Biogeographic variation in genetic variability, apomixis expression and ploidy of St. John's wort (*Hypericum perforatum*) across its native and introduced range. *Ann. Bot.* 113(3): 417–427. doi:10.1093/aob/mct268

Molloy, B.P.J. 2019. Apomixis in indigenous New Zealand woody seed plants and its ecological and wider significance: a working hypothesis. *N. Z. J. Ecol.* 43(1): 3357. doi:10.20417/nzjecol.43.4

Mondal, B. and R. Saha. 2014. Identification of zygotic and nucellar seedlings of *Citrus reticulata* and *Citrus aurantifolia* using RAPD. *Intern. J. Adv. Biotechnol. Res.* 5: 25–30.

Monteiro, R. 1980. Taxonomia e biologia da reprodução da *Stevia rebaudiana* Bert. MSc thesis, Universidade Estadual de Campinas, Campinas, Brasil.

Montemartini, L. 1899. Contributo all studio dell'anatonia de frutto e del seme delle opunzie. *Instituto Botanicao Della R. Universita di Pavia (Laboratorio Crittogamico Italiano)*. 1896: 59–67.

de Moraes, I.C.R., C.A.F. Pinto-Maglio and R.A. Lombello. 2009. Reproductive biology and cytology of *Hypericum brasiliense* (Hypericaceae). *Rev. Bras. Bot.* 32: 539–544.

Moran, G.F., L.A.J. Thomson, J. Grant and J.C. Bell. 1992. Distribution of genetic variation within two dry-zone *Acacia* species and implications for their genetic variation. In: House, A.P.N. and C.E. Harwood (eds.). *Australian dry-zone acacias for human food*, Proceedings of a workshop held at Glen Helen, Northern Territory, Australia, 7–10 August 1991. CSIRO Division of Forestry, Canberra, pp. 74–81.

Moreno, J., A. Terrones, M. Alonso, A. Juan and M. Crespo. 2016. *Limonium tobarrense* (Plumbaginaceae), a new species from the southeastern Iberian Peninsula. *Phytotaxa* 257(1): 61–70. http://dx.doi.org/10.11646/phytotaxa.257.1.4

Moskova, R.D. 1975. An electron microscopic study of the nucellus cells in *Bothriochloa ischaemum* L. *Caryologia* 28(3): 295–300. doi:10.1080/00087114.1975.10796619

Moskova, R.D. and M.S. Yakovlev. 1974. Apomixis of *Bothriochloa ischaemum* L. flowering, pollination, fertilization. In: Linskens, H.F. (ed.), *Fertilization in Higher plants*. North Holland Publishing, Amsterdam, pp. 341–349.

Moura, Y.A., A. Alves-Pereira, C.C. da Silva, L.M. Souza, A.P. de Souza and S. Koehler. 2020. Secondary origin, hybridization and sexual reproduction in a diploid–tetraploid contact zone of the facultatively apomictic orchid *Zygopetalum mackayi*. *Plant Biol. J.* 22: 939–948. doi:10.1111/plb.13148

Movsesyan, S.N. 1964. Fertilization process in *Rudbeckia maxima*. *Akad. Nauk. Armyanskoi SSR. Izv. Biol. Nauk.* 17(2): 77–83 (in Russian).

Movsesyan, S.N. and I.P. Kasparova. 1977a. Semigamia *Rudbeckia laciniata* after different types of pollination. *Bot. Zhur.* 62: 422–429.

Movsesyan, S.N. and I.P. Kasparova. 1977b. Some features of semigamy. In: Petrov, D.F. (ed.), *Cytogenetic Basis of Plant Breeding*. Novosibirsk, pp. 151–159.

Mráz, P., P. Zdvořák, M. Hartmann, M. Štefánek and J. Chrtek Jr. 2019. Can obligate apomixis and more stable reproductive assurance explain the distributional successes of asexual triploids in *Hieracium alpinum* (Asteraceae)? *Plant Biol.* 21(2): 227–236. doi:10.1111/plb.12930

Mráz, P. and P. Zdvořák. 2019. Reproductive pathways in *Hieracium s.s.* (Asteraceae): strict sexuality in diploids and apomixis in polyploids. *Ann. Bot.* 123(2): 391–403. doi:10.1093/aob/mcy137

Mráz, P. and Z. Szelag. 2004. Chromosome numbers and reproductive systems in selected species of *Hieracium* and *Pilosella* (Asteraceae) from Romania. *Ann. Bot. Fenn.* 41: 405–414.

Mráz, P., B. Šingliarová, T. Urfus and F. Krahulec. 2008. Cytogeography of *Pilosella officinarum* (Compositae): altitudinal and longitudinal differences in ploidy level distribution in the Czech Republic and Slovakia and the general pattern in Europe. *Ann. Bot.* 101(1): 59–71.

Mráz, P., J. Chrtek Jr. and B. Šingliarová. 2009. Geographical parthenogenesis, genome size variation and pollen production in the arctic-alpine species *Hieracium alpinum*. *Bot. Helv.* 119: 41–51. doi:10.1007/s00035-009-0055-3

Mráz, P., J. Chrtek Jr. and J. Kirschner. 2001. Genetic variation in the *Hieracium rohacsense* group (*Hieracium* sect. *Alpina*). *Phyton* (Horn, Austria) 41(2): 269–276.

Muarawski, D.A. and K. Bawa. 1994. Genetic structure and mating system of *Stemonoporus oblongifolius* (Dipterocarpaceae) in Sri Lanka. *Am. J. Bot.* 81(2): 155–160. doi:10.2307/2445628

Mukherjee, P.K. and S.G. Desai. 1980. Apomixis in *Eupatorium triplinervis*. *Third All India Bot. Conf. Abstract*, pp. 61.

Mukherjee, P.K. and S.G. Desai. 1990. Reduced and unreduced embryo sacs in *Solidago purpurea* L. *J. Indian Bot. Soc.* 69: 75–78.

Muller, H.J. 1964. The relation of recombination to mutational advance. *Mutat. Res.* 1(1): 2–9.

Müller, K. 1847. Beiträge zur Entwicklungsgeschichte des Pflanzen-Embryo. *Bot. Zeit.* 5: 737–742, 753–761, 769–778.

Mulligan, G.A. 1966. Chromosome numbers of the family Crucifereae III. *Can. J. Bot.* 44: 309–319.

Mulligan, G.A. 1996. Synopsis of the genus *Arabis* (Brassicaceae) in Canada, Alaska and Greenland. *Rhodora* 97: 109–163.

Mulligan, G.A. and J.N. Findlay. 1970. Sexual reproduction and agamospermy in the genus *Draba*. *Can. J. Bot.* 48(2): 269–270. doi:10.1139/b70-040

Mullins, M.G. and C. Srinivasan 1976. Somatic embryos and plantlets from an ancient clone of the Grapevine (cv. Cabernet-Sauvignon) by apomixis *in vitro*. *J. Exp. Bot.* 27(5): 1022–1030.

Muniyamma, M. 1977. Triploid embryos from endosperm *in vivo*. *Ann. Bot.* 41(175): 1077–1079. https://www.jstor.org/stable/42756486

Muniyamma, M. 1978. Variations in microsporogenesis and the development of embryo sacs in *Echinochloa stagnina* (Gramineae). *Bot. Gaz.* 139: 87–94.

Muniyamma, M. and J.B. Phipps. 1979. Cytological proof of apomixis in *Crataegus* (Rosaceae). *Am. J. Bot.* 66: 149–155.

Muniyamma, M. and J.B. Phipps. 1984a. Studies in *Crataegus:* A note on the occurrence of diplospory in *Crataegus dissona* Sarg. (Maloideae, Rosaceae). *Can. J. Genet. Cytol.* 26: 249–252.

Muniyamma, M. and J.B. Phipps. 1984b. Studies in *Crataegus* XI. Further cytological evidence for the occurrence of apomixis in North American hawthorns. *Can. J. Genet. Cytol.* 62: 2316–2324.

Munter, D.A., J.J. Luby and N.O. Anderson. 2018. Reproductive biology in Northern Prickly Ash. *J. Am. Soc. Hort. Sci.* 143(1): 72–83. doi:10.21273/JASHS04280-17

Müntzing, A. 1928. Pseudogamie in der Gattung *Potentilla*. *Hereditas* 11: 267–283.

Müntzing, A. 1931. Note on the cytology of some apomictic *Potentilla* species. *Hereditas* 15: 166–178.

Müntzing, A. 1933. Apomixis and sexual seed formation in *Poa*. *Hereditas* 24: 487–491.

Müntzing, A. 1958. Heteroploidy and polymorphism in some apomictic species of *Potentilla*. *Hereditas* 44: 280–329.

Müntzing, A. 1966. Apomixis and sexuality in new material of *Poa alpina* from middle Sweden. *Hereditas* 54: 314–337.

Murawski, D.A. and K.S. Bawa. 1994. Genetic structure and mating system of *Stemonoporus oblongifolius* (Dipterocarpaceae) in Sri Lanka. *Amer. J. Bot.* 81(2): 155–160. doi:10.2307/2445628

Murbeck, S.V. 1901. Parthenogenetische Embryobildung in der Gattung *Alchemilla*. *Lunds Univ. Årsskr.* 36(7): 1–45.

Murbeck, S.V. 1902. Über Anomalien im Bau des Nucellus un des Embryosackes bei parthenogenetischen Arten der Gatttung *Alchemilla*. *Lunds Univ. Årsskr.* 38(2): 1–11.

Murbeck, S.V. 1904. Parthenogenese bei den Gattungen *Taraxacum* und *Hieracium*. *Bot. Not.* 57(6): 285–296.

Murbeck, S. 1895. Skandinaviska former af *Alchemilla vulgaris* L. *Bot. Not.* 1895: 264–266.

Murbeck, S.V. 1897. Om vegetative embryobildning hos flertalet Alchemillor och den förklaring öfver formbestädigheten inom slägtet, som densamma innerbär. *Bot. Not.* 1897: 273–277.

Murthy, Y.S. and H.P. Sharma. 1977. Embryological studies in Compositae. II. *Bellis perennis* – a casual apomict. In: V.R. Dnyansagar et al. (eds.), *Recent Trends and contacts between cytogenetics, embryology and morphology*. Today & Tomorrow, New Delhi, pp. 257–260.

Murty, U.R. 1973. Polyploidy and apomixis in *Apluda mutica* var. *aristata* (L.) Pilger. *Cytologia* 38: 347–356.

Musiał, K. and M. Kościńska-Pająk. 2017. Pattern of callose deposition during the course of meiotic diplospory in *Chondrilla juncea* (Asteraceae, Cichorioideae). *Protoplasma* 254(4): 1499–1505. doi:10.1007/s00709-016-1039-y

Musiał, K. and M. Kościńska-Pająk. 2019. Callose is integral to meiotic diplospory of the *Taraxacum* type: new evidence from ovules of *Chondrilla brevirostris* (Asteraceae-Cichorioideae). *Bot. Lett.* 166(3): 274–282. doi:10.1080/23818107.2019.1573378

Musiał, K., M. Kościńska-Pająk, E. Sliwinska and A. Joachimiak. 2012. Developmental events in ovules of the ornamental plant *Rudbeckia bicolor* Nutt. *Flora* 207: 3–9.

Musiał, K., M. Kościńska-Pająk, R. Antolec and A.J. Joachimiak. 2015. Deposition of callose in young ovules of two *Taraxacum* species varying in the mode of reproduction. *Protoplasma* 252: 135–144.

Musiał, K., P. Górka, K.P. Maria and P. Marciniuk. 2013. Embryological studies in *Taraxacum udum* Jordan (sect. *Palustria*). *Botany* 91: 614–620.

Nagai, K. and T. Tanikawa. 1928. On citrus pollination. *Third Pan-Pacific Sci. Cong.* (Tokyo, 1928) *Proc.* 2: 2023–2039.

Nakamura, M., S. Nanami, S. Okuno, S.K. Hirota, A. Matsuo, Y. Suyama, H. Tokumoto, S. Yoshihara and A. Itoh. 2021. Genetic diversity and structure of apomictic and sexually reproducing *Lindera* species (Lauraceae) in Japan. *Forests* 12(2): 227. doi:10.3390/f12020227

Nakano, M., K. Kigoshi, T. Shimizu, T. Endo, T. Shimada, H. Fujii and M. Omura. 2013. Characterization of genes associated with polyembryony and in vitro somatic embryogenesis in *Citrus*. *Tree Genet. Genomes* 9: 795–803.

Nakano, M., T. Shimada, T. Endo, H. Fujii, H. Nesumi, M. Kita, M. Ebina, T. Shimizu and M. Omura. 2012. Characterization of genomic sequence showing strong association with polyembryony among diverse *Citrus* species and cultivars, and its synteny with *Vitis* and *Populus*. *Plant Sci.* 183: 131–142.

Nakano, M., T. Shimizu, H. Fujii, T. Shimada, T. Endo, H. Nesumi, T. Kuniga and M. Omura. 2008b. Marker enrichment and construction of haplotype-specific BAC contigs for the polyembryony genomic region in *Citrus*. *Breed Sci.* 58: 375–383.

Nakano, M., T. Shimizu, T. Kuniga, H. Nesumi and M. Omura. 2008a. Mapping and haplotyping of the flanking region of the polyembryony locus in *Citrus unshiu* Marcow. *J. Jpn. Soc. Hortic. Sci.* 77: 109–114.

Nakazawa, Y., K. Namai, A. Kojima, S. Kobayashi, K. Tasaki and M. Amagai. 2006. Estimating the inheritance for parthenogenesis in tetraploid Chinese leek (*Allium ramosum*, syn. *A. tuberosum*). *Breed. Res.* 8: 89–98.

Nannfeldt, J.A. 1937. The chromosome numbers of *Poa* sect. *Ochlopoa* A. & Gr. and their taxonomical significance. *Bot. Not.* 1937: 238–254.

Narayan, K.N. 1951. Cytogenetic studies in apomictic *Pennisetum* species. PhD thesis, University of California, Berkeley, CA, USA.

Narayan, K.N. 1955. Cytogenetic studies in apomixis in *Pennisetum: Pennisetum clandestinum* Hochst. *Proc. Indian Acad. Sci.* B 41: 196–208.

Narayan, K.N. 1962. Apomixis in some species of *Pennisetum* and in *Panicum antidotale*. In: P. Maheshwari (ed.), *Plant embryology*. CSIR, New Delhi, pp. 55–61.

Narayan, K.N. 1967–1968. Cytogenetical studies in Gramineae and Compositae. *J. Mysore Univ.* (N.S.) Sect. B. Golden Jubilee Vol. 144–156.

Narayanaswami, S. 1940. Megasporogenesis and origin of triploids in *Saccharum*. *Indian J. Agric. Sci.* 10: 534–551.

Narayanaswami, S. 1955. The structure and development of the caryopsis in some Indian millet. 4. *Echinochloa frumentacea*. *Phytomorphology* 5: 161–171.

Narayanaswami, S. and S.K. Roy. 1960. Embryo sac development and polyembryony in *Syzygium cumini*. *Bot. Not.* 1960: 273–284.

Nardi, F.D., Ch. Dobeš, D. Müller, T. Grasegger, T. Myllynen, H. Alonso-Marcos and A. Tribsch. 2018. Sexual intraspecific recombination but not *de novo* origin governs the genesis of new apomictic genotypes in *Potentilla puberula* (Rosaceae). *Taxon*. 67(6): 1108–1131.

Narmatha Bai, V. and K.K. Lakshmanan. 1982. Occurrence of polyembryony in *Toddalia asiatica* Lamb. *Curr. Sci.* 51:374

Nassar, N.M., A. Chaib and A.Y. Elsayed. 2011. Apomixis in different ploidy levels of cassava. *Hereditas* 148: 125–128.

Nassar, N.M.A., E. Dos Santos and S.R.O. David. 2000. The transference of apomixis genes from *Manihot neusana* Nassar to cassava, *M. esculenta* Crantz. *Hereditas* 132: 167–170.

Nassar, N.M.A. 1994. Development and selection for apomixis in cassava, *Manihot esculenta* Crantz. *Can. J. Plant. Sci.* 74: 857–858.

Nassar, N.M.A. 2002. Apomixis and cassava. *Genet. Mol. Res.* 1: 147–152.

Nassar, N.M.A. 2006. Cassava genetic resources: Extinct everywhere in Brazil. *Genet. Resour. Crop Evol.* 53(5): 975–983.

Nassar, N.M.A., M.A.R. Vieira, C. Vieira and D. Grattapaglia. 1998. Evidence of apomixis in cassava (*Manihot esculenta* Crantz). *Genet. Mol. Biol.* 21(4): doi:10.1590/S1415-47571998000400020

Natesh, S. and Rau, M.A. 1984. The embryo. In: B.M. Johri, (ed.), *Embryology of Angiosperms*. Springer-Verlag, Berlin, pp. 377–443.

Naumova, T.N. 1970a. The development of sexual and apomictic embryos in *Euonymus europaeus* L. (Celastraceae). *Bot. Zhu.* 5: 638–648.

Naumova, T.N. 1970b. Polyembryony in *Euonymus macroptera* Rupr. and *E. planipes* (Kochne) Kochne. *Bot. Zhu.* 9: 1270. (in Russian).

Naumova, T.N. 1971. Adventive embryony in genus *Euonymus* (Celastraceae). PhD Dissertation, Leningrad, Russia.

Naumova, T.N. 1978. Developmental characteristics of nucellar tissue and nucellar polyembryony in *Opuntia elata*, Cactaceae. *Bot. Zhu.* 63: 344–355.

Naumova, T.N. 1979. Embryology of representatives of the family Celastraceae. The significance of integumental embryony in the evolution of genus *Euonymus*. In: *Actualnye Voprosy Embryologi Pokrytosemennykh Rastenii*. Nauka, Leningrad, 46 pages.

Naumova, T.N. 1980. Nucellar polyembryony in the genus *Sarcococca* (Buxaceae). *Bot. Zhu.* 65: 230–240.

Naumova, T.N. 1993. *Apomixis in Angiosperms. Nucellar and integumentary embryony*. CRC Press, Boca Raton, FL.

Naumova, T.N. and M.T.M. Willemse. 1983a. Ultrastructural characteristic of nucellar polyembryony in *Sarcococca humilis* (Buxaceae). Nucellus and initial cells of nucellar proembryos. *Bot. Zhu.* 68: 1044–1054.

Naumova, T.N. and M.T.M. Willemse. 1983b. Ultrastructural characteristic of nucellar polyembryony in *Sarcococca humilis* (Buxaceae). Division of initial cell and nucellar proembryos. *Bot. Zhu.* 68: 1184–1195.

Naumova, T.N. and M.T.M. Willemse. 1995. Ultrastructural characterization of apospory in *Panicum maximum*. *Sex. Plant Reprod.* 8: 197–204.

Naumova, T.N., J. van der Laak, J. Osadtchiy, F. Matzk, A. Kravtchenko, J. Bergervoet, K.S. Ramulu and K. Boutilier. 2001. Reproductive development in apomictic populations of *Arabis holboellii* (Brassicaceae). *Sex. Plant Reprod.* 14: 195–200.

Naumova, T.N., A.P.M. den Nijs, and M.T.M. Willemse. 1993. Quantitative analysis of aposporous parthenogenesis in *Poa pratensis* genotypes. *Acta Bot. Neerl.* 42: 299–313.

Naumova, T.N., J.V. Osadtchiy, V.K. Sharma, P. Dijkhuis and K.S. Ramulu. 1999. Apomixis in plants: Structural and functional aspects of diplospory in *Poa nemoralis* and *Poa palustris*. *Protoplasma* 208: 186–195.

Naumova, T. N. and M.S. Yakovlev. 1972. Adventive embryony in angiosperms. *Bot. Zhu.* 57(8): 1006.

Nawashin, S.G. 1898. Resultate einer Revision der Bebrchtungsvorgange bei *Lilium martagon* und *Fritillaria tenella*. *Bull. Acad. Imp. des Sci. St. Petersburg* 9: 377–382.

Neal, P.R. and G.J. Anderson. 2004. Does the 'old bag' make a good 'wind bag'?: Comparison of four fabrics commonly used as exclusion bags in studies of pollination and reproductive biology. *Ann Bot.* 93(5): 603–607. doi:10.1093/aob/mch068

Nelson-Jones, E.B., D. Briggs and A.G. Smith. 2002. The origin of intermediate species of the genus *Sorbus*. *Theor. Appl. Genet.* 105: 953–963. doi:10.1007/s00122-002-0957-6

Németh, C., N. Papp, J. Nosková and M. Höhn. 2020. Speciation by triparental hybridization in genus *Sorbus* (Rosaceae). *Biol. Futura* 71: 209–222.

Németh, C.S. 2012. Two new *Sorbus* (Rosaceae) species from the Bakony Mts, Hungary. *Acta Bot. Hung.* 54(1–2): 131–144.

Ng, K.K.S., S.L. Lee and C.L. Koh. 2004. Spatial structure and genetic diversity of two tropical tree species with contrasting breeding systems and different ploidy levels. *Mol. Ecol.* 13(3): 657–669. doi:10.1046/j.1365-294X.2004.02094.x

Ngulube, M., J. Hall and J. Maghembe. 1998. Reproductive ecology of *Uapaca kirkiana* (Euphorbiaceae) in Malawi, Southern Africa. *J. Trop. Ecol.* 14(6): 743–760. doi:10.1017/S0266467498000546

Nielsen, E.L. 1946. The origin of the multiple macrogametophytes in *Poa pratensis*. *Bot. Gaz.* 108: 41–50.

Niemann, J., A. Wojciechowski and J. Janowicz. 2012. Identification of apomixis in the Kentucky bluegrass (*Poa pratensis* L.) using auxin test. *Acta Soc. Bot. Pol.* 81(3): 217–221. doi:10.5586/asbp.2012.020

Niemirowicz-Szczytt, K. 1984a. The results of intergeneric pollination of *Fragaria* × *ananassa* Duch. and *Fragaria virginiana* Duch. by *Potentilla* species. *Acta Soc. Bot. Pol.* 53: 443–454.

Niemirowicz-Szczytt, K. 1984b. Morphological and cytological evaluation of progeny obtained from pollination of *Fragaria* × *ananassa* Duch. with *Potentilla* spp. *Acta Soc. Bot. Pol.* 53: 455–468.

Niemirowicz-Szczytt, K. 1984c. Embryological studies on *Fragaria* × *ananassa* Duch. after pollination by three species of *Potentilla*. *Acta Soc. Bot. Pol.* 53: 469–484.

Nieto-Angel, R., W.B. Michal, M. Betancourt-Olvera and J. Martínez-Solís. 2013. Hawthorn (*Crataegus* spp.) as an ornamental plant. *Acta Hortic.* 977: 61–66. doi:10.17660/ActaHortic.2013.977.5

Niissalo, M.A., J. Leong-Škorničková, O. Šída and G.S. Khew. 2020. Population genomics reveal apomixis in a novel system: uniclonal female populations dominate the tropical forest herb family, Hanguanaceae (Commelinales). *AoB Plants* 12(6): plaa053. doi:10.1093/aobpla/plaa053

Nishimura, M. 1922. On the germination and the polyembryony of *Poa pratensis* L. *Shokubutsugaku Zasshi* 36(425) 47–54. doi:10.15281/jplantres1887.36.425_en47

Noack, K. 1939. Über *Hypericum*-Kreuzungen. VI. Fortpflanzungs-Verhältnisse und Bastarde von *Hypericum perforatum* L. *Z. Indukt. Abstamm. Vererbungsl* 76: 569–601.

Nogler, G.A. 1990. Simplified methods for embryological studies. *Apomixis Newsl.* 2: 56–58.

Nogler, G.A. 1971. Genetik der Aposporie bei *Ranunculus auricomus* s.l. W. Koch. I. Embryologie. *Ber. Schweiz. Bot. Ges.* 81: 139–179.

Nogler, G.A. 1972. Genetik der Aposporie bei *Ranunculus auricomus*. II. Endospermzytologie. *Ber. Schweiz. Bot. Ges.* 82: 4–63.

Nogler, G.A. 1973. Genetik der Aposporie bei *Ranunculus auricomus*. III. F2-Rückkreuzungsbastarde. *Ber Schweiz Bot. Ges.* 83: 295–305.

Nogler, G.A. 1984. Genetics of apospory in apomictic *Ranunculus auricomus*. V. Conclusion. *Bot. Helv.* 94: 411–422.

Nogler, G.A. 1995: Genetics of apomixis in *Ranunculus auricomus*. VI. Epilogue. *Bot. Helv.* 105: 111–115.

Nordborg, G. 1963. Studies in *Sanguisorba officinalis* L. *Bot. Not.* 116: 267–288.

Nordborg, G. 1967. Embryological studies in the *Sanguisorba minor* complex (Rosaceae). *Bot. Not.* 120: 109–119.

Norrmann, G.A. 1981. Citología y método de reproducción en dos especies de *Paspalum* (Gramineae). *Bonplandia* 5: 149–158 (in Spanish).

Norrmann, G.A., C.L. Quarin and B.L. Burson. 1989. Cytogenetics and reproductive behaviour of different chromosome races in six *Paspalum* species. *J. Heredity* 80: 24–28.

Norstog, K. 1963. Apomixis and polyembryony in *Hierochloe odorata*. *Am. J. Bot.* 50: 815–821.

Nosrati, H., A.H. Price and C.C. Wilcock. 2010. No evidence of apomixis in matroclinal progeny from experimental crosses in the genus *Fragaria* (strawberry) based on RAPDs. *Euphytica* 171: 193–202.

Novak, S.J. and R.N. Mack. 1995. Allozyme diversity in the apomictic vine *Bryonia alba* (Cucurbitaceae): potential consequences of multiple introductions. *Am. J. Bot.* 82(9): 1153–1162.

Novak, S.J. and R.N. Mack. 2000. Clonal diversity within and among introduced populations of the apomictic vine *Bryonia alba* (Cucurbitaceae). *Can. J. Bot.* 78(11): 1469–1481. doi:10.1139/b00-116

Novo, P.E., F. Espinoza and C.L. Quarin. 2014. An apomictic tetraploid *Paspalum chaseanum* cytotype and its cytogenetic relationship with *P. plicatulum* (Poaceae): taxonomic and genetic implications. *Aust. J. Bot.* 61: 538–543. doi:10.1071/BT13194

Nowaczyk, L. and P. Nowaczyk. 2006. Genetic analysis of tomato (*Lycopersicon esculentum* Mill.) twin forms. *Acta Biol. Cracov. Ser. Bot.* 48(1): 53–58.

Nowaczyk, P. 1987. Spontaneous and induced polyembryony in pepper *Capsicum annuum* L. *Genetica Pol.* 28: 109–125.

Noyes, R.D. 2007. Apomixis in the Asteraceae: Diamonds in the rough. *Funct. Plant Sci. Biotechnol.* 1: 207–222.

Noyes, R.D. and J.D. Wagner. 2014. Haploidy yields diploid apomicts and parthenogenesis in *Erigeron* (Asteraceae). *Am. J. Bot.* 101(5): 865–874. doi:10.3732/ajb.1400008

Noyes, R.D. 2000. Diplospory and parthenogenesis in sexual × agamospermous (apomictic) *Erigeron* (Asteraceae) hybrids. *Int. J. Plant Sci.* 161: 1–12.

Noyes, R.D. 2006. Apomixis via recombination of genome regions for apomeiosis (diplospory) and parthenogenesis in *Erigeron* (daisy fleabane, Asteraceae). *Sex. Plant Reprod.* 19: 7–18.

Noyes, R.D. 2008. Sexual devolution in plants: apomixis uncloaked? *BioEssays* 30(9): 798–801.

Noyes, R.D. and D.E. Soltis. 1996. Genotypic variation in agamospermous *Erigeron compositus* (Aseraceae). *Am. J. Bot.* 83: 1292–1303.

Noyes, R.D. and L.H. Rieseberg. 2000. Two independent loci control agamospermy (apomixis) in the triploid flowering plant *Erigeron annuus*. *Genetics* 155: 379–390.

Noyes, R.D., R. Baker and B. Mai. 2007. Mendelian segregation for two-factor apomixis in *Erigeron annuus* (Asteraceae). *Heredity* 98: 92–98.

Nybom, H. 1985. Pollen viability assessments in blackberries (*Rubus* subgen. *Rubus*). *Plant Syst. Evol.* 150: 281–290.

Nybom, H. 1988. Apomixis versus sexuality in black berries (*Rubus* subg. *Rubus*. Rosaceae). *Plant Syst. Evol.* 160: 207–218. doi:10.1007/BF0093608

Nybom, H. 1995. Evaluation of interspecific crossing experiments in facultatively apomictic blackberries (*Rubus* subgen, *Rubus*) using DNA fingerprinting. *Hereditas* 122: 57–65.

Nybom, H. 2007. Unique reproduction in dogroses (*Rosa* sect. Caninae) maintains successful and highly heterozygous genotypes. In: Hörandl, E., U. Grossniklaus, T. Sharbel and P. van Dijk (eds.), *Apomixis: Evolution, Mechanisms and Perspectives*. Ruggell, Liechtenstein, Gantner Verlag, pp. 281–297.

Nybom, H. and I.V. Bartish. 2007. DNA markers and morphometry reveal multiclonal and poorly defined taxa in an apomictic *Cotoneaster* species complex. *Taxon* 56: 119–128.

Nybom, H., G. Werlemark, D. G. Esselink and B. Vosman. 2005. Sexual preferences linked to rose taxonomy and cytology. *Acta Hortic.* 690: 21–28.

Nybom, H., G.D. Esselink, G. Werlemark and B. Vosman. 2004. Microsatellite DNA marker inheritance indicates preferential pairing between two highly homologous genomes in polyploid and hemisexual dogroses, *Rosa* L. Sect. *Caninae* DC. *Hereditas* 92: 139–150. doi:10.1038/sj.hdy.6800332

Nygren, A. 1946. The genesis of some Scandinavian species of *Calamagrostis*. *Hereditas* 32: 131–262.

Nygren, A. 1950a. A preliminary note on cytological and embryological studies in arctic Poae. *Hereditas* 36: 231–232.

Nygren, A. 1950b. Cytological and embryological studies in arctic Poae. *Symb. Bot. Upsaliensis* 10: 1–64.

Nygren, A. 1954a. Apomixis in the angiosperms II. *Bot. Rev.* 20: 577–649.

Nygren, A. 1954b. Investigations on North American *Calamagrostis* I. *Hereditas* 40: 377–397.

Nygren, A. 1967. Apomixis in the angiosperms. In: W. Ruhland (ed.), and co-editors, *Encyclopedia of Plant Physiology*, Vol. 18. Springer, Berlin, Heidelberg, pp. 551–596.

Nyléhn, J., E. Hamre and I. Nordal. 2003. Facultative apomixis and hybridization in arctic *Potentilla* section *Niveae* (Rosaceae) from Svalbard. *Bot. J. Linn. Soc.* 142: 373–381.

Ochogavía, A., G. Galla, J.G. Seijo, A.M. González, M. Bellucci, F. Pupilli, G. Barcaccia, E. Albertini and S. Pessino. 2018. Structure, target-specificity and expression of *PN_LNC_N13*, a long non-coding RNA differentially expressed in apomictic and sexual *Paspalum notatum*. *Plant Mol. Biol.* 96: 53–67. doi:10.1007/s11103-017-0679-4

Odiyi, A.C. and J.O. Uzo. 2001. Multiple seedlings in fluted pumpkin (*Telfairia occidentalis* Hook. f.): Evidence for the occurrence of polyembryony. *J. Appl. Chem. Agric. Res.* 7(1): 17–25. doi:10.4314/jacar.v7i1.41120

Ogawa, D., S.D. Johnson, S.T. Henderson and A.M.G. Koltunow. 2013. Genetic separation of autonomous endosperm formation (AutE) from the two other components of apomixis in *Hieracium*. *Plant Reprod.* 26(2): 113–123. doi:10.1007/s00497-013-0214-y

Okabe, S. 1930. Über Parthenogenesis bei *Houttuynia cordata*. *Jap. J. Genet.* 6: 14–19.

Okabe, S. 1932. Parthenogenesis bei *Ixeris dentata*. *Bot. Mag.* 46: 518–523.

Okabe, S. 1963. Apomixis in the genus *Boehmeria*. *Sci. Rep. Tôhoku Univ.* 29: 207–215.

Okada, T., Hu, Y., M.R. Tucker, J.M. Taylor, S.D. Johnson, A. Spriggs, T. Tsuchiya, K. Oelkers, J.C.M. Rodrigues and A. M. G. Koltunow. 2013. Enlarging cells initiating apomixis in *Hieracium praealtum*

transition to an embryo sac program prior to entering mitosis. *Plant Physiol.* 163: 216–231. doi:10.1104/pp.113.219485
Okada, T., K. Ito, S.D. Johnson, K. Oelkers, G. Suzuki, A. Houben, Y. Mukai and A.M. Koltunow. 2011. Chromosomes carrying meiotic avoidance loci in three apomictic Eudicot *Hieracium* subgenus *Pilosella* species share structural features with two monocot apomicts. *Plant Physiol.* 157(3): 1327–1341. doi:10.1104/pp.111.181164
Okoro, P., A.A. Shaibu, G. Ude and B.A. Olukolu. 2011. Genetic evidence of developmental components of parthenocarpy in apomictic *Musa* species. *J. Plant Breed. Crop Sci.* 3(8): 138–145.
Olden, E.J. 1953. Sexual and apomictic seed formation in *Malus sieboldii* Rehd. *Bot. Not.* 106: 105–128.
Olien, W.C. 1987. Apomictic crabapples ad their potential for research and fruit production. *Hort. Sci.* 22: 541–546.
Oliveira, P.E., P.E. Gibbs, A.A. Barbosa and S. Talavera. 1992. Contrasting breeding systems in two *Eriotheca* (Bombacaceae) species of Brazilian Cerrados. *Plant Syst. Evol.* 179: 207–219.
Olmedo-Monfil, V., N. Durán-Figueroa, M. Arteaga-Vázquez, E. Demesa-Arévalo, D. Autran, D. Grimanelli, R.K. Slotkin, R.A. Martienssen and J-P. Vielle-Calzada. 2010. Control of female gamete formation by a small RNA pathway in *Arabidopsis*. *Nature* 464: 628–632. doi:10.1038/nature08828
Olsen, R.T., T.G. Ranney and D.J. Werner. 2006. Fertility and inheritance of variegated and purple foliage across a polyploid series in *Hypericum androsaemum* L. *J. Amer. Soc. Hort. Sci.* 131: 725–730.
Olsson, Å.M.E. 1999. Morphometric and molecular variation in the Nordic dogroses (*Rosa* sect. *Caninae*, Rosaceae). PhD thesis, Lund University, Lund, Sweden.
Opperman, M.M. 1904. A contribution to the life history of *Aster*. *Bot. Gaz.* 37: 353–362.
Orellana, M.R., A.M. Rovira, C. Blanché and M. Bosch. 2005. Pollination and reproductive success in the gynodioecious endemic *Thymus loscosii* (Lamiaceae). *Can. J. Bot.* 83(2): 183–193. doi:10.1139/b04-166
Orr, M.Y. 1923. Polyembryony in *Sarcococca ruscifolia* Stapf. *Roy. Bot. Gard. Edinburgh. Notes* 14: 21–23.
Ortiz, J.P.A., C.L. Quarin, S.C. Pessino, C. Acuña, E.J. Martínez, F. Espinoza, D.H. Hojsgaard, M.E. Sartor, M.E. Cáceres and F. Pupilli. 2013. Harnessing apomictic reproduction in grasses: what we have learned from *Paspalum*. *Ann. Bot.* 112(5) 767–787. doi:10.1093/aob/mct152
Ortiz, J.P.A., F. Pupilli, C.A. Acuña, O. Leblanc and S.C. Pessino. 2020. How to become an apomixis model: The multifaceted case of *Paspalum*. *Genes* 11: 974. doi:10.3390/genes11090974
Ortiz, J.P., S.C. Pessino, O. Leblanc, M.D. Hayward and C.L. Quarin. 1997. Genetic fingerprinting for determining the mode of reproduction in *Paspalum notatum*, a subtropical apomictic forage grass. *Theor. Appl. Genet.* 95: 850–856.
Ortiz, J.P.A., O. Leblanc, C. Rohr, M. Grisolia, L.A. Siena, M. Podio, C. Colono, C. Azzaro and S.C. Pessino. 2019. Small RNA-seq reveals novel regulatory components for apomixis in *Paspalum notatum*. *BMC Genom.* 20: 487. doi:10.1186/s12864-019-5881-0
Ortiz, J.P.A., S. Revale, L.A. Siena, M. Podio, L. Delgado, J. Stein, O. Leblanc and S.C. Pessino. 2017. A reference floral transcriptome of sexual and apomictic *Paspalum notatum*. *BMC Genom.* 18: 318.
Osawa, J. 1913. Studies on the cytology of some species of *Taraxacum*. *Ark. Zellforsch.* 10: 450–469.
Osawa, J. 2012. Cytological and experimental studies in *Citrus*. *J. Coll. Agric. Tokyo* 4: 83–116.
Ostenfeld, C.H. 1906. Experimental and cytological studies in the Hieracia. I. Castration and hybridization experiments with some species of Hieracia. *Bot. Tidsskr.* 27: 225–248.
Ostenfeld, C.H. 1910. Further studies on the apogamy and hybridization of the Hieracia. *Z. Indukt. Abstamm. Vererbungsl.* 3: 241–285.
Ostenfeld, C.H. and C. Raunkiaer. 1903. Kastreringsforsog med *Hieracium* og andre Cichorieae. *Bot. Tidsskr.* 25: 409–413.
Overton, J.B. 1902. Parthenogenesis in *Thalictrum purpurascens*. *Bot. Gaz.* 33: 363–375.
Overton, J.B. 1904. Über Parthenogenesis bei *Thalictrum purpurascens*. *Ber. Deutsch. Bot. Ges.* 22: 274–283.
Ozias-Akins, P. and J.A. Conner. 2020. Clonal reproduction through seeds in sight for crops. *Trends Genet.* 36: 215–226. doi:10.1016/j.tig.2019.12.006
Ozias-Akins, P. and P.J. van Dijk. 2007. Mendelian genetics of apomixis in plants. *Ann. Rev. Genet.* 41: 509–537. doi:10.1146/annurev.genet.40.110405.09051
Ozias-Akins, P., D. Roche and W.W. Hanna. 1998. Tight clustering and hemizygosity of apomixis-linked molecular markers in *Pennisetum squamulatum* implies genetic control of apospory by a divergent locus which may have no allelic form in sexual genotypes. *Proc. Natl. Acad. Sci. U.S.A.* 95: 5127–5132.
Ozias-Akins, P., Y. Akiyama and W.W. Hanna. 2003. Molecular characterization of the genomic region linked with apomixis in *Pennisetum/Cenchrus*. *Funct. Integr. Genomics* 3: 94–104.
Pace, L. 1913. Apogamy in Atamasco. *Bot. Gaz.* 56: 376–394.
Pagni, P. 1954. Il gametofito tetramegasporiale di *Erigeron glaucum* Ker.-Gawl. *G. Bot. Ital.* 61(1): 67–79. doi:10.1080/11263505409431560

Pak, J-H. and S. Kawano. 1990. Biosystematic studies on the genus *Ixeris* (Compositae- Lactuceae). II. Karyological analyses. *Cytologia* 55: 553–570.

Pak, J-H., N.C. Kim, K. Choi, I. Kim, B.B. Seo and S.D. Song. 1997. Geographical distributions of diploid and polyploids of *Ixeris chinensis* complex (Asteraceae: Lactuceae) in East Asia. *Folia Geobot. Phytotax.* 48: 187–196.

Pal, N. 1960. Development of the seed of *Millettia ovalifolia*. *Bot. Gaz.* 122(2): 130–137.

Palacios, C. and F. González-Candelas. 1997a. Lack of genetic variability in the rare and endangered *Limonium cavanillesii* (Plumbaginaceae) using RAPD. *Mol. Ecol.* 6(7): 671–675.

Palacios, C. and F. González-Candelas. 1997b. Analysis of population genetic structure and variability using RAPD markers in the endemic and endangered *Limonium dufourii* (Plumbaginaceae). *Mol. Ecol.* 6(12): 1107–1121.

Palop-Esteban, M., J.G. Segarra-Moragues and F. González-Candelas. 2011. Polyploid origin, genetic diversity and population structure in the tetraploid sea lavender *Limonium narbonense* Miller (Plumbaginaceae) from eastern Spain. *Genetica* 139: 1309–1322. doi:10.1007/s10709-012-9632-2

Palop-Esteban, M., J.G. Segarra-Moragues and F. González-Candelas. 2007. Historical and biological determinants of genetic diversity in the highly endemic triploid sea lavender *Limonium dufourii* (Plumbaginaceae). *Mol. Ecol.* 16: 3814–3827. https://doi.org/10.1111/j.1365-294X.2007.03449.x

Pandey, S.N. and A. Chadha. 1997. *Plant anatomy and embryology*. Vikas publishing, Noida, India.

Pangsuban, S., N. Bamroongrugsa, K. Kanchanpoom and C. Nualsri. 2009. Facultative apomixis in *Garcinia atroviridis* (Clusiaceae) and effects of different pollination regimes on reproductive success. *Trop. Life Sci. Res.* 20(2): 89–108.

Pank, F., F. Matzk, U. Kästner, W.D. Blüthner, E. Foltys de Garcia, A. Meister, U. Ryschka and G. Schumann. 2003. Reproductive diversity and strategies for breeding in St. John's wort (*Hypericum perforatum* L.). *Euphytica* 134: 77–84. doi:10.1023/A:1026115508883

Pansarin, L.M., E.R. Pansarin and M. Sazima. 2008. Reproductive biology of *Cyrtopodium polyphyllum* (Orchidaceae): a Cyrtopodiinae pollinated by deceit. *Plant Biol.* 10: 650–659. doi:10.1111/j.1438-8677.2008.00060.x

Papanicolaou, K. 1984. Chromosome Number Reports LXXXII. *Taxon* 33: 126–134.

Parlevliet, J.E. and J.W. Cameron. 1959. Evidence on the inheritance of nucellar embryony in citrus. *Proc. Amer. Soc. Hort. Sci.* 74: 252–260.

Pattemore, D.E. 2017. Pollination. In: Thomas, B., B.G. Murray and D.J. Murphy (eds.), *Encyclopedia of Applied Plant Sciences*, Vol. 2, 2nd Ed., Academic Press, Heidelberg, pp. 309–320.

Paule, J. 2010. Evolutionary patterns and processes in the genus *Potentilla* L. (Rosaceae). PhD thesis, Ruperto-Carola University of Heidelberg, Heidelberg, Germany, 146 pages.

Paule, J., A. Scherbantin and Ch. Dobeš. 2012. Implications of hybridisation and cytotypic differentiation in speciation assessed by AFLP and plastid haplotypes–a case study of *Potentilla alpicola* La Soie. *BMC Evol. Biol.* 12: 132. doi:10.1186/1471-2148-12-132

Paule, J., F.G. Dunkel, M. Schmidt and T. Gregor. 2018. Climatic differentiation in polyploid apomictic *Ranunculus auricomus* complex in Europe. *BMC Ecol.* 18(1):16. doi:10.1186/s12898-018-0172-1

Paule, J., T.F. Sharbel and Ch. Dobeš. 2011. Apomictic and sexual lineages of the *Potentilla argentea* L. group (Rosaceae): cytotype and molecular genetic differentiation. *Taxon* 60: 721–732.

Paun, O., J. Greilhuber, E.M. Temsch and E. Hörandl. 2006a. Patterns, sources and ecological implications of clonal diversity in apomictic *Ranunculus carpaticola* (*Ranunculus auricomus* complex, Ranunculaceae). *Mol. Ecol.* 15(4): 897–910.

Paun, O., T.F. Stuessy and E. Hörandl. 2006b. The role of hybridization, polyploidization and glaciation in the origin and evolution of the apomictic *Ranunculus cassubicus* complex. *New Phytol.* 171(1): 223–236.

Pauzé, F. 1973. Developpement floral et embryologie d'*Ochna atropurpurea* DC. (Floral development and embryology of *Ochna atropurpurea* DC.). PhD Dissertation, McGill University, Montreal, Canada.

Pavlova, D. 1999. Mediterranean chromosome number reports 9 (1064–1065). *Fl. Medit.* 9: 356–359.

Pazy, B. 1998. Diploidization failure and apomixis in Orobanchaceae. *Bot. J. Linn. Soc.* 128(1): 99–103.

Pazy, B. and U. Plitman. 1996. Asynapsis in *Cistanche tubulosa* (Orobanchaceae). *Plant Syst. Evol.* 201: 271–273.

Peckert, T., J. Chrtek and I. Plačková. 2005. Genetic variation in agamospermous populations of *Hieracium echioides* in southern Slovakia and northern Hungary (Danube Basin). *Preslia* 77(3): 307–315.

Peel, M.D., J.G. Carman and O. Leblanc. 1997a. Megasporocyte callose in apomictic buffelgrass, Kentucky bluegrass, *Pennisetum squamulatum* Fresen, *Tripsacum* L. and weeping lovegrass. *Crop Sci.* 37: 724–732.

Peel, M.D., J.G. Carman, Z.W. Liu and R.R.C. Wang. 1997b. Meiotic anomalies in hybrids between wheat and apomictic *Elymus rectisetus* (Nees in Lehm.) A. Löve & Connor. *Crop Sci.* 37: 717–723.

Pegoraro, L., E.C. Baker, D. Aeschimann, M. Balant, R. Douzet, T. Garnatje, M.S. Guignard, I.J. Leitch, A.R. Leitch and L. Palazzesi, J-P. Theurillat, O. Hidalgo and J. Pellicer. 2020. The correlation of phylogenetics, elevation and ploidy on the incidence of apomixis in Asteraceae in the European Alps. *Bot. J. Linn. Soc.* 194(4): 410–422.

Pellicer, J., S. Clermont, L. Houston, T.C.G. Rich and M.F. Fay. 2012. Cytotype diversity in the *Sorbus* complex (Rosaceae) in Britain: Sorting out the puzzle. *Ann. Bot.* 110(6): 1185–1193. doi:10.1093/aob/mcs185

Pellino, M., D. Hojsgaard, E. Hörandl and T.F. Sharbel. 2020. Chasing the apomictic factors in the *Ranunculus auricomus* complex: Exploring gene expression patterns in microdissected sexual and apomictic ovules. *Genes* 11: 728. doi:10.3390/genes11070728

Pellino, M., D. Hojsgaard, T. Schmutzer, U. Scholz, E. Hörandl, H. Vogel and T.F. Sharbel. 2013. Asexual genome evolution in the apomictic *Ranunculus auricomus* complex: examining the effects of hybridization and mutation accumulation. *Mol. Ecol.* 22(23): 5908–5921. doi:10.1111/mec.12533

Pennell, R.I. and K. Roberts. 1990. Sexual development in the pea is presaged by altered expression of arabinogalactan protein. *Nature* 344: 547–549. doi:10.1038/344547a0

Persson, N.L., T. Eriksson and J.E.E. Smedmark. 2020. Complex patterns of reticulate evolution in opportunistic weeds (*Potentilla* L., Rosaceae), as revealed by low-copy nuclear markers. *BMC Evol. Biol.* 20: 38. doi:10.1186/s12862-020-1597-7

Peruzzi, L. and N.G. Passalacqua. 2006. On a new subspecies of *Adoxa moschatellina* (Adoxaceae), apoendemic in Calabria (S Italy). *Nord. J. Bot.* 24: 249–256.

Pessino, S.C., C. Evans, J.P.A. Ortiz, I. Armstead, C.B. do Valle and M.D. Hayward. 1998. A genetic map of the apospory-region in *Brachiaria* hybrids: Identification of two markers closely associated with the trait. *Hereditas* 128: 153–158.

Pessino, S.C., J.P. Ortiz, O. Leblanc, C.B. do Valle, C. Evans and M.D. Hayward. 1997. Identification of a maize linkage group related to apomixis in *Brachiaria. Theor. Appl. Genet.* 94: 439–444.

Peter, J. 1920. Zur Entwicklungsgeschichte einiger Calycanthaceen. *Beitr. Biol. Pflanzen* 14: 59–84.

Petrella, R., M. Cucinotta, M.A. Mendes, C.J. Underwood and L. Colombo. 2021. The emerging role of small RNAs in ovule development, a kind of magic. *Plant Reprod.* 34: 335–351. doi:10.1007/s00497-021-00421-4

Petrov, D.F. 1939. On the occurrence of facultative pseudogamy in a triploid variety of raspberries, Immer tragende (*R. idaeus*). *Comp. Rend. l'Acad. Sci. l'U.R.S.S.* 22: 352–353.

Petrov, D.F., N.I. Belousova and E.S. Fokina. 1979. Inheritance of apomixis and its elements in corn-*Tripsacum* hybrids. *Genetika* 15: 1827–1836.

Petrov, D.J., N.I. Belousova, E.S. Fokina, R.M. Yatsenko, L.I. Laikova and T.P. Sorokina. 1984. Transfer of some elements of apomixis from *Tripsacum* to maize. In D.F. Petrov (ed.), *Apomixis and its role in evolution and plant breeding*. Russian Translation Series 22. Rotterdam: AA Balkena. pp. 9–78.

Pfitzer, P. 1957. Chromosomenzahlen von Araceae. *Chromosoma* 8: 436–446.

Pfitzer, P. 1962. Apomixis und Embryosackentwicklung in der Gattung *Aglaonema* Exgl. Portugal. *Acta Biol.* 6: 279–293.

Phatak, V.G. 1971. Embryology of *Zygophyllum coccineum* L. and *Z. fabago* L. *Proc. Kon. Ned. Akad. Wet.* 74: 379–397.

Philipson, M.N. 1978. Apomixis in *Cortaderia jubata* (Gramineae). *N. Z. J. Bot.* 16: 45–60.

Philipson, M.N. 1986. A re-assessment of the form of reproduction in *Danthonia spicata* (L.) Beauv. *New Phytol.* 103: 231–243. doi:10.1111/j.1469-8137.1986.tb00611.x

Phillips, W., T. Ranney, D.H. Touchell and T.A. Eaker. 2016. Fertility and reproductive pathways of triploid flowering pears (*Pyrus* sp.). *HortScience* 51(8): 968–971.

Phipps, J.B. and M. Muniyamma. 1980. A taxonomic revision of *Crataegus* (Rosaceae) in Ontario. *Can. J. Bot.* 58: 1621–1699.

Piazzano, M. 1998. Números cromosómicos en Bignoniaceae de Argentina. *Kurtziana* 26: 179–189.

Picarella, M.E. and A. Mazzucato. 2019. The occurrence of seedlessness in higher plants: Insights on roles and mechanisms of parthenocarpy. *Front. Plant Sci.* 9: 1997. doi:10.3389/fpls.2018.01997

Pichot, C., B. Fady and I. Hochu. 2000. Lack of mother tree alleles in zymograms of *Cupressus dupreziana* A. Camus embryo. *Ann. For. Sci.* 57: 17–22.

Pichot, C., M. El Maâtaoui, S. Raddi and P. Raddi. 2001. Surrogate mother for endangered *Cupressus. Nature* 412: 39.

Piedade, L.H. and N.T. Ranga. 1993. Ecologia da polinização de *Galipea jasminiflora* Engler (Rutaceae). *Rev. Bras. Bot.* 16: 151–158 (in Portuguese).

Pignatti, S. 1972. *Limonium* Miller. In: Tutin, T.G., Heywood, V.H., Burges, N.A., Moore, D.M., Valentine, D.H., S.M. Walters and D.A. Webb (eds.), *Flora Europaea*, Vol. 3. Cambridge: Cambridge University Press, pp. 38–50.

van der Pijl, L. 1934. Über die Polyembryonie bei *Eugenia*. *Recueil Trav. Bot. Neerl*. 31: 113–187.

du Plessis, H. and J.J. Spies. 1992. Chromosome numbers in the genus *Pentaschistis* (Poaceae, Danthonieae). *Taxon* 41: 709–720.

Pliego-Alfaro, F. and T. Murashige. 1988. Somatic embryogenesis in avocado (*Persea americana* Mill.) *in vitro*. *Plant Cell Tiss. Organ Cult*. 12: 61–66. doi:10.1007/BF00043108

Plitmann, U. 2002. Agamospermy is much more common than conceived: A hypothesis. *Israel J. Plant Sci*. 50(supl): 111–117. doi:10.1560/4T32-TWGY-T5NP-JDAE

Ploompuu, T. 1999. Eesti kodumaised tuhkpuud [Native *Cotoneasters* of Estonia]. pp. 38–52, 53–54 In: Sander, H. (ed.), Dendroloogilised uurimused Eestis I [Dendrologi cal researches in Estonia]. Metsanduslik Uurimisinstituut [EAU Forest Research Institute], Tallinn.

Poddubnaja-Arnoldi, V.A. 1933. Geschlechtliche und ungeschlechtliche Fortpflanzung bei einigen *Chondrilla*-Arten. *Planta* 19: 46–86.

Poddubnaja-Arnoldi, V.A. 1964. *Cytoembryology of Angiosperms*. Nauka, Moscow (in Russian).

Poddubnaja-Arnoldi, V.A. 1976. *Cytoembryology of Angiosperms*. Revised edition. Nauka, Moscow (in Russian).

Poddubnaja-Arnoldi, V.A. and V. Dianova. 1964. Eine zytoembryologische Untersuchung einiger Arten der Gattung *Taraxacum*. *Planta* 23: 19–46.

Polegri, L., O. Calderini, S. Arcioni and F. Pupilli. 2010. Specific expression of apomixis-linked alleles revealed by comparative transcriptomic analysis of sexual and apomictic *Paspalum simplex* Morong flowers. *J. Exp. Bot*. 61: 1869–1883.

Polunina, N.N. 1958. Biology of flowering and embryology of *Callistemon lanceolatus* Sweet (in Russian). *Bot. Zhu*. 43: 1169–1178.

Popoff, A. 1935. Über die Fortpflanzungsverhältnisse der Gattung *Potentilla*. *Planta* 24: 510–522.

Poverene, M.M. and P.W. Voigt. 1995. Identification of apomictic and sexual *Eragrostis curvula* (Schrad.) Ness hybrids by isozyme analysis. *Mendeliana* 11: 29–36.

Povilus, R.A., P.K. Diggle and W.E. Friedman. 2018. Evidence for parent-of-origin effects and interparental conflict in seeds of an ancient flowering plant lineage. *Proc. Roy. Soc. B. Biol. Sci*. 285: 20172491.

Powers, L. and R.C. Rollins. 1945. Reproduction and pollination studies on guayule, *Parthenium argentatum* Gray and *P. incanum* H.B.K. *J. Amer. Soc. Agron*. 37: 96–112. doi:10.2134/agronj1945.00021962003700020003x

Prakash, N., L.L. Lim and R. Menurung. 1977. Embryology of Duku and Langsat variation of *Lansium domesticum*. *Phytomorphology* 27(1): 50–59.

Pratt, C. and J. Einset. 1955. Development of the embryo sac in some American blackberries. *Am. J. Bot*. 42(7): 637–645. https://www.jstor.org/stable/2485323

Pratt, C., J. Einset and R.T. Clausen. 1958. Embryology, breeding behaviour and morphological characteristics of apomictic, triploid *Rubus idaeus* L. *Bull. Torrey Bot. Club*. 85: 242–254.

Price, S. 1959. Critique on apomixis in sugarcane. *Econ. Bot*. 13(1): 67–74. https://www.jstor.org/stable/4288006

Pritchard, A.J. 1967. Apomixis in *Brachiaria decumbens* Stapf. *J. Austr. Inst. Agric. Sci*. 33: 264–265.

Pritchard, A.J. 1970. Meiosis and embryo sac development *Urochloa mosambicensis* and three *Paspalum* species. *Aust. J. Agric. Res*. 21: 649–652.

Proctor, M.C.F. and A.C. Groenhof. 1992. Peroxidase isozyme and morphological variation in *Sorbus* L. in South Wales and adjacent areas, with particular reference to *S. porrigentiformis* E.F. Warb. *Watsonia* 19: 21–37.

Ptak, K. 1989. Cyto-embryological investigations on the Polish representatives of the genus *Crataegus* L. II. Embryology of triploid species. *Acta Biol. Cracov. Ser. Bot*. 31: 97–112.

Pullaiah, T. 1976. Embryological investigations in compositae with some observations on Basellaceae. PhD thesis, Andhra University, Visakhapatnam, India.

Pullaiah, T. 1982. Studies in the embryology of Compositae. The Tribe Eupatorieae. *Indian J. Bot*. 5: 183–188.

Pullaiah, T. 1984. *Embryology of Compositae*. Today and Tomorrows Printers & Publishers, New Delhi.

Pullaiah, T. and G.N.V. Febulaus. 2000. *Embryology and apomixis in Grasses*. Regency Publications, New Delhi.

Pupilli, F., E.J. Martinez, A. Busti, O. Calderini, C.L. Quarin and S. Arcioni. 2004. Comparative mapping reveals partial conservation of synteny at the apomixis locus in *Paspalum* spp. *Mol. Genet. Genomics* 270: 539–548.

Pupilli, F., P. Labombarda, M.E. Cáceres, C.L. Quarin and S. Arcioni. 2001. The chromosome segment related to apomixis in *Paspalum simplex* is homoeologous to the telomeric region of the long arm of rice chromosome 12. *Mol. Breed.* 8: 53–61.

Puri, V. 1939. Studies in the order Parietales. 2. A contribution to the morphology of *Garcinia livingstonii*. *Proc. Indian Acad. Sci. B.* 9: 74–86.

Qu, X.L., Z.Q. Mao, Y.L. Hu, X. Shen and L. Guo. 2008. Apomixis and ploidy of several crabapple cultivars, *Malus* Mill. *Acta Hortic.* 769: 387–392.

Quarin, C.L., W.W. Hanna and A. Fernández. 1982. Genetic studies in diploid and tetraploid *Paspalum* species. *J. Hered.* 73(4): 254–256. doi:10.1093/oxfordjournals.jhered.a109634

Quarin, C.L. and B.L. Burson. 1991. Cytology of sexual and apomictic *Paspalum* sp. *Cytologia* 56: 223–228.

Quarin, C.L. and E.P. Lombardo. 1986. Niveles de ploidía y distribución geográfica de *Paspalum quadrifarium* (Gramineae). *Mendeliana* 7(2): 101–107.

Quarin, C.L. and M.H. Urbani. 1990. Cytology and reproduction of diploid and tetraploid cytotypes of *Paspalum coryphaeum*. *Apomixis Newsl.* 2: 44–46.

Quarin, C.L. and W.W. Hanna. 1980. Effect of three ploidy levels on meiosis and mode of reproduction in *Paspalum hexastachyum*. *Crop Sci.* 20: 69–75.

Quarin, C.L., G.A. Norrmann and F. Espinosa. 1998. Evidence for autoploidy in apomictic *Paspalum rufum*. *Hereditas* 129: 119–124.

Quarin, C.L., J.F.M. Valls and M.H. Urbani. 1997. Cytological and reproductive behaviour of *Paspalum atratum*, a promising forage grass for the tropics. *Trop. Grassl.* 31: 114–116.

Quarin, C.L., M.T. Pozzobon and J.F.M. Valls. 1996. Cytology and reproductive behaviour of diploid, tetraploid and hexaploid germplasm accessions of a wild forage grass: *Paspalum compressifolium*. *Euphytica* 90: 345–349.

Quiroga, M., K. Grunberg, A. Ribotta, E. López Colomba, E. Carloni, E. Tommasino, C. Luna and S. Griffa. 2013. Obtaining sexual genotypes for breeding in buffel grass. *S. Afr. J. Bot.* 88: 118–123. doi:10.1016/j.sajb.2013.04.016

Rabiger, D.S., J.M. Taylor, A. Spriggs, M.L. Hand, S.T. Henderson, S.D. Johnson, K. Oelkers, M. Hrmova, K. Saito, G. Suzuki, Y. Mukai, B.J. Carroll and A.M.G. Koltunow. 2016. Generation of an integrated *Hieracium* genomic and transcriptomic resource enables exploration of small RNA pathways during apomixis initiation. *BMC Biol.* 14: 86. doi:10.1186/s12915-016-0311-0

Raga, V., G.P. Bernet, E.A. Carbonell and M.J. Asins. 2012. Segregation and linkage analyses in two complex populations derived from the citrus rootstock Cleopatra mandarin. Inheritance of seed reproductive traits. *Tree Genet. Genomes* 8: 1061–1071.

Raghavan, T.S. 1951. The Sugarcanes in India: Some cytogenetic considerations. *J. Hered.* 42(4): 199–206.

Rai, Y. 2014. Seedling behaviour and growth status of endangered medicinal tree *Amoora rohituka* Roxb. in Meerut, U.P. India. *Elixir Appl. Bot.* 77: 28894–28898.

Raina, N.S. and T.N. Khoshoo. 1972. Cytogenetics of tropical bulbous ornamentals. IX. Breeding systems in *Zephyranthes*. *Euphytica* 21: 317–323. doi:10.1007/BF00036772

Raja, K. 2017. Polymorphism influences seed viability and vigour in polyembryonic Kaffir lime (*Citrus hystrix*) seeds. *Seed Sci. Technol.* 45(1) 189–197. doi:10.15258/sst.2017.45.1.15

Ramage, C.M., L. Sando, C.P. Peace, B.J. Carroll and R.A. Drew. 2004. Genetic diversity revealed in the apomictic fruit species *Garcinia mangostana* L. (mangosteen). *Euphytica* 136: 1–10.

Ramirez, N. and Y. Brito. 1990. Reproductive biology of a tropical palm swamp community in the Venezuelan Llanos. *Am. J. Bot.* 77(10): 1260–1271.

Ramsey, J. and D.W. Schemske. 1998. Pathways, mechanisms, and rates of polyploid formation in flowering plants. *Ann. Rev. Ecol. Syst.* 29: 467–501.

Ramulu, K.S., V.K. Sharma, T.N. Naumova, P. Dijkhuis and M.M. van Lookeren Campagne. 1999. Apomixis for crop improvement. *Protoplasma* 208: 196–205.

Rancken, G. 1934. Zytologische Untersuchungen an einigen wirtschaftlich wertvollen Wiesengräsern, mit besonderer Berücksichtigung von strukturellen Abweichungen in dem Chromosomenkomplement. *Acta Agral. Fenn.* 29: 1–92.

Randell, B.R. 1970. Adaptations in the genetic system of Australian arid zone *Cassia* species (Leguminosae, Caesalpinioideae). *Aust. J. Bot.* 18(1): 77–97.

Rangaswamy, N.S. 1959. In vitro studies on the ovules of *Citrus microcarpa* Bunge. PhD thesis, University of Delhi, New Delhi, India.

Rangaswamy, N.S. 1961. Experimental studies on female reproductive structures of *Citrus microcarpa* Bunge. *Phytomorphology* 11: 109–127.

Ranney, T.G., T.A. Eaker, N.P. Lynch, and R.T. Olsen. 2004. Reproductive pathways among flowering crabapples. SNA Research Conference - Vol. 49. Plant Breeding & Evaluation Section, pp. 575–579.

Rao, N.G.P., L.L. Narayana and R.N. Reddy. 1978. Apomixis and its utilisation in grain *Sorghum* -I. Embryology of two apomictic parents. *Caryologia* 31(4): 427–433. doi:10.1080/00087114.1978.10796762

Rathi S., M. Rani, A. Vashisth, N. Mittal, D. Kumar and A. Kumar Srivastava. 2020. Occurrence of apomixis in *Eleusine coracana*. *Flora* 265: 151575. doi:10.1016/j.flora.2020.151575

Raunkiaer, C. 1903. Kimdannelse uden Befrugtning hos Mælkebøtte. *Bot. Tidsskr.* 25: 109–140.

Raven, P.H., D.W. Kyhos and A.J. Hill. 1965. Chromosome numbers of spermatophytes, mostly Californian. *Aliso* 6: 105–113.

Ravishankar, S. and N. Gulati. 1985. Polyembryony in Symplocaceae. *Indian J. Bot.* 8: 13.

Ray, D.K., N.D. Mueller, P.C. West and J.A. Foley. 2013. Yield trends are insufficient to double global crop production by 2050. *PLoS One* 8: e66428.

Rebozzio, R.N., M.E. Sartor, C.L. Quarin and F. Espinoza. 2011. Residual sexuality and its seasonal variation in natural apomictic *Paspalum notatum* accessions. *Biol. Plant.* 55: 391–395.

Reddy, P.S. and R. D'Cruz. 1969. Mechanism of apomixis in *Dichanthium annulatum*. *Bot. Gaz.* 130: 162–165.

Reddy, R.N., L.L. Narayana and N.G.P. Rao. 1979. Apomixis and its utilization in grain *Sorghum*-II. Embryology of F_1 progeny of reciprocal crosses between R473 and 302. *Proc. Indian Acad. Sci.* B88: 455–461.

Reisch, C. 2004. Molecular differentiation between coexisting species of *Taraxacum* sect. *Erythrosperma* (Asteraceae) from populations in south-east and west Germany. *Bot. J. Linn. Soc.* 145(1): 109–117.

Reisch, C., T. Windmaißer, F. Vogler, F. Schuhwerk and N. Meyer. 2015. Genetic structure of the rare and endangered *Hieracium wiesbaurianum* group (Asteraceae) in Bavaria. *Bot. J. Linn. Soc.* 177: 112–123.

Renner, S.S. 1984. Pollination and breeding systems in some central Amazonian Melastomataceae. *Proc. 5th Intern. Symposium Pollination INRA Publ.* pp. 175–280.

Renner, S.S. 1987. *Sandemania hoehnei* (Melastomataceae: Tibouchineae): Taxonomy, distribution, and biology. *Brittonia* 39(4): 441–446. doi:10.2307/2807327

Renner, S.S. 1989. A survey of reproductive biology in Neotropical Melastomataceae and Memecylaceae. *Ann. Mo. Bot. Gard.* 76(2): 496–518. doi:10.2307/2399497

Ribaut, J-M. and D. Hoisington. 1998. Marker-assisted selection: new tools and strategies. *Trends Plant Sci.* 3(6): 236–239. doi:10.1016/S1360-1385(98)01240-0

Rich, T.C.G., E.J. McDonnell and M. Dolores Lledó. 2008. Conservation of Britain's biodiversity: the case of *Hieracium cyathis* (Asteraceae) and its relation to other apomictic taxa. *Bot. J. Linn. Soc.* 156: 669–680.

Richards, A.J. 1970. Eutriploid facultative apomixis in *Taraxacum*. *New Phytol.* 69: 761–774.

Richards, A.J. 1986. *Plant breeding systems*. Allen & Unwin, London.

Richards, A.J. 1990. Studies in *Garcinia*, dioecious tropical forest trees: The phenology, pollination biology and fertilization of *Garcinia hombroniana* Pierre. *Bot. J. Linn. Soc.* 103(3): 251–266.

Richards, A.J. 1997. *Plant breeding systems*. 2nd ed. Chapman & Hall, London.

Richards, A.J. 2003. Apomixis in flowering plants: An overview. *Philos. Trans. R. Soc. Lond. B Biol. Sci.* 358(1434): 1085–1093. doi:10.1098/rstb.2003.1294

Ritzerow, H. 1907. Über Bau und Befruchtung kleistogamer Blüten. *Flora* 98: 163–212.

Robert, H.S., L. Crhak Khaitova, S. Mroue and E. Benková. 2015. The importance of localized auxin production for morphogenesis of reproductive organs and embryos in *Arabidopsis*. *J. Exp. Bot.* 66(16): 5029–5042. doi:10.1093/jxb/erv256

Robertson, A. 2006. *Sorbus pseudomeinichii*, a new endemic *Sorbus* (Rosaceae) microspecies from Arran, Scotland. *Watsonia* 26: 9–14.

Robertson, A., A.C. Newton and R.A. Ennos. 2004. Multiple hybrid origins, genetic diversity and population genetic structure of two endemic *Sorbus* taxa on the Isle of Arran, Scotland. *Mol. Ecol.* 13: 123–143.

Robertson, A., T.C.G. Rich, A.M. Allen, L. Houston, C. Roberts, J.R. Bridle, S.A. Harris and S.J. Hiscock. 2010. Hybridization and polyploidy as drivers of continuing evolution and speciation in *Sorbus*. *Mol. Ecol.* 19: 1675–1690.

Robins, A. 1959. Note preliminare sur la polyembryonie dans *Bombax kimuenzae* de Wild. *Bulletin du Jardin botanique de l'État a Bruxelles* 29(1): 23–26.

Robinson, H., A.M. Powell, G.D. Carr, R.M. King and J.F. Weddin. 1989. Chromosome numbers in Compositae XVI: Eupatorieae II. *Ann. Mo. Bot. Gard.* 76: 1004–1011.

Robson, N.K.B. 1981. Studies in the genus *Hypericum* L. (*Guttiferae*) 2. Characters of the genus. *Bull. Brit. Mus. Nat. Hist. Bot.* 8: 55–226.

Roche, D., J.A. Conner, M.A. Budiman, D. Frisch, R. Wing, W.W. Hanna and P. Ozias-Akins. 2002. Construction of BAC libraries from two apomictic grasses to study the microcolinearity of their apospory-specific genomic regions. *Theor. Appl. Genet.* 104: 804–812.

Roche, D., P. Cong, Z.B. Chen, W.W. Hanna, D.L. Gustine, R.T. Sherwood and P. Ozias-Akins. 1999. An apospory-specific genomic region is conserved between buffelgrass (*Cenchrus ciliaris* L.) and *Pennisetum squamulatum* Fresen. *Plant J.* 19: 203–208.

Rodkiewicz, B. 1970. Callose in cell walls during megasporogenesis in angiosperms. *Planta* 93: 39–47.

Rodrigo, J.M., D.C. Zappacosta, J.P. Selva, I. Garbus, E. Albertini and V. Echenique. 2017. Apomixis frequency under stress conditions in weeping lovegrass (*Eragrostis curvula*). *PLoS One* 12: e0175852. doi:10.1371/journal.pone.0175852

Rodrigues, J.C., G.B. Cabral, D.M. Dusi, L.V. de Mello, D.J. Rigden and V.T. Carneiro. 2003. Identification of differentially expressed cDNA sequences in ovaries of sexual and apomictic plants of *Brachiaria brizantha. Plant Mol. Biol.* 53: 745–757. doi:10.1023/B:PLAN.0000023664.21910.bd

Rodrigues-Correia, M.C., C.R. Benevides and H.A. de Lima. 2011. Two cases of polyembryony in Clusiaceae: *Clusia criuva* Cambess. and *Clusia fluminensis* Planch. & Triana. *Hoehnea* 38(2): 321–324.

Róis, A.S., F. Sádio, O.S. Paulo, G. Teixeira, A.P. Paes, D. Espírito-Santo, T.F. Sharbel and A.D. Caperta. 2016. Phylogeography and modes of reproduction in diploid and tetraploid halophytes of *Limonium* species (Plumbaginaceae): Evidence for a pattern of geographical parthenogenesis. *Ann. Bot.* 117(1): 37–50. doi:10.1093/aob/mcv138

Róis, A.S., G. Teixeira, T.F. Sharbel, J. Fuchs, S. Martins, D. Espírito-Santo and A.D. Caperta. 2012. Male fertility versus sterility, cytotype, and DNA quantitative variation in seed production in diploid and tetraploid sea lavenders (*Limonium* sp., Plumbaginaceae) reveal diversity in reproduction modes. *Sex. Plant Reprod.* 25(4): 305–318. doi:10.1007/s00497-012-0199-y

Rojek, J., M. Kapusta, M. Kozieradzka-Kiszkurno, D. Majcher, M. Górniak, E. Sliwinska, T.F. Sharbel and J. Bohdanowicz. 2018. Establishing the cell biology of apomictic reproduction in diploid *Boechera stricta* (Brassicaceae). *Ann. Bot.* 122(4): 513–539. doi.10.1093/aob/mcy114.

Rollins, R.C. 1940. Notes on some crucifers of Texas, Mexico and South America. *J. Arnold Arbor.* 21: 392–394.

Rollins, R.C. and L. Rüdenberg. 1971. Chromosome numbers of Cruciferae II. *Contr. Gray Herb.* 201: 117–133.

Román-Palacios C., Y.F. Molina-Henao and M.S. Barker. 2020. Polyploids increase overall diversity despite higher turnover than diploids in the Brassicaceae. *Proc. R. Soc. B.* 287: 20200962. doi:10.1098/rspb.2020.0962

Ronceret, A. and J.P. Vielle-Calzada. 2015. Meiosis, unreduced gametes, and parthenogenesis: Implications for engineering clonal seed formation in crops. *Plant Reprod.* 28: 91–102.

Rosenbaumová, R. and F. Krahulec. 2015. Sexual reproduction as a source of ploidy level variation in the model agamic complex of *Pilosella bauhini* and *P. officinarum* (Asteraceae: Lactuceae). *Plant Syst. Evol.* 301: 279–290. doi:10.1007/s00606-014-1072-2

Rosenberg, O. 1906. Über die Embryobildung in der Gattung *Hieracium* (+1 lithographie). *Ber. Deutsch. Bot. Ges.* 24: 157–161.

Rosenberg, O. 1907. Experimental and cytological studies in the *Hieracium*. II. Cytological studies on the apogamy in *Hieracium. Bot. Tidsskr.* 28: 143–170.

Rosenberg, O. 1912. Über die Apogamie bei *Chondrilla juncea. Svensk Bot. Tidskr.* 6: 914–919.

Rosenberg, O. 1917. Die Reduktionsteilung und ihre Degeneration in *Hieracium. Svensk Bot. Tidskr.* 10: 145–206.

Rosenberg, O. 1927. Die semiheterotypische Teilung und ihre Bedeutung für die Entstehung verdoppelter Chromosomenzahlen. *Hereditas* 8: 305–338.

Rosenberg, O. 1930. Apogamie und Parthenogenesis bei Pflanzen. In: Baur, B. and M. Hartman (eds.), *Handbuch der Vererbungswissenschaft* 12: 1–66.

Rossa, J. 2013. *Potentilla rigoana* Theodor Wolf (Rosaceae) – Taxonomischer Status und systematische Charakterisierung. Mag. thesis, University of Vienna, Vienna, Austria, 75 pages.

Rosselló, J.A., L. Sáez and Á.C. Carvalho. 1998. *Limonium carvalhoi* (Plumbaginaceae), una especie nueva endémica de las Islas Baleares. *Anales Jard. Bot. Madrid* 56(1): 23–31.

Roth, F. 1906. Die Fortpflanzungsverhältnisse bei der Gattung *Rumex. Verh. Naturhist. Ver. Preuss. Rheinl. Westphalens* 63: 327–360.

Rotman, A.D. 1982. Los géneros *Calycorectes, Hexachlamys, Myrciaria, Paramyrciaria, Plinia* y *Siphoneugena* en la Flora Argentina (Myrtaceae). *Darwiniana* 24(1–4): 157–185.

Rotreklová, O., A. Krahulcová, D. Vaňková, T. Peckert and P. Mráz. 2002. Chromosome numbers and breeding systems in some species of *Hieracium* subgen. *Pilosella* from Central Europe. *Preslia* 74: 27–44.

Rotreklová, O., A. Krahulcová, P. Mráz, V. Mrázová, L. Mártonfiová, B. Šingliarová and T. Peckert. 2005. Chromosome numbers and breeding systems in some species of *Hieracium* subgen. *Pilosella* from Europe. *Preslia* 77(2): 177–195.

Rousi, A. 1956. Cytotaxonomy and reproduction in the apomictic *Ranunculus auricomus* group. *Ann. Bot. Soc. Zool.-Bot. Fennica "Vanamo"* 29(2): 1–64.

Rousi, A. 1973. Studies on the cytotaxonomy and mode of reproduction of *Leontodon* (Compositae). *Ann. Bot. Fenn.* 10(3): 201–215.

Roy, A., D.R. Malaviya, P. Kaushal, A. Chandra, U.P. Singh, S.K. Mahanta, R. Chauhan and R. Tiwari. 2009. Descriptors for *Dichanthium–Bothriochloa* Complex. Indian Grassland and Fodder Research Institute, Jhansi, India.

Roy, B.A. 1995. The breeding systems of six species of *Arabis* (Brassicaceae). *Am. J. Bot.* 82(7): 869–877. doi:10.2307/2445973

Roy, B.A. and L. Rieseberg. 1989. Evidence for apomixis in *Arabis*. *J. Hered.* 80: 506–508.

Roy, S.K. 1953. Embryology of *Eugenia jambos*. *Curr. Sci.* 22: 249–250.

Roy, S.K. 1961. Embryology of *Eugenia fruticosa*. *Proc. Natl. Acad. Sci. India* B. 31: 80–87.

Roy, S.K. and R. Sahai. 1962. The embryo sac and embryo of *Syzygium caryophyllifolium*. *J. Indian Bot. Soc.* 41: 45–51.

Rozenblum, E., S. Maldonado and C.E. Waisman. 1988. Apomixis in *Eupatorium tanacetifolium* (Compositae). *Amer. J. Bot.* 75: 311–322.

Rudenko, F.E. 1961. Apomixis in certain high mountain plants of the Ukrainian Carpathians. *Ukr. Bot. Zour.* 28(6): 24–31.

Rueda da Silva, C.T., E.M. Bonfim-Silva, T.J. de Araújo da Silva, E.A. Rodrigues Pinheiro, J.J. Vieira and A. Pereira Freire Ferraz. 2020. Yield component responses of the *Brachiaria brizantha* forage grass to soil water availability in the Brazilian Cerrado. *Agriculture* 10(1): 13. doi:10.3390/agriculture10010013

Ruiz, C., M. Paz Breto and M. Asíns. 2000. A quick methodology to identify sexual seedlings in citrus breeding programs using SSR markers. *Euphytica* 112: 89–94. doi:10.1023/A:1003992719598

Russell, S.D. 1979. Fine structure of megagametophyte development in *Zea mays*. *Can. J. Bot.* 57: 1093–1110. doi:10.1139/b79-134

Russell, S.D. 1993. The egg cell: development and role in fertilization and early embryogenesis. *Plant Cell* 5:1349–1359.

Rutishauser, A. 1943a. Konstante Art- und Rassenbastarde in der Gattung *Potentilla*. *Mitt. Naturf. Ges. Schaffausen* 18: 111–134.

Rutishauser, A. 1943b. Über die Entwicklungsgeschichte pseudogamer Potentillen. *Arch. Jul. Klaus-Stift.* 18: 687–691.

Rutishauser, A. 1943c. Untersuchungen über die Fortpflanzung und Bastardbildung apomiktischer Potentillen. *Ber. Schweiz Bot. Ges.* 53: 1–83.

Rutishauser, A. 1945a. Über die Fortpflanzung einiger Bastarde von pseudogamen Potentillen. *Arch. Jul. Klaus-Stift.* 20: 300–314.

Rutishauser, A. 1945b. Zur Embryologie amphimiktischer Potentillen. *Ber. Schweiz. Bot. Ges.* 55: 19–32.

Rutishauser, A. 1946. Überkreuzungsversuche mit pseudogamen Potentillen. *Arch. Klaus-Stift. Vererb. Forsch.* 21: 469–472.

Rutishauser, A. 1947. Untersuchungen über die Genetik der Aposporie bei pseudogamen Potentillen. *Experientia* 3: 204.

Rutishauser, A. 1948. Pseudogamie und Polymorphie in der Gattung *Potentilla*. *Arch. Jul. Klaus-Stift. Vererb. Forsch.* 23: 267–423.

Rutishauser, A. 1953. Die Entwicklungserregung des Endosperms bei pseudogamen *Ranunculus*-Arten. *Mitt. Nat. Ges. Schaffhausen* 25: 1–45.

Rutishauser, A. 1954. Entwicklungserregung der Eizelle bei pseudogamen Arten der Gattung *Ranunculus*. *Bull. Schweiz. Akad. Med. Wiss.* 10: 491–512.

Rutishauser, A. 1956. Cytogenetik des Endosperms. *Ber. Schweiz. Bot. Ges.* 66: 318–336.

Rutishauser, A. 1969. *Embryologie und Fortpflanzungsbiologie der Angiospermen*. Springer-Verlag: Vienna.

Rutishauser, A. 1982. *Introducción a la embriología y biología de la reproducción de las angiospermas*. 1ra Ed. Hemisferio Sur. Buenos Aires, Argentina.

Rutishauser, A. and H.R. Hunziker. 1950. Untersuchungen über die Zytologie des Endosperms. *Arch. J. Klaus-Stift. Vererb. Forsch.* 25: 477–483.

Ruzin, S.E. 1999. *Plant microtechnique and microscopy*. Oxford University Press, Oxford, UK.

Rychlewski, J. 1961. Cyto-embryological studies in the apomictic species *Nardus stricta*. *Acta Biol. Cracov. Ser. Bot.* 4: 1–23.

Rye, B.L. 1979. Chromosome number variation in the Myrtaceae and its taxonomic implications. *Aust. J. Bot.* 27: 547–573.

Sachar, R.C. and P. Murgai. 1958. Embryology of *Aerva tomentosa* Forssk. *Curr. Sci.* 27: 105–106.

Sachar, R.C. and R.N. Chopra. 1957. A study of the endosperm and embryo in *Mangifera* L. *Indian J. Agric. Sci.* 27: 219–228.
Sáez, L. and J.A. Rosselló. 1998. *Limonium inexpectans* (Plumbaginaceae) a new apomictic species from Mallorca (Balearic Islands). *An. Jard. Bot. Madr.* 54: 285–289.
Sáez, L. and J.A. Rosselló. 1999. Is *Limonium cavanillesii* Erben (Plumbaginaceae) really an extant species? *An. Jard. Bot. Madr.* 57(1): 47–55. doi:10.3989/ajbm.1999.v57.i1.188
Sáez, L., A. Curco and J.A. Rosselló. 1998. *Limonium vigoi* (Plumbaginaceae), a new tetraploid species from the Northeast of the Iberian Peninsula. *An. Jard. Bot. Madr.* 56(2): 269–278.
Sailer, C., B. Schmid and U. Grossniklaus. 2016. Apomixis allows the transgenerational fixation of phenotypes in hybrid plants. *Curr. Biol.* 26: 331–337.
Salma, I. and B. Razali. 1987. The reproductive biology of duku langsat, *Lansium domesticum* Corr. (Meliaceae), in Peninsular Malaysia. *MARDI Research Bulletin* 15: 141–150.
Salomão, A.N. and A.C. Allem. 2001. Polyembryony in angiospermous trees of the Brazilian Cerrado and Caatinga vegetation. *Acta Bot. Brasilica.* 15: 369–378. doi:10.1590/S0102-33062001000300007
Samaniego, F., F. Kolar, T. Urfus, A. Barrgan and K. Romoleroux. 2018. Determination of apomixis by flow cytometry in two species of *Lachemilla* (Rosaceae) in Ecuador. *Neotrop. Biodivers.* 4(1): 152–163.
Sampaio, D.S., N.S. Bittencourt Jr and P.E. Oliveira. 2013a. Mating in the pseudogamic apomictic *Anemopaegma acutifolium* DC: Another case of pseudo-self-compatibility in Bignoniaceae? *Plant Biol.* (Stuttgart) 15: 919–924.
Sampaio, D.S., N.S.B. Júnior and P.E. Oliveira. 2013b. Sporophytic apomixis in polyploid *Anemopaegma* species (Bignoniaceae) from central Brazil. *Bot. J. Linn. Soc.* 173: 77–91.
Sampson, D.R. 1969. Use of a leaf color marker gene to detect apomixis in *Malus* species and observations on the variability of the apomictic seedlings. *Can. J. Plant Sci.* 49: 409–416.
Sapkota, S., J.A. Conner, W.W. Hanna, B. Simon, K. Fengler, S. Deschamps, M. Cigan and P. Ozias-Akins. 2016. *In silico* and fluorescence *in situ* hybridization mapping reveals collinearity between the *Pennisetum squamulatum* apomixis carrier-chromosome and chromosome 2 of *Sorghum* and Foxtail Millet. *PLoS One* 11(3): e0152411. doi:10.1371/journal.pone.0152411
Saraiva, L.C., O. Cesar and R. Monteiro. 1996. Breeding systems of shrubs and trees of a Brazilian savanna. *Arquivos de Biologia e Tecnologia. Curitiba-Paraná: Inst. Tecnologia Parana* 39(4): 751–763.
Saran, S. and J.M.J. de Wet. 1970. Environmental control of reproduction in *Dichanthium intermedium* (Gramineae). *Bull. Torrey Bot. Cl.* 97: 6–13.
Saran, S. and O.N. Mishra. 1986. Apomixis in some *Pennisetum* spp. In: Manna, G.K. and U. Sinha (eds.), *Perspectives in Cytology and Genetics*. Kalyani Publishers, India, pp. 779–784.
Šarhanová, P., R.J. Vašut, M. Dančák, P. Bureš and B. Trávníček. 2012. New insights into the variability of reproduction modes in European populations of *Rubus* subgen. *Rubus*: How sexual are polyploid brambles? *Sex. Plant Reprod.* 25: 319–335. doi:10.1007/s00497-012-0200-9
Šarhanová, P., T.F. Sharbel, M. Sochor, R.J. Vašut, M. Dančák and B. Trávníček. 2017. Hybridization drives evolution of apomicts in *Rubus* subgenus *Rubus*: evidence from microsatellite markers. *Ann. Bot.* 120(2): 317–328. doi:10.1093/aob/mcx033
Sartor, M.E., C.L. Quarin, M.H. Urbani and F. Espinoza. 2011. Ploidy levels and reproductive behaviour in natural populations of five *Paspalum* species. *Plant Syst. Evol.* 293: 31–41.
Sartorius, R. and R. Stösser. 1997. On the apomictic seed development in the walnut (*Juglans regia* L.). *Acta Hortic.* 442: 225–230. doi:10.17660/ActaHortic.1997.442.33
Sato, H., N. Shimizu, H. Nakagawa and H. Matsuoka.1992. A new registered cultivar "Natsuyutka" of guineagrass, *Panicum maximum* Jacq. *Jpn. Agric. Res. Q.* 25: 259–266.
Satyamurthy, T.V.C. and V. Seshavathram. 1983. Apospory in *Eremopogon foveolatus*. *Curr. Sci.* 53: 1158–1159.
Savidan, Y. 1981. Nature et hérédité de l'apomixie chez *Panicum maximum* Jacq. Diploma thesis. Orsay: University of Paris-Sud, France, 169 pages.
Savidan, Y. 2000. Apomixis: Genetics and breeding. *Plant Breed. Rev.* 18: 13–86.
Savidan, Y. 2001. Transfer of apomixis through wide crosses. In: *The flowering of apomixis: From mechanisms to genetic engineering*. Savidan, Y., J.G. Carman and T. Dresselhaus (eds.), CIMMYT, IRD, European Commission OC VI (FAIR), Mexico City, Mexico, pp. 153–167.
Savidan, Y.H., L. Jank, J.C. Costa and C. Valle. 2004. Breeding *Panicum maximum* in Brazil. 1. Genetic resources, modes of reproduction and breeding procedures. *Euphytica* 41: 107–112.
Sax, H.J. 1954. Polyploidy and apomixis in *Cotoneaster*. *J. Arnold Arbor.* 35: 334–365.
Sax, K. 1949. The use of *Malus* species for apple rootstocks. *Proc. Amer. Soc. Hort. Sci.* 53: 219–220.
Sax, K. 1959. The cytogenetics of facultative apomixis in *Malus* species. *J. Arnold Arbor.* 40: 289–297.

Schacht, H. 1850. *Entwicklungsgeschichte des Pflanzen-Embryon*. PhD Dissertation, Amsterdam, the Netherlands.

Schallau, A., F. Arzenton, A.J. Johnston, U. Hähnel, D. Koszegi, F.R. Blattner, L. Altschmied, G. Haberer, G. Barcaccia and H. Bäumlein. 2010. Identification and genetic analysis of the APOSPORY locus in *Hypericum perforatum* L. *Plant J*. 62: 773–784. doi:10.1111/j.1365-313X.2010.04188.x

Schanderl, H. 1964. Untersuchungen über die Blütenbiologie und Embryonenbildung von *Juglans regia* L. *Biol. Zentralblatt* 83: 71–102.

Scheben, A. and D. Hojsgaard. 2020. Can we use gene-editing to induce apomixis in sexual plants? *Genes* 11: 781. doi:10.3390/genes11070781

Schinkel, C.C.F., B. Kirchheimer, A.S. Dellinger, S. Klatt, M. Winkler, S. Dullinger and E. Hörandl. 2016. Correlations of polyploidy and apomixis with elevation and associated environmental gradients in an alpine plant. *AoB Plants* 8: plw064. doi:10.1093/aobpla/plw064

Schinkel, C.C.F., E. Syngelaki, B. Kirchheimer, S. Dullinger, S. Klatt and E. Hörandl. 2020. Epigenetic patterns and geographical parthenogenesis in the alpine plant species *Ranunculus kuepferi* (Ranunculaceae). *Int. J. Mol. Sci*. 21: 3318. doi:10.3390/ijms21093318

Schkorbatow, L. 1912. Parthenogenetische und apogame Entwicklung bei den Blütenpflanzen. Entwicklungsgeschichtliche Studien an *Taraxacum officinale*. *Trav. Soc. Nat. Univ. Imp. Kharkow* 45: 15–25.

Schlimbach, H. 1924. Beiträge zur Kenntnis der Samenanlagen und Samen der Amaryllidaceen mit Berücksichtigung des Wassergehaltes der Samen. *Flora* 117(1–2): 41–54. doi:10.1016/S0367-1615(17)31058-3

Schmelzer, G.H. and J.F. Renno. 1997. Genetic variation in the agamic species complex of *Pennisetum* section *Brevivalvula* (Poaceae) from West Africa: ploidy levels and isozyme polymorphism. *Euphytica* 96: 23–29. doi:10.1023/A:1002974304592

Schmelzer, G.H. and J.F. Renno. 1999. Genotypic variation in progeny of the agamic grass complex *Pennisetum* section *Brevivalvula* in West Africa. *Plant Syst. Evol*. 215: 71–83. doi:10.1007/BF00984648

Schmidt, A. 2020. Controlling apomixis: Shared features and distinct characteristics of gene regulation. *Genes* 11: 329. doi:10.3390/genes11030329

Schmidt, A., M.W. Schmid, U.C. Klostermeier, W. Qi, D. Guthörl, C. Sailer, M. Waller, P. Rosenstiel and U. Grossniklaus. 2014. Apomictic and sexual germline development differ with respect to cell cycle, transcriptional, hormonal and epigenetic regulation. *PLoS Genet*. 10(7): e1004476. doi:10.1371/journal.pgen.1004476

Schmidt, E.D., F. Guzzo, M.A. Toonen and S.C. de Vries. 1997. A leucine-rich repeat containing receptor-like kinase marks somatic plant cells competent to form embryos. *Development* 124(10): 2049–2062.

Schmidt, H. 1964. Beiträge zur Züchtung apomiktischer Apfelunterlagen I. Zytogenetische und embryologische Untersuchungen. *Z. Pfl. Zucht*. 52: 27–102.

Schmidt, H. 1977. Contributions on the breeding of apomictic apple stocks. 4. On the inheritance of apomixis. *Z. Pfl. Zucht*. 78: 3–12.

Schmidt, J.M. and A.E. Antlfinger. 1992. The level of agamospermy in a Nebraska population of *Spiranthes cernua* (Orchidaceae). *Am. J. Bot*. 79: 501–507. doi:10.1002/j.1537-2197.1992.tb14585.x

Schneeweiss, G.M., T. Palomeque, A.E. Colwell and H. Weiss-Schneeweiss. 2004. Chromosome numbers and karyotype evolution in holoparasitic *Orobanche* (Orobanchaceae) and related genera. *Am. J. Bot*. 91(3): 439–448. doi:10.3732/ajb.91.3.439

Schranz, M.E., C. Dobeš, M.A. Koch and T. Mitchell-Olds. 2005. Sexual reproduction, hybridization, apomixis, and polyploidization in the genus *Boechera* (Brassicaceae). *Am. J. Bot*. 92(11): 1797–1810.

Schranz, M.E., L. Kantama, H. de Jong and T. Mitchell-Olds. 2006. Asexual reproduction in a close relative of *Arabidopsis*: a genetic investigation of apomixis in *Boechera* (Brassicaceae). *New Phytol*. 171(2): 425–438.

Schrick, K., D. Nguyen, W.M. Karlowski and K.F.X. Mayer. 2004. START lipid/sterol-binding domains are amplified in plants and are predominantly associated with homeodomain transcription factors. *Genome Biol*. 5: R41. doi:10.1186/gb-2004-5-6-r41

Schuhwerk, F. 2002. Some thoughts on the taxonomy of *Hieracium*. *Ber. Bayer. Bot. Ges*. 72: 193–198.

Schürhoff, P.N. 1923. Zur Apogamie von *Calycanthus*. *Flora* (Jena) 116(1–2): 73–84. doi:10.1016/S0367-1615(17)31133-3

Schürhoff, P.N. 1924. Zytologische Untersuchungen in der Reihe der Geraniales. *Jahrb. Wiss. Bot*. 63: 707–759.

Schürhoff, P.N. and F.J. Jüssen. 1925. Nuzellarembryonie bei *Spathiphyllum patinii* (Hogg.) N.E.Br. *Ber. Deutsch. Bot. Ges*. 43: 454–456.

Schwertschlager, J. 1915. Beobachtungen und Versuche zur Biologie der Rosenblüte und Rosenbefruchtung. *Ber. Bayer, bot. Ges. in München* Bd. 15.

Scott, R.J. 2007. Polyspermy in apomictic *Crataegus*: yes and no. *New Phytol*. 173: 227–230.
Scott, R.J., M. Spielman, J. Bailey and H.G. Dickinson. 1998. Parent-of-origin effects on seed development in *Arabidopsis thaliana*. *Development* 125: 3329–3341.
Sears, C.J. and J. Whitton. 2016. A reexamination of the North American *Crepis* agamic complex and comparison with the findings of Babcock and Stebbins' classic biosystematics monograph. *Am. J. Bot*. 103(7): 1289–1299. doi:10.3732/ajb.1600057
Sears, P.B. 1917. Amitotic parthenogenesis in *Taraxacum vulgare* (Lam.) Scrk. and *Taraxacum laevigatum* (Willd.) DC. *Ohio J. Sci*. 17(3): 97–100.
Sears, P.B. 1922. Variation in cytology and gross morphology of *Taraxacum*. *Bot. Gaz*. 72: 308–325.
Secor, D. and S. Russell. 1988. Megagametophyte organization in a polyembryonic line of *Linum usitatissimum*. *Am. J. Bot*. 75(1): 114–122. doi:10.2307/2443909
Seefeldner, G. 1912. Die Polyembryonie bei *Cynanchum vincetoxicum* (L.) Pers. *Sitzungsberichte d.kais. Akad. d.Wiss. Math. Nat*. 121: 273–296.
Seilova, L.B. 1996. *Apomiksis u sakharnoy svekly I ego ispo''-zovanie v prakticheskoy selektsii*. Na pravakh rukopisi. Kazakhskaya Akademiya Se"skokhozyaystvennykh Nauk, Almaty.
Selva, J.P., D. Zappacosta, J. Carballo, J.M. Rodrigo, A. Bellido, C.A. Gallo, J. Gallardo and V. Echenique. 2020. Genes modulating the increase in sexuality in the facultative diplosporous grass *Eragrostis curvula* under water stress conditions. *Genes* 11(9): 969. doi:10.3390/genes11090969
Selva, J.P., L. Siena, J.M. Rodrigo, I. Garbus, D. Zappacosta, J. Romero, J.P. Ortiz, S. Pessino, O. Leblanc and V. Echenique. 2017. Temporal and spatial expression of genes involved in DNA methylation during reproductive development of sexual and apomictic *Eragrostis curvula*. *Sci. Rep*. 7: 15092. doi:10.1038/s41598-017-14898-5
Sennikov, A. 2010. Cotoneasters: a comprehensive guide to shrubs for flowers, fruit, and foliage. *Nordic J. Bot*. 28(4): 509–512.
Sennikov, A.N. 2018. *Scandosorbus* (Rosaceae), a new generic name for *Sorbus intermedia* and its hybrid. *Ann. Bot. Fenn*. 55(4): 321–323. doi:10.5735/085.055.0413
Sennikov, A.N. and A. Kurtto. 2017. A phylogenetic checklist of *Sorbus* s.l. (Rosaceae) in Europe. *Memoranda Soc. Fauna Flora Fennica* 93: 1–78.
Seshagiriah, K.N. 1941. Morphological studies in the Orchidaceae I. *Zeuxine sulcata* Lindl. *J. Indian Bot. Soc*. 20: 357–365.
Seshavatharam, V. 1983. Embryological studies on apomixis in some forage grasses. ICAR project report. Mss. Andhra University, India.
Shamakumari, K. 1960. Cytogenetic investigations in Paniceae. Occurrence of apospory in a diploid species of *Panicum, P. antidotale*. *Curr. Sci*. 29: 191.
Shamrov, I.I. 1997. Ovule and seed development in *Paeonia lactiflora* (Paeoniaceae) (in Russian). *Bot. Zhu*. (Moscow and Leningrad) 82: 24–46.
Shanthamma, C. 1973. *Studies in Poaceae*. PhD thesis, University of Mysore, Karnataka, India.
Shanthamma, C. 1982. Apomixis in *Cenchrus glaucus* Mudaliyar et Sundarraj. *Proc. Indian Acad. Sci. (Plant Sci.)* 91: 25–36.
Shanthamma, C. and K.N. Narayan. 1976–1977. Studies in Poaceae (Graminae). *J. Mysore Univ. Sect*. B. 27: 302–305.
Shapiro, H.M. 2007. Cytometry and cytometers: Development and growth. In: Doležel, J., Greilhuber, J., Suda, J. (eds.), *Flow cytometry with plant cells: Analysis of genes, chromosomes and genomes*. Wiley-VCH Verlag GmbH & Co, KGaA, pp. 1–17.
Sharbel, T.F., M-L. Voigt, J.M. Corral, T. Thiel, A. Varshney, J. Kumlehn, H. Vogel and B. Rotter. 2009. Molecular signatures of apomictic and sexual ovules in the *Boechera holboellii* complex. *Plant J*. 58: 870–882.
Sharbel, T.F., M-L. Voigt, J.M. Corral, G. Galla, J. Kumlehn, C. Klukas, F. Schreiber, H. Vogel and B. Rotter. 2010. Apomictic and sexual ovules of *Boechera* display heterochronic global gene expression patterns. *Plant Cell* 22: 655–671.
Sharma, H.C., J.H. Crouch, K.K. Sharma, N. Seetharama and C.T. Hash. 2002. Applications of biotechnology for crop improvement: Prospects and constraints. *Plant Sci*., 163: 381–395.
Sharma, H.P. and Y.S. Murthy. 1977. Embryological studies in the Compositae. 3. Tribe Astereae. *Brachyscome iberidifolia*. A casual apomict. *J. Indian Bot. Soc*. 56: 321–328.
Sharma, K.K.and T.A. Thorpe. 1995. Asexual embryogenesis in vascular plants in nature. In: Thorpe T.A. (eds.), *In vitro embryogenesis in plants*. Current Plant Science and Biotechnology in Agriculture, Vol. 20. Springer, Dordrecht. doi:10.1007/978-94-011-0485-2_2

Sharma, R., R. Geeta and V. Bhat. 2014. Asynchronous male/female gametophyte development in facultative apomictic plants of *Cenchrus ciliaris* (Poaceae). *S. Afr. J. Bot.* 91: 19–31.
Sharma, V.K., Praveen, D.P. Uniyal and P. Bisht. 2004. Polyembryony in *Oroxylum indicum* Vent. *Indian For.* 130(6): 825–828.
Sherman, R.A., P.W. Voight, B.L. Burson and C.L. Dewald. 1991. Apomixis in diploid × triploid *Tripsacum dactyloides* hybrids. *Genome* 34: 528–532.
Sherman, W.B., R.J. Knight and T.E. Crocker. 1978. Peach and nectarine breeding and testing in warm parts of the world. *Proc. Tropical Reg. Amer. Soc. Hort. Sci.* 22: 103–107.
Sherwood, R.T. 2001. Genetic analysis of apomixis. In: Savidan, Y., J.G. Carman and T. Dresselhaus (eds.), *The flowering of apomixis: From mechanisms to genetic engineering*. CIMMYT, IRD, European Commission, DG VI, (FAIR), Mexico, DF, pp. 64–82.
Shi, Y., R.J. Gornall, J. Draper and C.A. Stace. 1996. Intraspecific molecular variation in *Hieracium* sect. *Alpina (*Asteraceae*)*, an apomictic group. *Folia Geobot.* 31: 305–313. doi:10.1007/BF02815375
Shibata, K. and K. Miyake. 1908. Über Parthenogenesis bei *Houttuynia cordata. Bot. Mag.* (Tokyo) 22(261): 141–144.
Shilpa, G. and R. Nadgauda. 2000. Induction of somatic embryogenesis in cashewnut (*Anacardium occidentale* L.). *In Vitro Cell. Dev. Biol.* 36: 41–46.
Shimada, T., T. Endo, H. Fujii, M. Nakano, A. Sugiyama, G. Daido, S. Ohta, T. Yoshioka and M. Omura. 2018. MITE insertion-dependent expression of *CitRKD1* with a RWP-RK domain regulates somatic embryogenesis in citrus nucellar tissues. *BMC Plant Biol.* 18: 166.
Shimatomai, N. 1935. Zur Kenntnis der Pseudogamie bei *Potentilla. Proc. Imp. Acad. Japan* 11: 338–339.
Shirasawa K., M.L. Hand, S.T. Henderson, T. Okada, S.D. Johnson, J.M. Taylor, A. Spriggs, H. Siddons, H. Hirakawa, S. Isobe, S. Tabata and A.M.G. Koltunow. 2015. A reference genetic linkage map of apomictic *Hieracium* species based on expressed markers derived from developing ovule transcripts. *Ann. Bot.* 115(4): 567–580. doi:10.1093/aob/mcu249
Shirke, N.S. 1963. Studies on Palms: A study of the genus *Caryota* L. in India. PhD thesis, University of Poona (now Savitribai Phule Pune University), Pune, India.
Shiryaeva, E.I. and G.I. Yarmoluk. 1983. Adventive embryony and polyembryony of *Beta vulgaris* L., in *Thesis of VIII Vsesojusnyi Simposiumpo embriologii rastenii*, FAN, Tashkent, 152.
Shishkinskaya, N.A. and A.V. Borodko. 1987. Apomixis in *Festuca drymeja. Biol. Nauki* (Moscow) 1: 84–89.
Shobha, J. 1988. *Cytoembryological studies in a few species of Panicoideae*. PhD thesis, University of Mysore, Karnataka, India.
Siena, L.A., J.P.A. Ortiz, O. Calderini, F. Paolocci, M.E. Cáceres, P. Kaushal, S. Grisan, S.C. Pessino and F. Pupilli. 2016. An apomixis-linked ORC3-like pseudogene is associated with silencing of its functional homolog in apomictic *Paspalum simplex. J. Exp. Bot.* 67: 1965–1978.
Siena, L.A., J.P.A. Ortiz, O. Leblanc, and S.C. Pessino. 2014. PnTgs1-like expression during reproductive development supports a role for RNA methyltransferases in the aposporous pathway. *BMC Plant Biol.* 14: 297. doi:10.1186/s12870-014-0297-0
Siena, L.A., M.E. Sartor, F. Espinoza, C.L. Quarin and J.P.A. Ortiz. 2008. Genetic and embryological evidence of apomixis at the diploid level in *Paspalum rufum* support recurrent auto-polyploidization in the species. *Sex. Plant Reprod.* 21: 205–215. doi:10.1007/s00497-008-0080-1
Silva, A.L., B. Cardoso Lopes and L.D. Ribeiro de Santiago Fernandes. 2013. Embriogênese adventícia em *Eugenia uniflora* L. (Myrtaceae). *Revista Tropica: Ciências Agrárias e Biológicas* 7(1): 97–105.
da Silva Araújo, F.D., A. do Nascimento Cavalcante, M.D.D.B. Sousa, C.V.R. de Moura, M.H. Chaves, S. Aued-Pimentel, M.S. Fernandes Caruso, L.J. Tozetto and S.K. Morais Chaves. 2017. Biodiesel production from *Bombacopsis glabra* oil by methyl transesterification method. *Energies* 10(9): 1360. doi:10.3390/en10091360
Silveira, É.D., Guimarães, L.A., D.M. de Alencar Dusi, F. Rodrigues da Silva, N. Florencio Martins, M. Mota do Carmo Costa, M. Alves-Ferreira and V. Tavares de Campos Carneiro. 2012. Expressed sequence-tag analysis of ovaries of *Brachiaria brizantha* reveals genes associated with the early steps of embryo sac differentiation of apomictic plants. *Plant Cell Rep.* 31: 403–416. doi:10.1007/s00299-011-1175-y
Simon, W. 1968. Chromosome numbers and B-chromosomes in *Listera. Caryologia* 21(2): 181–189. doi:10.1080/00087114.1968.10796298
Simonini, S. and L. Østergaard. 2019. Female reproductive organ formation: A multitasking endeavor. *Curr. Top. Dev. Biol.* 131: 337–371.
Simpson, C.E. and E.C. Bashaw. 1969. Cytology and reproductive characteristics in *Pennisetum setaceum. Am. J. Bot.* 56: 31–36.
Simpson, G.G. 1961. A basic work on evolutionary theory. *Evolution* 15(1): 112–113.

Sindhe, A.N. 1976. Reproductive behaviour of *Pennisetum squamulatum*. *Proc. Indo-Soviet Symp. on Embryology of Crop Plants*, Abstr. p. 41–42.
Singh, A.N. and A. Thakur. 2004. Polyembryony in *Dipterocarpus retusus*. *J. Trop. For. Sci.* 16(4): 475–476.
Singh, B. 1959. Studies in the family Malpighiaceae. I. The morphology of *Thryallis glauca* Kuntze. *Hortic. Adv.* 3: 1–19.
Singh, B.G., R. Sivalingam, N.P. Mahadevan, G. Chevanan, R.R. Warrier, R. Anandalakshmi and V. Sivakumar. 2005. High frequency polyembryony in *Saraca asoca* (Roxb.) de Wilde (Caesalpiniaceae) – A red listed medicinal tree. *MyForest* 41(4): 573–582.
Singh, B.P. and M.J. Soule Jr. 1963. Studies in polyembryony and seed germination of trifoliate orange (*Poncirus trifoliata*). *Indian J. Hort.* 20: 21–29.
Singh, D. 1955. Embryological studies in *Cucumis melo* var. *pubescens*. *J. Indian Bot. Soc.* 34: 72–78.
Singh, D. and A.S.R. Dathann. 1972. Development of generative and aposporic embryo sacs in *Cucumis metuliferus* E. Mey ex Schrad. *Curr. Sci.* 41: 33–35.
Singh, D.P., A.K. Singh and A. Singh. 2021. *Plant Breeding and Cultivar Development*. Academic Press.
Šingliarová, B., I. Hodálová and P. Mráz. 2011. Biosystematic study of the diploid-polyploid *Pilosella alpicola* group with variation in breeding system: Patterns and processes. *Taxon* 60(2): 450–470. http://www.jstor.org/stable/41317142
Skalińska, M. 1959. Embryological studies in *Poa granitica*, an apomictic species of the Carpathian range. *Acta Biol. Cracov. Ser. Bot.* 2: 91–112.
Skalińska, M. 1967. Cytological analysis of some *Hieracium* species, subgenus *Pilosella*, from mountains of southern Poland. *Acta Biol. Cracov. Ser. Bot.* 10: 127–141.
Skalińska, M. 1973. Further studies in facultative apomixis of *Hieracium aurantiacum* L. *Acta Biol. Cracov. Ser. Bot.* 16: 121–137.
Skalińska, M. 1976. Cytological diversity in the progeny of octoploid facultative apomicts of *Hieracium aurantiacum*. *Acta Biol. Cracov. Ser. Bot.* 19: 39–46.
Skalińska, M. and E. Kubien. 1972. Cytological and embryological studies in *Hieracium pratense*. *Acta Biol. Cracov. Ser. Bot.* 15: 39–50.
Skalińska, M. and R. Czapik. 1958. Badania cytologiczne nad rodzajem *Potentilla* L. Studies in the cytology of the genus *Potentilla* L. *Acta Biol. Cracov. Ser. Bot.* 1: 137–149.
Skawinska, R. 1962. Apomixis in *Hieracium alpinum*. *Acta Biol. Cracov. Ser. Bot.* 5: 89–96.
Slovak, M., B. Šingliarová and P. Mráz. 2007. Chromosome numbers and mode of reproduction in *Picris hieracioides* s.l. (Asteraceae), with notes on some other *Picris* taxa. *Nordic J. Bot.* 25(3–4): 238–244.
Smith, B.W. 1948. Hybridity and apomixis in the perennial grass *Paspalum dilatatum*. *Genetics* 33: 628–629.
Smith, G.L. 1963a. Studies in *Potentilla* L. I. Embryological investigations into the mechanism of agamospermy in British *P. tabernaemontani* Aschers. *New Phytol.* 62: 264–282.
Smith, G.L. 1963b. Studies in *Potentilla* L. II. Cytological aspects of apomixis in *P. crantzii* (Cr.) Beck ex Fritsch. *New Phytol.* 62: 283–300.
Smith, J. 1841. Notice of a plant which produces seeds without any apparent action of pollen. *Trans. Linn. Soc. (London)*: 18: 509.
Smith, M.W., M. Webb, D.L. Gultzow, T.K. Newman, D. Innes, N. Dillon, J. Owen-Turner and Q. Xu. 2019. Application of a MITE *Citrus* apomixis marker in the Australian rootstock breeding program. *Acta Hortic.* 1230: 1–6.
Smith, P.G. and J.B. Phipps. 1988. Studies in *Crataegus* (Rosaceae, Maloideae) XIX. Breeding behavior in Ontario *Crataegus* series *Rotundifoliae*. *Can. J. Bot.* 66: 1914–1923.
Snyder, L.A. 1957. Apomixis in *Paspalum secans*. *Am. J. Bot.* 44: 318–324.
Snyder, L.A., A.R. Hernandez and H.E. Warmke. 1955. The mechanism of apomixis in *Pennisetum ciliare*. *Bot. Gaz.* 116: 209–221.
Sobir, S. and R. Poerwanto. 2007. Mangosteen genetic and improvement. *Intl. J. Plant Breed.* 1(2): 105–111.
Sobir, S., R. Poerwanto, E. Santosa, S. Sinaga and E. Mansyah. 2013. Genetic variability of mangosteen, an apomictic *Garcinia*. *Acta Hortic.* 975: 155–164. doi:10.17660/ActaHortic.2013.975.16
Sobrevila, C. and M.T.K. Arroyo. 1982. Breeding systems in a montane tropical cloud forest in Venezuela. *Plant Syst. Evol.* 140: 19–37.
Sochor, M., B. Trávníček and J.C. Manning. 2018. Biosystematic revision of the native and naturalised species of *Rubus* L. (Rosaceae) in the Cape Floristic Region, South Africa. *S. Afr. J. Bot.* 118: 241–259.
Sochor, M., R.J. Vašut, T.F. Sharbel and B. Trávníček. 2015. How just a few makes a lot: Speciation via reticulation and apomixis on example of European bambles (*Rubus* subgen. *Rubus*, Rosaceae). *Mol. Phylogenet. Evol.* 89: 13–27. doi:10.1016/j.ympev.2015.04.007

Soejima, A., T. Yahara and K. Watanabe. 2001. Distribution and variation of sexual and agamospermous populations of *Stevia* (Asteraceae, Eupatorieae) in the lower latitudes, Mexico. *Plant Species Biol.* 16: 91–105.

Sokolova, D.V. 2020. Apomictic lines of sugar beet: development and studying. *Proc. Appl. Bot. Genet. Breed.* 181(4): 93–101. doi:10.30901/2227-8834-2020-4-93-101

Sokolowska-Kulzyca, A. 1959. Apomixis in *Leontopodium alpinum*. *Acta Biol. Cracov. Ser. Bot.* 2: 51–63.

Soler, J.X. and J.A. Rosselló. 1997. On the status of *Statice dianiae* Pau (Plumbaginaceae). *An. Jard. Bot. Madr.* 55(1): 9–15.

Soliman, M., F. Espinoza, J.P.A. Ortiz and L. Delgado. 2019. Heterochronic reproductive developmental processes between diploid and tetraploid cytotypes of *Paspalum rufum*. *Ann. Bot.* 123(5): 901–915.

Solntzeva, M. 1973a. Semigamy and fertilisation in *Rudbeckia laciniata*. *Bot. Zhur.* 58: 1261–1277.

Solntzeva, M. 1973b. Types of development of the embryo sac and morphology of embryo in *Rudbeckia laciniata* L. In: *Embryology of angiospermous plants*. Kishinev, Shtiitsa. pp. 47–56.

Solntzeva, M. 2003. About some terms of apomixis: Pseudogamy and androgenesis. *Biologia – section Botany* 58(1): 1–7.

Solntzeva, M.P. 1978. Apomixis and hemigamy as one of its forms. *Proc. Indian Natl. Sci. Acad.* Part B 44: 78–90.

Solntzeva, M.P. 1989. Classification of the embryo sacs of apomictic plants, including the structural basis of heterogeneity among embryo-sac nuclei. *Apomixis Newsl.* 1: 4–10.

Somlyay, L. and A.N. Sennikov. 2014. Atlas Florae Europaeae notes 23. The typification and revised taxonomic circumscription of *Sorbus bakonyensis* (Rosaceae), with a description of *Sorbus udvardyana*, a new apomictic species endemic to Hungary. *Phytotaxa* 164: 265–275.

Sonpipare, S.A. 1984. Studies in the Gramineae: Polyembryony in *Eragrostiella bifaria*. *Proc. 71st Indian Sci. Congr.* Part III. Abstract., p. 258.

Sood, S.K. and R.L. Ram. 1995. Co-occurrence of zygotic and adventives embryony in *Malaxis wallichii* (Orchidaceae). *Acta Bot. Indica* 23(1): 145–146.

Soreng, R.J., L.J. Gillespe and L.L. Consaul. 2017. Taxonomy of the *Poa laxa* group, including two new taxa from Arctic Canada and Greenland, and Oregon, and a reexamination of *P.* sect. *oreinos* (Poaceae). *Nord. J. Bot.* 35(5): 513–538.

Sorensen, A.M., D.T. Rouse, M.A. Clements, P. John and E. Perotti. 2009. Description of a fertilization-independent obligate apomictic species: *Corunastylis apostasioides* Fitzg. *Sex. Plant Reprod.* 22(3): 153–165.

Souèges, R. 1935. Polyembryonie chez le *Potentilla reptans* L. *Bulletin de la Société Botanique de France* 82(3): 381–384.

Souza, L.A., M.C. Iwazaki and I.S. Moscheta. 2005. Morphology of the pericarp and seed of *Tabebuia chrysotricha* (Mart. ex DC.) Standl. (Bignoniaceae). *Braz. Arch. Biol. Technol.* 48: 407–418.

Souza Ferreira, D.N., J.L. Camargo and I. Ferraz. 2017. Multiple shoots of *Carapa surinamensis* seeds: Characterization and consequences in light of post-germination manipulation by rodents. *S. Afr. J. Bot.* 108(1): 346–351.

Souza-Pérez, M. and G. Speroni. 2017. New apomictic pathway in Myrtaceae inferred from *Psidium cattleyanum* female gametophyte ontogeny. *Flora* 234: 34–40. doi:10.1016/j.flora.2017.06.010

Souza-Pérez, M., D. Mourelle, C. Trujillo, A. Borges and G. Speroni. 2021. Pollen grain performance in *Psidium cattleyanum* (Myrtaceae): A pseudogamous polyploid species. *Flora* 281: 151863. doi:10.1016/j.flora.2021.151863

Souza-Silva, S.C.S. 2000. Biologia reprodutiva e polinização em Melastomataceae no Parque do Sabiá, Uberlândia, MG. Brasília Dissertação de mestrado, Universidade de Brasília, Brazil.

Sparvoli, E. 1960. Osservazioni cito-embriologiche in *Eupatorium riparium*. II. Megasporogenesi e sviluppo del gametofito femminile. *Ann. Bot.* (Rome) 26: 481–504.

Spies, J.J. and H. du Plessis. 1985. The genus *Rubus* in South Africa. I. Chromosome numbers and geographical distribution. *Bothalia*. 15: 591–596.

Spies, J.J. and H. du Plessis. 1986. The genus *Rubus* in South Africa. III. The occurrence of apomixis and sexuality. *S.-Afr. Tydskr. Plantk.* 52: 226 – 232.

Spies, J.J., K.C. Klopper and B. Visser. 1999. Apomixis in the genus *Pentaschistis* (Arundineae). *Bothalia* 29: 203–209.

Spillane, C., M.D. Curtis and U. Grossniklaus. 2004. Apomixis technology development-virgin births in farmers' fields? *Nat. Biotechnol.* 22: 687–691.

Spongberg, S.A. 1971. A systematic and evolutionary study of North American arctic and alpine monocephalus species of *Erigeron* (Compositae). PhD thesis, University of North Carolina, Chapel Hill, NC, USA.

Sprecher, M.A. 1919. Étude sur la semence et la germination du *Garcinia mangostana* L. *Rev. Gén. Bot.* 31: 513–531.

Sprunck, S. 2020. Twice the fun, double the trouble: gamete interactions in flowering plants. *Curr. Opin. Plant Biol.* 53: 106–116.
Srivastava, A.K. 1982. Apomixis in *Paspalum paspaloides*. *Acta Bot. Indica* 10: 111–113.
Sriyot, N., A. Thammathaworn and P. Theerakulpisuti. 2015. Embryology of *Spathoglottis plicata* Blume: A reinvestigation and additional data. *Trop. Nat. Hist.* 5(2): 97–115.
Stace, H.M. 1978. Cytoevolution in the genus *Calotis* R.Br. (Compositae: Astereae). *Aust. J Bot.* 26: 287–307.
Staudt, G. 1970. Tests for apomixis in strawberries. *Hortic. Res.* 10: 148–150.
Stebbins, G.L. and J.A. Jenkins. 1939. Aposporic development in the North American species of *Crepis*. *Genetica* 21: 191–224. doi:10.1007/BF01508152
Stebbins, G.L. 1941. Apomixis in the Angiosperms. *Bot. Rev.* 7(10): 507–542. http://www.jstor.org/stable/4353257
Stebbins, G.L. 1932. Cytology of *Antennaria*. 2. Parthenogenetic species. *Bot. Gaz.* 94: 322–344.
Stebbins, G.L. 1949. Asexual reproduction in relation to plant evolution. *Evolution* 3: 98–101.
Stebbins, G.L. 1950. *Variation and evolution in plants*. Columbia University Press: New York.
Stein, J., C.L. Quarin, E.J. Martínez, S.C. Pessino and J.P.A. Ortiz. 2004. Tetraploid races of *Paspalum notatum* show polysomic inheritance and preferential chromosome pairing around the apospory-controlling locus. *Theor. Appl. Genet.* 109: 186–191. doi:10.1007/s00122-004-1614-z
Stein, J., S.C. Pessino, E.J. Martínez, M.P. Rodriguez, L.A. Siena, C.L. Quarin and J.P.A. Ortiz. 2007. A genetic map of tetraploid *Paspalum notatum* Flüggé (bahiagrass) based on single-dose molecular markers. *Mol. Breed.* 20: 153–166. doi:10.1007/s11032-007-9083-0
Steinbauer, G.P. 1943. Frequency of polyembryony in *Fraxinus* seeds. *Bot. Gaz.* 105(2): 285.
Stenar, H. 1932. Parthenogenesis in der Gattung *Calamagrostis*. Embryobildung bei *Calamagrostis obtusata Trin.* and *C. purpurea Trin. Arkiv. f. Wiss. Bot.* 25: 1–8.
Sterk, A.A., J.C.M. den Nijs and W. Kreune. 1982. Sexual and agamospermous *Taraxacum* species in the Netherlands. *Acta Bot. Neerl.* 31(3): 227–237.
Stevens, P.F. 2020. Angiosperm Phylogeny Website. Version 14, July 2017 [and more or less continuously updated since]. http://www.mobot.org/MOBOT/research/APweb/
Stevens, P.F. 1974. A review of *Calophyllum* L. (Guttiferae) in Papuasia. *Aust. J. Bot.* 22: 349–411. doi:10.1071/BT9740349
Stewart, S.L. 2007. Integrated conservation of Florida Orchidaceae in the genera *Habenaria* and *Spiranthes*: Model orchid conservation systems for the Americas. PhD thesis, University of Florida, Florida, FL, USA, 226 pages.
Stolze, A., A. Wanke, N. van Deenen, R. Geyer, D. Prüfer and C. Schulze Gronover. 2017. Development of rubber-enriched dandelion varieties by metabolic engineering of the inulin pathway. *Plant Biotechnol. J.* 15: 740–753. doi:10.1111/pbi.12672
Stonebloom, S.H. and H.V. Scheller. 2019. Transcriptome analysis of rubber biosynthesis in guayule (*Parthenium argentatum* Gray). *BMC Plant Biol.* 19: 71. doi:10.1186/s12870-019-1669-2
Štorchová, H., J. Chrtek Jr, I.V. Bartish, M. Tetera, J. Kirschner and J. Štěpánek. 2002. Genetic variation in agamospermous taxa of *Hieracium* sect. *Alpina* (Compositae) in the Tatry Mts. (Slovakia). *Plant Syst. Evol.* 235: 1–17.
Stork, M.A. 1920. Studies in the genus *Taraxacum*. *Bull. Torrey. Bot. Cl.* 47: 199–210.
Stort, M. and E. Dos Santos Pavanelli. 1985. Formation of multiple or adventive embryos in *Epidendrum nocturnum* Jacq (Orchidaceae). *Ann. Bot.* 55(3): 331–336. http://www.jstor.org/stable/42757491
Strasburger, E. 1877. Über Befruchtung und Zellteilung. *Jenaische Zeitschr. f. Naturw.* 11: 435–536.
Strasburger, E. 1878. Über Polyembryonie. *Jenaische Zeitschr. f. Naturw.* 12: 647–670.
Strasburger, E. 1879. *Die Angiospermen und die Gymnospermen*. G. Fischer, Jena.
Strasburger, E. 1905. Die Apogamie der Eualchimillen und allgemeine Gesichtspunkte, die sich aus ihr ergeben. *Jahrb. Wiss. Bot.* 41: 88–164.
Streetman, L.J. 1963. Reproduction of the lovegrasses, the genus *Eragrostis*. I. *E. chloromelas* Steud., *E. curvula* (Schrad.) Nees, *E. lehmanniana* Nees, and *E. superba* Peyr. *Wrightia* 3: 41–51.
Strother, J.L. 1989. Chromosome numbers in *Thymophylla* (Compositae: Tageteae). *Sida Contrib. Bot.* 13(3): 351–358.
Strydom, A. and J.J. Spies. 1994. Embryo sac development in some representatives of the tribe Cynodonteae (Poaceae). *Bothalia* 24: 101–105.
Subba Rao, A.M. 1937. A note on the development of the female gametophytes of some Malpighiaceae and polyembryony in *Hiptage madablota*. *Curr. Sci.* 6(6): 280–282.
Subba Rao, A.M. 1940. Studies in the Malpighiaceae. 1. Embryo sac development and embryogeny in the genera *Hiptage, Banisteria* and *Stigmatophyllum*. *J. Indian Bot. Soc.* 18: 145–156.

Subramanian, C.L. 1946. Cytological behaviour of certain parthenogenetic sugarcanes. Thesis submitted to the University of Madras, Chennai, India [cited from Price, 1959].

Subramanyam, K. 1942 Gametogenesis and embryogeny in a few members of the Melastomataceae. *J. Indian Bot. Soc.* 21: 69–85.

Subramanyam, K. 1948. An embryological study of *Melastoma malabathricum* L. *J. Indian Bot. Soc.* 27: 11–19.

Suessenguth, K. 1921. Beiträge zur Frage des systematischen Anschlusses der Monocotyledon. *Beih. Bot. Zbl.* 38B: 1–79.

Suessenguth, K. 1923. Über Pseudogamie bei *Zygopetalum mackayi* Hook. *Ber. Deutsch. Bot. Gesell.* 41: 16–23.

Sukatschew, W.N. 1940. Über Polymorphismus und Apomixis bei der Gattung *Adenophora* Fisch. *Bot. J.* 25: 297–303.

Sukhareva, N.B. 1978. Znacheniye nieredukcyi gamet u *Fragaria*. In: *Apomiksis u Rasteni i zhyvotnykh*. Nauka, Novosibirsk, pp. 118–128.

Sukhareva, N.B., S.O. Baturin and K. Trajkovski. 2002. Induction of apomixis in *Fragaria vescana. Acta Hortic.* 567: 231–234.

Sullivan, V.I. 1976. Diploidy, polyploidy and agamospermy among species of *Eupatorium* (Compositae). *Can. J. Bot.* 1907–1917.

Sun, M. 1997. Genetic diversity in three colonizing orchids with contrasting mating systems. *Am. J. Bot.* 84(2): 224–232.

Sun, M. and K.C. Wong. 2001. Genetic structure of three orchid species with contrasting breeding systems using RAPD and allozyme markers. *Am. J. Bot.* 88(12): 2180–2188.

Sundaresan, V. and M. Alandete-Saez. 2010. Pattern formation in miniature: the female gametophyte of flowering plants. *Development* 137(2): 179–189.

Swamy, B.G.L. 1943. Gametogenesis and embryogeny of *Eulophia epidendraea. Proc. Natl. Inst. Sci. India* 9: 59–65.

Swamy, B.G.L. 1946. The embryology of *Zeuxine sulcata* Lindl. *New Phytol.* 45: 132–136.

Swamy, B.G.L. 1948a. On the post-fertilization development *of Trillium undulatum. Cellule* 52: 5–14.

Swamy, B.G.L. 1948b. Agamospermy in *Spiranthes cernua. Lloydia* 11(3): 149–162.

Symon, D.E. 1956. Polyembryony in *Cassia. Nature* 177: 191.

Szkutnik, T. 2010. Apomixis in the sugar beet reproduction system. *Acta Biol. Cracov. Ser. Bot.* 52(1): 87–96. doi:10.2478/v1018 2-010-0011-y

Täckholm, G. 1920. On the cytology of the genus *Rosa*. A preliminary note. *Svensk Bot. Tidskr.* 14: 300–311.

Täckholm, G. 1922. Zytologische Studien über die Gattung *Rosa. Acta Horti Berg.* 7(3): 97–381.

Tahara, M. 1915. Parthenogenesis in *Erigeron annuus. Bot. Mag.* 29: 245–254.

Talent, N. 2009. Evolution of gametophytic apomixis in flowering plants: an alternative model from Maloid Rosaceae. *Theory Biosci.* 128(2): 121–138. doi:10.1007/s12064-009-0061-4

Talent, N. and T.A. Dickinson. 2007a. Endosperm formation in aposporous *Crataegus*. (Rosaceae, Spiraeoideae, tribe Pyreae): parallels to Ranunculaceae and Poaceae. *New Phytol.* 173: 231–249. doi:10.1111/j.1469-8137.2006.01918.x

Talent, N. and T.A. Dickinson. 2007b. Apomixis and hybridization in Rosaceae subtribe Pyrinae Dumort.: a new tool promises new insights. In: Hörandl, E., U. Grossniklaus, P. van Dijk and T.F. Sharbel (eds.), *Apomixis: evolution, mechanisms and perspectives*. Gantner Verlag, Ruggell, Liechtenstein, pp. 301–316.

Talent, N. and T.A. Dickinson. 2007c. The potential for ploidy level increases and decreases in *Crataegus* (Rosaceae, Spiraeoideae, tribe Pyreae). *Botany* 85: 570–584.

Taliaferro, C.M. and E.C. Bashaw. 1966. Inheritance and control of obligate apomixis in breeding buffelgrass *Pennisetum ciliare. Crop Sci.* 6: 473–480.

Tang, Q., G. Zang, C. Cheng, M. Luan, Z. Dai, Y. Xu, Z. Yang, L. Zhao and J. Su 2017. Diplosporous development in *Boehmeria tricuspis*: Insights from *de novo* transcriptome assembly and comprehensive expression profiling. *Sci. Rep.* 7: 46043. doi:10.1038/srep46043

Tang, Q., G. Zang, L. Zhao, C. Cheng, Z. Dong and C. Gao. 2016. Embryological and genetic evidence of amphimixis and apomixis in *Boehmeria tricuspis. J. Plant Biol.* 59: 114–120.

Tantravahi, R.V. 1965. Apomixis in Teosinte. *M.G.C. Newsletter* 39: 70.

Taşkin, K.M., K. Turgut and R.J. Scott. 2004. Apomictic development in *Arabis gunnisoniana. Israel J. Plant Sci.* 52(2): 155–160. doi:10.1560/L3DE-FMVY1XCQ-QRY5

Taşkin, K., K. Turgut and R. Scott. 2009. Apomeiotic pollen mother cell development in the apomictic *Boechera* species. *Biol. Plant.* 53: 468–474.

Tateoka, T. 1968. Notes on *Calamagrostis hakonensis* (Gramineae). *Bull. Nat. Sci. Mus.* (Tokyo) 11: 293–298.

Tateoka, T. 1969. A cytotaxonomic study of the *Calamagrostis arundinacea* complex (Gramineae) in Japan. *Bull. Nat. Sci. Mus.* (Tokyo) 12: 273–289.
Tateoka, T. 1973. A taxonomic study of the genus *Calamagrostis* in Japan. *Bot. Mag.* (Tokyo). 89: 99–114.
Tateoka, T. 1984. *Calamagrostis hakonensis* (Poaceae): Distribution and differentiation of cytotypes. *Bot. Mag.* (Tokyo) 97: 247–270. doi:10.1007/BF02488697
Taylor, R.L. 1967. The foliar embryos of *Malaxis paludosa. Can. J. Bot.* 45(9): 1553–1556.
Teixeira-Caixeta, E., L.F. Ventorim Ferrão, E. Maciel-Zambolim, and L. Zambolim. 2014. Molecular markers. In: Borém, A. and R. Fritsche-Neto (eds.), *Biotechnology and plant breeding: Applications and approaches for developing improved cultivars*. Academic Press, pp. 19–45.
Teppner, H. 1996. Adventitious embryony in *Nigritella* (Orchidaceae). *Folia Geobot. Phytotax.* 31(3): 323–331. doi:10.1007/bf02815377
Teppner, H. and E. Klein. 1985. Karyologie und Fortpflanzungsmodus von *Nigritella* (Orchidaceae-Orchideae), inkl. *N. archiducis-joannis* spec. nov. und zweier Neukombinationen. *Phyton* (Horn, Austria) 25(1): 147–176.
Teryokhin, E.S. 1994. Apomixis in *Potamogeton obtusifolius* Mert. et Koch (Potamogetonaceae). *Apomixis Newsl.* 7: 38–41.
Terziĭski, D. and A. Stefanova. 1990. Nature of apomixis in some Bulgarian varieties of walnut (*Juglans regia*). *Rasteniev'dni Nauki* 27 (6): 73–77.
Terziĭski, D., P. Jurukova-Grancharova, P. Robeva and T. Daskalova. 1995a. Contribution to the embryology of genus *Achillea* L. (Asteraceae). *Nauka-za-Gorata* 32: 36–42.
Terziĭski, D., P. Jurukova-Grancharova, P. Robeva and T. Daskalova. 1995b. Apomixis in the morphological complex of *Achillea millefolium* (Asteraceae). *C.R. de l'Academie Bulgare de Sciences* 48: 53–56.
Thapliyal, M. 2004. Twin seedlings in *Putranjiva roxburghii* Wall. *Indian J. For.* 27(1): 43–44.
Thomas, P.T. 1940. Reproductive versatility in *Rubus. J. Genetics* 40: 119–128. doi:10.1007/BF02982485
Thomas, S.C. 1997. Geographic parthenogenesis in a tropical forest tree. *Am. J. Bot.* 84(8): 1012–1015.
Thompson, M.M. 1995. Chromosome numbers of *Rubus* cultivars at the National Clonal Germplasm Repository. *HortSci.* 30(7): 1453–1456. https://journals.ashs.org/hortsci/view/journals/hortsci/30/7/article-p1453.xml
Thompson, S.L. and J. Whitton. 2006. Patterns of recurrent evolution and geographic parthenogenesis within apomictic Easter daisies (*Townsendia hookeri*). *Mol. Ecol.* 15: 3389–3400.
Thoppil, J.E. and J. Jose. 1996. Chromosome analysis as correlated with essential oil composition in two species of *Plectranthus. The Nucleus* 39 (3): 144–147.
Thurlby, K.A.G., C. Connelly, P.G. Wilson and M. Rossetto. 2011. Development of microsatellite loci for *Syzygium paniculatum* (Myrtaceae), a rare polyembryonic rainforest tree. *Conserv. Genet. Resour.* 3(2):205–208.
Tiagi, Y.D. 1956. Polyembryony in *Mammillaria tenuis* DC. *Bull. Bot. Soc. Univ. Sagar* 8: 25–27.
Tilley, D., S. Jensen, and L. St. John. 2012. Plant Guide for tapertip hawksbeard (*Crepis acuminata*). USDA Natural Resources Conservation Service, Aberdeen Plant Materials Center. Aberdeen, Idaho 83210.
Tinney, F.W. and O.S. Aamodt. 1940. The progeny test as a measure of the types of seed-development in *Poa pratensis* L. *J. Heredity* 31: 457–464.
Tinney, F.W. 1940. Cytology of parthenogenesis in *Poa pratensis. J. Agric. Res.* 60: 351–360.
Tisserat, B., B.E. Esan and T. Murashige. 1979. Somatic embryogenesis in Angiosperms. *Hortic. Rev.* 1: 1–78.
Titow de Tschischow, N. 1956. Número de cromosomas de algunas plantas chilenas. *Bol. Soc. Biol. Concepc.* 31: 145–147.
Tiwary, N.K. 1926. On the occurrence of polyembryony in the genus *Eugenia. J. Indian Bot. Soc.* 5: 124–136.
Tock, E. 1981. Embryology of *Sambucus racemosa. Acta Biol. Crac. Ser. Bot.* 22(2): 173–188.
Tohda 1971. Development of the embryo of *Lecanorchis japonica. Sci. Rep. Tôhoku Univ.* Ser. 4. 35: 245–251.
Tothill, J.C. 1968. Variation and apomixis in *Heteropogon contortus*, Gramineae. *Bot. Soc. Argent.* 12: 188–201.
Traub, H.P. 1939. Polyembryony in *Myrciaria cauliflora. Bot. Gaz.* 101: 233–234.
Tretjakow, S. 1895. Die Beteiligung der Antipoden in Fällen der Polyembryonie bei *Allium odorum* L. *Ber. Deutsch. Bot. Ges.* 13: 13–17.
Treub, M. 1911a. Le sac embryonnaire et l'embryon dans les Angiospermes. I. *Garcinia Kydia* Roxb., *Garcinia Treubii* Pierre. *Ann. Jard. Bot. Buitenz.* 24: 1–27.
Treub, M. 1911b. Le sac embryonnaire et l'embryon dans les Angiospermes. *Ann. des Sci. Nat. Bot.* (2) 9: 1–17.
Trueblood, C.E., T.G. Ranney, N.P. Lynch, J.C. Neal and R.T. Olsen. 2010. Evaluating fertility of triploid clones of *Hypericum androsaemum* L. for use as non-invasive landscape plants. *HortScience* 45(7): 1026–1028. https://journals.ashs.org/hortsci/view/journals/hortsci/45/7/article-p1026.xml

Tucker, M.R. and A.M. Koltunow. 2014. Traffic monitors at the cell periphery: The role of cell walls during early female reproductive cell differentiation in plants. *Curr. Opin. Plant Biol.* 17: 137–145.

Tucker, M.R., N.A. Paech, M.T.M. Willemse and A.M.G. Koltunow. 2001. Dynamics of callose deposition and beta-1,3-glucanase expression during reproductive events in sexual and apomictic *Hieracium*. *Planta* 212: 487–498.

Turesson, G. and B. Turesson. 1960. Experimental studies in *Hieracium pilosella* L. Reproduction, chromosome number and distribution. *Hereditas* 46: 717–736.

Turesson, G.1943. Variation in the apomictic microspecies of *Alchemilla vulgaris* L. *Bot. Not.* 1943: 413–427.

Turesson, G. 1956. Variation in the apomictic microspecies of *Alchemilla vulgaris* L. II. Progeny tests in agamotypes with regard to morphological characters. *Bot. Not.* 109: 400–404.

Tyler, T. 2006. Patterns of morphometric variation and a new supraspecific classification of apomictic taxa of *Hieracium* (Asteraceae) from Denmark and southern Sweden. *Plant Syst. Evol.* 261(1): 39–88.

Tyler, T. and J. Jönsson. 2013. Patterns of plastid and nuclear variation among apomictic polyploids of *Hieracium*: Evolutionary processes and taxonomic implications. *Ann. Bot.* 111(4): 591–609. doi:10.1093/aob/mct019

Tysdal, H.M. 1950. Apomictic interspecific hybrids are promising for rubber production from guayule1. *Agron. J.* 42: 351–355. doi:10.2134/agronj1950.00021962004200070008x

Uemura, A., N. Yamaguchi, Y. Xu, W.Y. Wee, Y. Ichihashi, T. Suzuki, A. Shibata, K. Shirasu and T. Ito. 2018. Regulation of floral meristem activity through the interaction of *AGAMOUS*, *SUPERMAN*, and *CLAVATA3* in *Arabidopsis*. *Plant Reprod.* 31(1): 89–105. doi:10.1007/s00497-017-0315-0

Uhlemann, I. 2001. Distribution of reproductive systems and taxonomical concepts in the genus *Taraxacum* F. H. Wigg. (Asteraceae, lactuceae) in Germany. *Feddes Repert.* 112: 15–35. doi:10.1002/fedr.4921120105

Uhríková, A. and J. Májovský. 1980. In Chromosome number reports LXIX. *Taxon* 29: 725–726.

Uhrinová, V., J. Zozomová-Lihová, D. Bernátová, J. Paule, L. Paule and D. Gömöry. 2017. Origin and genetic differentiation of pink-flowered *Sorbus* hybrids in the Western Carpathians. *Ann. Bot.* 120(2): 271–284.

Ulloa, C., P. Acevedo-Rodríguez, S. Beck, M.J. Belgrano, R. Bernal, P.E. Berry, L. Brako, M. Celis, G. Davidse, R.C. Forzza, S.R. Gradstein, O. Hokche, B. León, S. León-Yánez, R.E. Magill, D.A. Neill, M. Nee, P.H. Raven, H. Stimmel, M.T. Strong, J.L. Villaseñor, J.L. Zarucchi, F.O. Zuloaga and P.M. Jørgensen. 2018 onwards. Vascular Plants of the Americas (VPA) Website. Tropicos, Botanical Information System at the Missouri Botanical Garden, St. Louis, Missouri, USA. http://www.tropicos.org/Project/VPA Accessed: 30 September 2021.

Ulloa, C., P. Acevedo-Rodríguez, S. Beck, M.J. Belgrano, R. Bernal, P.E. Berry, L. Brako, M. Celis, G. Davidse, R.C. Forzza, S.R. Gradstein, O. Hokche, B. León, S. León-Yánez, R.E. Magill, D.A. Neill, M. Nee, P.H. Raven, H. Stimmel, M.T. Strong, J.L. Villaseñor, J.L. Zarucchi, F.O. Zuloaga and P.M. Jørgensen. 2017. An integrated assessment of the vascular plant species of the Americas. *Science* 358: 1614–1617.

Ulum, F.B., C. Costa Castro and E. Hörandl. 2020. Ploidy-dependent effects of light stress on the mode of reproduction in the *Ranunculus auricomus* complex (Ranunculaceae). *Front. Plant Sci.* 11: 104. doi:10.3389/fpls.2020.00104

Underwood, C. J., K. Vijverberg, D. Rigola, S. Okamoto, C. Oplaat, R.H.M. Op den Camp, T. Radoeva, S.E. Schauer, J. Fierens, K. Jansen, S. Mansveld, M. Busscher, W. Xiong, E. Datema, K. Nijbroek, E.-J. Blom, R. Bicknell, A. Catanach, S. Erasmuson, C. Winefield, A.J. van Tunen, M. Prins, E. Schranz and P.J. van Dijk. 2022. A *Parthenogenesis* allele from apomictic dandelion can induce egg cell division without fertilization in lettuce. *Nat. Genet.* 54: 84–93. doi:10.1038/s41588-021-00984-y

Urbani, M.H., C.L. Quarin, F. Espinoza, M.I.O. Penteado and I.F. Rodrigues. 2002. Cytogeography and reproduction of the *Paspalum simplex* polyploid complex. *Plant Syst. Evol.* 236: 99–105.

Ur-Rahman, H. 1997. Tissue culture and molecular studies on rejuvenation, somatic embryogenesis and gene transfer in apomictic *Malus* species. PhD thesis, University of Reading, Reading, England.

Ustinova, E.N. 1955. The apospory phenomenon in the sunflower. *Dokl. Akad. Nauk SSSR* 100(6): 1163–1166.

Ustinova, E.N. 1970. Apomixis in the sunflower. In: S.S. Khokhlov (ed.), *Apomiksis i selektsiya* (Apomixis and Breeding). Nauka, Moscow, pp. 110–116.

Vaclav, E. 1956. Vliv umeleho opylovani brizy na kvalitu osiva. Sborn. *Cesk. Akad. Zemedel. Ved., Rocnik* 29: 531–554.

Valdeviesso, T. 1990. Apomixis in Portuguese walnut varieties. *Acta Hortic.* 284: 279–283.

Valentin-Silva, A., M.A.S. Godinho, K.C. Cruz, S.M. Lelis and M.F. Vieira. 2016. Three psychophilous Asteraceae species with distinct reproductive mechanisms in southeastern Brazil. *N. Z. J. Bot.* 54(4): 498–510. doi:10.1080/0028825X.2016.1236735

Vamosi, J.C. and T.A. Dickinson. 2006. Polyploidy and diversification: A phylogenetic investigation in Rosaceae. *Int. J. Plant Sci.* 167(2): 349–358.

van der Hulst, R., T.H.M. Mes, T. Falque, P. Stam, J.C.M. Den Nijs and K. Bachmann. 2003. Genetic structure of a population sample of apomictic dandelions. *Heredity* 90: 326–335. doi:10.1038/sj.hdy.6800248

van Oostrum, H., A. Sterk and H. Wijsman. 1985. Genetic variation in agamospermous microspecies of *Taraxacum* sect. *Erythrosperma* and sect. *Obliqua*. *Heredity* 55: 223–228. doi:10.1038/hdy.1985.94

van Tieghem, P. 1885. Mémoire sur les canaux sécréteurs des plantes. *Ann. Sci. Nat. Bot.* 16: 1–96 [Clusiacées, 174–181].

Vašková, D. and V. Kolarčik. 2019. Breeding systems in diploid and polyploid hawthorns (*Crataegus*): Evidence from experimental pollinations of *C. monogyna, C. subsphaerica*, and natural hybrids. *Forests* 10: 1059. doi:10.3390/f10121059

Vašut R.J. and Ľ. Majeský. 2015. *Taraxacum pudicum*, a new apomictic microspecies of *T.* section *Erythrosperma* (Asteraceae) from Central Europe. *Phytotaxa* 227(3): 243–252. doi:10.11646/phytotaxa.227.3.3

Vašut, R.J., J. Štěpánek and J. Kirschner. 2005. Two new apomictic *Taraxacum* microspecies of the section *Erythrosperma* from Central Europe. *Preslia* 77: 197–210.

Veerasamy, S. 1982. Polyembryony and twin seedlings in *Borassus flabellifer* L. (Palmae). *Bot. J. Linn. Soc.* 85(2): 147–152.

Vegas, A., G. Trujillo, Y. Sandrea and J. Mata. 2003. Apomixis, poliembrionía somática y cigótica *in vivo* en lechosa. *Interciencia* 28(12): 715–718. http://ve.scielo.org/scielo.php?script=sci_arttext&pid=S0378-18442003001200009&lng=en&tlng=

Venkateswarlu, J. 1946. A case of polyembryony in *Daphne cannabina*. *Curr. Sci.* 15: 169.

Venkateswarlu, J. 1947. A case of polyembryony in *Poivrea coccinea* DC. (≡*Combretum coccineum* Lamk.). *Curr. Sci.* 16(3): 95–95.

Venkateswarlu, J. and M.K. Rao. 1966. Further studies on apomixis in *Coix aquatica*. *Maize Newsl.* 40: 166.

Venkateswarlu, J. and P.I. Devi. 1964. Embryology of some Indian grasses. *Curr. Sci.* 33: 104–106.

Venkateswarlu, J. and P.N. Rao. 1975a. Apomictic maternal diploids in tetraploid job's tars. *Theor. Appl. Genet.* 45: 274–276.

Venkateswarlu, J. and P.N. Rao. 1975b. Apomixis in *Coix aquatica* Roxb. *Ann. Bot.* 39: 1131–1136.

Ventura, M. 1934. Sulla poliembryonia di *Mallotus japonicus*. *Ann. Bot.* (Rome) 20: 568–578.

Verboom, G.A., H.P. Linder and N.P. Barker. 1994. Haustorial synergids: An important character in the systematics of danthonioid grasses (Arundinoideae: Poaceae)? *Amer. J. Bot.* 81:1601–1610.

Verduijn, M.H., P.J. van Dijk, J.M.M. van Damme. 2004. Distribution, phenology and demography of sympatric sexual and asexual dandelions (*Taraxacum officinale* s.l.): Geographic parthenogenesis on a small scale. *Biol. J. Linn. Soc.* 82(2): 205–218.

Vesque, J. 1893. Guttiferae. In: DC. Monogr. Phanerog. 8: 1–669.

Veyret, Y. 1965. Embryogénie comparée et blastogénie chez les Orchidaceae Monandrae. *ORSTOM*, Paris, 107 pages. (Mémoires ORSTOM; 12).

Veyret, Y. 1967. L'apomixie chez le *Cynosorchis lilacina* Ridley (Orchidaceae). *Compt. Rend. Acad. Sci. Fr.* Ser. D. 265: 1713–1716.

Veyret, Y. 1972. Etudes embryologiques dans le genre *Cynorkis* (Orchidaceae). *Adansonia* Ser. 2, 12: 389–402.

Veyret, Y. 1982. Apomixie et variation chez *Epidendrum nocturnum* Jacq. (Orchidaceae). *Bull. Mus. Natl. Hist. Nat., B, Adansonia* 4(3-4): 183–192.

Viana, M.L. 2018. Apomixia como estratégia reproductiva em *Microlicia* D. Don (Microlicieae, Melastomataceae). Dissertação de mestrado, Universidade de Brasília, Uberlândia, Brazil, 62 pages.

Viana, M.L., P.E. Oliveira, R. Romero and A.P.S. Caetano. 2021.The best of both worlds: Apomixis and sexuality co-occur in species of *Microlicia*, Melastomataceae. *Plant Species Biol.* 2021: 1–13. doi:10.1111/1442-1984.12330

Vielle-Calzada, J.R., C.F. Crane and D.M. Stelly. 1996. Apomixis: the asexual revolution. *Science* 274: 1322–1323.

Vijayaraghavan, M.R. and U. Dhar. 1978. Embryology of *Cyrilla* and *Cliftonia* (Cyrillaceae). *Bot. Not.* 131: 127–138.

Vijayaraghavan, M.R. and K. Prabhakar. 1984. The endosperm. In: B.M. Johri (ed.), *Embryology of Angiosperms*. Springer Verlag, Berlin, Germany, pp. 319–347.

Vijverberg, K., P. Ozias-Akins and M.E. Schranz. 2019. Identifying and engineering genes for parthenogenesis in plants. *Front. Plant Sci.* 10: 128. doi:10.3389/fpls.2019.00128

Vijverberg, K., R.G. van Der Hulst, P. Lindhout and P.J. van Dijk. 2004. A genetic linkage map of the diplosporous chromosomal region in *Taraxacum officinale* (common dandelion; Asteraceae). *Theor. Appl. Genet.* 108: 725–732

Vijverberg, K., S. Milanovic-Ivanovic, T. Bakx-Schotman and P.J. van Dijk. 2010. Genetic fine mapping of *DIPLOSPOROUS* in *Taraxacum* (dandelion; Asteraceae) indicates a duplicated *DIP*-gene. *BMC Plant Biol.* 10: 154. doi:10.1186/1471-2229-10-154

Vila, S., A. Gonzalez, H. Rey and L. Mroginski. 2009. Somatic embryogenesis and plant regeneration in *Cedrela fissilis*. *Biol. Plant.* 53(2): 383–386. doi:10.1007/s10535-009-0072-5

Vinogradova, G.Y. 2013. Polyembryony in *Allium schoenoprasum* L. (Alliaceae). Origin of embryos [Poliembrioniya u *Allium schoenoprasum* L. (Alliaceae). Proiskhozhdenie zarodyshei]. *Botanicheskii zhurnal* 98(8): 959–974. [In Russian]

Vinogradova, G.Y. and N.A. Zhinkina. 2021. Why does only one embryo sac develop in the *Paeonia* ovule with multiple archesporium?. *Plant Biol. J.* 23: 267–274. doi:10.1111/plb.13206

Visser, N.C. and J.J. Spies. 1994. Cytogenetic studies in the genus *Trilobium* (Poaceae: Danthonieae): II. A report on the embryo sac development special reference to the occurrence of apomixis in diploid specimens. *South Afr. J. Bot.* 60: 22–26.

Visser, N.C., J.J. Spies and H.J.J. Venter. 1999. Apomictic embryo sac development in *Cenchrus ciliaris* (Panicoideae). *Bothalia* 30(1):103–106.

Vít, P., M. Lepší and P. Lepší. 2012. There is no diploid apomict among Czech *Sorbus* species: a biosystematic revision of *Sorbus eximia*, and the discovery of *Sorbus barrandienica*. *Preslia* 84: 71–96.

Vitkova, A., A. Gavrilova, M. Delcheva, A. Trendafilova and M. Todorova. 2013. Cultivation of high antioxidant activity *Alchemilla* spp. (Rosaceae) for sustainable use. *J. Appl. Hortic.* 15(3): 166–172.

Voigt, P.W. 1971. Discovery of sexuality in *Eragrostis curvula* (Schrad.) Nees. *Crop Sci.* 11: 424–425.

Voigt, M.L., M. Melzer, T. Rutten, T. Mitchell-Olds and T.F. Sharbel. 2007. Gametogenesis in the apomictic *Boechera holboellii* complex: the male perspective. In: Hörandl, E., U. Grossniklaus, P. Van Dijk and T. Sharbel (eds.), *Apomixis: evolution, mechanisms and perspectives*. Ruggell, Liechtenstein, Gantner Verlag, pp. 235–258.

Voigt, P.W. and E.C. Bashaw. 1972. Apomixis and sexuality in *Eragrostis curvula*. *Crop Sci.* 12: 842–847.

Voigt, P.W. and E.C. Bashaw. 1976. Facultative apomixis in *Eragrostis curvula*. *Crop Sci.* 16: 803–806.

Voigt-Zielinski, M.L., M. Piwczyński, and T.F. Sharbel. 2012. Differential effects of polyploidy and diploidy on fitness of apomictic *Boechera*. *Sex. Plant Reprod.* 25: 97–109. doi:10.1007/s00497-012-0181-8

Vorobik, L.A. 1985. Hybridization and reproductive isolation between sympatric *Arabis* (Cruciferae) in Southwestern Oregon. 245. PhD Dissertation, University of Oregon, Eugene, USA.

Voronova, O.N. 2013. Development of female reproduction structures and apomixis in some *CMS* lines of sunflower. *HELIA* 36(58): 47–60. doi:10.2298/HEL1358047V

Vorster, T. and H. Liebenberg. 1984. Classification of embryo sacs in the *Eragrostis curvula* complex. *Bothalia* 15(1/2): 167–174. doi:10.4102/abc.v15i1/2.1122

de Vos, M.P. 1950. Die Ontwikkeling van die Saadknop en Saad by *Cyanella capensis*: 'n Gefal von Polyembryonie. *S. Afr. J. Sci.* 46: 220–226.

Vuille, C. and P. Küpfer. 1985. Aposporie chez le *Ranunculus parnassifolius*. L. I. Etude cytoembryologique. *Bull. Soc. Neuchl. Sci. Nat.* 108: 123–134.

Vuillemin, F. 1992. Origine du sympatrisme des cytotypes du *Ranunculus auricomus* L. s.l.: le cas de la population à Mauensee (canton de Lucerne, Suisse). *Bauhinia* 10: 85–90.

Waddell, M.G.B. 2006. Reproductive biology of relict *Trillium* (*Trillium reliquum*). MSc thesis, Auburn University, Auburn, AL, USA, 63 pages.

Wagner, V. 2016. A review of software tools for spell-checking taxon names in vegetation databases. *J. Vegetation Sci.* 27(6): 1323–1327. doi:10.1111/jvs.12432

Wakana, A. and S. Uemoto. 1987. Adventive embryogenesis in *Citrus* I. The occurrence of adventive embryos without pollination or fertilization. *Am. J. Bot.* 74: 517–530.

Wakana, A. and S. Uemoto. 1988. Adventive embryogenesis in *Citrus* (Rutaceae) II. Postfertilization development. *Am. J. Bot.* 75: 1033–1047.

Waki, T., T. Hiki, R. Watanabe, T. Hashimoto and K. Nakajima. 2011. The *Arabidopsis* RWP-RK protein RKD4 triggers gene expression and pattern formation in early embryogenesis. *Curr. Biol.* 21: 1277–1281.

Waldo, G.F. 1930. Fruit-bud development in strawberry varieties and species. *J. Agr. Res.* 40: 393–408.

Wanage, S.S., A.B. Mirgal, M.M. Naik, R.P. Gunaga, A.D. Rane, S.S. Narkhede and S.G. Bhave. 2010. A note on polyembryony in *Saraca asoca* (Roxb.) Wilde, a critically endangered medicinal tree species. *Karnataka J Agric Sci.* 23(4): 662.

Wang, C., Q. Liu, Y. Shen, Y. Hua, J. Wang, J. Lin, M. Wu, T. Sun, Z. Cheng, R. Mercier and K. Wang. 2019. Clonal seeds from hybrid rice by simultaneous genome engineering of meiosis and fertilization genes. *Nat. Biotechnol.* 37: 283–286. doi:10.1038/s41587-018-0003-0

Wang, L., D.G. Han, C. Gao, Y. Wang, X.Z. Zhang, X.F. Xu and Z.H. Han. 2012. Paternity and ploidy segregation of progenies derived from tetraploid *Malus xiaojinensis*. *Tree Genet. Genomes* 8: 1469–1476. doi:10.1007/s11295-012-0535-2

Wang, X., Y. Xu, S. Zhang, L. Cao, Y. Huang, J. Cheng, G. Wu, S. Tian, C. Chen, Y. Liu, H. Yu, X. Yang, H. Lan, N. Wang, L. Wang, J. Xu, X. Jiang, Z. Xie, M. Tan, R.M. Larkin, L.L. Chen, B.G. Ma, Y. Ruan, X. Deng and Q. Xu. 2017. Genomic analyses of primitive, wild and cultivated citrus provide insights into asexual reproduction. *Nat. Genet* 49(5): 765–772. doi:10.1038/ng.3839

Wang, Y.Q., Q-W. Zeng, X-L. Ye, Y.H. Liu and C-Y. Liang. 2004. Embryological studies on facultative apomixis in *Woonyoungia septentrionalis* (Dandy) Law. *Phytomorphology* 54: 247–252.

Wang, Z., B. Zhang, G. Xia, Y. Huang, D. Hong and J. Huang. 2010. Analysis of progeny of *Carya cathayensis* × *C. illinoensis* and the xenia effect. *J. Fruit Sci.* 27: 908–913 [in Chinese].

Warburg, E.F. and Z.E. Kárpáti. 1968. *Sorbus*. In: Tutin, T.G., V.H. Heywood, N.A. Burges, D.M. Moore, D.H. Valentine, S.M. Walters and D.A. Webb (eds.), *Flora Europaea*. Cambridge University Press, Cambridge, pp. 67–71.

Warmke, H.E. 1943. Macrosporogenesis, fertilisation and early embryology of *Taraxacum kok-saghyz*. *Bull. Torrey Bot. Cl.* 70: 164–173.

Warmke, H.E. 1952. Apomixis in grasses with special reference to *Panicum maximum*. *Proc. 6th Int. Grassland Cong*. pp. 209–215.

Warmke, H.E. 1954. Apomixis in *Panicum maximum*. *Am. J. Bot.* 41: 5–11.

Watanabe, K., P.S. Short, T. Denda, Y. Suzuki, M. Ito, T. Yahara and K. Kosuge. 1996. Chromosome number determinations in the Australian Asterae (Asteraceae). *Muelleria* 9: 197–228.

Webb, M.C. and B.E.S. Gunning. 1994. Embryo sac development in *Arabidopsis thaliana*. II. The cytoskeleton during megagametogenesis. *Sex. Plant Reprod.* 7: 153–163.

Webber, J.M. 1940. Polyembryony. *Bot. Rev*. 6: 575–598.

Weber, H.E. 1996. Former and modern taxonomic treatment of the apomictic *Rubus* complex. *Folia Geobot. Phytotax*. 31: 373–380.

Weber, J.E. and C.S. Campbell. 1989. Breeding system of a hybrid between a sexual and an apomictic species of *Amelanchier*, shadbush (Rosaceae, Maloideae). *Am. J. Bot.* 76: 341–347.

Wei, L. and X-J. Wu. 2012. Genetic variation and population differentiation in a medical herb *Houttuynia cordata* in China revealed by Inter-Simple Sequence Repeats (ISSRs). *Int. J. Mol. Sci.* 13(7): 8159–8170.

Weimarck, G. 1967a. Apomixis and sexuality in *Hierochloe australis* and in Swedish *H. odorata* on different polyploidy levels. *Bot. Not.* 120: 209–235.

Weimarck, G. 1967b. Apomixis in *Hierochloe monticola* (Gramineae). *Bot. Not.* 120: 448.

Weimarck, G. 1970. Apomixis and sexuality in *Hierochloe alpina* (Gramineae) from Finland nd Greenland and *H. monticola* from Greenland. *Bot. Not.* 123: 495–504.

Weismann, A. 1891. *Amphimixis, oder die Vermischung der Individuen*. Verlag Gustav Fischer, Jena.

Weiss-Schneeweiss, H., K. Emadzade, T-S. Jang and G.M. Schneeweiss. 2013. Evolutionary consequences, constraints and potential of polyploidy. *Cytogenet. Genome Res.* 140: 137–150. doi:10.1159/000351727

Wen, X.S., X.L. Ye, Y.Q. Li, Z.L. Chen and S.X. Xu. 1998. Embryological studies on apomixis in *Pennisetum squamulatum*. *Acta Bot. Sin.* 40: 598–604.

Werlemark, G. 1999. Genetic variability and reproductive strategies in Nordic dogroses, *Rosa* section *Caninae*. Doctoral thesis, Swedish University of Agricultural Sciences, Alnarp, Sweden, 33 pages.

Werlemark, G. 2000. Evidence of apomixis in hemisexual dog roses, *Rosa* section *Caninae*. *Sex. Plant Reprod.* 12: 353–359. doi:10.1007/s004970000028

Werlemark, G. and H. Nybom. 2003. Pollen donor impact on progenies of pseudogamous blackberries (*Rubus* subg. *Rubus*). *Euphytica* 133: 71.

Werpachowski, J.S., I.G. Varassin and R. Goldenberg. 2004. The occurrence of apomixis and parthenocarpy in some subtropical Asteraceae. *Rev. Bras. Bot.* 27: 607–613. doi:10.1590/S0100-84042004000300020

Wettstein, R. 1925. Fakultative parthenogenesis beim Hopfen (*Humulus lupulus*). *Flora* 118/119: 600–604. doi:10.1016/S0367-1615(17)32911-7

de Wet, J.M.J. and J.R. Harlan. 1968. Taxonomy of *Dichanthium* Section *Dichanthium* (Gramineae). *Bol. Soc. Argent. Bot.* 12: 206–227.

de Wet, J.M.J. and J.R. Harlan. 1970. Apomixis, polyploidy, and speciation in *Dichanthium*. *Evolution* 24(2): 270–277. doi:10.2307/2406803

de Wet, J.M.J., D.S. Borgaonkar and W.L. Richardson. 1963. Chromosome number and mode of reproduction in the Bothriochloininae. *Caryologia* 16(1): 47–55. doi:10.1080/00087114.1963.10796083

Wheeler, M., L.S. Lee and R.J. Henry. 2001. The reproductive biology and genetic diversity of *Potamophila parviflora* R. Br. *Genet. Resour. Crop Evol.* 48(5): 483–497. doi: 10.1023/A:1012020527094

Wheeler, Q.D. and R. Meier. 2000. Species Concepts and Phylogenetic Theory. A debate. Columbia University Press, New York. http://www.jstor.org/stable/10.7312/whee10142

Wickneswari, R., S.L. Lee and D. Mariam. 1995. Utility of RAPD markers in detecting genetic variation in multiple seedlings of *Hopea odorata* Roxb. (Dipterocarpaceae). *Malays. For.* 58: 42–50.

Wieners, R.R., S. Fei and R.C. Johnson. 2006. Characterization of a USDA Kentucky Bluegrass (*Poa pratensis* L.) core collection for reproductive mode and DNA content by Flow Cytometry. *Genet. Resour. Crop Evol.* 53: 1531–1541. doi:10.1007/s10722-005-7766-0

Wiger, J. 1930. Ein neuer Fall von autonomer Nuzellarpolyembryonie. *Bot. Not.* 83: 368–370.

Willemse, M.T.M., J.L. van Went. 1984. The female gametophyte. In: Johri BM (ed.), *Embryology of Angiosperms*. Springer-Verlag, Berlin, pp. 159–196.

Williams, J.H. and W.E. Friedman. 2002. Identification of diploid endosperm in an early angiosperm lineage. *Nature* 415: 522–526.

Williams, J.H. and J.B. Reese. 2019. Evolution of development of pollen performance. *Curr. Top. Dev. Biol.* 131: 299–336. doi: 10.1016/bs.ctdb.2018.11.012

Windham, M. D. and I. A. Al-Shehbaz. 2006. New and noteworthy species of *Boechera* (Brassicaceae) I: sexual diploids. *Harv. Pap. Bot.* 11: 61–88.

Windham, M.D. and I.A. Al-Shehbaz. 2007a. New and noteworthy species of *Boechera* (Brassicaceae). II. Apomictic hybrids. *Harv. Pap. Bot.* 11: 257–274.

Windham, M.D. and I.A. Al-Shehbaz. 2007b. New and noteworthy species of *Boechera* (Brassicaceae). III. Additional sexual diploids and apomictic hybrids. *Harv. Pap. Bot.* 12: 235–257.

Winkler, H. 1904. Die Parthenogenesis bei *Wikstroemia indica. Ber. Deutsch. Bot. Ges.* 22: 41.

Winkler, H. 1906. Über Parthenogenesis bei *Wikstroemia indica* (L.) C.A. Mey. *Ann. Jard. Bot. Buitenz.* 2 Ser. 5(20): 208.

Winkler, H. 1908. Über Parthenogenesis und Apogamie in Pflanzenreich. *Progr. rei Bot.* (Jena) 2: 293–454.

Wirth, M. and C.L. Withner. 1959. Embryology and development on the Orchidaceae. In: Withner, L.C. (ed.) *The Orchids: A scientific survey*. Ronald Press, New York.

Wissemann, V. and F.H. Hellwig. 1997. Reproduction and hybridisation in the genus *Rosa*, section *Caninae* (Ser.) Rehd. *Bot. Acta* 110: 251–255.

Wolf, S.J. 1987. Cytotaxonomic studies in the genus *Arnica* (Compositae: Senecioneae). *Rhodora* 89: 391–400.

Wong, C.Y. 1939. The influence of pollination on seed development in certain varieties of *Citrus*. *Proc. Am. Soc. Hort. Sci.* 37: 161–164.

Woo, J.K., Y.C. Park, J.W. Lee, S.H. Yun, M. Kim, S. Park, Y. Lee, K.J. Song and H.B. Kim. 2019. Evaluation of polyembryony for genetic resources and efficacy of simple sequence repeat markers for the identification of nucellar and zygotic embryo-derived individuals in citrus. *Appl. Biol. Chem.* 62: 30. doi:10.1186/s13765-019-0437-1

Wood, C. and P. Adams. 1976. The genera of Guttiferae (Clusiaceae) in the Southeastern United States. *J. Arnold Arbor.* 57(1): 74–90.

Woodworth, R.H. 1929. Parthenogenesis and polyembryony in *Alnus rugosa* (Duroi). *Science* 70: 192–193.

Woodworth, R.H. 1930. Cytological studies in the Betulaceae. III. Parthenogenesis and polyembryony in *Alnus rugosa*. *Bot. Gaz.* 89: 402–409.

Worthington, M., C. Heffelfinger, D. Bernal, C. Quintero, Y.P. Zapata, J.G. Perez, J. De Vega, J. Miles, S. Dellaporta and J. Tohme. 2016. A parthenogenesis gene candidate and evidence for segmental allopolyploidy in apomictic *Brachiaria decumbens*. *Genetics* 203: 1117–1132. doi:10.1534/genetics.116.190314

Worthington, M., M. Ebina, N. Yamanaka, C. Heffelfinger, C. Quintero, Y.P. Zapata, J.P. Perez, M. Selvaraj, M. Ishitani, J. Duitama, J.F. de la Hoz, I. Rao, S. Dellaporta, J. Tohme and J. Arango. 2019. Translocation of a parthenogenesis gene candidate to an alternate carrier chromosome in apomictic *Brachiaria humidicola*. *BMC Genom.* 20: 41. doi:10.1186/s12864-018-5392-4

Worthington, M.L. and J.W. Miles. 2015. Reciprocal full-sib recurrent selection and tools for accelerating genetic gain in apomictic *Brachiaria*. In: Budak, H. and G. Spangenberg (eds.), *Molecular Breeding of Forage and Turf*. Springer, Cham, pp. 19–30. doi:10.1007/978-3-319-08714-6_3

Wright, T. 1937. Pollination and the seediness of Marsh grapefruit. *Trinidad and Tobago Agr. Soc. Proc.* 37: 51–60.

Wu, B-J., M-T. Xie, Y.P. Chen, H. Jiang and X-X. Guo. 1991. Cytologic and embryologic studies on apomicts discovered from male sterility line C 1001 in rice (*Oryza sativa* L.). *Sci. China Ser. B. Chem. Life Sci. Earth Sci.* 34: 823–831.

Wu, G-L., Y-Q. Song, J.A. Teixeira da Silva, Q-L. Liu, L. Ji, J-Q. Yang and P-F. Zhang 2007. Recent advances in the study of apomixis in Juglandaceae. *Fruit, Vegetable and Cereal Science and Biotechnology* 1(1): 53–59. Global Science Books.

Wu, S-B., Y-J. Shang, X-M. Han, J-X. Wang, T-T. Niu, F-Y. Zhang, Y-M. Wei, C-G. Meng, X-M. Yan and J-B. Zheng. 1994. Embryological study on apomixis in *Sorghum* line SSA-1. *Acta Bot. Sin.* 36: 833–837.

Xiang, C. and M.L. Roose. 1988. Frequency and characteristics of nucellar and zygotic seedlings in 12 citrus rootstocks. *Sci. Hortic.* 37(1-2): 47–59.

Xie, E., Y. Li, D. Tang, Y. Lv, Y. Shen and Z. Cheng. 2019. A strategy for generating rice apomixis by gene editing. *J. Integr. Plant Biol.* 61(8): 911–916. doi:10.1111/jipb.12785

Xu, Y., H. Jia, X. Wu, A.M.G. Koltunow, X. Deng and Q. Xu. 2021. Regulation of nucellar embryony, a mode of sporophytic apomixis in *Citrus* resembling somatic embryogenesis. *Curr. Opin. Plant Biol.* 59: 101984. doi:10.1016/j.pbi.2020.101984

Yahara, T. 1983. A biosystematic study of the local popullations of some species of the genus *Boehmeria* with special reference to apomixis. *J. Fac. Sci. Univ. Tokyo* 13: 217–261.

Yahara, T. 1986. Distribution of sexual and agamospermous populations of *Boehmeria sylvestrii* and its three relatives (Urticaceae). *Mem. Natn. Sci. Mus., Tokyo* 19: 121–132.

Yahara, T. 1990. Evolution of agamospermous races in *Boehmeria* and *Eupatorium. Plant Species Biol.* 5: 183–196.

Yakovlev, M.S. 1967. Polyembryony in higher plants and principles of its classification. *Phytomorphology* 17(1–4): 278–282.

Yamada-Akiyama, H., Y. Akiyama, M. Ebina, Q. Xu, S.I. Tsuruta, J. Yazaki, N. Kishimoto, S. Kikuchi, M. Takhara, T. Takamizo, S-I. Sugita and H. Nakagawa. 2009. Analysis of expressed sequence tags in apomictic guineagrass (*Panicum maximum*). *J. Plant Physiol.* 166: 750–761. doi:10.1016/j.jplph.2008.10.001

Yamashita, K., Y. Nakazawa, K. Namai, M. Amagai, H. Tsukazaki, T. Wako and A. Kojima. 2012. Modes of inheritance of two apomixis components, diplospory and parthenogenesis, in Chinese chive (*Allium ramosum*) revealed by analysis of the segregating population generated by back-crossing between amphimictic and apomictic diploids. *Breed Sci.* 62(2): 160–169. doi:10.1270/jsbbs.62.160

Yang, H. 1968. Fertilization and development of embryo on Satsuma orange (*C. unshiu* Marc.) and Natsudaidai (*C. natsudaidai* Hayata). *J. Japan. Soc. Hort. Sci.* 37: 102–110.

Yang, X-H. and Z.Q. Zhou. 2009. Anatomical characters of apomixis development in the gynoecia of *Malus xiaojinensis*. *Fruit Sci.* 26(1): 1–5.

Yankova-Tsvetkova, E., P. Yurukova-Grancharova, G. Baldjiev and A. Vitkova. 2016. Embryological features, pollen and seed viability of *Arnica montana* (Asteraceae) – a threatened endemic species in Europe. *Acta Bot. Croat.* 75(1): 39–44.

Yao, J.L., P-F. Yang, C-G. Hu, Y-D. Zhang and B-S. Luo. 2004. Embryological evidence of apomixis in *Eulaliopsis binata. Acta Bot. Sin.* 46(1): 86–92.

Yao, J.L., Y. Zhou and C.G. Hu. 2007. Apomixis in *Eulaliopsis binata*: characterization of reproductive mode and endosperm development. *Sex. Plant Reprod.* 20: 151–158. doi:10.1007/s00497-007-0051-y

Yao, L., Y. Zhang, C. Liu,Y. Liu, Y. Wang, D. Liang, J. Liu, G. Sahoo and T. Kelliher. 2018. *OsMATL* mutation induces haploid seed formation in indica rice. *Nat. Plants* 4: 530–533. doi:10.1038/s41477-018-0193-y

Young, B.A., R.T. Sherwood and E.C. Bashaw. 1979. Cleared-pistil and thick sectioning techniques for detecting aposporous apomixis in grasses. *Can. J. Bot.* 57: 1668–1672.

Young, D.P. 1952. Studies in the British *Epipactis*. IV. A revision of the *phyllanthes-vectensis-pendula* group. *Watsonia* 2: 259–276.

Yudakova, O.I. 2009. Abnormalities of female gametophyte development in apomictic bluegrass forms. *Russian J. Dev. Biol.* 39(3): 150–156.

Yudakova, O.I. and Shishkinskaya, N.A. 2008. *Osobennosti embriologii apomiktichnykh zlakov* (Embryological Patterns in Apomictic Cereals). Izd-vo un-ta, Saratov, Russia.

Yurukova Grancharova, P., M. Anchev and V. Goranova. 2004. Embryological study on diploid and triploid populations of *Iberis saxatilis* subsp. *saxatilis* (Brassicaceae) in the Bulgarian flora. *Phytol. Balc.* 10: 69–78.

Zagorcheva, L. 1985. Autonomous apomictic propagation of *Cucumis ficifolius* and *Cucumis anguria. Genet. Sel.* 18: 460.

Zaikovskaja, N.B., N.A. Peretjatjko, E.I. Shiryaeva and G.I. Jarmoluk. 1978. Apomixis in the sugar beet forms with genic sterility of pollen. In: *Apomixis and Cytoembryology of Plants*. Izd. Sarat. University, Saratov 39.

Zang, G. 1995. Plant apomixis and its application for breeding in ramie (*Boehmeria nivea* L.). *China's Fibre crops* 17(1): 16–19.

Zarubin, A.F. 1948. Possibility of parthenogenesis in walnut and black walnut. *Priroda* 10: 64–65.

Záveský, L., V. Jarolímová and J. Štěpánek. 2007. Apomixis in *Taraxacum paludosum* (section *Palustria*, Asteraceae): Recombinations of apomixis elements in inter-sectional crosses. *Plant Syst. Evol.* 265: 147–163. doi:10.1007/s00606-007-0518-1

Zeisek, V., J. Kirschner, P.J. van Dijk, J. Štěpánek, T. Černý, J. Kotek and P. Urbanovský. 2019. Analysis of wild *Taraxacum bicorne* Dahlst. (Compositae–Crepidinae) as a potential alternative natural rubber crop. *Genet. Resour. Crop Evol.* 66: 1341–1361. doi:10.1007/s10722-019-00788-4

Zeng, Q-W., D-X. Zhang, Z-Z. Gao and F-W. Xing. 2003. Facultative apomixis in an endangered dioecious species, *Woonyoungia septentrionalis* (Magnoliaceae). *Acta Bot. Sin.* 45(11): 1270–1273.

Zhang, B., Z.J. Wang, S.H. Jin, G.H. Xia, Y.J. Huang and J.Q. Huang. 2012. A pattern of unique embryogenesis occurring via apomixis in *Carya cathayensis*. *Biol. Plant.* 56(4): 620–627.

Zhang, G., L. Zhao and J. Sun. 1997. Cytoembryological studies on apomixis in *Boehmeria silvestrii*. *Acta Bot. Sin.* 39(3): 210–213.

Zhang, W.L. and J.Y. Gao 2018. High fruit sets in a rewardless orchid: a case study of obligate agamospermy in *Habenaria*. *Aust. J. Bot.* 66(2): 144–151. doi:10.1071/bt17182

Zhao, L-L., Y-F. Ru, M. Liu, J-N. Tang, J-F. Zheng, B. Wu, Y-H Gu and H-J. Shi. 2017. Reproductive effects of cadmium on sperm function and early embryonic development *in vitro*. *PLoS One* 12(11): e0186727. doi:10.1371/journal.pone.0186727

Zhgonti, L.P. 1978. Apomixis in genus *Paeonia*. *Vest. Gruz. Bot. Sada Acad. Nauk Gruz. SSR* 7: 113.

Zhong, J., F-C. Wu, P. Qiu and L-J. Dai. 2016. Correlation analysis of genetic diversity and population structure of *Houttuynia cordata* Thunb with regard to environment. *Genet. Mol. Res.* 15(3). doi:10.4238/gmr.15038780

Zhong, Y., C. Liu, X. Qi, Y. Jiao, D. Wang, Y. Wang, Z. Liu, C. Chen, B. Chen, X. Tian, J. Li, M. Chen, X. Dong, X. Xu, L. Li, W. Li, W. Liu, W. Jin, J. Lai and S. Chen. 2019. Mutation of *ZmDMP* enhances haploid induction in maize. *Nat. Plants* 5: 575–580. doi:10.1038/s41477-019-0443-7

Zhou, L-Z. and T. Dresselhaus. 2019. Friend or foe: Signaling mechanisms during double fertilization in flowering seed plants. *Curr. Top. Dev. Biol.* 131: 453–496. doi: 10.1016/bs.ctdb.2018.11.013

Zhou, Z., X. Guo, L. Xia and Z. Wang. 1995. Studies on apomixis in *Malus* Mill. *J. Southwest Agric. Univ.* 17(5): 378–380. https://europepmc.org/abstract/cba/286004

Zilli, A.L., C.A. Acuña, R.R. Schulz, E.A. Brugnoli, V. Guidalevich, C.L. Quarin and E.J. Martínez. 2018. Widening the gene pool of sexual tetraploid bahiagrass: Generation and reproductive characterization of a sexual synthetic tetraploid population. *Crop Sci.* 58: 762–772. doi:10.2135/cropsci2017.07.0457

Zilli, A.L., C.A. Acuña, R.R. Schulz, F. Marcón, E.A. Brugnoli, S.F. Novo, C.L. Quarin and E.J. Martínez. 2019. Transference of natural diversity from the apomictic germplasm of *Paspalum notatum* to a sexual synthetic population. *Ann. Appl. Biol.* 175: 18–28. doi:10.1111/aab.12507

Zilli, A.L., D.H. Hojsgaard, E.A. Brugnoli, C.A. Acuña, A.I. Honfi, M.H. Urbani, C.L. Quarin and E.J. Martínez. 2014. Genetic relationship among *Paspalum* species of the subgenus *Anachyris*: Taxonomic and evolutionary implications. *Flora* 209(10): 604–612. doi:10.1016/j.flora.2014.08.005

Zollikofer, C. 1930. Zur Fortpflanzung von *Poa alpina* L. *Jahrb. der St. Gallischen Gesellesch* 65: 99–117.

Zühl, L., C. Volkert, D. Ibberson, A. Schmidt and Z. Wilson. 2019. Differential activity of F-box genes and E3 ligases distinguishes sexual versus apomictic germline specification in *Boechera*. *J. Exp. Bot.* 70: 5643–5657.

Zuloaga, F.O., O. Morrone, M.J. Belgrano, C. Marticorena and E. Marchesi. (eds.) 2008. Catálogo de las plantas vasculares del Cono Sur. *Monogr. Syst. Bot. Missouri Bot. Gard.* 107(1–3): i–xcvi, 1–3348.

Zumajo-Cardona, C., N. Pabón-Mora and B.A. Ambrose. 2021. The evolution of *euAPETALA2* genes in vascular plants: From plesiomorphic roles in sporangia to acquired functions in ovules and fruits. *Mol. Biol. Evol.* 38(6): 2319–2336. doi:10.1093/molbev/msab027

Zweifel, R. 1939. Cytologisch-embryologische Untersuchungen an *Balanophora abbreviata* Blume und *B. indica* Wall. *Viertelj.schr. Naturf. Ges. Zürich* 84: 245–306.

Zymone, K., L. Raudone, R. Raudonis, M. Marksa, L. Ivanauskas and V. Janulis. 2018. Phytochemical profiling of fruit powders of twenty *Sorbus* L. cultivars. *Molecules* 23(10): 2593. doi:10.3390/molecules23102593

Index

Indexing of taxonomic names cited in the main text does not include synonyms.

In the subject indexing, pages in bold correspond to tables and pages in italics correspond to figures.

A

N

O

P